The QUEST for PURE WATER

The History of Water Purification From the Earliest Records To the Twentieth Century

by

M. N. BAKER

Associate Editor (Retired) *Engineering News-Record*

NEW YORK
THE AMERICAN WATER WORKS ASSOCIATION, INC.
1949

COPYRIGHT 1948 BY

THE AMERICAN WATER WORKS ASSOCIATION, INC.

First Printing, September 1948
Second Printing, July 1949

PRINTED IN UNITED STATES OF AMERICA BY
LANCASTER PRESS, INC., LANCASTER, PA.

M. N. BAKER

Books by M. N. Baker

The Manual of American Water Works (*Editor*)
1888–1897

Sewage Purification in America
1893

Sewage Disposal in the United States (with G. W. Rafter)
1894

Sewerage and Sewage Purification
1896

Potable Water
1899

Municipal Engineering and Sanitation
1901

British Sewage Works
1904

Notes on British Refuse Destructors
1905

The Quest for Pure Water
1948

Publisher's Foreword

The late Lord Acton, distinguished historian at Cambridge University, commented once that "the recent past contains the key to the present time." M. N. Baker no doubt was impressed by the same axiom when he initiated many decades ago his collection of rare historical volumes on water supply. Fortunately for the water works practitioner, Mr. Baker, earlier than most in a professional career, acquired "historical mindedness," that rare mark of the matured worker. When he crystallized, in the present volume, the distilled wisdom of the ages, which he had been gathering for over a half century, he captured a permanent place in the list of distinguished workers in sanitary engineering.

For thousands of years, the search for "pure" water has been persistently pursued by man. Criteria of purity have become more complex, more quantitative, perhaps even more rigid, but principles, methods and materials for purifying water have remained remarkably similar from the earliest recorded data of 2000 B.C. down to the present time. To trace the practices through almost 4,000 years would at first sight appear to be an impossible task for one man, no matter how endowed. But the task has been accomplished. In this one field, M. N. Baker has "relived the entire life of Mankind as a single imaginative experience" and his diary is at hand!

The student of water treatment will find in the chapters which follow an inexhaustible treasury of authenticated information. He will be enabled to trace the continuing demand by the consumer, and the consequent search by the practitioner, for improved quality of water. The search is not ended—more than likely it is never ending! The intuitive avoidance centuries ago of "dirty water," the subsequent use of the chemical test, the great progress in the biological era, the more recent esthetic demands—all are indices of the endless character of this "Quest."

Neither the consumer nor the professional should be disconcerted by this endless adventure in search of "pure" water. It is the natural concomitant of advancing knowledge and expanding horizon. Mr.

Baker makes clear that, although evidence of water purification effort dates back thousands of years and great advances in the art were made in the last half of the eighteenth century, it was not until the first half of the nineteenth century that treatment methods were applied to the water supplies of whole cities and, from the late 1890's, rule-of-thumb has been increasingly replaced by the making and application of the results of scientific research by thousands of engineers, chemists and biologists.

More disconcerting perhaps than the endless adventure is the conclusion which may be drawn from the present volume that the history of "conditioning or treatment" of water is *largely* a history of empiricism, a history of an art and not of a science. Mr. Baker would probably be the first to agree that his task has been best accomplished by making it clear that history is a challenger to the future, that "history made and history making are scientifically inseparable," and that the water works man of today has still a world to conquer!

Abel Wolman
Consulting Sanitary Engineer
Baltimore, Maryland
May 1948

Author's Preface

During my forty-five years on the editorial staff of *Engineering News* and *Engineering News-Record* from 1887 to 1932, much of my work was devoted to public water supplies. I began by gathering and compiling descriptions of all the water works of the United States and Canada. These descriptions first formed a volume of more than 600 closely printed pages entitled *The Manual of American Water Works, 1888*. Three revisions and extensions of the work followed, bringing the data down to 1896. In addition to statistical data, these manuals contained brief descriptions of such water purification plants as existed or of those which had been built and later abandoned.

In 1893–1896, I wrote for *Engineering News* a series of articles on "Water Purification in America." These reports described the very few slow sand filters and the comparatively many rapid or mechanical filtration plants then in use. Also, before, during and after that period, I wrote or edited hundreds of articles on water purification in America and abroad.

In the early 1930's I was asked to write a few pages on the history of water purification for either a proposed revision of the American Water Works Association's *Water Works Practice Manual* or a short monograph. The outcome was a sketch entitled "The History of Water Purification," which appeared in the *Journal of the American Water Works Association* in July 1934.

Having become deeply interested in the historical background of current water works practices through the study undertaken in preparing the article, I began what proved to be nine years of gathering and digesting data from libraries, rare-book dealers and correspondence with engineers and librarians the world over. After much rewriting and painful condensation—often cutting to a few sentences what would have made a long chapter—I delivered to the American Water Works Association on my seventy-ninth birthday (January 26, 1943) manuscript, illustrations and an extensive bibliography for the present volume. War and postwar conditions have delayed publication for the five years past.

AUTHOR'S PREFACE

A large amount of source material, chronologically arranged, of which much has been reduced to monograph form and all has been typewritten, and a collection of books and pamphlets, many of which are rare, have been deposited with the Association and, by it, transferred to the Library of the United Engineering Societies, New York City, where the material will be readily accessible for reference.

My thanks are given to librarians and to a host of water works superintendents, engineers, chemists and bacteriologists in America and Europe, all of whom have contributed unstintingly to my researches for this volume. Thanks are also given to the American Water Works Association for its cooperation in my efforts, for financial aid and for publishing my book.

M. N. Baker.

M. N. Baker
Upper Montclair, N.J.
May 1948

Table of Contents

CHAPTER		PAGE
I.	From the Earliest Records Through the Sixteenth Century ...	1
II.	Seventeenth Century	9
III.	Eighteenth and Early Nineteenth Centuries	19
IV.	Four Centuries of Filtration in France	29
V.	British Contributions to Filtration	64
VI.	Slow Sand Filtration in the United States and Canada	125
VII.	Inception and Widespread Adoption of Rapid Filtration in America	179
VIII.	Upward Filtration in Europe and America	248
IX.	Multiple Filtration: Seventeenth to Twentieth Centuries	253
X.	Drifting-Sand Rapid Filters	265
XI.	Natural Filters: Basins and Galleries	273
XII.	Plain Sedimentation	286
XIII.	Coagulation: Ancient and Modern	299
XIV.	Disinfection	321
XV.	Distillation	357
XVI.	Aeration in Theory and Practice	361
XVII.	Algae Troubles and Their Conquest	391
XVIII.	Softening	415
XIX.	Cause and Removal of Color	440
XX.	Iron and Manganese Removal	445
XXI.	Taste and Odor Control	449
XXII.	Medication by Means of the Water Supply	456
	Epilogue	465
	Bibliography	469
	Index	511

List of Illustrations

FIG.		PAGE
	M. N. Baker	*Frontispiece*
1.	Egyptians Siphoning off Water or Wine Clarified by Sedimentation	2
2.	Bacon's Experiment on Filtration	10
3.	Porzio's Multiple Filter	12
4.	Sculptured Bronze Curb to Venetian Filter-Cistern	14
5.	Cross Section of Venetian Filter-Cistern	15
6.	Water Power and Clarification Plant for Seventeenth Century Paper Mill at Auvergne, France	16
7.	Pages From the Works of Joseph Amy, Filtration Pioneer	32
8.	Amy's Machine to Purify Water	33
9.	Amy's Sextifold Filter for Army Garrisons	34
10.	Fonvielle Pressure Filter at Paris Sales Fountain, 1856	50
11.	Souchon Pressure Filter at Paris Sales Fountain, 1856	50
12.	Thirty-six Souchon Filters at Nantes, France	56
13.	Darcy's Upward-Flow Mechanically Cleaned Filter of 1856	59
14.	Cross Section of Lancashire Filter	65
15.	Old Filter at Close Bleachworks, Radcliffe, England	66
16.	First British Water Filter Patent	68
17.	Title Page of Peacock's Promotional Pamphlet	70
18.	Peacock's Diagram to Illustrate Reasons for Arranging Filter Media in Layers of Decreasing Size	72
19.	Three-Tank Form of Peacock's Upward-Flow Backwash Filter	74
20.	Single-Tank Form of Peacock's Upward-Flow Backwash Filter	75
21.	First Known Filter to Supply an Entire City With Water, Completed at Paisley, Scotland, in 1804, by John Gibb	78
22.	Thomas Telford (1757–1834)	82
23.	Robert Thom (1774–1847)	92
24.	Thom's Self-cleaning Filter at Paisley, 1838	96
25.	Frontispiece and Title Page of *The Dolphin*	100–101

LIST OF ILLUSTRATIONS

xiii

FIG.		PAGE
26.	The Dolphin, or Intake, of the Southwark Water Works Co.	102
27.	James Simpson (1799–1869)	104
28.	James Simpson's Experimental Filter of 1827–1828	108
29.	Cross Section of Simpson's One-Acre Filter for Chelsea Water Works Co., 1829	112
30.	Albert Stein (1785–1874)	126
31.	James P. Kirkwood (1807–1877)	134
32.	Allen Hazen (1869–1930)	138
33.	Miniature Slow Sand Filter at Marshalltown, Iowa, 1876	142
34.	Covered Slow Sand Filters at St. Johnsbury, Vt.	159
35.	Birdsill Holly (1820–1894)	180
36.	John Wesley Hyatt (1837–1920)	182
37.	Isaiah Smith Hyatt (1835–1885)	184
38.	Evolution of Rapid Filter by Clark and Hyatt	186
39.	Hyatt Sand-Transfer Wash Filter	187
40.	George F. Hodkinson (1868–)	192
41.	John E. Warren (1840–1915)	196
42.	Section of Warren Gravity Filter	198
43.	National Filter at Terre Haute, Ind.	200
44.	Albert R. Leeds (1843–1902)	206
45.	The American Filter	210
46.	Blessing Horizontal Duplex Filter	214
47.	The Three Jewells—Father and Sons	218
48.	The Staff at the Louisville Experiment Station in 1896	232
49.	Upward-Flow Filter at Burlington, Iowa, 1878	250
50.	J. D. Cook (1830–1902)	254
51.	Multiple Filters and Aerators at Glasgow	256
52.	Puech-Chabal Multiple Filters and Aerators	260
53.	Bollmann Combined Reverse-Flow and Sand-Ejector Wash Filter	267
54.	Sand Traps, Extractor Pipes and Washers in Toronto Drifting-Sand Filters	268
55.	Gore-Ransome Drifting-Sand Filter at Toronto	270
56.	Filter Basin and Galleries at Toulouse, France	276
57.	Adam Anderson (Early 19th Century)	278
58.	Filter Gallery at Nashville, Tenn., 1880	284

LIST OF ILLUSTRATIONS

FIG.		PAGE
59.	Rain Water Storage and Settling Reservoirs at Ancient Carthage	288
60.	Piscana on Roman Aqueduct Virgo	291
61.	Baffled Settling Reservoir in Swiss Stream	293
62.	Dominique Francois Arago (1786–1853)	306
63.	Coagulation and Sedimentation Basins and Lateral-Flow Filter at Vicksburg, Miss.	312
64.	Jewell Chlorine Gas Generator	333
65.	George A. Soper (1868–1948)	344
66.	Still and Wick Siphon	358
67.	Diagram and Explanation of Littlewood-Hales Aerator for Stinking Water	364–365
68.	Russian Aerator, Circa 1860	368
69.	Multiple Jet Aerator at Highland Reservoir, Rochester, N.Y.	372
70.	Forced Aeration at Norfolk, Va., and Brockton, Mass.	382
71.	First Biological Laboratory on an American Water Works System, Chestnut Hill Reservoir, Boston Water Works	399
72.	Inside View of Chestnut Hill Reservoir Laboratory	400
73.	Dr. Francis Home of Edinburgh	416

The QUEST for PURE WATER

CHAPTER I

From the Earliest Records Through the Sixteenth Century

The quest for pure water began in prehistoric times. Into those recordless millenniums only speculation can enter. Sanskrit medical lore and Egyptian inscriptions afford the earliest recorded knowledge of water treatment.

"It is good to keep water in copper vessels, to expose it to sunlight, and filter through charcoal." So wrote Francis Evelyn Place (1) from India in 1905, crediting the quotation to *"Ousruta Sanghita*—a collection of medical lore in Sanskrit, probable date 2000 B.C., Chap. XIV, verse 15." In the same letter, Place cites, as from another Sanskrit source "of about the same date, . . . it is directed to heat foul water by boiling and exposing to sunlight and by dipping seven times into it a piece of hot copper, then to filter and cool in an earthen vessel. The direction is given by the god who is the incarnation of medical science." This citation is attributed to "the *Naghrund Bhuson*—a collection of medical maxims from the *Ayura Veda,* the earliest Sanskrit work on medicine extant . . . in the chapter on water, in the last *sloka* but two." Both citations were translated by Kundan Jugendia Pal Singh, of Jaipur, India, where Place was located in 1905.*

The *Sus'ruta Samhita,* a body of medical lore said to date from 2000 B.C., but not known to have been put into manuscript form until 400 A.D. (2), declares: "Impure water should be purified by

---

* Being unable to find an English translation of the works named by Place, I appealed to the Library of Congress for aid. Miss Gerda Hartman, Division of Indic Studies, replied that Place's *Ousruta Sanghita* must be the same as the *Sus'ruta Samhita* (see below) and that *Naghrund Bhuson* may be corrupted from *Nighantu Bhusana,* "nighantu" being part of the name of several medical glossaries, probably of recent date, possibly one of the latest of these. *Ayura Veda* may have been intended for the traditional fifth Veda, or "Veda of Longevity," the preservation of which has not been proved. The passages cited by Place have been repeated again and again in journals and books, sometimes with and sometimes without credit to him, but with no attempt to authenticate the material. Place's chapter and verse citations were probably derived from a contemporary translator, Singh, who, Miss Hartman suggests, may have drawn from one of numerous extant manuscripts.

being boiled over a fire, or being heated in the sun, or by dipping a heated iron into it, or it may be purified by filtration through sand and coarse gravel and then allowed to cool." Other methods of purifying water include the use of "Gomedaka"—a kind of stone— and a number of vegetable substances, most notably seed of *Strychnos potatorum*. These statements may be regarded as a summary of methods of water purification known and used by the Aryan and Indic

Fig. 1. Egyptians Siphoning off Water or Wine Clarified by Sedimentation

Pictured on wall of tomb of Amenophis II at Thebes, 1450 B.C.
(From Wilkinson's *Manners and Customs of the Ancient Egyptians*, 1879)

priests and physicians down to 400 A.D. They include some of the methods cited by Place as having been taken from Sanskrit literature of a much earlier date.

Egyptian Customs

The earliest known apparatus for obtaining clarified liquids was pictured on Egyptian walls in the fifteenth and thirteenth centuries B.C. The first picture, in a tomb of the reign of Amenophis II, represents men siphoning off either water or settled wine (3); the second, in the tomb of Rameses II, shows assorted sizes of wick siphons in an Egyptian kitchen (4).

"In porticoes of Egyptian temples," says Hero of Alexandria, "revolving wheels . . . are placed for those who enter to turn around, from an opinion that bronze purifies [water]." The apparatus was connected behind an entrance pillar. It included a wheel attached to a perforated horizontal tubular axis placed beneath and in contact with the perforated bottom of a [bronze?] water container. When the wheel was revolved, the holes in the tube and the axis came opposite each other and water was sprayed upon the worshipper through a small tube extending through the wheel. This was not a device invented by Hero himself, nor does he say how long it had been in use. He probably lived in the last half of the first century A.D. (5).

Bible Lore

In view of the well-deserved fame of the sanitary and hygienic code of the early Hebrews, directions for water treatment might be looked for in the Old Testament. None appears. But three incidents may be cited as examples of the quest for pure water in dry or alkaline lands: At Marah, Moses is said to have sweetened bitter waters by casting into them a tree shown him by the Lord (6). During the forty years' wandering in the wilderness, he is said to have brought forth water by smiting a rock (7). Much later, Elisha is said to have "healed unto this day" the spring water of Jericho by casting "salt" into it (8).

Greek and Roman Records

For convenience there are assembled here, out of chronological order, a number of similar later instances of water conditioning. Diophanes (first century B.C.) advised putting macerated laurel into rain water (9). Paxamus (first century A.D.) proposed that bruised coral or pounded barley, in a bag, be immersed in bad-tasting water—evidently to cure taste due to mineral salts (9). Vitruvius (15 B.C.) recommended that cisterns be constructed in two or three compartments and the water transferred from one to another of them, thus allowing the mud to settle, and insuring clearness and limpidity. Otherwise, he wrote, it would be necessary to clarify the water by adding "salt" (10). Pliny (c. 77 A.D.) said that polenta, a kind of food, added to nitrous or bitter water would render it potable in two hours, and that a similar property is possessed by chalk of Rhodes and the argilla of Italy (11). This is the first mention found of lime and aluminous earth as precipitants. Referring to well water, Palladius (fourth or fifth century A.D.), in his rhymed work, *On Husbondrie,* wrote that "If water be lymous or infest admystion of salt wol it correct" (12).

A drinking cup which hid badly colored water from the sight of the drinker and caused mud to stick to its side was devised by Lycurgus, the Spartan lawgiver and reformer (ninth century B.C.), according to Plutarch (13). This is the earliest recorded instance of an attempt by the Greeks to purify water. More credible is the earliest specific instance of treating water for human consumption found on record. Herodotus tells us that Cyrus the Great, King of Persia (sixth century B.C.), when going to war, took boiled water in silver flagons, loaded on four-wheeled cars drawn by mules (14). A later Greek writer, Athenaeus of Naucratis (third and second centuries B.C.), says that the water was boiled in these instances to make it keep—which reflects the opinion if not also the custom of his time (15).

That the use of wick siphons was well known in the days of Socrates (469–399 B.C.) and Plato (427–347 B.C.) is shown by the simile in Plato's *Symposium,* wherein Socrates is represented as saying it would be well if wisdom would flow from a person filled with it to one less wise, as water flows through a thread of wool from a fuller to an emptier vessel (16). Common knowledge of filtration through porous vessels is indicated by Aristotle (384–322 B.C.) in his *De Generatione Animalium* (17). He says that the nutriment which produces flesh is deduced through the veins and pores in the same way as water through earthen vessels not thoroughly baked.

Hippocrates (460–354 B.C.), the father of medicine, wrote in *Air, Water and Places*—the first treatise on public hygiene—that "whosoever wishes to investigate medicine properly should consider the seasons of the year, the winds and the waters in relation to health and disease." As the "qualities of the waters differ from one another in taste and weight," he said, "so they differ much in their [other] qualities." One should "consider the waters which the inhabitants use, whether they be marshy and soft, or hard and running from elevated and rocky situations, and then if saltish and unfit for cooking . . . for water contributes much to health." Hippocrates' discussion of the qualities of water centers on the selection of the most health-giving sources of supply rather than on rectifying the waters that were bad. The only exception to this is an assertion that rain waters should be "boiled and strained for otherwise they would have a bad smell" and cause hoarseness. For straining he mentions a cloth bag which became known in later ages as "Hippocrates' sleeve" (18).

In 168 B.C., a thousand years after the siphons were depicted on the walls of Egyptian tombs, Athenaeus of Naucratis wrote his account of the voyage of Antiochus on the Nile (15). On this journey, says the author of *The Deipnosophists,* water from the river was exposed to sun and air; strained; settled overnight in jars the outsides of which were kept cool by being kept wet by slaves; and then, in the morning, drawn off and placed in chaff for use as needed. Thus the water was made clear and healthful.

When Caesar began taking possession of Alexandria, in 47 B.C., it is said that he found the city underlaid with aqueducts bringing water from the Nile to cisterns in which it became clarified by sedimentation for the use of the masters and their families. The Egyptians cut off the river water and turned in salt water from the other side of the city. Whereupon, Caesar, firmly believing that all seashores naturally abounded with fresh springs, ordered his centurions to lay aside all other works and dig wells day and night. The very first night an abundance of water was found. This incident, related with zest by Hirtius (19), is doubly significant: it affords an early instance of clarification by sedimentation and shows that Caesar did not suppose, as was almost universally assumed for centuries before and afterwards, that salt water could be freshened by percolation through sand.

Not all the citizens, soldiers and slaves of Alexandria were supplied with water brought underground to cisterns. Athenaeus of Attilia, writing about 50 A.D. on *Purification of Water* (20), mentioned Alexandria as a city where water was purified by jars called "stacta." He also cited a single, double, and even triple filtration as producing water of the greatest purity. Channels were dug along the sea to draw off water or along lakes to produce water free from mud and leeches. These channels may be regarded as prototypes of the infiltration galleries of the nineteenth century in England, France and Germany.

Pliny credited to his contemporary, the Emperor Nero (reigned 54–68 A.D.), the device of pouring boiled water into glass and incasing it in snow, adding that it was generally admitted that water was more wholesome when boiled. Pliny also said that sea water may be freshened by filtration through argillaceous earth. Perhaps most remarkable of Pliny's assertions is that water in which iron had been plunged was useful as a potion in many diseases, particularly dysentery (11).

None of the writers thus far cited was an engineer. None of them mentioned the treatment of public water supplies. But Sextus Julius Frontinus, who in 97 A.D. became *curator aquarium* or water commissioner of Rome, deserves to be called an engineer and he was the author of a treatise on public water supply. He had previously filled many important civil and military positions and, on assuming a new office, he wrote, he always considered that "the first and most important thing to be done [was] to learn thoroughly what I have undertaken." Accordingly, in 98 A.D., he produced *De Aquis Urbis Romae Libri II* [*Two Books on the Water Supply of Rome*], the first known detailed description of a water works system. It was made available to English readers by the noted American hydraulic engineer, Clemens Herschel, in 1899 (21).*

Of particular interest is Frontinus's description of a settling reservoir at the head of one of the aqueducts supplying Rome and of the *piscanae* or ingeniously designed pebble catchers built into most of the aqueducts.

Albert Neuburger, in *The Technical Arts and Sciences of the Ancients,* mentions the "piscana mirabilis" at Baiae, Italy, which "served the double purpose of storing and clarifying the water simultaneously" (22).

In the writings of Galen, the noted Greek physician (130–200 A.D.), we find that the Egyptians used water filtered through earthen jars (23).

* The study of Frontinus was Herschel's avocation for many years. He made a "pious journey" from New York to Cassino, Italy, and there had photographed the oldest known manuscript of Frontinus. Besides reproducing this in facsimile, Herschel printed on facing pages the text in Latin and his own translation into English. These versions were supplemented by explanatory chapters based on wide library research and on personal inspection of the remains of the aqueducts of Rome.

Paulus Aegineta, a Greek physician of the seventh century A.D., a great traveler and a writer of works on medicine that were chiefly compilations and commentaries, noted that water containing impurities or having a fetid smell might be made fit to drink by boiling or by mixing with judiciously selected wine (24). Marshy, saltish or bituminous waters, he said, were benefited by straining.

Arabia and Persia

We come next to the highly specialized treatise on distillation by the noted Arabian alchemist, Geber, of the eighth century A.D. (25). He described not only various stills for water and other purposes but also a filter leading from one stone vessel to another, which must have been a wick siphon. This also he called a still. The object of distillation, by whatever means, Geber said, was the "purification of liquid matter from its turbulent feces, and conservation of it from putrefaction." The "invention of pure water," he said, was for "the imbibition of spirits and clean medicines."

Avicenna, an early eleventh century Persian physician of the Arabian school, gathered in his *Canon of Medicine* what he thought were the most important and practical means of water purification (26). Needs of travelers were his chief concern. He did not mention filtration but he advised travelers to strain all their drinking water through cloth. Boiling he thought more efficient than distillation or sedimentation in overcoming objectionable matter. If a thick water were left to stand for a long time, he said, hardly anything would be deposited, while boiling produced an abundant precipitate, rarifying the water and leaving it light and clear. Aeration by the force of gravity in aqueducts, or by agitation resulting from lifting water from wells, he deprecated, not realizing that water flowing through closed aqueducts instead of being aerated loses its free oxygen and that when well water is bad it needs *more* aeration than the little that is given by drawing it to the surface. Particularly significant was Avicenna's declaration that acetous water added to rain water corrects putrefaction and provides immunity from the possible ill effects on human beings.

Another Engineer

In 1591, some fifteen centuries after Frontinus, an engineer again appears in the quest for pure water. This is the Italian military engineer and soldier of fortune, Federigo Gianibelli, who, while in the service of Queen Elizabeth, floated "fire-ships" down the Scheldt, blew up a bridge and a thousand Spaniards, and thus relieved the besieged city of Antwerp. Afterwards, while at London in 1591, and still in Elizabeth's service, Gianibelli prayed Lord Burghleigh to make known to the queen an "invention" for cleaning certain "filthy ditches round about the city" and displacing their contents with "plenty of wholesome clean water for the use of the inhabitants" and for fighting fire. Strype, in his 1720 revision and extension of Stow's *A Survey of the Cities of London and Westminster,* does not say how the waters were to be cleansed (27). Later writers have asserted that filtration was proposed but they cite no authority in support of their statement.

CHAPTER II

Seventeenth Century

Next to be noted in the quest for pure water is the philosopher Sir Francis Bacon. In his *Sylva Sylvarum, or a Natural History in Ten Centuries,* published in 1627, a year after his death, ten of the thousand "experiments" recorded dealt with water purification (1). Percolation or filtration, boiling, distillation, and clarification by coagulation were briefly reviewed. He repeated the notion that by digging a pit on the seashore, fresh water could be obtained from salt water by percolation through the sand. Caesar had been wrong, Bacon held, in believing that the water he obtained by digging pits at Alexandria (see Chap. I) came from springs fed from landward. Bacon had read that an experimenter who passed salt water through ten vessels had failed to freshen it, but that another person had succeeded by using twenty vessels. Ignoring the alleged success with twenty vessels, and accepting the old claims of success with the pit on the seashore, Bacon said that what man did not accomplish nature might and he pointed out that the water passed downward through the many vessels while it passed upward in the pits by the seaside. In another "experiment," he noted that clarifying water tends to improve health and increase the pleasure of the eye, and described clarification as "effected by casting and placing pebbles, at the head of a current; that the water may strain through."

Sir Kenelm Digby, naval commander, philosopher and author, said in his *Nature of Bodies* (1644) (2) that filtration was a form of motion familiar to "alchymists." He then described and explained the action of the wick siphon. In his better known *Powder of Sympathy* (3), in setting forth the various forms of "attraction," he again described the wick siphon.

Johann Rudolf Glauber, a German chemist and physician, who in 1658 produced the commercial salts bearing his name by treating sodium chloride with sulfuric acid, wrote (4) of the use of salt (presumably sodium sulfate) in ships "against thirst and also the Scurvy." Earlier (1651?) he had described methods of precipitating salt from sea water by passing it through a singular sand which not only drives

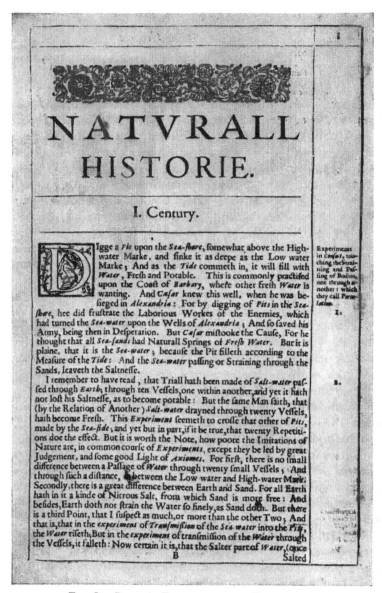

Fig. 2. Bacon's Experiment on Filtration

(Half-title and opening paragraph of *Bacon's Sylva Sylvarum, or a Natural History in Ten Centuries*, first published in 1627)

salt downward but also "all Phlegm, Sordes and Impurities" so that although the water should be "like to a Fen or Dunghill," in a few moments it would become clear and free from odor and taste. Glauber also indorsed the notion that passing sea water through sand would freshen it, because the sand would imbibe the salt, "for those two have a mutual communion and communication, seeing that both are generated of Water . . ." (4).

Catch channels to intercept the runoff from hillsides, with catch pits covered with grates to retain mud or filth carried by the water are described in Rapin's Latin poem *Hortorum* or *Of Gardens* (1665–66) (5).

The advent of the patent era had now been reached. In 1675, William Walcott and, in 1683, Robert Fitzgerald were granted British patents on methods for making salt water fresh (see Chap. XV).

Porzio's Multiple Filter

The first known illustrated description of sand filters was published in 1685. Its author was Luc Antonio Porzio (Lucas Antonius Portius), an Italian physician who had gained distinction at Rome and Venice and gone to Vienna in 1684. The Austro-Turkish war of 1685 led him to write a book on conserving the health of soldiers in camps, probably the earliest published treatise on mass sanitation (6).

Multiple filtration through sand, preceded by straining and sedimentation, was proposed by Porzio. The filters could be placed in the hull of a boat or on land, depending on whether surface or ground water was to be purified. For a boat, he showed three pairs of filters, each pair consisting of a downward-flow filter and an upward-flow filter —the larger the boat and the more filters, the better the filtrate.

Water entered the first or settling compartment of the boat through a perforated plate acting as a strainer. The top or clearest water in the settling compartment flowed through two funnels in the top of the first partition, passed down through the first filter, out from it through oblong openings in the bottom of the second partition and up through the second filter of the first pair. The course of the water was the same in succeeding pairs of filters. Pebbles were placed "near the funnels of each partition." The funnels were diagonally across from each other in each pair of filters to increase total travel of the water between the raw-water and clear-water basins. "The smaller

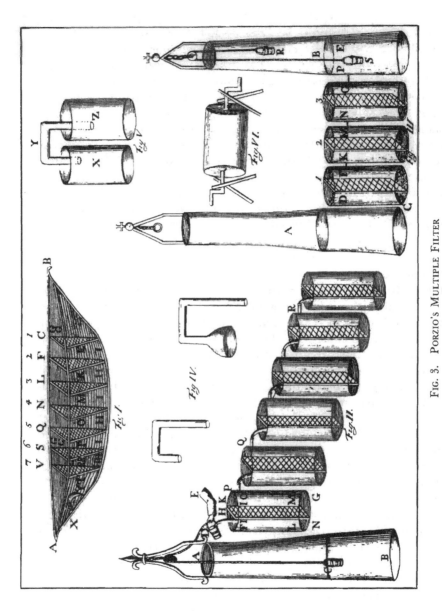

Fig. 3. Porzio's Multiple Filter

Three types of filters are shown: one floating; two on land, for well water

(From *Militis in Castro Sanitate Tuenda*, 1685. Translated into English as *The Soldier's Vade Mecum*, 1747)

the pebbles and the larger the sand," Porzio said, "the better they are; but there is no necessity for being over-nice in this respect, since 'tis sufficient that they be both clean."

Porzio said his plan was an imitation of Nature's method of passing water through the "Bowels of the Earth," producing at last "Fountains whose Waters are good and salutary." One statement by Porzio (6) seems to indicate that he had built multiple filters and certainly indicates that he had constructed a simpler type of sand filter. He said, "We have also made use of the same Means employ'd by those who built the Wells in the Palace of the Doges in Venice and in the Palace of Cardinal Sachetti at Rome."

Filter-Cisterns at Venice

Venice, Queen of the Adriatic, occupies a unique position in the quest for pure water. Built on a hundred islands, she depended for 1,300 years primarily on catching and storing rain water in cisterns. Her only means of supplementing this was to bring water in boats from the mainland. Other cities largely dependent on rain from the skies above them could build large storage reservoirs, as was done at ancient Carthage. This could not be done on the sea-encircled isles of Venice, and therefore hundreds of cisterns were built. Many if not all of these were surrounded with sand filters. The cisterns must have dated from the founding of Venice in the fifth century A.D. but how soon and how generally filters were provided cannot be stated. Probably it was soon enough to put Venice first among cities to be largely supplied with filtered water.

Most famous of the Venetian filter-cisterns were two in the great courtyard of the Ducal Palace, which was founded early in the ninth century. After being several times destroyed, rebuilt and extended, it attained its "perfected" state in 1550 (7). There remained to be brought into harmony with the palace, however, the curbs protecting the cisterns. To that end "well-heads" of richly sculptured bronze were provided—one, in 1556, by Nicolo de Conti, Director of the Foundries of the Republic, the other in 1559, by the sculptor Alfonso Alborchetti. Although the dates and the designers of the well-heads are known those of the filter-cisterns are not.

A half century after these well-heads were completed, they were seen by Thomas Coryat and described in *Coryat's Crudities,* his book of travels published in 1611 (8). No filters were mentioned by Coryat.

Fig. 4. Sculptured Bronze Curb to Venetian Filter-Cistern

In courtyard of the Ducal Palace, Venice. Curb placed in the 1560's
(From a water color by Birket Foster made about 1868. Reproduced from Cundall's *Life of Birket Foster*, 1906)

In the part of his book relating to the Ducal Palace at Venice he wrote:

———in the middest of the court there are two very goodly wels which are some fifteen paces distant, the upper part whereof is adorned by a very faire work of bronse that incloseth the whole Well, whereon many pretty images, clusters of grapes, and of Ivy berries are very artificially carved. There is a fair ascent to each of these wels by three marble greeses [steps?]. They yield very pleasant water. For I tasted it. For which cause it is so much frequented in the Sommer time, that a man can hardly come thither at any time in the afternoone, if the sunne shineth very hote, but he shall finde some company drawing of water to drinke for the cooling of themselves. (8)

More detailed were the observations of Porzio, made at Venice in 1683 and related in his treatise on military sanitation (6). In the very dry summer of 1683, said Porzio, when vast numbers of Venetians

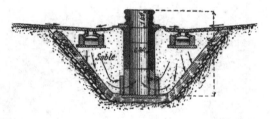

FIG. 5. CROSS SECTION OF VENETIAN FILTER-CISTERN
Accompanying text also described filter-cisterns in the Orient
(From Edouard Imbeaux's *L'Alimentation en Eaux*, 1902)

flocked to the wells at the Ducal Palace and they soon became dry, the water was "salutary and pure" because there was no mud or nastiness at the bottom.

The water came into the wells "perfectly purify'd" because all around the wells was "a large quantity of Sand, which the Venetians call the Spunge of the Wells." Surrounding the sand was "a kind of Fence" of "fat earth . . . which hinders the salt Water from penetrating into the Well." Rain water, or water brought "from adjacent Rivers in Boats," was conveyed in pipes or canals to the sand, flowed through it and the sides of the well, falling into it "clear and pure." Many of the bad qualities of the water were corrected by passing through the sand, especially "their Taste and Smell of Pitch and Tar" resulting from four or five hours of contact in the "small pitch'd Boats" used as carriers. This correction, Porzio said, led him

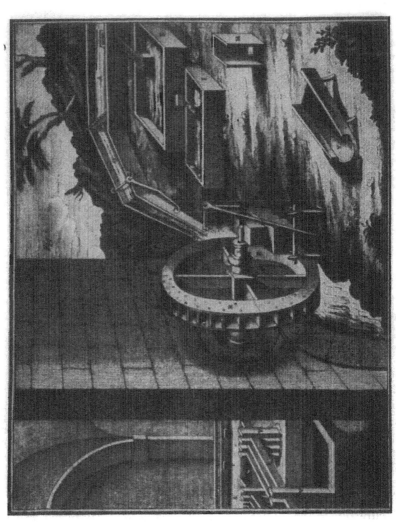

Fig. 6. Water Power and Clarification Plant for Seventeenth Century Paper Mill at Auvergne, France. Water for use in mill is drawn from power flume through two settling tanks working in series, passing through screens en route; M is detail of stilling box, shown in reverse

(From de la Landes' *Art de Faire de Papier*)

SEVENTEENTH CENTURY

to believe that filtration was "a very efficacious Method of correcting the bad Qualities of Water" (6).

In 1835, Matthews' *Hydraulia* (9) gave the depth of the excavation for the filter-cistern at the Ducal Palace as 24 ft. and said the bottom was shaped like a parabola 6 or 7 ft. in "diameter."

In 1855, the Imperial Gazetteer (10), in its article on Venice, said that in the inner court of the larger houses of the city a rain-water cistern was never wanting. In addition there were then 160 public cisterns.

A comprehensive article on the water supply of Venice as it existed shortly before a modern water works system was built appeared in *The Practical Mechanics' Journal* in 1863 (11). The land area of Venice was 5,200,000 sq.m., or nearly 12.85 acres. The rainfall in an average year was 32.3 in. Nearly all of the rainfall was collected in 177 public and 1,900 private cisterns. Their joint storage capacity was 7,160,600 cu.ft. or 1,002,484 gal. (U.S.), affording a daily average supply of about 4.2 gal. per capita. This low average was due in part to the absence of sewers and to the practice of washing clothes in the Lagoon.

Describing what had probably been standard practice for some time, the article went on, in substance: The cisterns were seldom deeper than 10 to 12 ft. In building them the earth was excavated to the shape of a truncated inverted pyramid. A timber form was then built against the sides of the pit. On this an inclined wall of well-puddled clay was built. A flat stone was placed in the bottom of the pit. On this as a base a cylindrical wall was built from brick molded to curves. The bricks were laid with open joints. Those in the three or four lower courses were perforated with radial holes. The space between the wall and the inclined puddle lining was filled with sand. If the cisterns were to receive water from roofs and interior courtyards of a palace, the pavement was sloped gently toward the cistern. At each corner above the sand forming the filter, the pavement was dished and connected with perforated stone blocks. The stone blocks were connected with perforated square drain tiles, discharging the water into the filter sand. Pitchers were let down by cords to draw filtered water from the cistern. The water was always fresh and cool, with a temperature of about 52°F. Venice, the article stated, "is a city with nearly no soot and scarcely any dust." Whether because of this or of circumstances not explained, the filter-cisterns were used for years, "without renewal or washing of sand."

In or about the sixteenth century the Senate of Venice ordered a canal built to carry water from the River Brenta to lagoons at the edge of the mainland from which it was taken to the city in barges and from them discharged by buckets into conduits leading to the filter-cisterns. In 1884, the Compagnie Générale des Eaux de Paris completed works to supply Venice with filtered water from the River Brenta (12). In 1924, a company supplied water to Venice from tubular wells on the mainland, eighteen miles distant (13).

Water Purification for Industrial Use

The first known designs for an industrial water purification plant are found in an engraving dated 1698 and repeated later in a French treatise on papermaking (14). The engraving shows the interior of a paper mill at Auvergne, France, with a flume leading to a water wheel and with a branch flume conveying water through grate screens and two small settling basins into the mill. Whether or not this was the beginning of water treatment for paper mills is a question that Dard Hunter, historian of papermaking and curator of the paper museum at Massachusetts Institute of Technology, cannot answer. He reproduced the engraving without giving its date and origin in his *Paper Making Through Eighteen Centuries* (15).

CHAPTER III

Eighteenth and Early Nineteenth Centuries

The quest for pure water in the eighteenth century was keen but the quarry was small. Many and various ideas and experiments on water treatment in Great Britain, France, Germany and Russia, and a few instances of practical achievements, are reviewed in this chapter. French adventures in patenting and promoting filters appear in Chap. IV, "Four Centuries of Filtration in France." Late eighteenth century developments in England and Scotland are found in Chap. V, "British Contributions to Filtration." America did not join the hunt, except incidentally, until the nineteenth century, and did not get into full cry until after the Civil War. Specialized means of water treatment appearing for the first time or in new form in the eighteenth century are given only passing mention in this chapter. They appear under their proper heads in later chapters.

In 1703, the Parisian scientist La Hire (1) presented before the French Academy of Sciences a plan for providing a sand filter and rain-water cistern in every house. He proposed an elevated cistern covered with matting to prevent freezing and also to exclude light that would otherwise foster "a greenish kind of moss" on the water surface. Rain water passed through river sand and stored underground, he said, would keep for years without spoiling and was usually the best water for drinking, washing, bleaching and dyeing, because "it is not mixed with any salt of the earth as all spring waters usually are." This is perhaps the earliest written recognition that spring waters, in their long passage through the earth, may take up objectionable mineral salts.

The long-held theory that sea water could be freshened by filtration was exploded in 1711 by the Italian oceanographer, Marsigli (2), when he reported that sea water was not made drinkable by filtering it fifteen times through superimposed vessels filled with clay and sand. Rochon (3), who cites the experiment of Marsigli, also reports an experiment made by Nicollet and Réaumur. They used a zigzag glass tube 1,000 fathoms or 6,000 ft. long, filled with sand but, says Rochon, the water came out as salty as when it was put in.

Boerhaave, noted Dutch physician and chemist, in *Elements of Chemistry* (4), first published in Latin in 1724, said that putrid water could be cured by boiling, with the addition of acid, and that spirit of vitriol would prevent putrefaction. Only one boiling of putrid water, he thought, would destroy the animals that generated spontaneously in the water, sending them to the bottom with other impurities. Then by adding a small quantity of strong acid the water would be made fit for use. This method, Boerhaave said, had given excellent service under the equator and between the tropics "where the waters putrefy horridly" and breed large quantities of insects, yet must be drunk.

Dr. Stephen Hales, writing in 1739 (5), reported experiments he had made to determine the ratio of vitriol necessary, as a small excess might be dangerous. He found that three drops of oil of sulfur in a wine quart ($\frac{1}{4}$ U.S. gal.) preserved "water from stinking for many months . . . the purer the water the less the acid spirit" [required]. He had been informed that the Dutch, to prevent water from stinking on long voyages, put into it spirit of vitriol. He also noted a statement in the *History of the Academy of Sciences* (1722) that fresh water had been kept from putrefying and breeding insects by fuming the cask with burning brimstone.

Various Reports on Filtration

Advanced ideas on water purification were recorded by the scientist Plüche in a popular work called *Spectacle de la Nature*, first published in 1732 (6). He reported that rivers fouled by the filth of cities were improved by sedimentation and by exposure to the sun, and said it was customary to let muddy river water stand in earthenware vessels a few days, in which time it settled and became clear as crystal. "Several" used copper cisterns with sand in them "through which by an artificial kind of filtration [the water] clarifies sooner and with equal safety, provided the sand it drains through be often wash'd, and the vessel [be] well tinned within to secure it from verdigrease." This is the first clear-cut recognition found on record of the slowness of sedimentation compared with filtration. It is also the first contemporary record of the need for occasional cleansing of sand in filters and for protecting copper containers from the verdigris that gave

Amy, a few years later, so much concern. It shows, too, the customary use of sedimentation and the less common use of sand filtration in 1732. A passing remark by Amy, in 1750, carries the use of household sand filters in Paris back another two centuries, or to about 1550.

Extensive use of filtering stones early in the eighteenth century is shown in a paper by Dr. Abraham Vater (7) describing the filtering stone of Mexico and comparing it with other stones then in use.

The earliest mention of filters in English and French encyclopedias is in the fifth edition of Chambers (1741) (8), which describes filtration through sand and pulverized glass and by ascent through cotton. The seventh edition of Chambers (1751–52) is a little more explicit.

In striking contrast to Chambers, Diderot's *Encyclopédie* (9), which featured science and industry (volumes pertinent to this discussion published in 1756–57), took up filters for use in pharmacy and manufacturing (*filtration en grande*); described stone filters used extensively in Japan; and devoted a page to *fontaines domestiques*, as the household filters then widely used in Paris were called. Although in 1757 the household filters were of many kinds, most of them appear to have been of sand, inclosed in copper vessels, the sand resting on and being covered with perforated copper plates. The article illustrated in detail and highly commended Amy's filter (see Chap. IV).

Ten years later Croker's *Complete Dictionary of Arts and Sciences* (10) defined "filter" as a strainer of filtering paper or bag of woolen or linen cloth. A page was devoted to depuration in pharmacy by decantation, by despumation or percolation through strainers of flannel, linen, cloth or paper and "by the attraction of a pendant thread." The first edition of the *Encyclopedia Britannica* (1771) had only this sentence on filtration: "Filter or filtre, in chemistry, a strainer commonly made of bibulous or filtering paper in the form of a funnel, in order to separate the gross particles from water and render it limpid."

Six British Questers for Pure Water

In the brief period, 1753–69, a half dozen English and Scotch questers for pure water came on the scene: Doctors Lind, Home, Butler, Rutty, Heberden, Percival and Buchan.

Lind on Supplying Water for Seamen.—James Lind, Scotch physician and naval surgeon, gave much attention to the purification of water for use on sea and land in *A Treatise on Scurvy* (11), first published in 1753, and in *An Essay on the Most Effectual Means of Preserving the Health of Seamen in the Royal Navy* (12), which appeared four years later. Both these books were pioneers in their field.

One method of at least delaying putrefaction of water at sea mentioned by Lind was that of fuming the casks with burning brimstone; another was that of adding a little oil of vitriol. A common and a good practice of curing putridity was that of throwing a little salt into water while it is being boiled and removing the "thick feculant unwholesome scum" that rises. This should always be done, he asserted, when peas or oatmeal are boiled. Another method of curing water that is putrid and stinking was that of taking the bungs out of the casks, shaking the casks or pouring the water from one vessel to another. Safer for common use than applying vitriol or salt is adding juice or extract of lemon. Acid, he explained, will precipitate the earthy particles in the water together with "various animalcules with their sloughs, now destroyed by boiling."

The stone filter sometimes used on ships to soften and clarify water, Lind said, was "very proper" if the water did not abound with "vitriolic or marine salts" but it could never yield enough water for a ship's company. "Sand," he added, "is the finest body for separating these heterogeneous particles from water." There followed a description of a multiple filter from "the ingenious essay of Sir Francis Home on the Dunse Spaw" (published in 1751). This seems to have been based on the filter described by Porzio in his book of 1685 (see Chap. II).

Lind elaborated on filtration through sand, and several other methods, in *An Essay on the Most Effectual Means of Preserving the Health of Seamen in the Royal Navy* (1757) (12). Bad water, next to bad air, he said, is a frequent cause of sickness, especially near the torrid zone. Where the water found on shore is bad, the ship's casks should be filled with rain water, if possible; or, if fuel is available on shore, sea water may be distilled, "which will prove as wholesome as that of the Thames." He gives, in a footnote to the revised edition of 1762, the substance of a paper read before the Royal Society, dated April 26, 1762. He advised digging a pit on shore to find water. If the water thus found were foul or impure, the pit should be made deep and large, its bottom and sides lined with large stone and then sand and gravel thrown in. By this means the water "will often become in 24 hours clear, soft and wholesome." If not, then a filter could be constructed by placing a small cask inside a large one, both headless; then putting sand and gravel in the small cask and in the concentric space between the two casks. Water poured on the sand of the inner cask will pass down through the inner cask and up through the outer cask where it may be drawn off through a cock.

When the surface sand of the inner filter became "loaded with the gross impurities of the water," it could be replaced with fresh sand. For private use, a large funnel could have a bit of sponge inserted in its narrow part, sand placed above that, then a piece of flannel and finally sand again. The materials had to be changed when they became fouled.

Other practices mentioned by Lind were: Toasted biscuit put into the water of the St. Lawrence River prevented fluxes in Sir Charles Saunders' fleet—about 4 lb. of burnt biscuit were used to a hogshead of water. Powdered ginger for the same purpose was mixed with water for troops in Canada, with resulting benefit. Vinegar in small quantity "is an excellent corrector of unhealthy waters"; also cream of tartar. Most notable of all is an early instance of the use of lime in water purification described by Lind:

> At Senegal, where the Water is extremely unwholesome, unquenched Lime has been used to purify it. . . . But water cannot be thus purified in a Ship, because I find [note the personal pronoun] that it must be exposed in a very wide mouthed Vessel for many Days, and sometimes Weeks, before it looses the Taste of the Lime: much of it also is expended, by daily removing the Scum; and it will sometimes require boiling. (12)

Home and the Appearance of Water Softening.—Dr. Francis Home, Scotch physician, experimenter and writer, contributed greatly to the knowledge of water softening in *Experiments on Bleaching* (13). His book was published in 1756 but apparently the experiments were made earlier. His studies appear to have included the first scientific experiments on water softening ever made and he seems to have been the first to suggest that softening be applied to city water supplies.

Butler and Soap Lye for Purification.—Dr. Thomas Butler (14) wrote in 1755 that if a wine glass of the strongest soap lyes were added to 15 gal. (Imp.) of sea water, and the water distilled, 12 gal. of fresh water generally would be produced. To keep fresh water sweet, he recommended putting ¼ lb. avoirdupois of "fine clear white pearl ashes" into 100 gal. of fresh water.

Rutty's Mineral Waters.—Dr. John Rutty, in his thick quarto treatise, *A Methodical Synopsis of Mineral Waters* (1757) (15), has a chapter on "Common Waters." He notes the "great importance in building a town to chuse a proper situation with regard to the quality of springs." Common spring waters, if not immoderately hard, may be softened by standing a few days, those at Henly, England, and divers others requiring only two days' exposure to become "fit to wash with." One method of softening was to put into the water an alkaline salt. Rutty mentions adding two or three drops of oil of sulfur per quart of water to keep river water from putrefying, and cites Dr. Butler's plan of using pearl ashes, and a proposal by a Dr. Alston to put a pound of quicksilver in a hogshead of water. In a long chapter on "Alum as an Ingredient in Mineral Water," Rutty says he had "found many spring waters to become more limpid upon the admixture of alum." In Kent, muddy pond water was cleared by throwing into it a little alum. Dr. Rutty quotes Churchill's *Collection of Voyages and Travels* as relating that a certain yellow or red river water in China was cleared and made drinkable by putting a little alum into a jar of it and shaking it about.

Heberden's Pump Waters.—Dr. William Heberden, in his discussion of London pump waters (1767) (16), took up softening, sedimentation and coagulation followed by filtration and distillation. Softening could be effected by boiling, which would free water of "most of its unneutralized limestone and selenite" but increase "saline matter." Limestone, either "loose" or "united to the acids," could be precipitated by adding salt of tartar—10 to 15 grains per pint! Heberden said that, although Prospero Alpino reported (*De Plantis Aegyptis Liber,* Venice, 1592) that powdered almonds were rubbed on the inside of jars containing Nile water as an aid to sedimentation in Egypt, he could not find it of any use. The common people of England, Heberden stated, successfully used alum to purify muddy water, 2 or 3 grains dissolved in a quart of water making the dirt very soon flocculate and then slowly precipitate. Subsequent filtration made the water fit for use. Such a small addition of alum "will hardly be supposed to make the water unfit for any purpose." No earlier suggestion for coagulation followed by filtration nor concern over the use of alum has been found. Distillation was given more space by Dr. Heberden than any other method of water treatment. To free the distillate from burnt taste, it could be let stand or boiled in an open vessel or air could be forced into it, as suggested by Dr. Hales. The wholesomeness of distilled water could hardly be doubted, Heberden thought, if it were recollected that Nature distilled all the fresh water in the world.

Percival's Well Waters.—Thomas Percival described (1769) (17) some experiments on treating hard and in some cases polluted well waters at Manchester, England. Hard water, first well boiled, then filtered through stone, was rendered "tolerably pure, potable and salutary and at the same time better adapted to a variety of culinary uses."

Buchan and Purification of Municipal Supplies.—In *Domestic Medicine* (1769) (18), by Dr. William Buchan of Edinburgh, is found the first positive statement that water should be purified, if need be, before it is supplied to great towns. After saying that many diseases may be caused or aggravated by bad water, and that, once a supply had been procured at great expense, people were unwilling to give it up, he wrote that the common methods of rendering water clear by filtration, or soft by exposing it to the air, etc., were so generally known that it was unnecessary to spend time in explaining them. What the common methods of filtration so well known in 1769 were, Dr. Buchan did not tell. So far as has been found, no city water supply had been filtered up to that time.

Infiltration

In a French book on rural economy (1767), Jean Bertrand, pastor at Orbe, Switzerland (19), discussed retention of water in ponds and its aeration and natural filtration through rocks and sandy earth. He said he had no doubt that nature could be imitated by passing water through an artificial bank of sand. Filtration, he believed, would be of inestimable advantage to a city supplied with water that caused goiter or had other defects. A suitable means of correcting chalky and viscous waters, he reported, was to pass them through green branches of fir trees pressed into a pond above or just below its outlet.

Pits near the tidal reaches of the Senegal River, West Africa, were seen by J. P. Scotte, M.D., about 1776 (20). During the rains, Dr. Scotte wrote, the water of the river is fresh, but very thick and troubled. In dry months the river is salt, because the low stage of the river permits the backing up of water from the sea. Where a pit is dug into the sand, water "filtrates from all sides and gathers up to the level of the river." Although brackish, this infiltrated water was used by the garrison and the inhabitants of the town.

Lowitz's Studies of Charcoal

The efficacy of powdered charcoal in preventing or removing bad tastes and odors from water, and incidentally in clarifying it, was established experimentally by Johann Tobias Lowitz in 1789–90 and detailed in a paper read before the Economics Society of St. Petersburg in 1790 (21).

Lowitz (1757–1804) was born at Gottingen, Holland, son of the German astronomer, George Moritz Lowitz, who became director of the Gottingen Observatory. The son was an eminent member of the Imperial Academy of St. Petersburg and a professor of chemistry.

In 1785, Lowitz showed that charcoal would decolorize brandy. The studies on charcoal centered on water for use on long voyages, but spring waters were also considered.

Addressing himself first to prevention rather than cure of tastes and odors, Lowitz said water free from heterogeneous particles is not subject to putrefaction, but, owing to its dissolving powers, it is difficult to preserve it long in a pure state. To keep it pure on shipboard it should be stored in vessels of glass or earthenware. This being impracticable, wooden casks were used. Such casks impart a great quantity of mucilaginous and extractive matters, which, in a state of division, furnish innumerable living creatures, the decomposition of which causes putrefaction. Clean casks are essential because the smallest quantity of corrupted matter left in them acts as a ferment which causes the water to putrefy. Therefore he advised that the casks be washed with hot water and sand, or any other substance capable of removing the mucilaginous particles. The clean casks having been filled with water, a small amount of vitriolic acid should be poured in, then powdered charcoal stirred in to nullify the acid taste. Six drams of powdered charcoal would deodorize and clarify three pints of water, provided 24 drops of vitriolic acid were also used. One cask of powdered charcoal would deodorize and clarify 34 casks of water; but more would be required to cure bad taste. Water drawn from the cask should be passed through a filtering bag, apparently filled with charcoal.

Spring waters having a hepatic or liver taste, said Lowitz, should be filtered through a bag half filled with powdered charcoal, but unless it was loaded with mucilaginous particles, use of acid would be unnecessary.

The influence of Lowitz's paper was strong for many decades. His suggestions on the use of charcoal for filters, either alone or with other media, gained particular eminence.

Filter Ships

Porzio's plan for filter boats adjacent to army camps, published in Vienna in 1685 (see Chap. II), was simplified and put into effect during the siege of Belgrade on the Danube, in 1790. The German technical encyclopedist Feldhaus (22) credits the Austrian Surgeon General, Mederer von Wuthwehr, with transforming putrid into potable water by boring holes in the bottom of old transport ships, putting in successive layers of cannon balls or coarse gravel, coarse sand, fine wood ashes and fine sand. It was claimed that the water issuing through these filters was as clear as that from artesian wells.

Cavallo on Water Purification

Tiberius Cavallo, F.R.S., who was born in Naples in 1749 but settled in England before 1775, published a treatise on natural philosophy in 1803 (23) and an essay on preserving and purifying water in 1807 (24). In the treatise he described filtration "as a finer species of sifting [through] the pores of paper, or flannel, or fine linen, or sand, or powdered glass, or porous stones," and said that filtration would remove suspended but not chemically combined matters—that is, mud but not salt—from water.

In the essay he stated that animal, vegetable and earthy substances might be eliminated from water by sedimentation, either before or after fermentation. Thames water, after standing in casks a few days, fermented and "stinks intolerably." It became covered with scum, yielding inflammable gas which might be lighted by a candle. But when fermentation had decomposed animal and vegetable substances, which took a few days, the components flew off in the air or fell to the bottom of the casks, leaving the water sweet and clear. He taught that most stagnant waters may be purified by filtration and water putrid with animal or vegetable matter may be purified by agitating in it powder of freshly made charcoal, then filtering it. Charcoal mixed with sand (in filters) purifies muddy and stinking waters, but the charcoal powder must be renewed frequently. Or, he suggested, a pound of quicklime may be added to 1 gal. of water, letting it stand for six to eight hours, stirring at intervals, and then filtering and exposing it to the atmosphere, with frequent agitation, for a few hours. According to Cavallo, sea water may be made drinkable by freezing or by distillation, but never by passing it into wells dug on the seashore—"no human art could ever effect it by filtration." Distillation will purify hard water; but if the water contains "volatile ingredients, and expecially the putrid effluvia of animal and vegetable bodies," quicklime should be added.

Cavallo described a filtration "machine" which was either a cask, open at both ends, or an earthenware chimney pot, placed in a tub or cistern; both vessels were filled to about three-quarters of their height with sand which had been repeatedly washed in boiling water to free it from clayey or saline particles. Muddy water poured upon the sand in the inner vessel would descend through it, then rise up through the sand outside. But double filtration, downward then upward, as has been seen, was not a new method. Municipal supplies are not mentioned by Cavallo.

CHAPTER IV

Four Centuries of Filtration in France

Filtration in France has been practiced for at least four centuries. During the first half of that period the art was limited to filters for household and probably small-scale industrial use. The filter containers were vessels of metal or pottery, containing sand resting on a perforated plate or false bottom. Raw water was poured or dropped into the top and filtered water drawn from beneath the false bottom.

In the eighteenth century several attempts were made to commercialize filtration—the first, about 1750, by patenting, manufacturing and selling filters, chiefly for household use. In the 1760's and 1780's, two promoters obtained patents or licenses for filter plants. One of these promoters built a filter plant and organized a city-wide carrier system of delivering water in any amount, as frequently as desired, in sealed containers.

At the very close of the century a patent was taken out in France for a filter on which many others were modeled during the next 60 years—although with important modifications. A unique purification plant, large for the time, was installed at Paris in 1806 and another a few decades later. Then came others, of a new type of filter, at public fountains in Paris. At Nantes, 36 notable filters were included in water works completed in 1855.

Most of the water filtered in Paris up to the end of the nineteenth century was sold on the cash-and-carry plan or else by licensed porters. Only a little was piped to houses. At Nantes, the product of the filters of 1855 was conveyed to houses by a separate distribution system.

Sponge, then charcoal, then wool were promoted as filtering media in France for more than a century: sponge from 1745 onward; charcoal from shortly after Lowitz presented his famous essay at St. Petersburg in 1790; wool for a short period in the 1840's; then sponge and wool in separate layers. Sponges were occasionally used as prefilters over a period of years. These media were generally used in combination with sand, crushed sandstone or gravel.

French persistency in using sponge as a filtering medium was remarkable; even more so was the later use of waste wool. Sponge met with official protest at the middle of the eighteenth century and patents then being sought were modified either by omitting the sponge or giving it a minor function, as a prefilter. The English encyclopedist, Tomlinson (1), commented adversely in 1852 on the use of any organic matter, except charcoal, as filtering media. When such material was kept wet constantly it decomposed and "imparted impurities to the water which it was intended to purify and make potable." Tomlinson may not have known that both sponge and wool were given special treatment before use in some of the nineteenth century filters.

English slow sand filters have found little favor in France, partly because water from springs or "natural filters" (galleries) is preferred to that from any kind of artificial filter. Paris continued to use spring water for domestic consumption, or in other words did not attempt to filter its main supply from the Seine and Marne, until the close of the nineteenth century, when it adopted multiple filters working in series. These were composed of two or more *dégrossisseurs* or roughing filters of gravel, a prefilter of coarse sand and a final filter more or less of the English slow sand type. Ably engineered and promoted, this system came into wide use throughout France after its adoption at Paris. Multiple filtration (see Chap. IX) had been proposed from time to time and place to place since 1685 and had been used in several installations, notably in Scotland and England.

The outstanding innovation of French filter designers was the use of high pressure in closed tanks. This practice began in the 1830's, a half century in advance of the American rapid filter. Efforts to introduce the earliest makes of American rapid filters to municipal supplies in France were unsuccessful, as they also were in Germany.

Amy's Filter Patents

Midway in the eighteenth century Joseph Amy was granted the first water filter patent issued by any country. Also, he published the first book on filters to appear anywhere in the world and founded the first-known filter manufactory. His patent application was filed in 1745. His book, *Nouvelles Fontaines Domestiques* (2), appeared in 1750. This was followed by a pamphlet in 1752 (3), a second book in 1754 (4), and a second pamphlet in 1758 (5). Altogether a thousand pages were printed to promote Amy's filter and to expose the menace of copper poisoning from the filter containers and other household vessels then widely used in Paris. No such amount of publicity material was ever issued by any other filter promoter.*

Amy was from southern France, advocate in the Provence Parliament. During a visit to Paris in 1742, he drank without ill effects water from the Seine that had stood in jugs of earthenware instead of copper. Returning to Paris in 1745, he drank water from copper containers, had symptoms of poisoning and had a long illness. He concluded that it was not the water from the Seine but verdigris on the vessels containing it that made strangers in Paris sick.

Introduction of Sponge.—Household filters of sand in copper containers, wrote Amy, had been used in Paris for two centuries (since about 1550). Believing that sand as a filter medium was objectionable, and copper a menace, Amy decided to substitute sponge for sand and some other material than copper as a container.

In his patent application of 1745, Amy showed a design for both a large and a small filter. His large filter was composed of many sponges pressed into holes in the sides of an open-topped wooden box floating in the water to be filtered. The filtrate passed through a flexible tube to a clear-water boat-shaped trailer. If desired, multiple filtration could be had by means of sponges inserted in crosswise partitions in the trailer. For smaller quantities of water a vessel of lead or earthenware, with sponges inserted in holes in its sides, was used.

On the strength of a favorable committee report from the French Academy of Sciences, royal assent to the granting of Amy's patent application was given on June 15, 1746, but subject to the approval of certain Paris functionaries. Difficulties followed. To overcome these, Amy supplied one of his filters to the scientist Réaumur. After trial in his house, Réaumur expressed preference for the sponge filter

* All four of Amy's book and pamphlets, collected by the author, are now deposited in the Engineering Societies Library, New York. The volume of 1750 carries the bookplate of Marquis de Bauffremont, Knight of the Golden Fleece and Lieutenant General of the Armies of the King, to whom the book was dedicated. The second book was dedicated to Senac, Councillor of State and First Physician to the King. The first pamphlet was from the library of Cardinal de Brieure, Minister to Louis XVI. The pamphlets are original folded sheets, unstitched. So far as is known, there are no other copies in America. None is listed in the Union Catalogue of rare books in certain libraries of the United States, compiled by the Library of Congress.—M.N.B.

NOUVELLES
FONTAINES
DOMESTIQUES,
APPROUVÉES
PAR L'ACADEMIE ROYALE
DES SCIENCES.,

A PARIS,
Chez { J. B. COIGNARD, Imprimeur du Roi.
       A. BOUDET, Libraire-Imprimeur.

MDCCL.
AVEC APPROBATION ET PRIVILEGE DU ROI.

NOUVELLES
FONTAINES
FILTRANTES
*APPROUVÉES PAR L'ACADÉMIE
ROYALE DES SCIENCES*

En plusieurs rencontres, dont quelques-unes sont présentées dans ce Livre,
**POUR LA SANTÉ DES ARMÉES
DU ROI,**
Sur Mer & sur Terre, & du Public,
Le tout accompagné de Figures expliquées par lettres indicatives.
Par M. AMY, *Avocat au Parlement de Provence.*

A PARIS,
Chez ANTOINE BOUDET, Imprimeur du Roi, rue saint Jacques.

M. DCC. LII.

SUITE DU LIVRE
INTITULÉ
NOUVELLES
FONTAINES
FILTRANTES.

FIGURE IX.

Fontaine publique pour les Villes de Garnison & autres, où les porteurs d'eau vont puiser dans les rivieres.

Ette figure représente une Fontaine pour purifier l'eau dans une Ville de Garnison, dont la situation oblige les soldats d'aller puiser dans une Riviere trouble par intervalles, comme la Seine à Paris, par les pluies & les fontes de neige, ou dans une Citerne, ou dans une Marre, ou dans
A

APPROUVÉES PAR L'ACADÉMIE. 289

1745.
N°. 469.

MACHINE
POUR PURIFIER L'EAU,
INVENTÉE
PAR M. AMY, AVOCAT.

CEtte machine consiste en une caisse AB, dont les côtés AC, CD sont percés de plusieurs trous, dans lesquels on fait entrer à demeure autant de cornets EEE, &c. que l'on bouche avec des éponges; chaque cornet, comme on le voit par le profil FG, est fixé de maniere que sa plus grande base E, qui est de 6 pouces, est extérieure, & la petite base I de 3 pouces est en-dedans de la caisse; on peut aussi substituer à la place de cette grande caisse, un bateau que l'on percera, & auquel on appliquera les mêmes cornets dont le nombre & l'arrangement est arbitraire. On joint à cette caisse un bateau KR, qui reçoit l'eau clarifiée par le moyen d'un sanche de cuir L attaché aux tuyaux M, N, & pour une plus parfaite purification, on partage la capacité du bateau KR en 3 ou 4 compartimens, par les cloisons O, P, Q, qui sont percées de même que la caisse AB d'une quantité de trous, qui portent de semblables cornets que l'on bouche aussi avec des éponges, de sorte que l'eau qui sera dans la partie QR aura été filtrée quatre fois.

L'inventeur a proposé de joindre la caisse AB, au bateau KR par deux tuyaux de fonte ST; mais en ce cas il faudroit que les deux tuyaux fussent assez solides pour qu'un des corps flottant pût imprimer son mouvement.

Rec. des Machines. TOME VII. Oo

FIG. 7. PAGES FROM THE WORKS OF JOSEPH AMY, FILTRATION PIONEER
Top: Title pages of first known books on filtration, 1750 and 1752
Bottom Left: A page from a sequel published in 1754
Bottom Right: Description of Amy's machine to purify water in *Machines et Inventions Approuvées par L'Académie Royale des Sciences*, 1745

rather than the commonly used sand filters in copper containers. The copper, he wrote, *"engendre un ver-de-gres très redoutable."* He had found that when his domestics neglected to keep the sponge covered with water, the first filtrate, after replenishing the water, had a spongy or marshy taste. When brought to Amy's attention this fault was remedied by a change in design. Réaumur also noted that he had tried cotton, wool, silk and sand as filter media. All gave very good water but he concluded that sand was the best. These are the first known comparative tests of filter media.

Next, a second committee of the Academy of Sciences, with Réaumur as one of its two members, was appointed. In a few days it reported approval of Amy's filter, subject to the changes Amy had proposed.

On July 9, 1749, an order was issued for the registration of the filter patent, as amended, giving Amy the exclusive right to make and sell

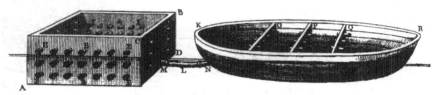

FIG. 8. AMY'S MACHINE TO PURIFY WATER

Approved by the French Academy of Sciences in 1745. Shown are the floating filter box and clear-water trailer; sponges were inserted in sides of box and, if needed, in partitions of trailer

(From *Machines et Inventions Approuvées par L'Académie Royale des Sciences*, one page of which is shown in Fig. 7, facing)

his filter until 1766. The containers might be composed of lead, pewter or earthenware, none of which, the grant said, was subject to verdigris. The use of filter boats had been renounced by Amy. Sponge was sanctioned as a filtering material but the patent included the use of sand also. It was noted that the sand and sponge could be washed in place and that washing was necessary to prevent the water from contracting a bad taste. The sand was to be packed firmly between two plates, the lower one to serve as a false bottom to the filter, the upper to prevent disturbance of the sand when water was poured into the vessel.

Promotion Literature.—The simplest and smallest of the many designs for filters described and illustrated in Amy's books and pamphlets merely provided for the passage of water through one or more sponges inserted in a perforated plate. A larger and more elaborate design, intended for a military garrison, showed three pairs of filters, working in series, the first of each pair filtering downwards and the

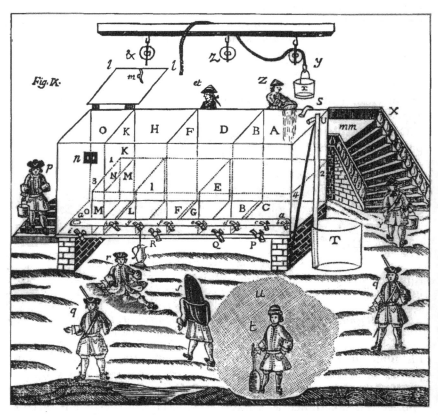

FIG. 9. AMY'S SEXTIFOLD FILTER FOR ARMY GARRISONS

Three pairs of down-up filters; water passed through 18 ft. of sand; tank of lead-lined wood or masonry, 18 x 3 ft. in plan, 6 ft. deep

(From Amy's book of 1754 for which see Fig. 7, p. 32)

second upwards, the water passing through 18 ft. of sand. This design so closely resembles the one described by Porzio in his book on military camp sanitation, published in 1685, that it suggests Amy had copied Porzio's design (see Chap. II). Another filter, designed to supply officers and men of a garrison, had several compartments, with valves for drawing filtered water from the bottom of each. Water could be supplied to the filter by buckets or hand pumps. Sand could be shoveled out for washing.

Amy's extensive work seems to have ended in 1757 or 1758 after he had written the larger part of a 160-page promotion booklet (5). In this he excoriated attackers of his filters, presented illustrated descriptions of various styles of filters, with costs of installing and servicing, and cleaning directions for those wishing to do their own servicing. Long lists of yearly purchasers from 1750 to 1757 were given. These included high government officials of France and other countries, cardinals, royal and other physicians, engineers, plain bourgeois, and the *"première de chambre de Madame le Marquis de Pompadour."* An installation with a capacity of 30 *voies* (approx. 160 gal.) * for a chateau was cited.

The Montbruel—Ferrand Project

An elaborate plan for supplying Paris with filtered, aerated and bottled water was given provisional royal approval in 1763. The patentees were Jean Baptiste Molin de Montbruel and Nicolas Ferrand. The source of supply, wrote Belgrand, Director of the Paris water works, a century later (6), was the Seine at Pont-a-l'Anglais, above the entrance of the turbid Marne. There one or more boats were to be moored into which water was to be pumped, then agitated, filtered through sand, then through sponge, then passed to a reservoir exposed to a current of air. The first filter was to be a 2-ft. layer of "good sand," 4 x 80 ft. in plan. The sponge was to be fine grained, well cleaned, passed through spirit of wine, then placed in 1,280 one-inch cylinders and covered with fine-meshed fabric.

The city bureau to which the patent was referred objected to the use of sponge, as it had done twenty years earlier in the case of Amy's patent. The patentees renounced the use of sponge, holding that sand filtration was sufficient. Thereupon the bureau recommended (July 7, 1763) that the tentative letters of patent of June 2 should be registered.

* Eugène Belgrand (6), in 1877, said the *voie* was 18 to 20 liters; John B. Hawley, who translated parts of Belgrand for use here, wrote in 1937 that a *voie* was formerly 29 liters; later, 23; and averages about 25 liters. It is sometimes loosely given as two bucketfuls.

Belgrand expressed the belief that Montbruel and Ferrand did not utilize the privilege granted to them (6). Four promotion pamphlets (7, 8) show the progress of the enterprise up to a few days before the date announced for the delivery of water to customers, January 30, 1764.

The earliest of the four pamphlets contains a favorable report on the project by three members of the Paris Faculty of Medicine, submitted October 1, 1763. The fourth pamphlet, containing a second report, dated January 11, 1764, from the same committee, stated that water would be delivered to subscribers on January 30. The committee certified that the reservoirs and filters contained no sponge, metal or other material injurious to water. It declared that the output of the plant was clear, limpid and fresh, and free from all foreign tastes.

The fourth pamphlet also listed fourteen bureaus of subscriptions. Water was to be delivered to subscribers in sealed containers, in either 60- to 80-pint jars or 6- to 8-pint bottles, the latter in wicker baskets holding six bottles. Delivery would be daily or at lesser intervals, as desired. The price was *six deniers la pinte, mesure de Paris*—not renderable with certainty in American units, but probably $2\frac{1}{2}$ cents per U.S. gallon.

Reasonably conclusive proof that Montbruel and Ferrand put their project into operation is afforded by Girard, Director of the Paris water works, who wrote in 1812 (9) that the sale of filtered water was not what the promoters expected. This he attributed to the preference of Parisians for spring water delivered by aqueducts. A more potent reason for lack of patronage may have been that so elaborate a plan for supplying filtered water was too far ahead of the demands of Parisians in 1764. Certainly the project did not fail for lack of promotion.

Charancourt's Filter Boats in the Seine

Baron François-Grégoire de Bourbon Charancourt, an engineer, on January 18, 1781, applied for permission to place pumps on boats in the Seine at Paris and, on its banks, fountains to be supplied with purified water. He had observed, he said, that the river was always charged with heterogeneous particles carrying with them the principles of corruption. His purification process, he said, had been successfully used at Toulouse (date not given). Subsequently he had improved the mechanism and had demonstrated its efficiency both at Versailles and at Paris. He had proved that the process did not rob the water of its natural beneficial properties. The process was simple and inexpensive. He wished protection for it before making its nature public.

Each fountain, wrote Belgrand (6), was to have thirteen valves for drawing water, of which twelve would be for the use of the bucket carriers and one for free service to the public. In addition, many of the fountains would have two valves for venders who carried the water in casks.

The application was denied because a city bureau objected that the project had not been approved by the Faculty of Medicine, the boats would impede navigation, and the fountains and their patrons would be in the way. Late in 1781, Charancourt renewed his application. He urged that a trial of his process had shown it to be good. On May 18, 1782, the patent was granted, but for only three boats and six fountains.

Belgrand (6) in 1877, and Girard (9), writing 65 years earlier, agree that the process was kept a secret, but Belgrand says it was never utilized at Paris, while Girard says its success at Toulouse had been certified by the city officials and at Versailles by Monsieur Lassone in a procès-verbal and that, moreover, Charancourt added to the pumping plants in use at Paris. A. Gury (10), former head of the Paris water bureau of today, says that Charancourt opened a horse-driven pumping plant in 1783 and two others in 1784 (thus making up the three which he was allowed to establish). Their locations and the dates of their acquisition by the city are given by Girard. Gury also says that in 1768 there existed a Compagnie Dufaud which had established at Pont Ile Saint Louis, near the Arsenal, water purification works that Charancourt took up again in 1782. Water from this plant was delivered in casks decorated with the arms of the king and of the city. The drivers wore uniforms and announced their approach by blowing trumpets (10).

Mirabeau's Account of Filter Fountains

Data given by three one-time chief engineers of the Paris water works—Girard, Belgrand and Gury, writing in 1812, 1877 and 1939—have been reviewed. A contemporary of Charancourt, the Comte de Mirabeau, also wrote about the filter fountains in Paris (11), three years after final approval was given to Charancourt's project. In what seems to have been the first attack on a private water company for alleged financial abuses, Mirabeau comments favorably on three filter fountains then in use. These seem to have been fountains established by Charancourt. After saying that one would assume that the water supplied by the company under attack was as superior to other water procurable in Paris as that of the New River is to all other waters one can drink in London, Mirabeau declares: "And this is precisely what it is not. . . . The *fontaines épuratoires*, which no one encourages at all, supply to Parisians a limpid water at the same price as the filthiest water. . . . There are as yet only three of these fountains; they ought to be established on the two banks of the Seine in all the length it runs through Paris." A stove in the bottom of the fountain and serpentine pipes, wrote Mirabeau, kept the water from freezing and thus interrupting service in winter.

From the combined testimony of Girard, Gury and Mirabeau it appears that commercially filtered water was made available to Parisians by Montbruel and Ferrand in 1764, by Compagnie Dufaud in 1768 and by Charancourt in 1782—Belgrand to the contrary notwithstanding.

The Smith-Cuchet-Montfort Filter

In 1800, the basic Smith-Cuchet-Montfort patent was granted by France and, in 1806, the Quai des Celestins filters, which operated for a half century or more, were established in Paris.

James Smith, a gunsmith from Glasgow, for a short time helped Richard Younger of Edinburgh, formerly a brewer, to assemble filters, the manufacture of which Younger began in or about 1795. These filters, wrote John Wilson (12), in 1802, were the most remarkable of the devices proposed up to that time to purify water by the use of charcoal, in accordance with the proposals of Lowitz (see Chap. III) and others.

Smith, having brought the Lowitz process to the attention of the French Minister of Marine "as an important secret," says Rochon (13), was sent to Brest. Numberless experiments were made there in the presence of twelve representatives of different branches of the Marine Department. An official report on the experiments was made in 1798. Smith went to Paris and, with others, took out a filter patent.

Younger's filters are thus described by Wilson: Into a cask a false bottom of flannel was driven. On this were placed: a thin layer of sand; garden mold; a thick layer of charcoal dust and lumps, mixed;

and fine sand. The filter was covered with flannel on a frame, driven into the cask. Filtration was downward. Smith's filter, says Wilson, differed from Younger's in that Smith used sponge prefilters and a device to prolong contact of the water with the charcoal.

Smith's process of water purification, according to a French article of 1804 (14), was the object of many experiments by "Citizen" Barry, a former Commissioner-General of Marine. They were conducted at Brest by order of the Minister of Marine. Apparently Barry ran away with the show, for in the article a detailed description is given of the "Barry filter," with data on preparation of filtering media and on equipping a "Barry filter" suitable for ships of "three bridges" and smaller.

The article also stated that Smith and "Citizen" Cuchet had set up works in Paris to make filters to purify and disinfect the very fetid water of the Seine. How closely these filters resembled those of Barry, the writer of the article would not attempt to say, but declared that each attained the same end perfectly. Nothing has been found to show whether either filter was ever used by the French Navy.

Filtres inaltérables tirés des trois règnes de nature * was the high-sounding title given by Smith, Cuchet and Denis Montfort to their French patent of July 23, 1800. The containers could be made of wood, stone or terra cotta. On a perforated false bottom was placed a web of wool. Then came 2 in. of crushed sandstone; then 12 in. of coarse powdered charcoal and either very fine well-washed crushed sandstone or fine river sand, mixed and strongly compressed so that the water would be in contact with the charcoal for a long time; then 12 in. of sand or crushed sandstone. On top of the unit thus formed was a plate of earthenware or stone, pierced with three or four holes to the inch. In each of these holes was placed a mushroom molded from crushed sandstone, the sides of which were covered with sponges, to be washed from time to time. An air vent was provided.

Five kinds of filters were described in the patent specifications: domestic, tonneau or cask; portable; marine; and soldier, to be carried at the top of a bayonet or gun. All but the last were illustrated in detail. The marine filter worked by upward flow and was suspended to free it from the motion of the ship.

---

* Unchangeable filters drawn from the three kingdoms of nature: animal, vegetable, and mineral. In the later official reprint the title was given as *Divers appareils à filtrer l'eau.*

Clarification at the Beginning of the 19th Century

In a comprehensive essay on clarification of liquids, published at Paris in 1801, "Citizen" Parmentier (15) asserted that in filtration the minute pores of the media allow the fluid to pass through, but intercept all suspended particles. His list of media included woolen, linen and cotton cloth, carded cotton, sponge, sand, earths, pounded glass and porous stone. Sand, he wrote, was commonly used to clarify water for domestic purposes—the more layers the better—and experience proved that the sand must be either renewed or washed from time to time. It must be confessed, said Parmentier, that filtration removes the superabundance of air with which water is sometimes impregnated and which gives it lightness and sharpness. After saying that neither "spontaneous clarification," meaning plain sedimentation, nor any kind of filter can give liquids perfect limpidity, he named as agents for use where filtration is inadequate, "albuminous and gelatinous matter, the acids, certain fats, lime, cream and blood." The four pages on coagulation are the most comprehensive discussion of that subject found published up to 1801.

The Plant on the Quai des Celestins

In 1806, two years after Gibb (see Chap. V) put his Paisley filters into use, Monsieur Happey opened a much larger plant in Paris. It was located on the Quai des Celestins, a well-known wharf on the Seine, near the Hôtel de Ville or city hall. The filters were modeled on the Smith-Cuchet-Montfort patent of 1800, which had expired. The plant was notable for its size, elaborateness, number of employees and its continued operation for a half century. So far as has been found, it was operated sixteen years before a description of it was published. Early in 1822, a brief but highly appreciative account appeared in a London journal, in the form of a letter to the editor from a friend in Paris (16).

The letter describes a "patent Institution" in Paris for purifying the water of the Seine for domestic use. The plant employed about 200 persons. Before filtration, the water was settled twelve hours. The sponge prefilters were renewed every hour. The main filters were composed, from the top down, of coarse river sand, clean sand, pounded charcoal and clean Fontainebleau sand. These units were "renewed [surface scraped?] every six hours." The pumps were driven by four horses working in three shifts. Steam power was not used because of the "dearness of fuel" (16a).

Slightly condensed, the letter from Paris was promptly reprinted in a second London journal (16b), with the following cryptic or crabbed editorial comment: "We apprehend this process would not suit London tastes"—strange words in view of the fact that most Londoners were then supplied with water from the turbid and sewage-polluted Thames, none of which was filtered by the water companies.

More specific data on the Celestins Filters as of 1823 are available. There were six rows of filters. These produced 2,000 hectoliters (52,840 gal.) of water daily out of 195,198 hectoliters supplied to Paris from all sources. The filtered water was delivered by 130 men using 75 casks and 109 horses, making two or three trips a day to different parts of the city. The charge for the perfectly limpid filtered water was the same as that for unclarified water delivered by 1,338 porters— 10 centimes per *voie* or 23 l. [about 0.4 cents for 5 gal.]. Inside the plant, 70 men were employed (17).

The most complete description of the Quai des Celestins plant, apparently written after personal inspection, appeared in 1826 (18) and told the following story:

The plant was founded and was still owned by Monsieur Happey. Water taken from the Seine through a pipe 300 ft. long was lifted by a set of three horse-driven pumps into three wooden settling tanks, each 15 ft. in diameter and 12 ft. high, holding 350 *muids* (about 23,800 gal.). Each tank was filled in about three hours and stood full. They were decanted in succession.

The most remarkable part of the plant, it was said, was the filter room on the second floor, measuring 87 x 32 ft. Settled water was delivered here by a second trio of pumps, driven by the same wheel as the first set. After discharging in a cascade facing the entrance door and falling over two more cascades, the overflow from the third receiving basin passed into conduits of which some ran all around the room and some were in the middle. The water passed from these conduits through horizontal bottle-shaped leaden vessels each containing a sponge which retained a large part of the matter in suspension. These sponges were changed every two or three hours and carefully washed, one workman being constantly employed for this purpose.

From the sponges the water fell into prismoidally shaped, lead-lined filter tanks, each tank fed by four or five pipes. The filters followed the Smith-Cuchet-Montfort model. On a perforated false bottom there was placed a 1-in. layer of gravel, then a thick layer of charcoal mixed with fine sand, topped with another inch or two of gravel.

"All the work at the Quai des Celestins establishment," the account concludes, "is carried out with the greatest exactness." Regulations posted on the premises combined the firmness of a master resolved to be obeyed with the paternal affection of a father. On the one hand was a list of offenses and fines, on the other were sickness benefits for the employees (18). These were the first regulations for the operation of a water purification plant ever posted.

Important supplementary data on the Quai des Celestins plant were given by Genieys in an essay written about 1835 (19). After quoting in full the *Dictionnaire* description, Genieys said that when he visited the establishment there were 34 filters, each 3.25 x 0.65 m. in plan, giving a total area of 71.82 sq.m., or 773 sq.ft. They yielded 1,000 *voies* or 230 kl. a day, a rate of 230 l. per sq.m. per 24 hr. Their total daily output was 60,766 gal., or 786 gal. per sq.ft., or about 3.42 mgd. per acre. Considering that the water had been presettled and prefiltered, this acre rate was about what was to become the standard for English slow sand filters.

A British and an American book, both published in 1835, briefly described the Quai des Celestins filters, each adding its mite to the earlier descriptions. Matthews' *Hydraulia* (20) said that "this useful concern affords important accomodation to the numerous *restaurateurs* of Paris, as well as the residents of the *Palais Royale,* and its environs, for their various purposes." Dr. Robley Dunglinson, in *Human Health,* the first known American book on private and public hygiene, said that after the water had passed through sponge and powdered charcoal, it was "made to fall, under the form of rain, from a height, into a large wooden reservoir, 14 or 15 ft. broad" (21). This was done "to restore the air it had lost during filtration."

Henry Darcy (22), French engineer, writing after the Quai des Celestins filters had been in use a half century, questions whether bone charcoal would not be more efficient than the "coals of the baker" which were used. It appears, he states, that the filters were washed six to seven times a month, when the charcoal was exposed to the air for several days. Such exposure, he said, apparently drawing on a report made at an unstated date for the company promoting the Fonvielle filters, was insufficient to restore the charcoal to its original capacity for absorbing organic matter.

The latest available data on the Quai des Celestins filters appeared in a memoir on the water supply of Paris, dated July 16, 1858, written by G. E. Haussmann, senator, prefect of the Seine, and rebuilder of Paris (23). After noting various sources of water supply for Paris, he said that an industrial company had established, on the Quai des Celestins, filtering works supplied directly from the Seine. The water thus clarified was carried to houses by a daily service of casks of 8 to 9 hectoliters capacity (about 225 gal.) and sold at 10 centimes per *voie* (the same rate as given in earlier descriptions). For the calendar year 1857 the gross yearly revenue from this service was about 6,000,000 francs—no mean sum when a franc was a franc.

Ducommun's Gravity Filters

Fifteen years after Happey installed the Quai des Celestins plant, another notable group of filters was put into use at the Boule-Rouge sales fountain, also in Paris. Belgrand (6), ignoring the Quai des Celestins filters, says the Boule-Rouge plant was installed by Ducommun in 1821 and was the first serious attempt at filtration made in Paris.

J. Ducommun of Paris took out a French patent on a gravity filter on January 28, 1814 (No. 1,072),* perfecting the Smith-Cuchet-Montfort patent of 1800. The media, from the top down, were: sponge; pulverized sandstone; charcoal; and coarse sand. The sponge, before use, was macerated daily with fresh water to remove the taste and odor of the sea. It was indicated that charcoal from oak wood should be used because it compacted more readily than other charcoal and afforded more material in a given space. The sand, the patent said, should be washed five or six times to remove the lime that would otherwise unite the grains. Two classes of public filters were described: *speculations* and *grands*. The former may have been for the sales fountain. They seem to have provided double filtration in open tubs, the first being *dégrossisseurs* or roughing filters. The

* An American filter patent was granted in 1816 to J. Du Commun, New York City. This may have been a duplicate of the French patent. No copy of the American patent is available. All copies of early American patents were destroyed by fire.

grands filtres were applicable to canals. They were rectangular, in pairs, above a clear water chamber and worked by downward flow.

Genieys, writing in or soon after 1826 (19), says the filters at Boule-Rouge could be used to treat water from the Seine or the Canal de l'Ourcq. There were 72 filters, each 0.487 x 0.975 m. in plan, giving a combined area of 34.2 sq.m. or 368 sq.ft., and delivering 400 to 500 *voies* of water a day.

H. C. Emmery, who edited and annotated Genieys's essay (19), says that in 1833 there were at Boule-Rouge 74 filters on the second floor treating l'Ourcq water and discharging it into nine circular or elliptical wooden tubs on the first floor, each holding 1,000 hectoliters (24,620 gal.), from which casks were loaded for distribution to consumers. The filtering media, as given by Genieys, were, from the top down: small gravel; sifted grit; sifted pulverized charcoal; sifted grit; small gravel. The depth of the charcoal was given as 3 to 6 in., according to the character of the water—which seems to have been of general application—and each of the other layers was always 2 in. The total depth would therefore be only 11 to 14 in. The plant was operated, in 1833, by an overseer and two workmen.

Ducommun Filters at Gros-Cailloux.—Genieys also describes similar filters at the artificial mineral water establishment of Planche at Gros-Cailloux, near the early *pompe à feu* [steam-driven pump]. Here, water was pumped from the Seine into a "vast reservoir" [filter?] 60 ft. above the river. After passing through layers of sand and charcoal, the water flowed to the second "reservoir" [filter?] by means of a siphon of hemp [hose?]. Filtration was upward. Before use in the mineral water "machines" such of the water as was not to be acidulated passed through a third filter.

Arago, in his report of 1837 (24) on the Fonvielle filter (see below), in writing of Ducommun's claim that the Fonvielle filter was copied from his, pointed out that Fonvielle's filter worked under pressure while Ducommun's did not, even at the mineral water plant where pressure was available. (See reference to litigation below (25).)

Pressure Filters by Another Ducommun

A few years later Théophile Ducommun of Paris took out a French patent for a lateral-flow pressure filter (No. 5,977 July 25, 1838), to which he added on August 25, 1840, a claim for a vertical-flow reversible or turn-over filter. The main patent called for a long hermetically sealed cast- or wrought-iron tank, divided vertically by perforated plates into five compartments. At either end of the tank was a narrow water compartment, receiving raw or discharging filtered water, according to the direction of flow. Then came, near each end, a compartment filled with sand and crushed sandstone, for clarification. The central and larger compartment was filled with charcoal, for purification. The filter could be backwashed by reversing the direction of flow and wasting the dirty water from openings near the bottom of the end of the tank. No evidence has been found to show that either type of T. Ducommun's pressure filter was put into use.*

Natural Filters at Toulouse

After years of deliberation during which several plans for "artificial filters" were considered and rejected, the city of Toulouse built a "natural filter" in or about 1825. First, an open pit or infiltration basin was dug. After much trouble with organic growths, the basin was converted into a filter gallery—the second on record, the first one having been built at Glasgow, Scotland, in 1810. Filter galleries were constructed by a number of other cities in France (see Chap. XI).

C.-F. Mallet's Proposal for a Parisian Water Works

The first complete plan for supplying Paris with filtered water was perfected in 1826 by C.-F. Mallet (27). In view of a proposal by English capitalists to build water works under a concession, Mallet had been ordered, in August 1824, to visit the water works of London, Glasgow and other cities in Great Britain to learn the best practice there. In March 1825, he outlined a project. This he was instructed in the following August to elaborate, with plans and estimates on which competitive bids for a franchise could be based. In February 1826, Mallet submitted to the municipal council a comprehensive plan, with 34 designs for various parts of the works. These included pumps, filters, distributing reservoirs and pipes in every street for providing water from the Canal de l'Ourcq, which was already being drawn upon (20).

* Delbrück (26) describes what he terms two Ducommun patents, but does not say that the first was granted to J. Ducommun in 1814 and the second to Théophile Ducommun in 1838 and perfected in 1840—these covering the lateral-flow stationary and the vertical-flow turn-over filters.

Instead of adopting Mallet's report, the Paris council referred it back and forth for three years. Various modifications were suggested. The last of these, made by Mallet in conjunction with Girard, who had for some years been engineer for a Canal de l'Ourcq project, called for a dual supply: domestic, from the canal; general, from the Seine. A review of all the schemes, with plans and estimates, was prepared by Mallet in 1829 and published in 1830 (27).

A remarkable feature of Mallet's plan, says Matthews (20) (apparently meaning his independent plan of 1826), was "two filters in which the water was to have an *ascending* motion," the filters so piped that either one could supply the city while the other was being cleaned or repaired.

None of these plans set forth in Mallet's summary of 1829 resulted in the construction of water works on the rigorous concession or franchise plan that had been drawn up by the Paris authorities.

Lees and Taylor's Plan

In 1834, Lees and Taylor, says Matthews (20), presented to Louis-Philippe a plan for supplying Paris with settled and filtered water taken from the Seine above the confluence of the Marne. The water was to be conveyed by tunnel to a point beneath the hill of Ivry, then pumped 150 ft. to settling and filtering basins. Nothing has been found to show that the Paris authorities gave this plan serious consideration. A number of small filters were installed by their promoters at sales fountains in several parts of the city.

Fonvielle's High-Pressure Filters

Arago, Gay Lussac and two lesser-known members of a special committee of the French Academy of Sciences reported most favorably, August 14, 1837, on the "filtering apparatus" of Henri de Fonvielle (24). Most of their report was devoted to a review of the quality of water from various natural sources, why some waters needed treatment and the various means of purification employed. Sedimentation was ruled out as too slow. Hastening it by coagulation with alum was banned as adding foreign matter to water. Filtration was given first place, but its capital and operating cost as practiced at London were held to be prohibitive. No artificial filter can be successful, said Arago, reporter for the committee, "unless prompt, economical and certain means are at hand, of cleaning and renewing the

FRENCH FILTRATION PRACTICE

filters." As compared with the James Simpson filter of the Chelsea Water Works Co. at London, the filter of Robert Thom had the advantage of a "self-cleaning operation" by means of reverse-flow wash.

In France, filtration had not yet been introduced on a large scale. "In several valuable establishments in Paris," many small open boxes, containing a layer of charcoal between two layers of sand, were used as filters—modeled on the Smith-Cuchet-Montfort patent of 1800. When the water of the Seine and Marne was heavily charged with silt, it was necessary to remove the media, or at least the upper strata, every day and even twice a day. The yield of these filters was only 3,000 l. (nearly 800 gal.) per sq.m. per day (about 3.25 mgd. per acre). To supply a town needing 1,000 in. of water a day, 7,000 filters with an area of 1 sq.m. each would be required.

Following this introductory explanation, Arago declared that by hermetically closing these little boxes and causing the water to pass through them under strong pressure instead of merely by its own weight, the yield of the filters would be greatly increased. This had been accomplished by Henri de Fonvielle in his filter at the Hôtel-Dieu (chief Paris hospital). Although the area of the filter was less than a square meter, it

——yields daily, by a pressure of 88 cm. [34.6 in.], 50,000 l. at least of clarified water [nearly 54 mgd. per acre]—seventeen times greater than by the methods commonly used.

It would seem, at first view, that de Fonvielle's filter, working at so high a rate, would have to be cleaned hourly. Not so. It requires no more attention than ordinary filters. The explanation is simple. Under a feeble pressure, a filter acts, as it were, only at its surface; the mud scarcely penetrates it; under great pressure, it must sink deeper, but being disseminated through a greater depth, the permeability will not be decreased—the cleaning will merely be more difficult. (24)

This difficulty was met in the Fonvielle filter "in the action of the two countercurrents—in the shock and sudden shaking and stirring which result from them." At the Hôtel-Dieu, "cocks of the tubes which connect the bottom and top of the apparatus with the elevated reservoir, or with the body of the feeding pump, [are opened] suddenly and almost simultaneously." As a result, the filter is "tumultuously agitated" and the foreign matter detached from the gravel. After the Fonvielle filter had been cleaned by the ascending current used in the Thom system, the water came out limpid. "As soon as the two other cocks were opened, the water rushed out from the filter in a very filthy condition . . . the patients who witnessed the operation, expressed their great surprise at seeing, after an interval of a few seconds, the same fountain furnish, first a yellow mass as thick as soup, and then water as clear as crystal."

For more than eight months (or since early December 1836), said the report, the Fonvielle filter at the Hôtel-Dieu had been operated, with the same layer of sand of about a square meter area, without renewal or intermission. Although the Seine had been extremely foul meanwhile, at least 12,000 cu.m. had been passed through the filter (about 53 mgd. per acre). In conclusion, the committee declared, "we do not hesitate to say that in showing the possibility of clarifying large quantities of water with a very small apparatus, M. Henri de Fonvielle has made an important advancement in the arts."

Fonvielle's Patents

Sieur Louis-Charles-Henri de Fonvielle took out a French patent on November 27, 1835, to which he made additions in the next three years. His original patent was on a system of filtration which he named *filtre à grande courantes*. This he said was adapted to both public and domestic use. Claims for novelty were operation under pressure in a hermetically sealed vessel and cleaning by opposing reverse-flow currents.

The filtering media named by Fonvielle in his main patent were, from the bottom up: river pebbles on a false bottom; *grès* [grit, presumably coarsely powdered sandstone]; iron filings [an innovation]; crushed charcoal. This assemblage was "closed" by double plates of perforated zinc. At a distance of 6 cm. (2.4 in.) above the zinc plates, there were inverted troughs holding sponges through which the raw water passed, leaving nine-tenths of the mud it carried. Apparently, the sponges were removed for washing, after which the filter was cleaned with filtered water from a reservoir. A tube was provided by means of which the filter could be saturated by atmospheric air, thus accelerating filtration, but just how this was achieved is not stated.

An addition, dated March 4, 1836, to the original patent covered many filters placed one above another in a single closed tank. These filters were in pairs, with a chamber between each two filters into which the raw water entered from a central tube, then passed up through one half and down through the other. Reverse flow produced the counter currents for washing on which Arago laid so much emphasis. As no filtering material was specified, it is assumed that it was the same as had been described in the original patent. It was stated that a square meter of surface, working under a 45-ft. head, would filter 12 hectoliters of water an hour (about 31 mgd. per acre).

In a third addition to the patent, it is stated that cleaning could be effected, or at least aided, by introducing air by a *jeu de pompe* [play of pump].

First Known Filter Patent Litigation.—The French Filtering Co., which had been formed to promote Fonvielle's filter, soon had to defend his patent—the first filter-patent infringement suit found (25).

Arago alluded to this controversy in his report of 1837. Several persons, he said, among them Ducommun, had recently claimed "the invention of filtering by increased pressure." This might be substantiated "in mathematical exactness, but no one before Fonvielle had proposed filtration in hermetically sealed vessels, so arranged that strong pressure could be applied without deranging or confusing the different layers of media." *

"In France," Arago continued, "we find everywhere, and especially at the beautiful mineral water establishment at Gros-Cailloux, a fine disposal of high pressure, entirely neglected. We see, in fact, M. Ducommun, whose name is honorably known in this department of the arts, using at the Hôtel-Dieu three cisterns to clarify 15 hectoliters of water in 24 hours, while a single one of these cisterns, modified by de Fonvielle, yields in the same time, agreeably to the report of M. Desportes, steward of the hospital, 900 hectoliters of water, perfectly filtered, in lieu of 15" (24).

The defendants seem to have been successful (26). About this time Ducommun took out the patent on pressure filters already noted.

* Arago overlooked or ignored several earlier designs for filters under pressure in hermetically sealed tanks.

In 1815 Graf von Real patented a *Zwangsfilter* (force filter) for which head was supplied by a reservoir. Delbrück (26) gives no data, but says that Cadet von Graffincourt, a distinguished chemist, suggested that pressure be supplied by a pump attached to the bottom of the filter. The mechanic Hoffman in Leipzig, adds Delbrück, invented an aerostatic process, operating by compressed air.

On Aug. 11, 1819, Henry Tritton took out an English patent on a closed filter with a clear water chamber beneath it, combined with an air pump attached to the chamber to produce a partial vacuum and thus a negative head. This is the first known proposal for negative head in a filter.

On May 11, 1827, J.-F.-E. Quarnier of Paris was granted a French patent on four hermetically sealed filters working in series under pressure applied by a column of water.

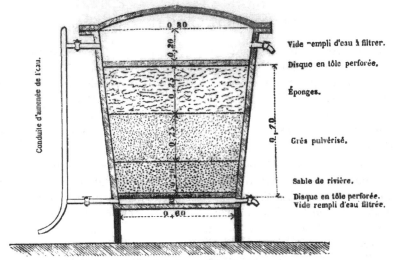

Filtre de la Compagnie française.

FIG. 10. FONVIELLE PRESSURE FILTER AT PARIS SALES FOUNTAIN, 1856
Compressed sponge, pulverized sandstone and river gravel, separated by grills;
covered and supported by perforated iron plates

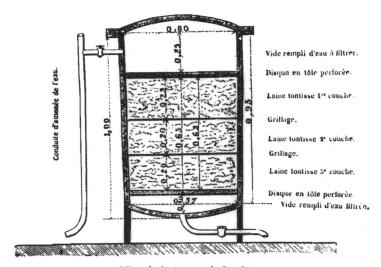

Filtre de la Compagnie Souchon.

FIG. 11. SOUCHON PRESSURE FILTER AT PARIS SALES FOUNTAIN, 1856
Three layers of wool cloth clippings obtained from tailor shops
[Both drawings from Belgrand's *Les Travaux Souterrains de Paris*, 1877;
dimensions in both are in meters]

Mareschal's Improvements to the Fonvielle Filters.—Mareschal et Compagnie, of Paris, were granted a patent, May 31, 1838, on improvements to the Fonvielle filter. This was followed by four *additions et perfectionnements* in 1839 and 1840 (the entire five *brevets* being numbered 11,006). Elaborate specifications were included. They named a variety of filter media plus devices for holding them in place. As a clincher the patentees claimed all means, known or unknown, for applying pressure. Emphasis was placed on washing by opposing currents and on economy of space by stacking filters in a single closed tank. The Fonvielle filter, as thus perfected, is described not only in the patents but also in Delbrück's long article of 1853 (26).

Other sources of information, cited below, seem to show that during the twenty years that the Fonvielle filter was in use at the Paris sales fountains a simple design was followed, using pressure which was sometimes high, but without stacking the filters or using conflicting currents for washing.

The Fonvielle and Souchon Filters at Paris Sales Fountains

In a sketchy review of filtration of water at sales fountains of Paris in the nineteenth century, Belgrand (6) discusses the filters of Fonvielle and of his rival, Souchon.

Ignoring the patent of 1835, the Hôtel-Dieu filter of 1836–37 and Arago's report, Belgrand says that the first trial of the Fonvielle system was made on June 2, 1838, in the presence of the French Ministers of the Interior and of Commerce; and that, on April 5, 1839, the city council of Paris "authorized the establishment of the first filters of this type at St. Denis gate fountain." This filter was promoted by the Compagnie Français.

The first trial of Souchon's filters, says Belgrand, was made at the basin of the Notre Dame Bridge, where it continued in use until 1847. The first "*tender* of the use of this device" was made December 10, 1846, by Bernard, who exploited the Souchon patent of 1839—presumably he proposed to rent it to the city. These two types of filters, says Belgrand, were a great success. They were applied to all sales fountains. The city paid the filter companies 6 centimes per cu.m. (264 gal.) and sold the filtered water to porters and to householders at 90 centimes.

The Fonvielle filter used at the fountains was in the form of an inverted truncated cast-iron cone, with a slightly domed cast-iron cover bolted on. There were three layers of filtering material with a total depth of 0.7 m. From the top down, these were: 0.25 m. of sponge; 0.25 m. of pulverized sandstone; and 0.20 m. of very clean river sand. The sand rested on a perforated iron plate, below which was a filtered water chamber only 0.03 m. deep. An outside vertical pipe brought raw water under pressure for admission, at will, either above or below the filter unit. The water surface was 0.8 m. in diameter, while the bottom of the unit was 0.6 m. and the filter area was 0.62 m.

The Fonvielle filters were cleaned every eight days when the water was muddy, but when it was "merely turbid" once in fifteen days was sufficient, and when clear (in summer) once a month. "Almost always," said Belgrand, "they were content merely" to remove, wash and replace the sponge, and return the filter to service. When the pulverized sandstone commenced to get muddy, it was treated similarly; the bottom or sand layer was cleaned the same way every two or three months, according to season (6).

The Souchon filter was entirely of wool-cloth clippings resting on and covered by perforated cast-iron plates. Two wire grills divided the unit into three layers of nearly equal depth, giving 0.62 m. total depth of wool. The waste wool was compressed by a screw, acting on the upper diaphragm. The container was a cast-iron cylinder, with bottom and top slightly curved. The filtering surface had an area of 0.57 m.*

The Souchon filters were cleaned daily when the water was turbid, but ordinarily once in three or four days. At each cleaning a "certain thickness" of wool was removed from the upper layer without being replaced until that layer was exhausted.

Output Tests.—In 1856 Belgrand (6) made official tests of the output of filters at four of the Paris sales fountains—the first tests of the kind found on record. A Fonvielle filter was tested at one fountain and Souchon filters at three others. The filters had areas of 0.25 to 0.5 sq.m. (2.3 to 4.6 sq.ft.). They worked under heads of 12 to 22.4 m. (39.4 to 74.7 ft.).

* Belgrand does not always make clear whether he is describing the first installations of the two makes of filters or those tested by him for output in 1856. He says that the Souchon filter at Notre Dame Bridge was open at the top and delivered water directly into the distribution system. But those tested in 1856, at other locations, worked under considerable heads.

The Fonvielle filter, located at the Arcade fountain, was tested February 23, using water from the Seine under a 15-m. head and, at another time, using the clearer water of the Canal de l'Ourcq under a head of 16.28 m. The yields were at the rate of 758 and 950 cu.m. per sq.m. per 24 hr., respectively. Reduced to a common head of 1 m., said Belgrand, these rates would be 51 and 58 cu.m. per sq.m.

Belgrand expressed the belief that no one ever obtained such high unit rates with any filter before. So far as is known, this was true. But, as he pointed out, the filters were "perfectly clean." In present-day American terms, the rates were 70.5 mgd. for the Seine and 105 mgd. for the Ourcq, under heads of 49.2 and 53.4 ft., respectively.

Omission of the layer of sponge from a Fonvielle filter located at the Pantheon fountain did not notably increase its yield—"a very important fact," wrote Belgrand, "since the greater part of the materials in suspension were stopped by the sponges." What would have been the result during a long run, Belgrand did not say.

A freshly-cleaned Souchon wool filter, located at the Boule-Rouge fountain, was tested by Belgrand on March 19, 1856. The average yield of three runs under a head of 12 m. was 974, and under a 14-m. head, 1,076 cu.m. per sq.m. per 24 hr.

At the Sèvres fountain, on March 16, 1856, a Souchon filter, working under 21-m. head, was tested "under ordinary clean conditions." That is, the first layer had not been cleaned for three days; the second, for 21 days; and the third for 45 days. The average output for two runs was 538 cu.m. per sq.m. per 24 hr. In another experiment, the same filter, under 22.4-m. head, but with the two upper layers of wool freshly renewed and the third one in use 45 days, gave an average discharge for two runs of 235 cu.m. per sq.m.

Amalgamation of the Fonvielle and Souchon Filters

The French (Fonvielle) and Vedel * Souchon filter companies joined forces on June 16, 1861, and made their two filters into one. The sponge of the Fonvielle filter was put above a single layer of the wool cloth clippings of the Souchon filter. The sponge, says Bel-

---

* Vedel of Paris was granted a French patent May 17, 1853 (No. 9,698) for a filter composed, from the top down, of sponge, crushed sandstone and finally river gravel mixed with wood charcoal, resting on a false bottom. It was claimed that a filter 0.8 m. in diameter and 1 m. high, working under a head of 6 to 8 m., would yield 100 to 120 l. per minute. The filter was cleaned by reverse-flow wash.

grand, served as a *dégrossisseur* or roughing filter, retaining all muddy and insoluble matter. The wool thus protected had to be cleaned once every month or two. It was much more easily cleaned than the pulverized sandstone beneath the sponge of the Fonvielle filter.

The price paid to the companies for filtered Seine water between July 1, 1853, and December 31, 1858, was 0.9 francs per cu.m. for water supplied to porters who delivered it in casks. Householders who came for the water paid 0.025 francs per *voie,* a bucketful of 18 to 20 l. For watering a horse the charge was 0.05 francs. After the suburbs were annexed to Paris, the price was raised to 1 franc per cu.m., but was paid by the Compagnie Générale des Eaux.

With the introduction of the practice of piping water to houses, says Belgrand, the number of water porters decreased. In 1833, these had numbered 1,216, delivering water either in buckets suspended from shoulder yokes or else in casks on horse-drawn carts. In 1859, before annexation of suburbs, the number of venders had fallen to 972 within the city. It then rose to 1,378 for the larger area. In 1876, the water venders had decreased to 710, of whom 431 used hand-drawn and only 41 horse-drawn carts.

When Belgrand wrote (1876), so large a part of the domestic water supply of Paris was piped from the Vanne and Dhuis springs that the only filters then in use were at three sales fountains: Sèvres, University and "Reservoirs." These were of the sponge-above-wool type, introduced in 1861. The number and capacity of these filters are not given by Belgrand (6). This seems to have been the virtual end of small-scale filtration at Paris after more than a century of promotion.

The Thirty-Six Souchon Filters at Nantes

Souchon filters were included in water works built in 1855 at Nantes, France. So far as can be found, these were the only Souchon filters installed outside of Paris.

The Nantes water works were built under a concession granted to the Compagnie Générale des Eaux calling for water pumped from the Loire: 2,000 cu.m. daily of settled and filtered water for house and factory use; 4,000 cu.m. daily of raw water for street washing, public fountains and other "communal" purposes, and for householders who did not subscribe for filtered water. The raw water was pumped directly into the mains.

FRENCH FILTRATION PRACTICE

Raw water was pumped to an open high-service reservoir and passed through submerged filters to a covered clear-water basin. This ensemble was semi-circular in plan with the clear-water basin inserted on the inner side. The settling reservoir was in three compartments. The filters were placed at the bottom of the reservoir alongside the wall between the reservoir and the basin. There were 36 filters, twelve in each compartment of the reservoir. Each filter had an area of 1 sq.m. (10.76 sq.ft.) and contained about 3 ft., in depth, of media.

Darcy, who described the works in his book of 1856 (22), using information supplied by the engineers, Jegau and Watier, says the filter material was of compressed wool. Kirkwood (28), who visited Nantes, April 25, 1866, says the media were sponge, sand, pebbles and broken stone, all of which were taken out once a month, washed and replaced. Instead of sponge, "a preparation of wool from the workings of the woolen factories" was often used, and was "more thorough in its operations than sponge." Whichever was employed was covered with a perforated plate and compressed with a screw. The filter boxes were covered with a watertight plate. Water was drawn from the top of the settling reservoir through a rubber tube supported by an india-rubber ball and was delivered into the top of each filter. The filtrate was then passed through the perforated false bottoms of the filters into a drain serving six filters and discharging into the clear-water basin.

When Kirkwood was at Nantes one of the compartments of the settling reservoir was used to store river water. The others were used alternately, giving twelve to eighteen hours of presedimentation. They were "cleaned at short intervals by flushing off about 2 ft. of the bottom water and sediment; the latter being stirred up and brushed off the bottom by men using sweeps and brooms."

The amount of water filtered in March 1866 was about 400 cu.m. or 0.105 mgd. This low figure obtained because many of the householders took raw water from the public fountains or duplicate mains.

Kirkwood concluded that the filters were too small and that whatever clarification was obtained was effected chiefly by sedimentation. It may be added that the 36 filters had a total area of only 387 sq.ft., or $\frac{1}{112}$ acre. With all at work, the rate of filtration, in March, 1866, would have been about 11.8 mgd., with no allowance for either settling reservoirs or filters out of use for cleaning. Darcy (22) states that the engineers for the plant assumed that the filters would produce 100 cu.m. of water per 24 hours, working under an average head of 2.5 m. or 8.2 ft. This would be about 105 mgd. per acre, with no allowance for cleaning.

The estimated population of Nantes in 1866, says Kirkwood, was

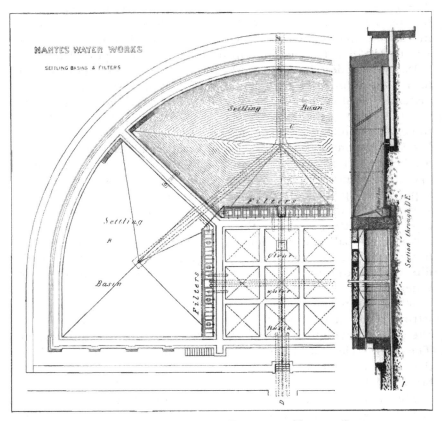

FIG. 12. THIRTY-SIX SOUCHON FILTERS AT NANTES, FRANCE
Placed 12 each at bottom of three settling basins along walls between settling basins and covered clear-water basin
(From Plate XXIII of Kirkwood's *Report on the Filtration of River Water*, 1869)

112,000. The total amount of water pumped in March was 1.524 mgd., of which only one-fifteenth was filtered. At the public fountains water was available only from 4 to 8 P.M., during which time half of the entire supply was carried away by consumers. About

50,000 gal. of "clarified" water was "sold about the streets in casks" by a second company (28).

How long the Nantes filters were used after Kirkwood saw them in 1866 is unknown. Imbeaux's *Annuaire* (29) says the works built in 1854 supplied raw water from the Loire but mentions no purification at Nantes before Puech multiple filters were installed there in 1900–01. (The promoters give the date as 1902–03.)

Reverse-Flow Wash Filters in Three Cities

At Marseilles, Tours and Dunkirk, filters washed by upward reverse-flow were built between 1842 and 1870. These and Nantes are the only four cities in France known to have been provided with "artificial" filters for their entire water supply before 1890.

A canal 52 mi. long was built in 1839–47 to bring water from the Durance to Marseilles. After debating whether to clarify the muddy water of the river by sedimentation or filtration, both methods were adopted. Filters were put in use in 1842 (29, 30). Dams across several gorges formed settling reservoirs along the canal. Near the city a filter with an area of 92,336 sq.ft., in two compartments, was constructed. It was of masonry, roofed and floored by arches. Beneath it was a clear-water reservoir. The filter unit was 23 in. deep. From the top down, the media were: sand of increasing size; small gravel; broken stone; and small stones. The filtrate passed into the clear-water reservoir, which was generally full, through 0.04-m. (1.57-in.) pottery tubes. These also served to backwash the filter.

Darcy (22) gives the yield of the filter under heads of from 0.4 to 0.8 m. (16 to 32 in.). When conditions were favorable, backwashing cleaned the filter but, with high turbidity, scraping the surface was necessary. When Kirkwood (28) visited Marseilles in March 1866, the filter was not being used because a large percentage of the settling-reservoir capacity had been lost by silting. Clemence, writing much later, states that filtration was given up in 1863 and the space devoted to storage; also that the original settling capacity was partly restored by desilting and a huge new settling reservoir was built (31).

At Tours, an ill-planned attempt to filter turbid water from the Cher without presedimentation, using a filter of inadequate size, soon ended in failure. The filter was put into use in 1856. When Kirkwood was at Tours early in 1866 the supply was taken directly from the river (28). Here also the head seems to have been too small for backwashing the filter.

An upward backwash filter, apparently modeled after Thom's Scottish filters of 1827–42, was a part of water works completed in 1870 for Dunkirk, with Monsieur Pauwells as engineer. As described by William Humber in 1876 (32), the filter was in four compartments, each 26¼ x 52½ ft. in area. The filter media from the top down were: sand, 8 in.; washed coke dust, 6 in., in equal layers—very fine, fine, moderately coarse; and Calais pebbles, 5 in. The total is 19 in. The filtering material rested on perforated tile 1.8 in. thick, set in portland cement on brick placed edgewise. Under a head of 16 to 32 in. the yield was 15 gal. per sq.ft. per hour or about 16 mgd. per acre. Imbeaux's *Annuaire* (29) states that the supply was from the Boubourg River but not being potable was given up in 1893 for an underground supply. It is not clear whether the change was due to pollution of the canal or to a rising standard of quality.

Darcy's Hydraulics of Filtration and His Mechanical Filter Patent

Henry Darcy, the far-seeing French hydraulician, delved into the hydraulics of filtration nearly a century ago, described and analyzed the principal filters of Great Britain and France, patented a mechanical filter that included the leading features devised by Scotch, English and French predecessors and anticipated in all but one important particular—coagulation—the American mechanical rapid filter.

Evidence that Darcy merits recognition as a filtration engineer is found in his book of 1856 on the newly built water works of Dijon (22) and also in his filter patent of the same year. The book contains an exposition of the hydraulics of filtration. The patent gives some of the same laws to elucidate the principles on which his filter was based.

In his British patent (Filtering Water on a Large Scale, September 19, 1856 [No. 2,196]), Darcy stated that filters for town water supply in Great Britain produced a daily average of only 4 cu.yd. per sq.yd. of area or 1.45 mgd. (U.S.). Therefore "very large establishments were required." The filter area could be lessened by increasing the head on, or decreasing it beneath, the filter (using "negative head"). The consequent reduction in volume of filtering material could be further reduced by using layers of less depth—30 cm. of filtering sand and 10 to 20 cm. of supporting gravel (12 in. of sand and 4 to 8 in. of gravel).

A new method of keeping the filter clean was proposed: (a) the mud deposit on the sand was diminished by keeping the suspended matter in suspension and discharging to waste that part of it within 50 cm. (20 in.) of the surface of the sand, without stopping filtration;

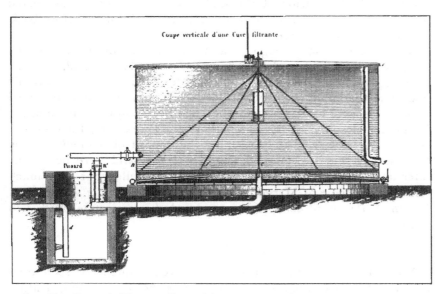

FIG. 13. DARCY'S UPWARD-FLOW MECHANICALLY CLEANED FILTER OF 1856
Thin layer of sand on gravel; raw water admitted concentrically to inside of tank to keep dirt in suspension and permit extraction at four equidistant points; supplementary reverse-flow wash aided by power-driven revolving broom; negative head used if conditions require
(From Plate 25, Atlas, Darcy's *Fontaines Publiques de la Ville de Dijon*, 1856)

(b) the filter surface was swept with a mechanically driven revolving broom; (c) at intervals determined by experience, reverse-flow wash was used. Despite all this, it was occasionally necessary to scrape off 2 to 3 cm. (0.8 to 1.2 in.) of sand.

To provide filtered water for a town of 100,000 requiring 38 gpd. (U.S.) per capita, Darcy assumed an open filter about 7 m. (23 ft.) in diameter and in height. Raw water was admitted to the tank about 1 m. (40 in.) above the surface of the sand. To prevent surface disturbance and to lessen the deposit of mud upon it, raw water was discharged horizontally along the inside of the tank. This gave a circular motion to the incoming water which kept a part of the suspended matter in suspension. Bottom water loaded with dirt was discharged through four orifices located just above the surface of the sand, into horizontal pipes leading to a main waste pipe concentric with the outside of the tank. This drainage system served also to draw down the water to the surface of the sand before backwashing and to discharge the wash water.

Reverse-flow wash through a false bottom had been patented by Peacock in 1791 and used by Thom at Greenock, Scotland, in 1827. Scraping was a feature of Simpson's Chelsea filter of 1829. Fonvielle and others had used heads of various magnitudes—Fonvielle in closed tanks. Negative head, Darcy implies, had been employed by Simpson in the Thames filter of the Lambeth Water Works Co., London, put into use in January 1840. Lessening the deposit of sediment on the filter surface and sweeping the surface of the sand were new. His filter units had a much greater area than Fonvielle's but were much smaller than those at Marseilles. Darcy's *tout ensemble* was unique. No one before him seems to have applied the laws of hydraulics to filtration.*

The Non-submerged Filter

Noteworthy as an illustration of the French quest for something new and perfect in the art of filtration was the non-submerged filter. Although laboratory studies of it began in the early 1890's and were in progress ten years later, and a plant was installed at a small works in 1907, this type of filter was soon forgotten except by a few chroniclers.

The essential principle of the non-submerged filter is expressed by its name. Water was showered upon a filter so drained that it would neither stand on or in the sand. To all intents and purposes, this

---

* Jules Dupuit, in the second (1865) edition of his treatise on water supply (33), devotes 56 pages to "Divers Processes of Filtration in Use." He cites Darcy on the hydraulics of filtration and carries Darcy's study further. Strangely enough, the bibliography in Hazen's *Filtration of Public Water Supplies* (34), includes no reference to Darcy, Dupuit or the earlier writer, Genieys (19); in fact it ignores all the French books and essays dealing wholly or in part with filtration. It contains many references to German writings on the subject. Hazen, it should be said, was chiefly concerned with slow sand filtration.

water filter was the same as the sewage trickling or sprinkling filter perfected during the 1890's in England.*

For more than ten years, according to a communication presented at Paris on July 18, 1904, P. Miquel and H. Mouchet had been studying the bacterial purification of water by means of non-submerged fine sand (36). A paper presented January 22, 1939 (37), by Marboutin, "recalls the studies of M. L. Janet, Ingenieur en Chef des Mines, and of MM. Miquel and Mouchet," who perfected "this type of filter," and gives "results of comparative researches of these last-named bacteriologists." Neither paper makes clear who conceived and designed the filter but apparently chief credit for the studies belongs to Miquel and Mouchet.

The first two papers describe a filter substantially like the typical English slow sand filters in composition, except that (1) the top layer of fine sand, 1 to 1.3 m. (40 to 53 in.) deep, was "tamped down and wetted," and when raw water low in earthy suspended matter was to be treated, the fine sand was covered with coarse gravel to prevent disturbance by the oncoming water; (2) if the raw water was turbid and loaded with organic detritus the fine sand was covered with "sieved sand of moderate size to retain the impurities which can ultimately be removed without interference with the fine sand." Ourcq water [Canal de l'Ourcq supply to Paris?] was "perfectly clarified" [original turbidity not given]; its dissolved oxygen increased by about 20 per cent; and its dissolved organic matter reduced 10 to 20 per cent, according to the rate of filtration. Ordinary microbes in the filtrate never exceeded 50 to 80 per ml., these being largely due to aftergrowths in the sand. Raw Ourcq water had up to 200,000 microbes per ml. In general terms, without stating turbidity or other conditions of the raw water, the rate of filtration is given as 2 cu.m. per day per sq.m. of filter surface—about 50 gpd. per sq.ft., or about 2.2 mgd. per acre.

Although it was stated in July 1904 (38) that studies had been under way for more than ten years, the paper dealt with laboratory tests of the preceding two years. During that time, it stated, there had been no appreciable choking of the filter nor difficulty in obtain-

* "One of the earliest" of the sewage sprinkling filters "was constructed at Salford, England, by Joseph Corbett [city engineer] about 1893, the inspiration for the design being furnished" by reports of the Lawrence Experiment Station of the Massachusetts State Board of Health (35).

ing an even distribution of water over the filter. Apparently no study had been made of the distribution of raw water over a large surface.

Marboutin's paper of January 1909 (37) states that the ordinary sand filter reduced bacteria "to 2 to 5 one-thousandths of those in the raw water" but might "fail to retain certain bacteria, especially the pathogens." In contrast, "the non-submerged filter [let] none of the bacteria in the raw water pass through." He adds that it is "logical to attribute a role of great importance to physical phenomena of molecular attraction in the filtration process."

Marboutin mentions non-submerged filters having an area of 250 sq.m. then newly installed at Chateaudun, on the River Loir. The medium was Loir sand passing a 1.5-mm. mesh. These filters yielded 800 cu.m. per 24 hours in summer, 500 in winter, but hours in use were different in the two seasons.

Imbeaux's *Annuaire* of January 1, 1930 (29), states that the Chateaudun filters were of the "système Miquel," established in 1907 by Baudet, Mayor of Chateaudun, had an area of 254 sq.m. (2,733 sq.ft.) and treated 800 cu.m. a day (about 3.35 mgd. per acre). The water came from two deep wells, sunk in 1893, and was apparently under suspicion before the filters were installed. The population of Chateaudun in 1930 was about 6,500.

So far as is known, the non-submerged filters at Chateaudun are the only ones of the kind ever installed for city use.

Frank Hannan, of Toronto, Ont., who kindly translated the French articles for use here, also supplied citations from later German authors:

Tillmanns [*] mentions the excellence of the filtrate but says no plants of this type had been installed in Germany.

August Gärtner,[†] under Regenfilter (rain filter), gives details of eighteen tests made by the Conseil Supérieur d'Hygiène de France between September 1905 and May 1906 on a 16-sq.cm. (172-sq.ft) experimental filter. No coliform organisms or putrefactive bacteria were found in the filtrate. The French government was very favorably impressed and specified the filter for municipal use. Gärtner did not consider the filter well suited for turbid waters but did not say why.

Eugen Goetz [‡] describes the non-submerged filters in considerable detail. He says the sand chokes up, is difficult to clean, and has to

---

[*] *Wasserreinigung u. Abwasserbeseitigung.* 1912.
[†] *Die Hygiene des Wassers.* 1915.
[‡] *Wasserversorgung* (Weyl's *Handbuch der Hygiene*). 1919.

be renewed from time to time; algal growth has to be checked by darkness; dirty water has to be prefiltered. He discounted the advantages of this type of filter and thought it must cost more to operate than the ordinary slow sand filter.

Opposition of French to Slow Sand Filters

Preference for water from springs rather than rivers, combined with hostility to filtration, says Imbeaux (36), retarded construction of filters in France until the 1890's. Even then filtration was preceded by some other treatment.

Anderson revolving purifiers were constructed ahead of the earliest slow sand filters in France. They were at Libourne, where slow sand filters were installed as part of water works built in 1890–92 (29, 36). Immediately afterward the General Water Co. began treating by the same method the water supplies of suburbs of Paris and of the city of Nice. In 1896–97, the city of Paris established Anderson "purifiers" and slow sand filters to treat the water of the Marne and in 1897–98 it did the same for the water of the Seine. Although these "purifiers" merely produced comminuted metallic iron and mixed it with the raw water on its passage to slow sand filters, the promoters applied the name "Anderson Process" to both coagulation and filtration (see Chap. XIII).

Puech multiple filters were installed by Paris before its slow sand filters on the Marne and Seine. Additions brought the daily capacity of each of these plants up to 300,000 cu.m. (79.2 mgd.) within a few years. By 1935 the Puech-Chabal system had been installed in 125 French cities, far outranking any other method of water purification in France. As changed from time to time it has become a succession of decreasingly rapid filters followed by slow sand filters (see Chap. IX).

Trailigaz filters, so far as is known, are now the only proprietary rivals of the Puech-Chabal system in France. In the period 1934–39, the Trailigaz filter had been installed in seventeen cities of France (39). It is of the rapid type, cleaned by both reverse-flow wash and by ejecting the sand from the bottom to the top of the unit.

Ozonation to supplement filtration has probably been used more extensively in France than in any other country. Before September 1939, it is reported, Paris had installed two large ozonation plants and contracted for a third (see Chap. XIV).

CHAPTER V

British Contributions to Filtration

England and Scotland divide the honors for pioneer work in filtration. The Lancashire filter appears to have been a crude forerunner of the slow sand filter. Its earliest development is uncertain but may have been before 1790. Two centuries before, the British patent office had started issuing patents on distillation, chiefly of salt water for use on shipboard. Not until 1790 was a patent relating to filtration granted, the first being one for the composition and manufacture of household filters of earthenware.

In 1791, James Peacock was granted the most remarkable filter patent issued in England (1). Two years later he published an expository pamphlet (2) which deserves a high and lasting place in the annals of filtration.

The first filter to supply water to a whole town was completed at Paisley, Scotland, in 1804, but the water it supplied was carted to consumers. At Glasgow, in 1807, filtered water was piped to consumers by one water company and immediately after by a rival. In 1810, the first of these companies built the earliest recorded filter gallery. Altogether the two companies built a half dozen filter plants within fifteen or twenty years; none of them was a success.

In 1827, slow sand filters designed by Robert Thom were put into use at Greenock, Scotland, and similar filters designed by James Simpson were completed at London in 1829. Both were slow sand filters. Thom's were cleaned by reverse-flow wash; Simpson's by surface scraping. The Simpson design became the model for English slow sand filters throughout the world, and it still is the model wherever that type of filter is continued in use.

Thom's filter design was followed in only a few places, most of them in Scotland; however, two of its main elements—false bottom and reverse-flow wash—were and are principal features of the rapid filter, developed in the United States during the 1880's. The rapid filter has largely supplanted the slow sand filter in most countries of the world.

British workers contributed little to filter design after the days of Simpson, but added much to the knowledge of the reduction of bacteria by slow sand filtration. They also demonstrated the importance of presedimentation, although they were slow to accept coagulation as an aid to sedimentation and rapid filtration.

The Lancashire Filter

References to the "Lancashire Filter" are numerous but vague in publications of the second quarter of the nineteenth century. Apparently the earliest of these filters were used for industrial water supplies, and some may have been installed before 1790. In the light of meager evidence that has been found they may be considered as primitive slow sand filters.

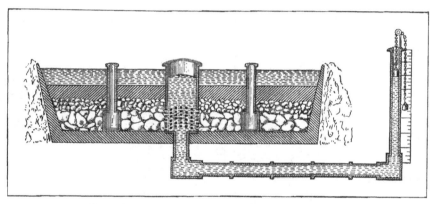

FIG. 14. CROSS SECTION OF LANCASHIRE FILTER
(From Thomas Graham's *Elements of Chemistry*, 1850 edition)

In the second edition of *Elements of Chemistry*, Thomas Graham (3) describes and illustrates a water filter, "as it is usually constructed for public works in Lancashire." It was placed in an excavation about 6 ft. deep, lined with well-puddled clay. On the bottom was a layer of large stones, while above this were smaller stones, then coarse sand and gravel. From the bottom layer of stones the filtrate found its way to a central iron cylinder, the lower part of which was perforated. Two air-vent pipes and a water-level gage were provided. The central collecting well and absence of underdrains suggest a primitive design.

Graham did not mention any of the municipal filters in Great Britain, some of which had been in use twenty years. But in both editions of his book (1842 and 1850) he describes large-scale filters as being composed of gravelly sand, for which there might be substituted

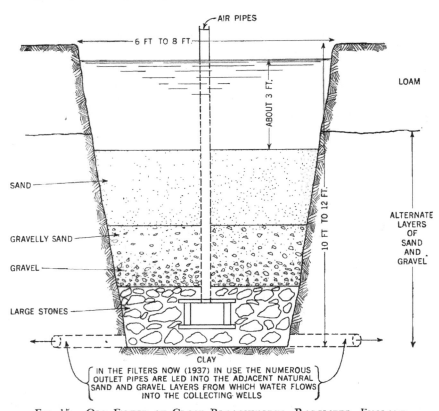

FIG. 15. OLD FILTER AT CLOSE BLEACHWORKS, RADCLIFFE, ENGLAND
James Simpson visited a filter at "calico works" near Manchester on his 2,000-mile inspection trip in 1827
(From sketch by W. F. Creber, Chief Engr., Manchester Corporation Waterworks)

crushed cinders or furnace clinkers. The function of any of these media, he said, was to support "finer particles of mud or precipitate where first deposited" on the surface, and "form the bed that really filters the water." When the sand became clogged, an inch or two of the surface was removed by scraping.

An appeal for information made in 1937 to W. F. H. Creber (4), then chief engineer of the city water works of Manchester, Lancashire, brought a hypothetical sketch of one of several very old filters at bleach works at Radcliffe near Manchester. These were 6 to 8 ft. wide, 50 to 150 ft. long and 10 to 12 ft. deep. On the natural clay base a collecting drain of brick or stone ran the length of the unit. This was surrounded by a layer of stone. Six air-vent pipes were provided for each unit. Turbid water was brought from the river to one end of each filter. The collecting drains led to circular brick pump wells. Although the earliest record of filters at the bleachery dates from 1878, Creber states that "there is little doubt that the filters have been in use for upwards of 150 years."

Johanna Hempel's Domestic Filter

The first evidence found of the manufacture of household filters in England is the grant of a British patent, on October 16, 1790, to Mrs. Johanna Hempel, a potter of Chelsea. The patent was for a composition of materials and for a means of manufacturing it into vessels "having the power of filtering water and other liquids in a more cheap, easy and convenient manner" than they could before be filtered. The principal materials were tobacco-pipe clay and sand in ratios varying with the size of the vessels. Mrs. Hempel is the only woman inventor and manufacturer of filters whose name has been found in the annals of filtration.

Peacock's Upward-Flow Filter With Reverse-Flow Wash

James Peacock, a London architect of note in his day, was granted the first British patent on a process and apparatus for water filtration (December 23, 1791, No. 1,844) (1). In 1793, Peacock published a promotion pamphlet (2) setting forth the need for filtration and the principles that should guide the choice, preparation and placing of filtering media, showing sketches of filters of different sizes and design. It includes a diagram showing superimposed spheres of diminishing size, illustrating a mathematical exposition of the reasons why coarse filtering material should be placed at the bottom of a filter with layers of material of regularly decreasing size above it. Peacock's exposition brings to mind the Wheeler filter bottom designed more than a cen-

XXXIII. *Specification of the Patent granted to Mr. James Peacock, of Finsbury-square, in the Parish of St. Luke, in the County of Middlesex, Architect; for his Invention of a new Method for the Filtration of Water and other Fluids, which would be of great public and private Utility.*

Dated December 23, 1791.

To all to whom these presents shall come, &c. Now know ye, that in compliance with the said proviso, I the said James Peacock do hereby declare, that the nature of my said invention for filtration of water and other fluids, applicable to heads of water, of various magnitudes or extents, for public service, reservoirs, or cisterns, for private use, and for other purposes of filtration, and in what manner the same is to be performed, is described as follows; that is to say: My method for the filtration of water and other fluids, is by impelling the ascent of the fluid through the filtering medium, instead of the common method by descent.

* * * * * * *

The filters will be cleansed, by drawing out the head or body of water or fluid; by which the water or fluid will descend in the filter, and carry with it all foul and extraneous substances. In witness whereof, &c.

FIG. 16. FIRST BRITISH WATER FILTER PATENT
Issued on December 23, 1791, to James Peacock; opening and closing paragraphs are shown
(From *Repertory of Arts* (London), pp. 221, 226 (1799))

tury afterwards (5). No such thesis had appeared before Peacock's day and none surpassing it has appeared since.*

Peacock opens his pamphlet by declaring: "The Poet's maxim, that 'God never made his works for man to mend,' if not generally false, is however pretty glaringly so, in many important particulars upon this atom of a universe." † Peacock continued:

> Among the various subjects evidently designed by Providence to ask amendment at the hands of men, there is one of immense importance, which has not yet received it in the degree it is capable of, and that is WATER.
> This element, necessarily of such universal use, and particularly in food and medicine, is suffered to remain laden with a great diversity of impurities, and is taken into the stomach, by the majority of mankind, without the least hesitation, not only in its fluid state, however turbid it may happen to be: but also in the forms of bread, pastry, soups, tea, medicines, and innumerable other particulars.
> Medical gentlemen can readily point out the probable advantages towards the preservation of health, and extending the period of human life, which would result from the use of soft water, cleared from the earthy, and the living, dead and putrid animal and vegetable substances, with which it is always, more or less, defiled and vitiated. (2)

Because of the "indelicacies of turbid soft water," many are "driven to the use of hard water, although they are not unapprized of the probable danger to their health, from its petrifying quality, or from the metallic, or other mineral, taints, too frequently suspended or concealed therein."

Peacock deprecates the use of natural "filtering stones," which may "contain copper, or other metallic, or mineral substances, dissoluble

---

* Through the kind aid of Sir William Paterson of the Paterson Engineering Co., London, a photostatic copy of Peacock's pamphlet has been supplied for use in this book by the British Museum (see Fig. 16). The only known copy of the pamphlet in the United States is in the Library of Congress. Extensive inquiries by the author failed to locate any other copies in the United States, England or Scotland, although appeals were made to many dealers in rare books.

† Like many other detached quotations this one has a different significance when considered in context. Dryden, whom Peacock did not name as the author, wrote:

> Better to hunt in field for health unbought
> Than fee a doctor for a nauseous draught
> The wise for cure on exercise depend;
> God never made his works for man to mend. (6)

The entire passage, applied to water supply today, would mean: search the fields and mountains for pure water rather than attempt purification of what is unfit; or in latter-day parlance, "innocence is better than repentance."

A
SHORT ACCOUNT
OF A
NEW METHOD
OF
FILTRATION BY ASCENT;

WITH

EXPLANATORY SKETCHES, UPON SIX PLATES;

By JAMES PEACOCK,

OF FINSBURY-SQUARE, ARCHITECT;
Author of OIKIDIA; or NUTSHELLS, SUPERIOR POLITICS, &c.

Adde quod e parvis ac levibus et elementis
Nec facile est tali naturæ obsistere quicquam
Inter enim fugit ac penetrat per rara viarum.

LUCRETIUS.

LONDON:
PRINTED FOR THE AUTHOR;

AND

SOLD BY LACKINGTON AND Co. CHISWELL-STREET,

1793.

FIG. 17. TITLE PAGE OF PEACOCK'S PROMOTIONAL PAMPHLET
(From a photostat of the copy in the British Museum; obtained for use here by William Paterson, Paterson Engineering Co., London)

by water" and render the filtrate "somewhat suspicious." As to "artificial productions" made of clay in which combustible material has been placed to be burned out in firing, rendering "the mass porous," Peacock remarks that "the ingenious Mr. Wedgwood [the famous English potter] informed the writer hereof that he had caused some of this kind to be made, but that their effects were so trifling, and temporary, that he did not think proper to continue the manufacture of them."

Whether Peacock knew of Mrs. Hempel's patent of 1790 on earthenware filters is not apparent, but like all the capable promoters he disparaged the products of rivals, both stone and artificial filter vessels, as follows:

> Neither of these kinds of filters will afford clear water in any considerable quantity, and notwithstanding the repeated brushing and cleansing applied to the surfaces of their concavities, the pores, beyond the reach of the brush, will, sooner or later, clog up; and the stones become entirely useless. (2)

Having set forth the need for water treatment and the inadequacy of the filter stones and vessels then in use, Peacock remarks, with the confidence and benevolent spirit of the inventor-promoter:

> To supply, therefore, the inhabitants of this great metropolis and its environs with more than a sufficiency of perfectly clear soft water from the inexhaustible sources contained in the noble rivers in its vicinity, has been the writer's study for several years past. He has viewed the subject with much attention; and has made a very great variety of experiments, in order to arrive, as near as possible, to the simplicity and perfection of nature, in her process of percolation, by using the same medium and the same mode, taking away, by human art, her hurtful and disgusting redundances only; how far he has succeeded herein, the impartial public will best judge. (2)

Peacock's Design.—The novelty of Peacock's invention, he declared in his patent, was filtration by ascent instead of the common method of descent. This could be applied under any head, in any quantity and for public as well as private use. A further novelty, far more significant, was cleaning the filter by reverse flow, the descending water carrying with it "all foul and extraneous substances."

To put his innovation into effect, Peacock proposed either three tanks, or one tank with three compartments; one for turbid water, one for the filter and one for clear water. The filter was fed from the bottom of the raw-water vessel, which discharged into a small chamber beneath the filter, the latter being supported on a false bottom of slats with spaces between them, arranged to form a flat cone.

The raw water passed through the false bottom and up through the filter. The filtrate passed from the water space above the unit into a clear-water tank or chamber.

Filter media were sand, sandy gravel, and broken glass or other material which could be graded into various sizes. A material was prepared for use by repeated washings until the wash water ran clear, spreading it to dry, and then grading it into various sizes by means of a set of superimposed sieves actuated in unison until the remaining particles were as small as possible, after which, if necessary, trituration or pulverization was used. When washed and graded, the media

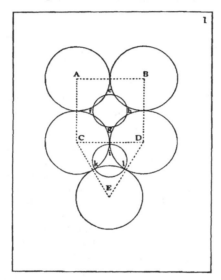

Fig. 18. Peacock's Diagram to Illustrate Reasons for Arranging Filter Media in Layers of Decreasing Size

Superimposed spheres with diameters in each layer about double those in the layer below would increase interstitial spaces in geometric ratio: 1, 2, 4, 8, 16...

(From Peacock's promotional pamphlet (Fig. 17))

were put in place in layers of size in nearly "subduple ratio to each other," with the largest at the bottom. The finest or top layer served as "the main agent of percolation." While these filter media were being arranged, pure water was run through the strata to condense the particles.

An air-vent pipe was placed in the center of the filter unit, extending from the top of the false bottom to the level of the top of the sand. This pipe was filled with media arranged as in the filter except that the coarsest bottom layer and the finest top layer were omitted.

The theory of filtration by ascension was that gravity would cause some of the sediment to be deposited in the chamber beneath the false bottom and that the remainder would be intercepted by the increasingly fine material; also that reverse-flow wash would cleanse the filter and the settling chamber. To "counterbalance and resist any disturbance from the [upward] pressure of the column of turbid water," there was to be placed on and above the filtering material a second series of materials, arranged in reverse order, but omitting the finest material.

The raw-water vessel was kept full by a pipe-and-ball cock, discharging into a bag or strainer. The latter, particularly in summer, intercepted "innumerable green filaments" abounding in the waters of some streams, which "coalesce and form a tough mucus" giving rise to "disagreeable effects." Peacock said that this did not occur unless the water was exposed to the sun. Perhaps this is the earliest statement of the kind recorded.

Four designs for filters were illustrated in Peacock's pamphlet. The first made use of three cylindrical glass containers side by side, and was apparently for household use. The second one showed a single cylindrical tank, divided by curved vertical partitions into three compartments. The third design, for "sea, camp or garrison service," showed three wooden casks, with wooden slats for the false bottom of the filter and a wooden grating at the top to compress the filter medium and hold it in place in land transit or on shipboard. The fourth design was intended to serve filtered water to a community of any size, at a small annual charge. In such a water works, three masonry-lined wells would be made in the ground at any convenient distance from a pond, ditch or river. If the body of water were large enough, two wells would give a constant supply, one for the filter, the other for the filtrate. The drawing showed a building with open sides above the three-well purification plant.

Peacock's Influence.—Of the many published comments on Peacock's filter, from 1795 to 1929, the first was the only one containing an adverse judgment (7). It was, however, the philosophy expressed in the pamphlet that was criticized and not the filter itself, which was called ingenious. Partly in defense of unfiltered water the reviewer said:

> The petrifying quality of hard water no philosopher, we believe, now regards as connected with the origin of nephritic complaints. That the ordinary qualities of sweet and soft water are prejudicial to health has never, so far as we know, been demonstrated, nor rendered probable: . . . We shall

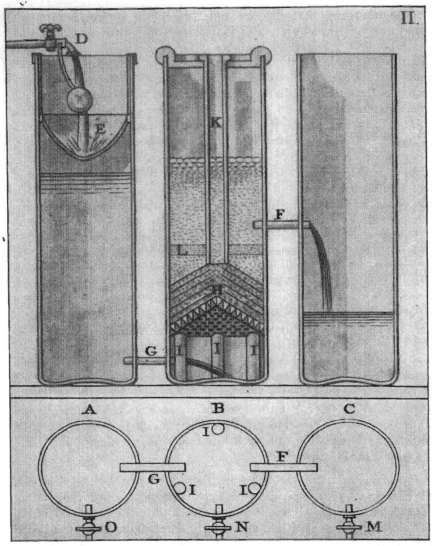

Fig. 19. Three-Tank Form of Peacock's Upward-Flow Backwash Filter
A—Raw-water tank with float valve and strainer; *B*—Filter, with media supported on inverted conical bottom composed of slats with spaces between; *C*—Clear-water tank
(From Peacock's promotional pamphlet (Fig. 17))

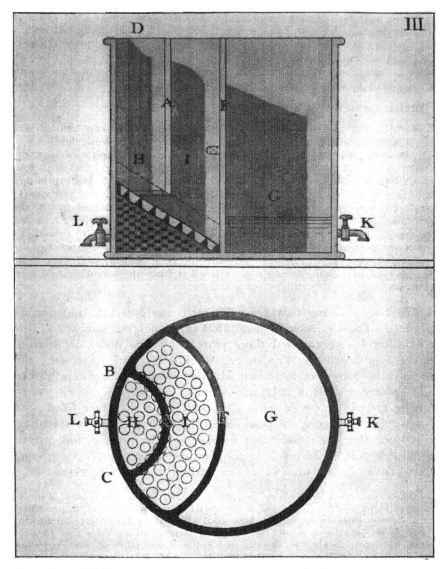

FIG. 20. SINGLE-TANK FORM OF PEACOCK'S UPWARD-FLOW BACKWASH FILTER
The same three elements as are shown in Fig. 19, combined into a
single tank unit
(From Peacock's promotional pamphlet (Fig. 17))

not deny that Mr. Peacock's process is highly desirable in point of delicacy; and on this score we wish him success, because he appears to deserve it; but we need not at any time apologize for exposing what we take to be mistakes or gratuitous suppositions; in the present case, we consider it humane to prevent, as far as our influence extends, a false alarm on account of their health from spreading amongst the drinkers of unfiltered water.

Referring to the filter's deficiencies, the critic said:

Now, by this contrivance, such matters only as are diffused through the water, and not such as it holds dissolved, would be separated and left behind. Mr. Peacock tried this with brine, and found the salt still remaining in the water, as any smatterer in chemistry would have predicted. [And as to Peacock's statement, in answer to an inquirer, that he had not yet found whether his filter would "sweeten putrid water," the reviewer declared that Peacock] might have answered in the negative, for he seems not to have heard of the purifying quality of charcoal; of which, we apprehend, he might take advantage.* We rather wonder that Mr. Peacock did not think of submitting his manuscript to the inspection of some person acquainted with recent philosophical discoveries. [The reviewer, for instance?—*M.N.B.*] A very little of this kind of knowledge might have freed it from the unauthorized assertions which it contains. (7)

The only evidence that Peacock's filter was ever put into use appears in a French article of June, 1804 (8), which stated that the Peacock filter had been tried three years before by order of Admiral Parker, on board the *Vengeance, Magnificent* and *Lancaster*. On these ships, according to reports of their captains, the filters yielded 2,880 *pintes* of water in 24 hours.†

The French article describes Peacock's filter as a box filled with washed gravel or sand. A plate shows the filter as a cube-shaped box, tilted on edge, with raw water entering at the bottom and filtered water drawn from the top. An air vent and a force-and-suction pump were provided, the latter for use in washing the filter by reverse-flow.

* When Peacock published his pamphlet in 1793 he may not have heard of Lowitz's paper of 1790 on the use of charcoal to sweeten putrid water (see Chap. III). Subsequently he appears to have used charcoal in the filters installed on naval vessels, mentioned below. It was in the days of sailing ships that "putrid" water was most troublesome. Lowitz demonstrated the value of powdered charcoal added to "stinking" water rather than the use of charcoal in filters. But the passage of years showed that sand and gravel were the best filtering media, thus justifying Peacock's earliest conceptions.

† The old French *pinte* seems to have been about equal to an English quart or 0.3 U.S. gal. On that basis the yield of the filters (on each ship?) would be about 860 U.S. gal. a day.

In special cases, the filter might contain a mixture of powdered wood charcoal and limestone to disinfect the water.

Peacock lived until 1814. He had a hand in designing many buildings, some of which were important and may well have been equipped with his filters at a time when London and vicinity were being supplied with turbid and filthy water from the Thames.

Potentially, Peacock's contribution to the art of filtration was great; but there is no way of learning how many of his successors in the field profited by his patent and pamphlet. His filter cleaning by reverse-flow was one of the basic elements of the later mechanical filter. Upward filtration was a delusion and a snare that caught the fancy of many, including some engineers, during the ensuing century. The false bottom and air vent were old. Sand had been used for centuries; but Peacock's specific directions for preparing sand for use and placing it in graded layers containing particles of decreasing size were both new and thorough.

The First Filtration Plant for City-wide Supply

The quest for pure water entered a new phase when John Gibb decided to supply filtered water to his bleachery at Paisley, Scotland, and cart it to "almost every door" in town. His is the first known filter for city-wide supply installed anywhere in the world. It was probably put into use in midsummer of 1804.

Famous for its shawls and threads, Paisley was early in the field with cotton mills, bleacheries and other industries. As was the case in other manufacturing towns of Britain and America, the industries at Paisley quickly monopolized and shamelessly polluted all sources of water supply within the town.

A contemporary description of the Paisley filters, written by the Rev. Robert Boog, first minister of the Abbey Church in Paisley, is found in Sir John Sinclair's *Code of Health and Longevity* (9). It was published in 1807 to show how the inhabitants of a town of 20,000, "who were formerly in a distressed state from the unwholesomeness of the water, are now plentifully supplied with that valuable article in great perfection."

The idea of supplying filtered water to Paisley, wrote Boog, occurred to a bleacher as an accessory to plans for improving his bleaching grounds. These grounds lay along the River Cart, a little above Paisley. The water of that stream was often muddy. It brought down wastes from print fields and from lime, copperas and alum works and so it was unfit for bleaching. This suggested filtration to the bleacher, "an operation," wrote Boog, "not uncommon but perhaps nowhere so carefully executed as here." If filtration were not uncommon, what a pity, it may be interjected, that Boog did not give the location and nature of the other filters he had in mind!

Transformation of the muddy, industrial waste-laden water of the Cart was effected by a roughing filter, sedimentation and subsequent double filtration. The flow was lateral throughout. The final filtrate occupied a central circular well, surrounded concentrically by the main filters and the settling chamber—an arrangement used more than a century later in the so-called Morse filter at Burnt Mills, Md., and elsewhere. Robert Morse used steel; Gibb used masonry.

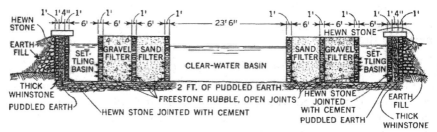

Fig. 21. First Known Filter to Supply an Entire City With Water, Completed at Paisley, Scotland, in 1804, by John Gibb

Water passed through stone-filled trench to ring-shaped settling chamber, then through two lateral-flow filters to central clear-water chamber; delivered to tank on a hillside from which it was carted to consumers

(From description in Sinclair's *Code of Health and Longevity*, London, 1807)

Water from the River Cart flowed to a pump well through a roughing filter about 75 ft. long, composed of "chipped" freestone, of smaller size near the well than at the upper end. This stone was placed in a trench about 8 ft. wide and 4 ft. deep, covered with "Russian matts" over which the ground was leveled.

A small steam engine placed over the well lifted the water to an "air-chest" about 16 ft. higher than the river, from which it was forced to the settling chamber through about 200 ft. of 3-in. bore wooden pipe of Scots fir. The settling basin, main filters and clear-water basin were formed by concentric masonry walls carried up 10 ft. above a puddled-earth bottom, the top of the latter being 2 ft. above the original ground surface.

The ring-shaped settling basin and the two filters nested within it were each 6 ft. wide. The outer filter was composed of coarse gravel, the inner of very fine gravel or sand. The depth of the sand is not given. The clear-water basin was 23.5 ft. in diameter. The outside wall of the settling basin was double, filled with 16 in. of puddled earth, with a coping stone over the whole. All the other walls were of open-jointed masonry, each about 1 ft. thick. Water passed laterally through the joints in the walls and through the filters into the clear-water chamber. All the stone in the roughing filters and in the walls that were in contact with the water was carefully chosen from "quarries perfectly free from any metallic tinge"—this may have been chiefly for the benefit of the bleachery.

From the clear-water basin a pipe extended about an eighth of a mile to a declivity where the filtered water could be discharged into a cask holding about 480 wine gallons. Such casks were placed on carts and two carts so loaded went seven times a day through the town—thus delivering about 6,700 gal. a day. The water was sold at first for a halfpenny (1 cent) a gallon. Later, to meet the cost, the price was increased to three farthings (1.5 cents), "but if any considerable quantity is bought," wrote Boog, "some gallons are allowed in addition." Commenting on the cost and value of the service, Boog said: "This is some addition to the family expenses; but, for pure water, all who value health will willingly pay at this rate; and, as it is brought to almost every door, to those who are at a distance from wells or river, there is considerable saving of time and labour. This plan is susceptible of improvements; but it is sufficient to demonstrate, that no town near a river, need be destitute of good water" (9).

The air-chest, wrote Boog, "is a contrivance employed for extinguishing fire. The water is driven into a receptacle, by a forcing pump, and its return prevented by a valve opening inward. A pipe is inserted into the top or side of the chest with its mouth near the bottom. The compressed air acting on the surface of the water forces it through the pipes" (9). Thus did the Paisley bleacher anticipate the apparatus widely used decades later to supply isolated buildings with water under pressure.

Boog described the Paisley filters as in use but did not say when they were put into operation nor give the name of their builder. The missing information was found in the Boulton & Watt collection preserved in the Reference Library of the city of Birmingham, England. A search made under the direction of H. M. Cashmore, City Librarian, disclosed letters showing that John Gibb, of Paisley, ordered an engine from Boulton & Watt early in 1803; and that in May, 1804, the engine "with some appendages of pumps [was] nearly done" and that settlement of the balance due on account would be made in a few days. This correspondence does not mention filters, but in a letter dated January 23, 1810, written by Boulton & Watt to the Glasgow Water Works Co., specific mention is made of "filters erected by Mr. Gibb at Paisley," which completely purified the muddy water of the River Cart (10).

How long water from Gibb's filters was carted to consumers is unknown. In 1838 a water company began to pipe water through the streets of Paisley from a reverse-flow-wash sand filter designed by Robert Thom (see below). A recent letter from James Lee, Water Engineer of Paisley (11), leads to the conclusion that the Gibb filters, with later duplications, continued to supply bleach works at least until 1861. An ordnance map of that date shows three groups of concentric circles, designated "filtering tanks," close by the "Linside Bleach Works," near the River Cart. The outside diameters of these circles, Lee says, were approximately 65, 50 and 40 ft.

Thirteen Decades of Filtration at Glasgow

Glasgow, Scotland, was the third city in the world to have a filtered water supply. Unlike its predecessors, Paisley and Paris, where filtered water was carted to consumers, Glasgow was supplied by pipes. At Glasgow, two rival companies began to introduce water from the Clyde, the Glasgow Water Works Co. in 1807 and the Cranston Hill Water Works Co. in 1808. The earliest filters of the first company were failures. They were immediately followed by others, but these were likewise unsuccessful and were supplemented by filter galleries. The Cranston Hill company, after various misfortunes, also built a filter gallery. Subsequently the second company was absorbed by the first. The galleries and at least some of the filters built by the two original companies continued in use until the city introduced a gravity supply from Loch Katrine, in 1859. Just before that, the city had acquired the property of the consolidated water company, and had also bought the property of the Gorbals Gravitation Water Co., which had completed works in 1848 to supply a suburb afterwards annexed to the city. This third company had unique filters which were remodeled by the city and were still in use in 1936 (12). Thus, for thirteen decades Glasgow has had filtered water piped to consumers—a record unparalleled.

The Glasgow Water Works Company

Thomas Telford, who later founded and served as first president of the Institution of Civil Engineers, was engineer for the Glasgow Water Works Co. Correspondence between him and Boulton & Watt (13) affords meager data regarding his plans for the earliest filter at Glasgow. In a letter dated May 25, 1806, he said that "if there is any difficulty in getting the water [from the Clyde] to subside or filtrate so as to be perfectly good—then instead of one reservoir 6 ft. in depth, it will be advisable to have two of 3 ft. in depth each—and each one acre in superficial area."

About forty years after the works were completed, Donald Mackain, engineer of the company then supplying water to Glasgow (14), described how Telford proposed that water be pumped from the Clyde at a point two miles above the city to three reservoirs each holding a day's supply. These reservoirs were to be so placed, wrote Telford, in a report no longer available, "that the water in passing from one to another shall be filtrated." Telford's plan was followed, says Mackain, but in times of flood the river brought down alluvial matter that did not soon subside, followed by water from sources higher up which had a deep brown color. Telford's filter yielded water differing little from that of the river.

Again what a pity that Telford and Mackain made only vague references to filters built so early. Neither Telford in his autobiography (15) nor Sir Alexander Gibb in his recent biography of Telford (16) mentions Telford's filters at Glasgow.

James Simpson, in a discussion (17) of Mackain's paper, describes Telford's filters as "a series of cells, filled with sand" through which the water passed in succession. When the water was at its worst it was little changed after passing through the first filter, but at times the filters worked satisfactorily.

More specific data are given by C.-F. Mallet (18) who, as chief engineer of a projected water supply for Paris, visited Glasgow in 1824. He reports that, years before, an attempt had been made "to unite sedimentation and filtration." Three ponds were built, each 120 x 30 m.

FIG. 22. THOMAS TELFORD (1757–1834)
Engineer for Glasgow Water Works Co. and designer of its first filters, 1807
(From a painting by Henry Raeburn, R.A., 1812, reproduced in Sir Alexander Gibb's
The Story of Telford, London, 1935)

and 1.5 m. deep (394 x 98 x 4.9 ft.), on levels 1.5 m. apart. The ponds were "separated by filters in which the water circulated alternately before reaching the last reservoir; but the wind agitating the water, the sedimentation upon which they had counted" was imperfect, "and the water left the sand without having experienced a notable change from its original condition."

Dr. Ure's Unsuccessful Filters.—Immediately after the failure of Telford's settling reservoirs and filter cells the Glasgow Water Works Co. offered premiums for filter plans. Amazingly, 22 plans were submitted. The highest premium was awarded to Dr. Andrew Ure, then a lecturer at Glasgow University. Filters were built under Ure's direction, says Mackain (14). For a short time and under favorable conditions these filters "yielded a sufficient supply of pure water; but they were too small . . . became clogged with silt deposited by flood waters," and so water had to be taken directly from the river, regardless of its condition.

Doubtless this was the first filter-plan competition ever held. No description of the plans can be found. Dr. Ure does not mention even his own plan in any of the books he wrote subsequently. Apparently he, like Telford, felt that the less said about his unsuccessful filters the better.

Alterations in both the Ure and Telford filters, says Mackain (14), were "unavailing." The market price of the stock of the company fell owing to the character of its water and "other causes." A change in the source of supply was imperative. Chief among the "other causes" was the competition of the Cranston Hill company, whose filtered water seems to have been in high favor when introduced in 1808.

Pioneer Filter Gallery in 1810.—To meet these conditions the Glasgow company decided to go across the Clyde to a peninsula of sand and gravel which it believed would yield "a large quantity of water, either filtered through the sand bed from the river or from natural springs which might be obtained by sinking wells and connecting them by tunnels or culverts." Tests for yield being successful, "tunneling was commenced in 1809" (14). Contrary to tradition, James Watt did not propose the filter gallery nor design the flexible-jointed pipe laid beneath the Clyde from the filter gallery to the existing pumping station. He may have suggested the basic idea for the flexible joint, which was said to have been like the joint of a lobster. When Boulton & Watt sent the design for the submerged pipe on January 23, 1810, they disclaimed approval of the filter gallery, and suggested building a third set of filters (10). The submerged pipeline, following the Boulton & Watt drawing, appears to have been put into use in midsummer of 1810, and with it the filter gallery. Each was the first of its kind.

The yield of the filter gallery, says Mackain (14), "was pure as to its origin and pure to the eye," and became almost exclusively used by the inhabitants—to the distress of the rival Cranston Hill company.

Cranston Hill Water Works Company

A calico printer named Richard Gillespie, having a "printfield" in the neighborhood of Cranston Hill, Glasgow, was the founder of the Cranston Hill Water Works Co. Like the Paisley bleacher, he wanted to supply his works, and at the same time serve private consumers with water pumped from the Clyde, below Glasgow, to a reservoir on Cranston Hill. Robertson Buchanan, author of *Mill Work*, became engineer of the company. The "prudence" of drawing water "below the drainage of the city" having been questioned, says Mackain (14), Thomas Simpson, Chief Engineer of the Chelsea Water Works Co., London, was consulted. Simpson reported that the supply of the London Bridge Water Works Co. was preferred to that of the other companies taking water from the Thames, "being by the filth" it received "materially improved." Strange doctrine! But the London companies were already on the defensive and were soon to be attacked for supplying grossly polluted water.

Although the Glasgow "doubters were silenced," Scotch skepticism led the company to install a settling basin of several days' capacity to remove the grosser particles of sediment and a filter to remove the finer. The filter was composed of several feet of sand and gravel with "tunnels" below, leading to a pure-water basin. "This being the first experience of artificial filtration on a large scale," wrote Mackain, many trials were made to determine the requisite depth of sand and also the yield of the filter under different circumstances. This suggests that the filter was built before the Telford filter cells were put into operation.

Charles Dupin, a French naval officer, wrote two descriptions of the Cranston Hill filter, apparently as seen by him during two visits to Glasgow (19, 20). Between these visits the filter seems to have been changed from upward- to downward-flow operation. Both descriptions indicate a ridge-and-valley surface. In the first, settled water discharged into pits or wells in the ridges, flowed into many conduits beneath the valleys, rose through open joints in the conduits, passed up through the stone or gravel and sand of the filter into the valley and flowed thence to the pure-water basin (19). The second description says that water from the settling reservoir was discharged through four iron pipes into four long channels in the filters. About 4 ft. beneath the bottoms of these channels was a layer of paving stone rising toward the center of the channel. Beneath the ridges were small open-jointed masonry drains. Both the drains and the pavements were covered with small pebbles, above which was sand (20). The channels were 14 ft. wide on top and 10 ft. apart. Whether or not Dupin's first description applied to the filter as completed in 1808 he does not say.

The Cranston Hill company began supplying filtered water in 1808, just at the time the Telford filter cells of the Glasgow company had proved themselves to be a failure. Immediately afterwards, the second or Ure filter of the Glasgow company met a like fate. The market value of its stock went down. The Cranston Hill company, elated by the favorable reception given to the product of its filter, put a premium on its shares in 1809. The elation was short-lived. In 1810, the Glasgow company completed its filter gallery and the market price of its shares went up. Contrariwise, the Cranston Hill company, however satisfactorily its filter seemed to work, was confronted with more and more objections to "water previously contaminated with drainage." Its revenue fell below expenditure. The use of its old intake and apparently of its filter on Cranston Hill was continued, but its consumers were chiefly industrial (14).

Cranston Hill's Unsuccessful Filter Gallery.—After some years the directors of the Cranston Hill company decided to build a filter gallery up the river near the one built by its rival ten years before. To their amazement, when this gallery was put into service in 1820, instead of yielding "brilliant" water, like that from the near-by gallery of its rival, the water was so heavily charged with iron that it was unpleasant to domestic consumers and unfit for manufacturing calico. This is the first instance found in the literature of trouble arising from iron in ground water.

William Mylne, Engineer of the New River Water Co., London, was called to Glasgow. He assumed that the trouble came from the water mains instead of the water, and spent many days, says Mackain (14), in an effort "to ascertain how the currents flowed in the pipe" but not until he got back to London was his attention "directed to the real seat of the trouble." He then furnished plans for a filter which Mackain does not describe. Instead of following Mylne's plans, however, the company started a tunnel between the filter gallery and the river on the assumption that the iron-impregnated water could be excluded from the gallery by back pressure.

Second Filter of Cranston Company.—Although this device promised to be a success, the company decided to follow the advice of its manager, a Glasgow architect named Weir. His plan was to build a filter near the unsuccessful filter gallery, similar, wrote Mackain, to the one that had done so well on Cranston Hill, except that the water, instead of being pumped to the filter, would flow from the river "into the hollows between the tunnels"—meaning underdrains.

Mallet, who saw the filter soon after it was completed, says that on the bottom of an excavation, 20-in. cylindrical "galleries" were built of wedge-shaped brick, laid dry. The galleries were 23 ft. apart. A layer of sand was placed on the bottom of the excavation and extended over the underdrains. Thirteen valleys were thus created, making a unit some 300 ft. across. Each underdrain discharged into a main drain of stone, laid dry, leading to a pump well. To clean the sand, wrote Mallet (18), "a very ingenious method was used." A cast-iron pipe was laid above the main drain, and from this pipe "branches descended to the bottom of the valleys, which were paved and on a level with the galleries." To clean the filter, wooden stoppers in the ends of the branch pipes "were removed, the pump started, and the water passed along each valley in a direction contrary to that employed when the filter was in use, thus carrying away the sediment that had been deposited." But "the water, charged with this sediment, quickly put the pump out of order; . . . it was longer out of repair than in use; they had to give up this means; the consumers diminished and the Cranston Hill company was placed in a difficult position." This implies that a pump was used to remove the wash water.

A different explanation of the failure of this filter is given by Mackain (14). Under some conditions, he says, the filter improved the quality of the water but whenever the river was swollen above mean summer level, iron-bearing water from the substratum rose into the underdrains.

When Mackain became chief engineer of the Cranston Hill company in 1829, he made alterations in the works which, he says, led to an increased demand for water (14). These changes he does not describe. Mallet, in his report of 1830 (18), says that Mackain sent him a plan of his scheme, probably executed meanwhile. The plan was to build a number of wells placed checkerboard-wise and connected by pipes. The wells were 10 ft. in diameter, 6½ ft. high, of brick laid dry and covered with plank. Matthews, in his book of 1835 (21), mentions that raw water was delivered to the wells, passed up through a thick stratum of sand and flowed to a reservoir. There is no direct evidence bearing on the success of this filter. According to Mackain, soon after he made the changes described, the two companies had a rate competition but in 1833 stopped it. In 1838, the Cranston Hill company was absorbed by the Glasgow company (14).

The Consolidated Company.—Under Mackain as engineer of the consolidated company, says J. M. Gale (22), "filters upon the Lancashire principle were constructed at both works." What that principle was, Gale, like many other writers of the period, did not think it necessary to say (see Lancashire Filter, above).

When Glasgow took over the works of the consolidated companies just before it introduced a gravity supply from Loch Katrine, maps were drawn showing the old river works of each company, as of October 26, 1856. Copies of these maps, supplied for consideration here by John Cochrane, Engineer and Manager of the present Glasgow Water Works, show a maze of river intakes, pipes, filter galleries, settling basins, filters, clear-water reservoirs, drains and pumping stations crowded into a small space. At the site of the works of the Glasgow company there were two acres and at the Cranston Hill river works 1.6 acres of typical slow sand filters.

The Gorbals Gravitation Water Company

Three-stage upward-flow filters were built in 1846–48 by the Gorbals Gravitation Water Co., a part of works to supply water to a suburb that was soon annexed to Glasgow. Multiple filters were not new but William Gale, Engineer of the Gorbals Co. (22), made an innovation in their construction. Instead of superimposing different sizes of filtering material in one bed, as had Peacock in 1791, Thom in 1827 and Simpson in 1829, he put each in a separate unit: the coarsest material was in the first, which had the greatest depth; the medium grade was in the second, which had a larger area and a lesser depth than the first; and the finest grade was in the third, which had a still larger area but lesser depth than the second. Each unit had a false bottom of perforated flat tile, supported by brick on edge, as in Thom's filters, described below. Water was cascaded from a reservoir to the first filter, then from one to another unit and finally to a clear-water basin. Each of the first three cascades had a fall of 9 in.; the fall of the fourth was 12 in. The depth of water on each filter was only 4 in.

A Glasgow bleacher named Stirrat, testifying before a committee in 1850 (23), said that the third or sand filter worked for six to eight weeks before removal and washing of the sand was required. Plans and a description supplied to the committee by William Gale (24) at the same time, show that there were two sets of these filters, each narrow, placed end to end except for a channel between them to carry off the dirty wash water. Six years later, Darcy (25) wrote that these filters were washed by upward flow but that once a month a layer of sand about 1 cm. or 0.4 in. deep was removed.

William Gale anticipated, in his general design, the filters built by Walker in 1890 at Reading, England, by Armand Puech a little later at Paris, and by Puech and Chabal throughout France (see Chap. IX).

After the Gorbals works were bought by Glasgow, in 1856, Mackain made over the filters "upon the Lancashire principle," says James M. Gale (22). According to a description of the Gorbals filters as they stood in the early 1870's, they consisted of sand, then perforated tile $1\frac{1}{2}$ in. thick, resting on gravel and stone (26). That is, the tile had been moved up between the gravel and sand. This arrangement was adhered to in a design for an additional filter made in 1881, in which the sand was supported by 3-in. fire-clay slabs resting on a checkerboard arrangement of brick underdrains. The efficiency of this remodeled type of filter, wrote Cochrane in 1936 (12), is shown by its continued use on all additions to the Gorbals filters, including those then under construction.

Filtration Proposals—1783–1825

Tantalizing in their vagueness are several early references to proposed filters. Data of great historical value might have been recorded then had its future importance been recognized.

At Glasgow, in 1783, David Young proposed construction to obtain a filtered water supply from the Forth and Clyde Canal. The volume must have been small, for the estimated cost of "a house at the canal, with a reservoir and filtering apparatus" was only £150. In 1804, a little before the two water companies built the first of the filters there, a supply from the Clyde, which was to be filtered, was proposed (27).

At London, also in 1804, Ralph Dodd, a British civil engineer of enough importance to be included in the *Dictionary of National Biography*, made reports to subscribers to projected London water companies, in which filtration was proposed as if it were too well known to need description. For projected works to supply South London he planned to draw water from that part of the Thames undisturbed by shipping and "throw" it into a reservoir by "tidal flow" for its "perfect purification." Here the water would be "sufficiently settled and clarified" for delivery to consumers. In a later report (November 2, 1804) to "subscribers to the intended East London Water Works Co.," he mentioned filtration, but did not say by what means. In this project, also, he proposed to divert water to large reservoirs at high tide. From these it would flow to lower reservoirs from which, after "settling and filtering," it would be forced to a summit reservoir for delivery to consumers (28).

At Manchester, a citizens' committee reported, on February 2, 1809, in favor of a filtered water supply from either the Irwell or the Tame (29). The former could be filtered through "beds of gravel and sand, either natural or artificial, at small expense." The better plan, it was suggested, would be to store flood water of the River Tame in reservoirs made in land of little value and convey it through the Ashton Canal to a reservoir near Manchester, "where it may be filtered and rendered pure." It would then be delivered into the highest apartments of any house in Manchester. Apparently the committee did not state the nature or estimate the cost of filters or give data on the adequacy of filtration. The committee was against a water supply proposed by "private adventurers," both because it believed in public ownership and because the proposed company intended to take water from a source that would cut off many springs and feeders which then supplied large printing, bleaching and dye works, and afforded condensing water for the steam engines of numerous cotton mills and other works. The committee's report was approved by a public meeting of February 2, 1809, but the town did not build water works. Instead, a company began supplying unfiltered water in 1809.

Many references to Lancashire filters in various later writings, and the fact that the Manchester water committee took it for granted that readers of its report needed no description of the filters it had in mind, suggest that filters were already in use at industrial works in the Manchester district in 1809 (see Lancashire Filters, above).

At Edinburgh, in 1811, Dr. Thomas C. Hope dismissed construction of "a proper filtering bed of sand" as too costly under local conditions and proposed settling reservoirs instead. Much of the imperfection in the water delivered to the city, the report stated, came from muddiness in two large ponds, following heavy rain or snow or high winds. Part of the yield of these ponds came from springs. Dr. Hope suggested that when the water of the ponds was muddy, the flow of springs into them should be diverted through earthen pipes into the conduit leading to the city; also that two settling reservoirs, each with a capacity equaling the water consumption of 48 hr., be constructed for use alternately. Formerly, said Dr. Hope, "Edinburgh was celebrated for the excellent quality of the water; but of late, it had become in an equal degree conspicuous for the badness of it" (30).

A reference to filtration which might fruitfully have gone into more detail, but one which is highly significant for what it does say, appeared in an article published in 1825 (31). The article is all the more interesting because it was printed in a Glasgow magazine conducted by a committee of civil engineers and practical mechanics. In that city both upward- and downward-flow filters had been used, while at near-by Paisley lateral-flow filters had been constructed twenty years earlier. None of these filters was specifically mentioned.

The object of the article was to advocate "A New Mode of Forming Artificial Filters" that would avoid the failure of earlier types. Ignoring the fact that a filter had been put into use at Paisley in 1804, the article noted that the inhabitants of that town were then much interested in finding the best mode of filtering the water from the River Cart. After showing a predilection for filter galleries, the article said that if Paisley could not get a "natural filter," great care should be taken in "forming an artificial one," since both Glasgow and Cranston Hill companies had wasted labor and expense in such undertakings.

Asserting that "we have seen not a few artificial filters," the article declared that "no matter how apparently well planned such filters had been, after a short time all [became] equally useless; got dirty and choked, and ceased to purify water as they once did. . . ." In view of this, the article urged that where natural filtration is not possible artificial filters should be built in the bed and bank of a river where they would be kept clean by water flowing over them.

The general plan of filter construction was to build large and expensive tanks through which the water flowed laterally, with no means of cleaning them provided or even possible. The editors were aware that many downward-flow filters had been built. They condemned alike downward-, upward- and lateral-flow filters that they had seen, and all of which, they declared, could be cleaned only by "emptying the filter entirely and removing the impurities which it had gathered." They said that downward-flow was correct, but the filter should be kept clean by the constant natural flow of a stream across the surface of the sand (31).

This discussion is the first critical review of filters for city water supply that has been found. It points out that the main cause of failure in filters was lack of means for cleaning them in place. It rightly asserts that both lateral- and upward-flow filters were wrong and downward-flow right in principle. But the method of cleaning suggested, natural lateral surface wash, was wrong. What these early editors failed to foresee was that the proper method for cleaning a filter was either by reverse-flow wash of the material in place or else scraping off, removing, washing and replacing the top layer of the sand. The time was coming when Robert Thom was to put into large-scale use the first of these methods, patented by Peacock in 1791; and when Simpson was to adopt the second method.

Thom's and Simpson's Filters

In the late 1820's, Robert Thom in Scotland and James Simpson in England blazed the trail for mechanical and for slow sand filtration. Thom's first municipal filter was put in use at Greenock late in 1827; Simpson's at London early in 1829. Each profited in his own way from past failures. Each based his design upon small-scale experiments.

Thom, with knowledge of several filters at and near Glasgow, concluded that their complete or partial failure was due to surface clogging, difficult, uncertain and costly to prevent, and so he devised a self-cleaning filter, washed by reverse-flow.

FIG. 23. ROBERT THOM (1774–1847)
Engineer from Ascog, Scotland; designer of reverse-flow-wash filters for Greenock (1827) and Paisley (1838), Scotland
(From *Centenary of Shaws Water Company's Works, Greenock, Scotland, 1827–1927*)

Simpson journeyed 2,000 miles to see filters in the north of England and south of Scotland. He concluded that a filter at a calico works near Manchester was better than the filter then used by one of the Glasgow water companies because it had coarser material at the bottom than at the top, thus facilitating the passage of water. Observing in his experimental work, as did Thom, that the dirt retained in filters was at and just below the surface, Simpson concluded that the best way to clean the filter was to scrape off the thin dirty top layer, remove, wash and restore it, and replace it at intervals.

Both Thom and Simpson used coarse material at the bottom of their filters and successively finer layers until the top was reached. Whether or not either the Scotchman or the Englishman knew it, careful arrangement of filter media, successively smaller from bottom to top, was called for in a British patent granted to Peacock in 1791 and justified scientifically in a pamphlet published by him in 1793 (see above). Peacock's filter, like Thom's, was cleaned by reverse-flow wash, and in using this method, each man was ahead of his time. Simpson set the model for the typical slow sand filters that are still dominant in conservative England. In Scotland, Thom's self-cleaning filter, either in its original or a modified form, was used for many decades in several towns and was still in use in at least one town in 1940.

Thom's Self-cleaning Filter

Like many other good enterprises, the water works system of Greenock, Scotland, in which Thom's first municipal filter was included, had its inception on a golf course. While Thom was "coursing" near his Rothesay cotton mills, in 1820, a companion asked whether the scarcity of water in Greenock could not be alleviated by an aqueduct similar to that by which Thom supplied water and power to his mills (32). After considering information regarding a stream near Greenock, Thom said he believed that water from the Shaws River could be brought to Greenock by gravity. Although invited to look into the matter the next year, Thom was too much occupied with his mills to do so. Early in 1824, he made a survey which showed that not only was a water supply for the town practicable, but also feasible was a large water-power development the capacity of which would exceed all the steam power then being used in and about Glasgow.

On the strength of this the Shaws Water Joint Stock Co. was incorporated in 1825. The first water for power was delivered April 16,

1827. Filtered water was supplied in the latter part of 1827, more than a year before Simpson's London filter went into service on January 14, 1829.

True, Simpson began studying filtration a few years before the date just mentioned. Apparently Thom did so still earlier. This is shown by Thom's letter of March 20, 1829, to a leading promoter of the Shaws water company (32). The letter was a 3,000-word exposition of Thom's self-cleaning filter, written by request and appended to Thom's final report of 1828, which barely mentions the filters.

Although filtration on a small scale had long been practiced, said Thom, all previous attempts to render turbid water pure on a scale sufficient to serve large cities had failed, the yield growing less and less until it entirely or nearly ceased. As examples, Thom cited the failure at near-by Glasgow of "artificial filters" built "by an eminent engineer" (Telford's name graciously omitted), later unsuccessful attempts on a different plan, followed by filter galleries ("natural filters") of gradually falling yield (see Glasgow, above). These failures Thom attributed "to the lodgment of sediment between the particles of sand." Solution of the problems thus presented had long occupied his attention. Finding that the grosser particles of extraneous matter lodged at the surface, the stirring or harrowing of which was not a permanent cure, and that frequent removal and renewal of a small quantity of surface material, while better than harrowing, was troublesome, expensive and incomplete, and that various other contrivances generally failed, he devised a self-cleaning filter. On a small scale this was tried for several years and was uniformly successful. Tried on a large scale at Greenock it gave equally satisfactory results.

The Greenock Filter.—In the permanent filter at Greenock, said Thom, water was made to pass, either downward or upward, at will, through about 5 ft. of very fine, clean, sharp sand. When the yield of pure water declined through lodging of sediment, the flow was reversed for a few minutes—carrying the sediment out over the top or down through the sand to the bottom, according to the direction of filtration. No evidence of upward filtration in any of the filters designed by Thom has been found.

In his paper of 1840 (33), Thom said that his filter at Greenock and the later one at Paisley removed not only suspended matter but also color due to moss water, thereby rendering the treated water similar to spring water. This he accomplished by means of a species of traprock or amygdaloid common in the hills about Greenock, broken down to the size of peas and smaller, and mixed with fine sharp sand. These filter media, he said, were rather expensive and in time became saturated and had to be replaced; therefore, great care was taken to exclude moss water from the filters by using a separate reservoir to supply the filters and diverting the moss water to the power reservoir.

"Cesspools for the deposit of sediment" were built at intervals in the aqueduct leading from the supply reservoir to the filters. There were three filters, each 12 x 50 ft. in plan, with walls 8 ft. high. There was also a clear-water reservoir with a capacity equal to a day's consumption.

The Greenock works were acquired by the town in 1836. Since then, both slow sand and rapid or mechanical filters have been built (34). Writing late in 1936, James MacAlister, Superintendent of Water Works in Greenock (35), said that he could not find out when or why Thom's filters were abandoned. There were no plans or other particulars regarding them in his office. A lease of land dated 1861 mentioned three 12 x 50-ft. filters, presumably Thom's filters of 1827.*

Second Paisley Filters.—Eleven years after completing his self-cleaning filters at Greenock, Thom built similar but larger filters for the near-by town of Paisley. These also treated an impounded gravity water supply, first delivered July 13, 1838. Twenty-four years earlier, the first known filters on a municipal supply had been put into use at Paisley (see above).

In June, 1843, in evidence before a Royal Commission, Thom said he had "created self-cleaning filters at Greenock, Paisley and Ayr" (41).

* In the years immediately after the Greenock filters were completed they received as much if not more attention in print than did Simpson's filters at Chelsea. C.-F. Mallet, Water Engineer of Paris, who in 1825 had proposed upward-flow filters for that city, translated Thom's expository letter of 1829 into French and published it in 1831 (36). J. C. Loudon, who saw the Greenock filter in 1831, described it briefly the following year in his *Gardeners Magazine* (37) and sent Thom's pamphlet to the *Mechanics Magazine*, where it was given extended notice in 1832 (38). In the United States Loammi Baldwin, in a report of 1834 on a new water supply for Boston (39), inserted notes on the Greenock filters taken from the *Mechanics Magazine*. Not having a copy of Thom's pamphlet, Baldwin translated back into English a goodly part of Mallet's French version of Thom's letter of 1829. In 1835, Baldwin's one-time pupil, Charles S. Storrow, described Simpson's Chelsea or London filter of 1829 at some length and Thom's Greenock filter briefly (40), crediting Mallet for the latter. Thus Thom got publicity for his self-cleaning filter in Scotland, England, France and the United States. Publicity for his Paisley filters is noted later.

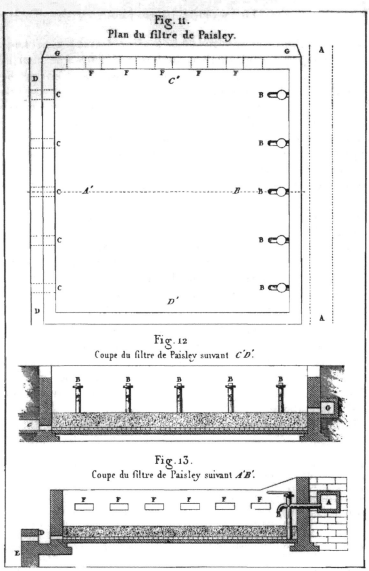

FIG. 24. THOM'S SELF-CLEANING FILTER AT PAISLEY, 1838
Media supported on false bottom of perforated flat tile; cleaned by reverse-flow wash; filter enclosed by rectangular masonry walls. Design similar in principle to rapid filters of 20th century, but rate of operation about same as that of Simpson's hand-scraped filters
(From "Atlas" of Darcy's *Les Fontaines Publiques de la Ville de Dijon*, Paris, 1856)

Thom's description of the filter at Paisley, which in essence was like that put in use in 1827 at Greenock, shows how carefully he designed his self-cleaning filter and how much it is resembled in form and in some details by the American mechanical filter.

The Paisley filter was 100 x 60 ft., with three compartments which could be used separately or together. The walls were of masonry in cement backed with puddle. The bottom was paved, had cement joints, and was supported by puddle. Much attention was given to the design of the underdrain and washing system. There was a false bottom of flat perforated tiles, similar to those used in oat-drying kilns—an example of how one industrial art builds on an earlier one. The holes were more than 0.1 in. in diameter and were very near each other. Fire brick on edge, resting on the paved bottom of the filters, supported the false bottom and formed underdrains 1 ft. wide and 5 in. high. These bricks were laid end to end, with $\frac{1}{4}$-in. open joints. Their upper edges were little more than 1 in. wide, so there would be little or no space in the false bottom without holes, and thus "nothing to prevent the water spreading equally over every part of the bottom of these drains," which "is particularly necessary," said Thom, "when the filters are cleaned by upward motion of the water."

The filtering materials from the bottom upward were: about 1 in. of 0.3-in. clean gravel, placed on the perforated flat tiles; five layers of gravel, each layer about 1 in. thick and of lesser size than the one below, the fifth described by Thom as coarse sand; and 2 ft. of "very clean, sharp, fine sand, similar to that used in hour glasses, but a very little coarser." Mixed with the upper 6 or 8 in. of this sand, in the ratio of 1 to 8 or 10, was animal charcoal, ground to about $\frac{1}{16}$ in. diameter. This gave a total depth of 36 to 38 in., mostly sand. The charcoal was used to decompose "any vegetable matter with which the water may be impregnated."

Although Thom claimed either upward or downward filtration in his report of 1829 on the Greenock filter, he says cleaning at Paisley was effected by manipulating the valves to change from downward to upward flow and wasting the dirty water. Cleaning was facilitated by stirring with a fine-toothed rake and admitting a little water through the raw-water conduit so it would flow over the surface.

The cost of the Paisley filter was given by Thom as £600 ($2,900) and the average quantity of water produced 106,320 cu.ft. [nearly 0.8 mgd. (U.S.)]. As the available area was under $\frac{1}{7}$ acre, the plant must have worked at an average rate of nearly 6 mgd. per acre or double the nominal rate of slow sand filters of that and later dates; and the rate was even faster when a compartment was being cleaned.

Thom's Influence.—The only known adverse criticisms of Thom's self-cleaning filters were uttered in support of filters patented by Fonvielle in France in 1835 and by Maurras in England in 1842 (see Arago's report and Sloper's testimony, below). The Maurras filter, which does not appear to have gone beyond the promotion stage, also employed reverse-flow wash, but under higher pressure than used by Thom. B. G. Sloper, probably an agent for the Maurras filter, testified before the Commissioners for Inquiring Into the State of Large Towns and Populous Districts in 1843: "We find that the return current of water [in Thom's filter, even under a head of 26 ft.] does not remove one-tenth of the dirt" (42). No supporting data were given. Presumably the Thom and Maurras filters were tested side by side at or near London.

Favorable comment was made by John Horsley in 1849 (43), who mentioned Thom's "self-depurating arrangement" used in "a modification of what has been called the Lancashire filter" (see Lancashire Filters, above).

Henry Darcy, in a book published in 1856 (25), says that besides being washed by reverse flow the Paisley filters were cleaned by removing 1 cm. of surface sand from time to time, and replacing it at longer intervals. He does not say how long this supplementary cleaning had been practiced.

The Paisley filter was used until 1874 when it was remodeled, wrote James Lee, Water Engineer of Paisley, in 1936 (11). It was abandoned in 1887 for new filters adjacent to one of the storage reservoirs. No plans of Thom's filters were available in the Paisley water office in 1936.*

A memoir of Thom published in 1848 (46) contains no details of his filters not already given here. It says that after his connection

---

* A plan and sections of the Paisley filter were shown on a folding plate accompanying Thom's testimony of 1843 before the Commissioners for Inquiring Into the State of Large Towns and Populous Districts (41). The plate was in the folio but not in the octavo edition. A section of the filter was shown in Tomlinson's *Cyclopedia* of 1852 (44) and in *Allgemeine Bauzeitung*, 1853 (45). Plate 24 of the Atlas to Darcy's *Fontaines de Dijon* (25) shows a plan and two sections of the filter, presumably from the Report of the Commissioners for Inquiring Into the State of Large Towns and Populous Districts.

with the Rothesay cotton mills ceased in 1840 he retired to his estate at Ascog, intending to pass his remaining days at leisure, but he was induced to lend his advice and assistance on water supply to several towns in the United Kingdom and to consider many foreign schemes. In all these "he adhered to the gravitation principle" and used with success the large filters described in the Greenock pamphlet. These towns are not named in the memoir.

After Thom's death in 1847 filters more or less like his were built at several Scotch towns. A false bottom of perforated flat tiles supported by bricks set on edge was used in a slow sand filter completed in 1850 for the borough of Kilmarnock, Scotland. The filter was designed by James M. Gale, Water Engineer to the city of Glasgow (47). A filter modeled after Thom's design was put in use at Dunkirk, France, in 1870 (48). In some filters in Scotland, notably in the Gorbals works at Glasgow (see above), perforated flat tiles were placed between the gravel and sand instead of being used as false bottoms.

James Simpson and the Chelsea Water Works Company

Best known of all the filtration pioneers is James Simpson. He was born July 25, 1799, at the official residence of his father, who was Inspector General (engineer) of the Chelsea Water Works Co. The house was on the north bank of the Thames, near the pumping station and near what was to become the site of the filter that was copied the world over. At the early age of 24, James Simpson was appointed Inspector (engineer) of the water company at a salary of £300 a year, after having acted in that capacity for a year and a half during the illness of his father. At 26, he was elected to the recently created Institution of Civil Engineers. At 28, he made his 2,000-mile inspection trip to Manchester, Glasgow and other towns in the North, after designing the model for a working-scale filter to be executed in his absence. On January 14, 1829, when Simpson was in his thirtieth year, the one-acre filter at Chelsea, commonly known as the first English slow sand filter, was put into operation.

Of the eight water companies supplying Metropolitan London in the 1820's, five, including the Chelsea until early in 1829, served raw water from the always polluted and sometimes turbid Thames, taken within the tidal reach of the stream into which numerous sewers discharged. The Chelsea Water Works Co., probably led by James Simpson, was the first to give official attention to this deplorable con-

THE DOLPHIN OR GRAND JUNCTION NUISANCE.

1.—The Dolphin 3.—Grand Junction Water Works
2.—Grand Common Sewer 4.—Chelsea Hospital
5.—Minor Common Sewers.

Geo Cruikshank 1827

THE

DOLPHIN;

OR,

𝕲𝖗𝖆𝖓𝖉 𝕵𝖚𝖓𝖈𝖙𝖎𝖔𝖓 𝕹𝖚𝖎𝖘𝖆𝖓𝖈𝖊:

PROVING THAT

SEVEN THOUSAND FAMILIES,

IN

WESTMINSTER AND ITS SUBURBS,

ARE SUPPLIED WITH

WATER,

IN A STATE, OFFENSIVE TO THE SIGHT,

DISGUSTING TO THE IMAGINATION,

AND

DESTRUCTIVE TO HEALTH.

"There is such a thing as Common Sense!"
Abernethy.

LONDON:
PRINTED FOR T. BUTCHER, 108, REGENT STREET.

1827.

FIG. 25. FRONTISPIECE (*left*) AND TITLE PAGE (*above*) OF *The Dolphin* (Reproduced from George Cruickshank's signed copy)

FIG. 26. THE DOLPHIN, OR INTAKE, OF THE SOUTHWARK WATER WORKS CO.

An illustration which accompanied a broadside entitled "Royal Address of . . . Water King of Southwark," which was "a satire on the pollution of the Thames by the Walbrook sewer and other outlets, founded on the report of the Commissioners of Inquiry in 1828"

(From colored print by George Cruikshank (Reid's No. 1464; Cohn's No. 1952)) dition and the first to build a filter. In this the company may have been stimulated by a project, launched late in 1824 by the Thames Water Co., to bring "pure and unpolluted water" from a point upriver on the Thames in the vicinity of Richmond and Brentwood and to deliver it wholesale to the London companies for distribution.

The Dolphin.—Grievances against the companies had been growing since 1810 when, promising better service and better water, the companies divided the areas served between them and increased the rates. This resulted in complaints of monopoly and high rates, and in mounting criticism of the bad character of the water supplied. On March 15, 1827, a veritable explosion occurred with the publication of a thick pamphlet called *The Dolphin or Grand Junction Nuisance* (49). A frontispiece engraving showed the dolphin or water intake of the Grand Junction Water Works Co., and near it the outlets of a large sewer and several small ones, one of which came from the Chelsea Fever Hospital. The pamphlet led to commission hearings and reports on the quality of the London water and projects for new sources of supply (50, 51).

Simpson's Working-Scale Filter.—It is to the credit of the Chelsea Water Works Co. that its first steps toward filtration were taken a year before *The Dolphin* appeared. Undoubtedly, the pamphlet spurred on both the company and Simpson. His filter-inspection trip and working-scale filter came after *The Dolphin* had been published and about the time of the creation of a royal commission. His permanent filter was already under construction when he testified before the commission.

Simpson's Inspection Tour.—Hope of finding that a detailed report by Simpson on his filter-inspection trip still exists was ended late in 1935 when Lt.-Col. J. R. Davidson, then Chief of the London Metropolitan Water Board, wrote that an exhaustive search of documents in the Board's Muniment House disclosed no evidence that a printed or written report had ever been made (52).* Fortunately, unpublished

---

* Similar disappointment resulted from an appeal made in 1937 to Clement P. Simpson, grandson of James Simpson. In answer to a request for data for use here, he wrote that as executor of the wills of James Simpson Jr., and Charles Liddell Simpson, a grandson of James Simpson Sr., he had access to all papers and documents belonging to them. He remembers no material which would be of assistance here. The journey of his grandfather to the North of England and Scotland has "always been rather obscure and so far as I know there is little documentary evidence of detail" (53).

FIG. 27. JAMES SIMPSON (1799–1869)
Engineer of Chelsea (London) and designer of hand-scraped filters for the Chelsea Water Works Co., 1829
(From a painting by Sir William Boxall, R.A., exhibited at the Royal Academy, London, 1856; photograph supplied by E. Graham Clark, Secretary, Institution of Civil Engineers)

Minutes of the Chelsea Water Works Co. contain several brief references not only to the trip but also to preceding and subsequent events. Extracts from the Minutes (54) were kindly supplied for use here by Col. Davidson. From these, from Simpson's testimony of 1828 before two Parliamentary Commissions (50, 51), from material supplied by Simpson for a lecture by William Thomas Brande (55), and from data supplied by Simpson for Telford's *Autobiography* (15), a consecutive story of all of Simpson's early filtration activities has here been constructed for the first time.

Rearranged chronologically, the minutes of November 1, 1827 (54), showed that filtration on a large scale had occupied Simpson's attention for the two previous years and that he had made many filtration experiments at the water works. (Elsewhere (15) Simpson said that the experiments were begun in the spring of 1826.) In January 1827, Simpson received permission from the directors of the company to make "experiments on a larger scale than he had been able to do privately." In August, 1827 (five months after the appearance of *The Dolphin* (49) and a few weeks after the public-protest meeting inspired by this pamphlet), "the directors manifested some impatience," and "directed" Simpson "to turn his whole attention to the subject; and having heard that a filter bed was working at Glasgow, he received permission to proceed there." A brief report on his northern trip was submitted to the directors on November 1, 1827, whereupon "he was ordered to make certain further experiments which he had proposed and report the results to the next Board."

Apparently, that was too short a time, for on November 7, Simpson submitted instead another account of his filter-inspection trip. This, as spread on the minutes, is so brief and yet is so important in the history of filtration, that it is given here in full:

During the journey to Manchester and Glasgow, which he undertook with permission of the Court [directors], he saw several large filter beds at work, and from the information he obtained, he has no doubt of being able to filter the quantity of water the Company requires. Filtration of water through the simple medium of sand and gravel, possesses so many advantages compared with reservoirs, that he feels assured that it will be the best method of purifying the water.

He takes the liberty of stating that from results of reservoirs in other Water Works, where various sums of from £10,000 to £50,000 have been expended, the improvement in the quality and appearance of the water has not in a single instance equalled his expectations. The River Thames is often affected by land floods, particularly during about 20 days every year when reservoirs are of very little use, but filtration will improve this water very much and render the loamy appearance scarcely perceptible. (54)

A little additional information regarding what he saw on his trip was given by Simpson in testimony before a committee of the House of Commons in July 1828 (51):

——he travelled over Britain, and examined many plans in operation for filtering water; he had travelled 2,000 miles, and in Lancashire, Lincolnshire * and Scotland, he had seen many manufactories and some waterworks supplied by filtered water. The filter beds he had seen had been in operation for various periods, some for four months, and others for longer periods up to 16 years.

Asked by the committee whether his proposed filter for the Chelsea company was like that of the Cranston Hill Water Works Co. at Glasgow, Simpson replied:

The plan I have adopted is partly like that; but I consider it improved by making use of a process [element in design] which they use in the neighborhood of Manchester, and that is by having a lower stratum of gravel, that the water may pass freely off.

[Asked] "Has the plan at Glasgow or at Manchester perfectly succeeded?" [Simpson replied:] "The plan at Manchester is used in the calico works, and they have been at work many years."

Out of the obscurity of a century ago, the foregoing notes are all that can be found on what Simpson saw on his inspection tour, the second journey of the kind on record, Mallet of Paris having visited Britain to study filtration, and water works generally, about 1824.

What a pity that Simpson did not put on record descriptions of the filters he saw, the names of their designers and the dates they were put into operation! Even the location of the filters and an indication of whether they were treating water for industrial works or municipal supply would be welcome information. A phrase in one of the passages cited—"many manufactories and some water works"—indicates that most of the filters which he inspected were for industrial supplies.

* No record can be found of filters in Lincolnshire as early as 1827. The water engineers of Lincoln (56) and Boston (57), the largest communities in the county, report that sand filters were included in the first works for general water supply, built in 1847 for each community. They could find no evidence of earlier filters, either municipal or industrial, in their part of the county. A county gazetteer, published about 1812, mentions water works for only one town—Stamford—to which water was brought in cast-iron pipes from the Walthorpe springs (58).

FILTRATION IN BRITAIN

Completion of Experimental Filter.—At a meeting of the directors of the water company on November 15, 1827, Simpson reported that the experimental filter was completed November 7; the "water had been cleaning the sand every day since." On November 8, after 27 hours work, the filtrate was cloudy, but "as pure as the Water taken from Hyde Park Basin, when it had been at rest 84 hours." On November 9, the cloudy appearance had gone off and after the following day the water was quite clear. To show "that this process of filtration did not render the water at all vapid but pure and brilliant—he laid a sample thereof before the Court" (54).

Convinced thus of the efficacy of the working-scale filter, the directors, on the same day, ordered the construction of a permanent filter "upon the Plan proposed by the Engineer"—who had just submitted cost estimates.

By far the most complete description of the large experimental filter was that supplied by Simpson to William Thomas Brande (55). He told Brande that the "pond" containing the filter was 44 ft. square at the top, 26 ft. square at the bottom and 6 ft. deep. The filter had a top surface of 1,000 sq.ft. and a depth of 4 ft. After the "pond" had been made watertight, a drain was laid to a clear water well and open-jointed branch drains of brick were laid. The filter media, from the bottom up, were gravel, graduated from coarse to fine, 2 ft., and sand, graduated from coarse to fine, 2 ft. Both the gravel and the sand were selected with great care and well washed. Two settling reservoirs, each 32 ft. square at the top, 20 ft. at bottom and 4 ft. deep, were provided. Their low-water line was level with the high-water line of the filter. The reservoirs worked alternately, regulated to filter 12,000 cu.ft. or 90,000 gal. (U.S.) per 24 hr. This would be 90 gal. per sq.ft. or about 3.9 mgd. per acre. The method of cleaning the sand and the underlying principle of filtration as seen by Simpson when he had operated his experimental filter some two months deserve to be given in his own words:

The silt which was stopped on the sand, was regularly cleaned off with a portion of the sand every fourteen days; the principle of the action depends upon the strata of filtering material being finest at the top, the interstices being more minute in the fine sand than the strata below; and the silt, as its progress is arrested, (while the water passes from it) renders the interstices between the particles of sand still more minute, and *the bed generally produces better water when it is pretty well covered with silt than at any other time.* [Italics mine.] Silt has never been found to penetrate into the sand more than 3 in., the greatest portion being always stopped within the top half-inch of the sand; and in cleaning the silt off, it has never been found necessary to scrape any more of the sand off with the silt than the first half-inch depth and sometimes only half that depth was removed. The small air-pipes from the drains are to prevent injury to them or the filtering materials by condensation or otherwise. (54)

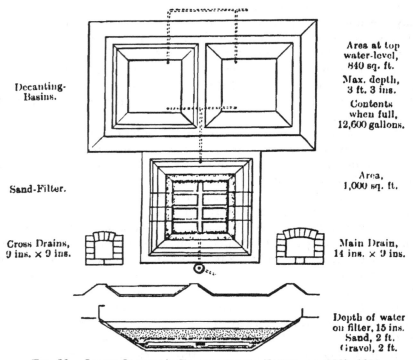

FIG. 28. JAMES SIMPSON'S EXPERIMENTAL FILTER OF 1827–1828
(From drawing in possession of Charles Liddell Simpson; previously reproduced in *Proc. Inst. M.E.*, April 1916, p. 300)

The experimental filter was continued in use at least eight months, for in his evidence before the Select Committee on the supply of Water to the Metropolis, given July 7, 1828 (51), Simpson said that the filter "is now in work." In answer to questions Simpson, on the same day, said that the head on the test filter "varies according to the state of the material; when the material is clean, it will go off very well with four inches head, and the head water increases as the material becomes clogged." This was once in 14 days, whereupon, "the sand was made dry, and men were sent in with common spades and [they] scraped off the surface."

Skepticism as to the wholesomeness of filtered water in 1828 and Simpson's reassurances on the subject are amusing today. At the hearing before the Royal Commission (50) a member asked whether any persons had been in the habit of drinking the water filtered on a small scale. "Yes," answered Simpson. Had they complained of the water "being insalubrious, giving them cholic or any other complaints?" To this, the engineer replied that none of the more than 100 men working on the ground (presumably on the permanent filter) had complained of the filtered water but there had been complaints of the "land spring-water being injurious." Fish, the commission was assured, did not die in the filtered water. Simpson willingly admitted that "water may contain so many ingredients chemically dissolved, that filtration will not purify it." Asked whether the discharge from King's Scholars Sewer could be "so filtered as to be fit to drink," Simpson cannily said he had never tried it. Asked whether filtration would remove bad taste from water, Simpson replied that "Thames water has a taste, according to season, of animal and vegetable matter"; filtration "seems to deprive it of the whole of that, and we cannot discover it after it has passed the bed."

The Chelsea Filters.—After three years of experimentation, travel observations and construction, Simpson's permanent filter was put into use January 14, 1829. Within a few days of that epoch-making date in the quest for pure water, Simpson provided Brande with a description of the filter (55). It was even shorter than the description of the experimental filter already noted. Simpson said that the large filter had a surface of nearly an acre and was "constructed precisely on the same principle as the experimental bed," and added that the details of forming and operating the large filter had been "greatly improved and adapted to the enlarged scale." How, he did not say, but this he indicated later.

In January, 1829, the filter was "working with the greatest success during the inclement season," testified Simpson, "and although the water on the bed is this day covered with ice five inches thick, it does not impede the filtering process" (55). This is the earliest testimony on the effect of ice on filtration that has been found.

More information on the design of the large filter, with data also on his experiments, was supplied by Simpson for inclusion in Telford's *Autobiography* (15)—presumably shortly before Telford's death on September 2, 1834, when the filter had been in use five years.

Simpson's Researches and Studies.—When he began his preliminary studies, Simpson wrote to Telford, the art of filtration "upon a large scale was yet to be acquired, and improvements [were yet] to be made upon the works at Glasgow, Manchester, and other places, where it appeared that instances of failure, as well as of success, had occurred." Anticipating by many decades the conclusions of other engineers "that preliminary experiments were indispensable" before venturing on large capital outlay "several trials were made on superfices exceeding 1,000 sq.ft. to ascertain the most approved principle, and the fitness of the various materials proposed to be employed." Moreover:

> All the modifications of lateral and ascending filters proved disadvantageous; difficulties were encountered in preserving the various strata in their assigned position, according to the sizes of their component particles; and effectual cleaning could not be accomplished without the removal of the whole mass of the filtering medium. All devices by currents, reactions of water, and other means, also proved either inefficient or inconvenient and expensive. (15)

Mention of "lateral" and of "ascending" filters suggests that Simpson may have seen or heard of the Paisley filter of 1804 and that he talked with Robert Thom about his "self-cleaning filter, worked by ascent or descent" (see above), put into use just before Simpson visited Scotland late in 1827. At Glasgow, he may have seen or heard of four different filters and extensive filter galleries built by the two water companies in the previous twenty years.* The only Glasgow filter mentioned by Simpson was one of three built by the Cranston Hill Water Works Co. at various times. Of that, Simpson says only that it was something like his Chelsea filter, having a ridge-and-furrow surface. This may have been the Glasgow filter that Simpson had in mind in testimony of 1828 (50, 51).

As a result of what he saw on his filter-inspection trip and learned from his experiments, Simpson decided on filtration "by descent"

---

* So far as has been found, no municipal filters had been built in England when Simpson made his filter journey of 1827, while in Scotland the only ones were those at Paisley (already abandoned?), Glasgow and Greenock.

through fine and coarse river sand, shells and pebbles, small and large gravel, and with the surface disposed in ridges, giving an undulated appearance. According to his letter to Telford:

> The first experiments by descent failed; sufficient care had not been taken in the selection and separation of the materials. Explosions of condensed air in the tunnels for collecting the filtered water deranged the strata occasionally, but were obviated by air drains. The filtration was, in one instance, stopped by the addition of fresh sand without having previously removed the old sand, which should be applied as the upper stratum; although in this case, the surface had been thoroughly cleansed previously. A film or puddle was formed on the original sand, and was sufficiently supported by the particles of sand to sustain five feet head of water, at first acting to impede, and eventually to stop the filtration. The process was greatly improved by the introduction of the small shells, such as are usually found at Shellness, the flat surfaces of which overlap, and assist in the great desideratum of separating the sand from the gravel, and thus tending to preserve the free percolation in the lower strata, which is essential for ensuring filtration sufficiently rapid for waterwork [sic] purposes. . . . The lower stratum of gravel contains the tunnels for collecting the filtered water. They are built up with cement blocks, and partially open-jointed, two spaces of an inch and a half on the bed and the heading joint of each brick being open. The fine gravel, pebbles and shells, and the coarse and fine sand are laid upon the large gravel. (15)

Water was let into the filter at nine places, discharged from pipes "fitted with curved boards to diffuse the currents of water and prevent the surface of the sand from being disturbed." Because the interstices of the fine top sand were smaller than those of the next lower stratum, the impurities were arrested near the surface. Careful examination showed that the sediment sometimes penetrated to a depth of 6 to 9 in., depending on the state of the land floods in the Thames. But it was never necessary to scrape off more than an inch of sand from the surface at one time, "the remainder tending rather to improve filtration by rendering the interstices between the sand still more minute." "From these observations," wrote Simpson, "it must not be inferred that the process is merely a fine mode of straining; for something more is evidently effected; an appearance resembling fermentation being discernible when water is in contact with the sand" (15). What was meant by "fermentation" is not apparent.

"The undulated surface" of the filter, wrote Simpson, "admits of parts of it being washed, and others drained; and it aids in cleansing, by admitting the grosser particles of the silt to slide down the ridges, and form a sediment easily manageable." This is not convincing today, but as Simpson ridged his later filters he must have continued to think it worth while.*

The amount of water being filtered when Simpson supplied data to Telford was from 2.25 to 3 mgd. (U.S.) and as the filter had an area of an acre that was also the filtration rate per acre. The period of sedimentation was probably short.

Reverse-flow wash and harrowing were used to supplement hand scraping of the London filters operated under Simpson's direction. This he stated in evidence given in July 1851, when asked if he had "adopted new means of cleaning filters" (59).

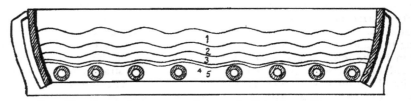

FIG. 29. CROSS SECTION OF SIMPSON'S ONE-ACRE FILTER FOR CHELSEA WATER WORKS CO., 1829

Media were: 1. fine sand; 2. loose sand; 3. pebbles and shells; 4. fine gravel; 5. large gravel, containing "brick tunnels" or underdrains. Similar undulating surfaces used by Simpson in several other filters and by Hiram F. Mills in filter at Lawrence, Mass., in 1893

(From communication by Simpson to Telford reproduced in latter's autobiography)

Water in relation to public health is barely mentioned by Simpson in anything he wrote on the Chelsea filter.

Simpson's Later Work

The Chelsea filter was continued in use until 1856, when the company began filtering water from an intake at Surbiton, higher up the Thames (52). These Surbiton filters were designed by Simpson, who

* The second of the three filters of the Cranston Hill Water Works Co. at Glasgow also had an undulating surface, raw water being passed longitudinally along the valleys. As designed, it was to be cleaned by discharging water through the valleys in a direction opposite to the normal flow, thus carrying away the sediment. For some reason not made clear by the available data this caused excessive wear on the pumps and the cleansing system had to be given up. (See "Glasgow," above.) A ridge-and-furrow surface was provided for the slow sand filters at Lawrence, Mass., put into use in 1893, but the stated object was to distribute the raw water without disturbing the surface of the layer.

was still Engineer of the Chelsea Water Works Co. He also designed filters completed in 1851 for the Lambeth Water Co. of London, which he likewise served as engineer for many years. In 1846, he advised the construction of filters for the water company at York, England, "in every respect except as to size, the same as those he had so successfully constructed at Chelsea." These were built soon afterward (61). It is also known that he designed filters built at Aberdeen, Scotland, in 1864 (62).

Simpson was president of the Institution of Civil Engineers in 1854–55. At the time of his death early in 1869 he was the oldest living member of the Institution.

Including Thomas Simpson, father of James, the family practiced engineering from late in the eighteenth until well into the twentieth century. James Simpson had three sons—James Jr., John, and Arthur Telford Simpson. All three were in some way connected with engineering and both James Jr. and John with water companies. All four sons of the second James were connected with both engineering and water companies—Charles Liddell, Thomas B., Edward P., and Arthur B. Simpson—the last still living in July 1937. John, son of the first James, had one son, Clement P. Simpson, writer of the letter from which these data are taken (53).

In 1851, James Simpson Jr. joined in the extensive engineering practice of his father. Six years later, he became a member of Simpson & Co., manufacturers of Woolf compound engines (63).

Charles Liddell Simpson, son of James Simpson Jr., and grandson of the first James Simpson, joined the firm of James Simpson & Co. in 1888, and became Managing Director in 1896. Later he became interested in the Worthington-Simpson Co., a combination of the Worthington Pump Co. and James Simpson & Co. (64). Keeping step with the march of engineering progress this grandson of the "father of the slow sand filter" used the Davis and Riddell type of American rapid filter when it was introduced in England in 1890 (65).*

The Maurras Filter

A multiple pressure filter much like Fonvielle's (see Chap. IV) was registered in the British patent office by André Eustache Gratien

---

* For help in obtaining data regarding James Simpson Jr. and Charles Liddell Simpson, I am indebted to Brig.-Gen. Magnus Mowatt, Secretary of the Institution of Mechanical Engineers.

Auguste Maurras * on November 15, 1842 (No. 9,520). Several filters were placed one above another in a closed tank. Water under pressure could be applied to either the upper or lower unit alternately, or from both directions at once. Pressure could be applied either from a reservoir at any desired elevation or by a pump. The filtering material could be cleaned in place.

In testimony submitted to the Commissioners for Inquiring Into the State of Large Towns and Populous Districts in 1844, B. G. Sloper, apparently agent for the Maurras filter, said he had examined 60 to 70 filter patents granted in England and on the Continent, besides inventions of many others who had not, according to him, thrown their money away on patents (42). Two recent inventions had seemed likely to become rivals—cotton and wool filters. Both failed because after a time they imparted impurities to the water. The compressed wool filter, Sloper said, was excellent so far as "minute porosity was concerned" and, he asserted, its inventor [Souchon] had influence enough to obtain a very favorable report on it by the French Academy of Medicine and its adoption for some of the public fountains of Paris, but in the last year [1843] the defects common to all organic filtering media had become manifest enough to overcome the influence employed to obtain its adoption, and its use in the fountains had been discontinued. [If some Souchon filters were abandoned, their use in Paris was resumed (see Belgrand tests of 1856, Chap. IV).]

Sloper claimed that a filter $5\frac{1}{2}$ x $5\frac{1}{2}$ ft. in plan, with a working surface of $60\frac{1}{2}$ sq.ft. [counting both faces] had a capacity of 180,000 gal. (U.S.) in 24 hours when working under a $12\frac{1}{2}$-ft. head. This capacity, he said, was based on continued working of a filter for three out of the four months it was in operation at New River Head [works of the New River Co., London]. Placing one unit over another would reduce the ground area required but increase the cost of operation. [The rate of filtration specified would equal 135 mgd. (U.S.) per acre with no allowance for time out for cleaning. No evidence of a permanent Maurras plant, big or little, has been found.] Illustrated descriptions of the Maurras filter are given in Tomlinson (44) and in Delbrück (45), apparently based on the Sloper article.

* The patent gave a London address for Maurras but searches made by the Reference Department of the New York Public Library indicate that he was a lawyer and *agent d'affaires* of Paris. He did not take out French or American patents.

Growing Concepts of Filtration; Early Nineteenth Century

At the beginning of the nineteenth century concepts of the nature and function of filtration were vague and sometimes contradictory. At its close they were well defined and generally in accord. Removal of suspended matter or turbidity was for many years the chief objective. Gradually more and more attention was given to organic matter. Whether it was harmful and, if so, how and why were moot questions.

Filtration was long regarded as a mere straining process, limited by the size of the interstices between the particles of media. Only charcoal, many held, had the power of removing, or at least transforming, organic matter in solution. Whatever the media, it was generally agreed that the finest-grained should be placed at the top of the filter, with coarser and coarser material below. One school was for upward-, the other for downward-filtration. The former believed that in upward flow through progressively finer material, gravity would carry most of the suspended matter to and below the bottom of the filter and that what remained lodged in or on the sand could be removed by reverse-flow wash. The downward-flow school, observing that most of the dirt was intercepted at the top of the filter, removed it by scraping off a thin layer of media. This school prevailed throughout Great Britain but in Scotland reverse-flow wash came into use just before scraping was adopted in England. At and after the close of the nineteenth century reverse-flow wash was employed wherever the American type of rapid filter was adopted.

Midway in the century, men of vision, aided by research, showed that organic matter, with possible harmful contents, was reduced by filtration. Toward the end of the century it was proved that slow sand filters, as perfected long before bacteria were more than dreamed of, were as efficient in removing bacteria as in effecting their original objective—clarification.

For convenience there have been assembled here notes and comments on methods and concepts of filtration, drawn from various British encyclopedias, a few books, government reports and papers before engineering and other scientific associations. The better to show progress these have been arranged chronologically.

Failure of English cyclopedias to keep pace with the quest for pure water during the first half or more of the nineteenth century is amazing. For the most part they did not go beyond generalities and mention of patented filters. Some of the books did better. In the last half of the century growing understanding and appreciation of filtration is noticeable.

Rees's *Cyclopedia* of 1819 (66) said that the filter most generally used was a porous stone basin. Its cost and liability to clogging had given rise to more simple filters, of sand, of powdered glass, or of charcoal. The last, because of its antiseptic qualities, corrected putrid water besides separating suspended matter. "Isaac Hawkins, of Twitchfield St.," made charcoal filters for use in the metropolis where the water, in general, required such treatment. Three small household filters were described.

The London Encyclopedia (1829) (67), although published the year Simpson's first sand filter was put into use and after the construction of the filters at Glasgow and Greenock, Scotland, does not mention a single filter for municipal supply.

Abraham Booth, who classed himself as "Operative Chymist," published a noteworthy little book on water in 1830 (68). In it he gives a comprehensive review of water purification as understood by a chemist familiar with what had been done on a small scale up to his day but with little knowledge or appreciation of accomplishments in the field of municipal filtration. Instead of describing any of the municipal filtration plants built within his time, Booth generalizes on filtration, noting that passage of water through filtering stones or through sand, gravel, or pounded glass would clarify water but would not remove "putrescent vapours," which could, however, be taken out by filtering through charcoal.

Matthews' *Hydraulia* (1835) (21), although written largely in defense of the London water companies, the qualities of whose supplies had recently been violently attacked, includes a wide-ranging review of water supply and purification in several countries. He describes "filtering reservoirs constructed at Chelsea, Glasgow and Greenock" and adds: "Probably at no very distant period, the practice of filtering the whole of the water supplied from rivers to the inhabitants of great towns for domestic use may be universally adopted."

The Penny Cyclopedia (1838) (69), says under "Filter" that within the previous few years various filters had been used for either domestic or culinary supplies, and were generally composed of sand or small pebbles and charcoal. In its final volume (1843) under "Water," it says that sand filtration will remove suspended matter. Charcoal is recommended to remove matter in solution or any taint from "putrid vegetable or animal substances." The *Cyclopedia* describes the Chelsea filter, London, with credit to its engineer, James Simpson.

Ure's *Dictionary of Arts* (1839) (70), strikes a new note by asserting that "agitation or vibration is of singular efficiency in quickening percolation," displacing particles and opening pores that have become closed. Dr. Ure sums up, as none of his predecessors do, the advantages of "hydrostatic or pneumatic pressure" to speed up filtration. In a closed vessel, this could be effected by piping water from a reservoir, by using a pump to exert pressure on the surface of the filter, by using an air or steam pump to create a partial void beneath the filter, or by means of a common siphon; however, pressure could not be pushed very far without chance of deranging the filter bed or making water muddy. Ure mentions no municipal filters.

The Encyclopedia Britannica (1842) (71) in its article on filters, takes up the Robins household filter, and even includes exterior views of three highly ornamental household filters of this manufacturer. The writer of the article disclaimed certain knowledge of the composition of the Robins filter but said he understood that it included sponge above "various strata of filtering material." He had great doubts of the validity of the patentee's claim for "voltaic action which decomposes soluble substances, and reduces the water equal in purity to distilled water," but he had "no doubt, from testimonials . . . that the filter is a very good one." This claim for voltaic action seems to be the first of a long series of attempts to win purchasers by alleging electrical action in water and sewage treatment. No Robins filter is included among British water filtration patents.

Cresy's *Encyclopedia of Civil Engineering* (1847) (72) makes available concise descriptions of Thom's self-cleaning filter at Greenock and Simpson's Chelsea hand-scraped filter, but without critical comment.

The Report on Supply of Water to the Metropolis, made in 1851 by three leading chemists (73), notes that filtration will not wholly remove turbidity and suggests coagulation. The committee was informed by one of the London water companies that seven grains of alum per Imperial gallon would clarify and deodorize water of the Thames.

Tomlinson's *Cyclopedia of Useful Arts* (1852) (44) contains a comprehensive review of water purification. It describes "repose in large reservoirs"; takes up coagulation at some length; outlines the principal filters thus far built for municipal service and describes chronologically many British and French filter patents. The article remarks on "the multitude of inventions and contrivances for domestic filters" and fire escapes. "In either class," it says, "examples are as numerous as attempts to solve the problem of perpetual motion, or to square the circle. Almost any kind of porous substance has been enlisted into the service of filters." All organic materials except charcoal were condemned by the writer because after they had been kept wet for a period of time they "underwent decomposition, and imparted impurities" to the water.

Samuel Hughes, a civil engineer, in what seems to have been the first comprehensive British book on water works design (74), describes several filters of his day (1856), which he classes as Scotch and English. In the Scotch filter, he says, "the various kinds of filtering material are placed in separate compartments," while in the English they "are placed in successive layers, one above the other." This is the only known credit to the Scotch for having evolved a distinctive type of filter—credit well deserved. But he errs in telling what the type was. In view of the pains taken by Hughes to present up-to-date descriptions of Scotch and English types of filters, his error is unfortunate.

Growing Concepts of Filtration; Later Nineteenth Century

New concepts of what filtration could do were manifested at the middle of the nineteenth century. At intervals these were enlarged until in the 1890's even the majority of the skeptics had recognized that, instead of being a mere straining process for the removal of suspended matter, filtration removed deadly germs of disease. For some time the broader views centered on the reduction of organic matter, both suspended and dissolved.

A pioneer in the broader concept of filtration was Dr. Angus Smith— now best known for his preservative coating for cast-iron pipe. His ideas on filtration were expressed in papers before the British Association for the Advancement of Science in 1848 and 1851. Formation of nitrates, he said in 1848 (75), is one of many ways in which water purifies itself from organic matters. "In large operations, carbon is also oxidized. A filter . . . as an oxidizing agent, acts in proportion to its cubic contents." Three years later (76) he expressed the belief that filtration was more potent than distillation in removing organic matter from water "and, more than any other known method, improves the taste and appearance."

Determined to learn the precise nature of the effects of filtration upon ordinary river water, Henry Witt, Assistant Chemist at the Government School of Applied Science, in 1855–56 made chemical analyses of water before and after filtration (77). Besides studying the efficiency of Simpson's slow sand filters at the Chelsea Water Works, London, he made laboratory tests on less polluted water taken higher up the Thames at the site chosen by the Chelsea company for new filters. The Chelsea tests appear to have been, if not the first, the most thoroughgoing investigations up till then made of the purification effected by a slow sand filter. Taking into account both the Chelsea and the laboratory studies, the latter including both sand and charcoal as filter media, Witt concluded that "sand, charcoal and probably other porous media, possess the very peculiar property of removing, not merely suspended impurities but even dissolved salts from solution in water." Although, of the two, charcoal was the more efficient in removing dissolved organic matter, sand was capable of doing so, but in less degree. These properties of porous media "have important bearings upon hygienic science," he said. The paper contained detailed figures significant in the history of chemical analyses of water.

Wholly different from Witt's conclusions were those expressed by Edward Byrne, in "Experiments on the Removal of Organic and Inorganic Substances in Water," a paper read in 1867 (78). The paper was perhaps not as significant as the lengthy discussion it elicited from engineers and others. The discussion indicated, however, the wide range of opinion on what filtration would do. Byrne, it should be understood, was a pronounced advocate of obtaining public water supplies that did not require purification—a doctrine afterwards expressed by the phrase, "innocence is better than repentance."

Much of Byrne's study was designed to determine whether vegetable matter, either nitrogenous or non-nitrogenous, is dissolved in water. To settle this point he evaporated bog or peaty water from an uninhabited area. Finding vegetable nitrogenous matter present, he concluded that it, like animal organic matter, could be decomposed into ammonia and nitric acid. This, he believed, disproved the conclusions of Dr. Edward Frankland (79) that, after deducting the nitrogen corresponding to the nitrates and nitrites, any remaining nitrogen must be due to sewage matter.

Next Byrne made laboratory experiments on the removal of organic matter from water by filtration through various kinds of charcoal, using water from a garden well in Dublin. His final conclusion was that filtration was valuable in removing suspended matter from water but for matter in solution it was "manifestly useless." Hence, "the inconsistency of bringing home foul water to undergo a delusive method of purification, instead of . . . procuring water which itself is naturally pure."

Outstanding in fifty pages of discussion of Byrne's paper were remarks by Thomas Hawksley, then Vice President of the Institution of Civil Engineers (78). Chemists, he says, had given valuable information on water purification but it did not enable engineers to make better water works. "Attempts should be made to understand how filters operated; whether the charcoal was necessary or unnecessary; and whether the common and ordinary sand filtration was sufficient." Twenty years of experience led him to believe that filters operated chemically as well as mechanically and that the chemical changes depended very much on the state of the organic matter in the water and on the admission of free atmospheric oxygen. He makes the significant statement that "the sand cleaned the water mechanically by the agency of the principle of the attraction of aggregation." During the slow passage of the water through the filter "the minute particles of matter suspended in it were attracted and held by the facets of the sand and adhered there, and the water became clear. . . . Scarcely in any filter did the water remain foul for more than a few inches from the surface." A filter 12 in. deep was as effective as one of 2 to 3 ft. The sand filter actually destroyed organic matter.

H. Shield, after considering analyses of Thames water which had been passed through various types of experimental filters (those studied by Witt, mentioned above) expressed the opinion (78) that the action of the sand on organic matter was due to adhesion of the impurities to the surface of the sand, and consequent neutralization of much of the organic impurities. "Possibly the portion of organic matter removed was that which was in a state of decomposition, and which alone was noxious to health."

Edwin Chadwick, social and sanitary reformer (78), said that water supplies from unpolluted sources might be contaminated en route or stagnate in open reservoirs, the latter condition giving rise to "rapid growth of vegetation, then animalcules, then decomposition of animal and vegetable matter, which subsequently public or private filtration only partially removed." He suggested a "competent examination, by scientific men, free from professional interest or bias, who would compare promises with results, in money as in quantity and quality, of the supplies given." Filters of sandy or other soil not containing "vegetation," said Chadwick, were little more than sieves while those containing "vegetation," as shown by chemical analyses made by Professor Way, removed much if not all the matter held in solution. Where filters containing "vegetation" could not be obtained it would be more economical to keep impurities out of water than to filter them out. Where water was derived from gathering grounds underlaid with granite or other primitive rocks, he suggested stripping off their commonly thin covering of peat or other vegetable matter down to bare and clean rock. The water should then be led in covered channels to covered reservoirs and thence direct to houses. This "large order" overlooked the magnitude of the task of stripping an entire water-collecting area for cities with even as low a consumption as prevailed in England in the sixties; but it anticipated the stripping of the sites of large storage reservoirs which was to be practiced for a time by some cities in the northeastern United States a few decades later.

The oxidizing power of carbon, in terms of albuminoid ammonia, was announced in 1872 by the *British Medical Journal* (80, 81). This conclusion was based on tests of a small commercial filter of silicated carbon made for the *Journal* by Professor J. Alfred Wanklyn, noted English chemist.

After decades of widespread belief that, unless charcoal was used, filtration removed only suspended matter from water, the contrary opinion, which had been voiced from time to time by the more progressive men, was reinforced in 1873 by William Corfield, Professor of Hygiene and Public Health at the University College, London (82), who stated that considerable chemical as well as mechanical action took place in a sand filter. Corfield, however, failed to mention what Hawksley had said in 1867, namely, that considerable chemical action is caused by the air held between the particles of sand coming in contact with the finely divided water passing through the filter. The resulting oxidation of organic matter and its transformation into "innocuous matter," Corfield says, is the "first important point to understand about filters," whether for water or sewage.

It might be well to interject here that while English contemporaries were centering their thought on the removal of dead organic matter from water, an investigator in Italy directed his attention to living organisms in sewage. "Microscopical observations" convinced Dr. Dario Gilbertini of Parma (83) that "germs of zymotic disease, especially cholera, could not be removed by filtration."

The efficacy of sand filtration was concisely summarized in the sixth report of the Rivers Pollution Commission (1874) (26). Mineral matters in suspension in water are almost always innocuous, it says, but "impart a repulsive appearance which often leads to the rejection of a wholesome water for a bright and sparkling though dangerous one." Slow filtration through sand almost always removes suspended matters whose separate particles are readily seen, but washings from clayey soils are very difficult to render bright by sand filtration. Organic matters in suspension have not only the objectionable quality of suspended mineral matters but in addition they are sometimes actively injurious and "always promote the development of crowds of animalculae." Finely divided organic matters in suspension cannot be entirely removed by filtration. Elsewhere in the report much space was given to "Propagation of Cholera by Water" in the light of deductions made and facts gathered concerning cholera epidemics of 1832, 1849, 1854 and 1866. Data are given showing the relatively light incidence of cholera after the introduction of slow sand filtration.

The Encyclopedia Britannica, which in earlier editions had not reflected progress in water purification, put itself nearly abreast of scientific progress in 1875 (84). It stated in its article on filters that putrescent organic matter may include "minute invisible disease germs" which should be removed from drinking water. Numerous outbreaks of "virulent disease, such as typhoid," had been "clearly traced to water so contaminated." It was pointed out too that the danger was much greater because "such water may be bright and sparkling, and peculiarly palatable."

To the astonishment of most of his listeners, Percy F. Frankland announced before a meeting of the Institution of Civil Engineers on April 16, 1886 (85), that filtration removed most of the bacteria from water. This had recently been proved by counts of the water supplied by the London water companies. The bacterial cultures were made by the recently devised "gelatin process" of Robert Koch. Besides his studies of the results of filtration at the large plants of the London companies, Frankland reported laboratory tests with various filter media and other tests to ascertain the effect of agitating water containing finely divided matter. He showed that storage alone greatly reduced the number of bacteria in water.

For the last four months of 1885, filtration reduced the average number of bacteria in the Thames 97.9 per cent while for the River Lee the reduction was 98.5 in November and 88.8 in December. Frankland's pronouncement deserves quotation; so does his tribute to engineers who in doing their best to accomplish known physical objectives incidentally achieved remarkably beneficial results in the realm of what was so long unknown:

Thus for the first time a definite conception has been obtained of the effect of sand-filtration upon these lower forms of life. Hitherto those who were acquainted with the size of these minute microscopic organisms on the one hand, and with the dimensions of the pores in a sand filter on the other, have believed that little or no barrier could be offered to these organisms by the comparatively spacious pores of the filter, and even the strongest advocate of sand filtration could not have reasonably anticipated that filtration through a few feet of material could effect the remarkable reduction in the number of micro-organisms to which the above table bears witness.

It is most remarkable, perhaps, that these hygienically satisfactory results have been obtained without any knowledge on the part of those who construct these filters, as to the conditions necessary for the attainment of such results. In the construction of filter beds, water works engineers have certainly never been guided by an acquaintance with the habits of micro-organisms and yet by carefully improving their methods, so as to secure the removal of visible suspended matter, they have hardly less successfully, although unconsciously, attacked the invisible particles, and reduced them to an extent that is surprising. (85)

The ability of storage to reduce bacteria is best utilized, asserted Frankland, by allowing water, when bad, to go on its way downstream instead of into the reservoir. Higher quality water may be stored, and the bacteria therein will be carried to the bottom of the reservoir with other forms of matter in suspension.

Frankland concluded that: Complete removal of micro-organisms demands the best filtering material, its frequent renewal, and reduction in the usual rate of filtration. Agitation of the water with certain finely divided solids may sometimes remove a large part of the organic matter in water, but that method is unreliable. Chemical precipitation will largely reduce bacteria in water.

Unfortunately, the Institution of Civil Engineers, apparently not fully realizing the importance of Frankland's paper, did not print it in full.

Rideal's Summary.—The evolution of nineteenth century concepts of what slow sand filters do and how they do it was aptly summed up early in 1902 by Samuel Rideal, noted English chemist (86).

Sand filters were first regarded simply as strainers; and the fineness and cleanness of the sand was the most important point. Analyses later proving that the soluble constituents were considerably affected, an explanation was sought in surface action. Afterwards from the fact that nitrates and carbonic acid were formed, a chemical theory of simple oxidation arose. Three discoveries, however, threw new light on the process: (1) The size of the finer mineral particles is only about 1/1,000,000 inch . . . and that of most bacteria 1/25,000 inch, or larger, but both are smaller than the interstices between the grains of even fine sand, consequently it follows (a) that the cleaning is not accounted for by simple straining, (b) that the organisms would be retained first. (2) Piefke in Berlin, about 1886, found that sterilized sand effected hardly any purification and did not retain microbes. It had previously been noticed that sand filters did not become efficient for several days after re-laying. (3) When the oxygen of the air, and the water, were sterilized little or no oxidation of organic matter occurred.

It was proved, therefore, that for the proper mechanical and chemical effects the action of organisms is essential. It must be remembered that some organisms have long flagella, while a large number, such as diatoms and bacteria, are normally surrounded by a gelatinous envelope which greatly increases their size, and enables them to adhere to surfaces, so that in a short time the sand of a new filter becomes covered with a living slimy layer which entangles suspended matters and effects the main part of the purification. This is called *schmützdecke*. (86)

Subsequently, less emphasis was put on *schmützdecke*, a term and idea taken over from German writers. It should be understood that Rideal's summary related to slow sand filters and was written in England at a time when rapid or mechanical filters were but little used there.

CHAPTER VI

Slow Sand Filtration in the United States and Canada

Filtration in the United States made a bold and early but unsuccessful start in 1832 with an upward-flow backwash filter at Richmond, Va. This was only five years after Thom's filter of much the same type was put in use at Greenock, Scotland, and three years after the completion of Simpson's downward-flow, manually cleaned filter at London. Not until 1855 was there another municipal filtration venture in the United States and that was a small charcoal, sand and gravel filter or strainer at Elizabeth, N.J.

In Canada, a lake-intake filter crib was built in 1849 at Kingston, Ont., and in 1859 an infiltration basin was built on the lake shore at Hamilton, Ont. Neither of these, however, would qualify as successful filters.

Up to the end of 1860 there had been constructed only 136 water works in the United States and ten in Canada. A large percentage of these supplied water from springs or other sources free from turbidity and at least relatively free from pollution. Although slow sand filtration was thoroughly established in England and Scotland, and to a lesser extent in Continental Europe, before the American Civil War, no such plants were in operation on this side of the Atlantic. The Civil War put a damper on water works construction in both the United States and Canada.

After the Civil War, water works construction in America was resumed at a rapid rate but for many years nearly all attempts at filtration were utterly inadequate.

America made three most notable contributions to filtration: (1) The rapid filter was introduced by inventors and promoters in the 1880's and early 1890's and put on a sound engineering basis by working-size scientific experiments divorced from proprietary interests, then further advanced by various elements of mechanical equipment and by filter operators. (2) Improvements were made in slow sand filters, beginning with the studies at the Lawrence Experiment Station of the Massachusetts State Board of Health and carried for-

Fig. 30. Albert Stein (1785–1874)
Engineer who designed first municipal filter in America for Richmond, Va., 1832, applying principles of upward flow with reverse-flow cleaning
(From painting by John Neagle, Philadelphia, 1837, in possession of Thomas Stein, Toulmanville, Ala.)

ward at other testing stations and at working installations, the latter contributing notable improvements in methods of operation. (3) Chlorination was initiated, early in the twentieth century, first as a bactericidal adjunct to rapid filtration, and then in conjunction with sand filtration.

The First Filters in America

The first American city to attempt water filtration was Richmond, Va. Early in 1832, a water works which included a small upward-flow filter of gravel and sand was completed. When this proved a failure, a filter operated by downward flow was soon constructed and it too failed. In both cases the filters were of small area and unable to cope with the highly turbid water of the James River. The city then waited a century before it obtained a complete purification plant. Much of this period, however, involved a wait for full development of the means to purify very turbid water, in this case contaminated with troublesome industrial wastes.

Richmond's first water works was designed by Albert Stein, a German-American engineer, who arrived in Richmond in the spring of 1830. At that time the city depended on wells and springs for its water, pipes having been laid in a few streets. Stein aroused interest in water works and was engaged to make preliminary plans and estimates, which were approved at a freeholders' meeting and later at a special election. The council then appointed a "Watering Committee" which engaged Stein as designing and constructing engineer at a fee of $6,500, to be paid on completion and acceptance of the works.

At sunrise, July 24, 1830, the Watering Committee met on the "Canal Bank near the little Arch," approved the sites of the dam and reservoir selected by Stein, and instructed Stein to begin construction of the pump house. For a time after construction began, the committee continued meeting at sunrise every Friday, to inspect the work.

On January 7, 1832, Stein reported to the committee that the works were completed. The total cost to that date, including material on hand, was $76,861, against the original estimate of $92,600. Neither figure included Stein's fee of $6,500, not yet paid. On February 17, the Watering Committee reported to the Council that it had "inspected the works in all their parts" and had made "full experiments with all the machinery, at the pump house as well as at the reservoir, and on and along the whole line of pipes, and that it had that day come to Resolutions:

That Mr. Stein, our Engineer, has faithfully and to our entire satisfaction performed the duties required of him, by the contract in the premises.—That the works ought to be received of him in discharge of his contract and that the same be and they are received by your Committee, *as competent to furnish an abundant supply of sweet and pure water for the use of our city* [Author's italics] (1).

Tap No. 1, for the first house supplied with water, was made March 5, 1832. On May 1, 1833, there were 295 water subscribers.

Stein described the new works in a report submitted to the committee on January 7, 1832 (2). Water was lifted to a reservoir by a 0.4-mgd. pump, driven by a water wheel. The reservoir was 194 ft. long, 104 ft. wide, 10 ft. 8 in. deep and held nearly 1 mil.gal. It was divided into four "apartments, of which two were for filtering. All the apartments were connected at the bottom by 10-in. cast-iron pipes, with gates attached." Each filter was $22\frac{1}{2}$ ft. long and 16 ft. wide. This would give an area of only $\frac{1}{60}$ acre for both filters against a pump capacity of 0.4 mgd.

The filter is rather vaguely described as "a body of gravel and sand through which the water percolates upwards," the gravel being at the bottom and the material becoming "finer and finer toward the top." When the "quantity of pure water falls short by lodgment of sediment among the gravel and sand, the water is made to enter at the top, and in passing downward with considerable force carries along with it the sediment into the reservoir [below the filter] from which it is carried off through the ascending main by means of a branch pipe with a stopcock attached to it."

Evidently the filter rested on a "floor," or false bottom [not described], 3 ft. above the bottom of the filter basin, thus affording "sufficient space to remove the sediment, which may remain at the bottom after the body of gravel and sand has been cleaned." From this it may be inferred that the filter was cleaned by reverse-flow wash.

In a semicentennial paper on the Richmond works, James L. Davis, Water Superintendent, stated that the filters were 5 ft. deep (3).

Stein expressed doubt that the filter was large enough "to produce the required amount of pure water." If an increased demand should make a second filter necessary, he said, it would be advisable to place it near the second reservoir—apparently then projected below the first reservoir and filters.

Stein declared that the Richmond filter was the first one he had "formed upon a large scale and I believe it is the only one formed in the United States for the purpose of producing pure water for a town."

So far as can be learned, the description thus condensed from Stein's report is the first and only locally recorded contemporary mention of Stein's filter. Loammi Baldwin's report on a new supply for Boston, dated October 1, 1834, cites a letter from the city clerk of Richmond, saying that a reservoir equal in size to the first, with a filter between the two, was being constructed "with a view of clearing the water, which at times has been too muddy for use. The first filter does not seem to have had much effect in purifying the water. The second differs from it, in filtering water downwards instead of ascending, and it is expected to render the water fit for use at all times, with the aid of settlement in the New Reservoir" (4).

Charles E. Bolling, a later superintendent, stated in 1889 that the filter was pronounced a failure and its use abandoned in 1835. Presumably he meant the second unit (5).

In view of the immediate failure of Stein's filter to provide "an abundant supply of good and pure water," and since he questioned the capacity of the filter immediately after its completion, why did he not put in a larger filter? It is possible that he did design and build the second filter to meet his obligation.

Stein was right in believing that the United States afforded no precedent for a municipal filter. But few were in use anywhere in the world. His contemporary, Baldwin of Boston, said of the Richmond filters: "This reversing the course of water through the filter appears to be like the plan adopted by Mr. Thom at Greenock" (4). The Greenock filter was put into use in 1827, some years after Stein came to America. It was described by Thom in a pamphlet published in Scotland in 1829 (see Chap. V). It is conceivable, but unlikely, that Stein saw a copy of this pamphlet before he designed the Richmond filters. Or he may have known of Peacock's British patent of 1791 and pamphlet of 1793 on an upward-flow filter, washed by reverse flow. Whether or not Stein knew of the Peacock and Thom filters, the one he built for Richmond was absurdly small, particularly so for a water much more turbid than that in any city supply previously subjected to filtration.

It was a bold venture for Richmond to adopt Stein's plans for water works, including a pumping plant and filters, for at the close of 1830 only 44 cities in the United States had public water supplies, mostly small gravity works, none of which included filters.

What was the background of this engineer—leader of this bold adventure? Albert Stein was born in Düsseldorf, Prussia, December 9, 1785 (6). After being educated as a civil engineer, he began work on a topographical survey of the Rhenish Provinces. In 1807, he was appointed hydraulic engineer by Murat, then Grand Duke of Berg by the favor of Napoleon I, whose cavalry had been led by Murat. After the fall of Napoleon and the cession of the duchy to Prussia, Stein resigned his position and came to America. He reached Philadelphia in 1816, where he seems to have had some relation with Frederic Graff, Chief Engineer of the Philadelphia Water Works. In 1817, Stein submitted plans for water works at Cincinnati. About that time, also, he made surveys for a canal from Cincinnati to Dayton. For a few years beginning in 1824 he was engineer for deepening the tidal section of the Appomattox River at and below Petersburg, Va. He was engineer for water works at Lynchburg, Va., in 1828–30. While building the Richmond works, Stein designed for Nashville, Tenn., water works which were completed in 1832. In the period 1834–40, Stein was at New Orleans, building a reservoir for the water works there, a canal from the city to Lake Pontchartrain, and making a survey and plan for the improvement of the Southwest Pass of the Mississippi. In 1840 he leased a small, privately owned water works system at Mobile, Ala., which he improved and operated. He died July 26, 1874, on his estate at Spring Hill near Mobile.

Although Richmond did nothing effective to improve its water supply until well into the twentieth century, settling basins were proposed from time to time. In 1860, the city council asked the superintendent, Davis, and its city engineer, W. Gill, to make plans for a new reservoir "with a proper filter." They proposed filters cleaned by reverse flow (1). A new reservoir was put in use January 1, 1876. Later, under Superintendent Charles E. Bolling, and the health officer, Dr. E. C. Levy, two narrow settling basins, about a mile long, with provision for drawing off the sediment alternately, were provided. On December 22, 1909, large coagulation basins were added. Chlorination with hypochlorite was begun June 26, 1913, on Levy's recommendation, following a few cases of typhoid fever in Richmond. In 1914, apparatus for applying liquid chlorine was installed. But not until August 29, 1924, was a complete purification plant available, with coagulation basins, mechanical filters, aerators and a clear-water basin, the whole of 30-mgd. capacity (1).

There remained a trouble that had been increasing for a quarter century. Dr. Levy had reported the discharge into the river of deleterious matter from sulfite pulp mills at Covington. These wastes seriously interfered with water purification. Although the owners spent large sums in alleviating pollution, other mills appeared on the river banks and nullified the benefits. "For the last few years," wrote Whitfield in 1930, "this condition has become serious" (1).

Carrying on the story, Marsden C. Smith, Engineer of Water Works, stated in 1934 that not only had the pollution been reduced but improved methods of preparing water for filtration had increased the effective capacity of the plant 50 per cent and in addition had produced a vastly better effluent at reduced cost (7). Among other improvements were continuous instead of seasonal treatment for algae control in the raw water settling basin; pH control of the raw water by the addition of lime or acid just before applying the coagulant; mechanical mixers or flocculators to improve coagulation; taste and odor control by activated carbon "fed in batch at the beginning of a filter run directly onto the filters"; ammonia and chlorine combined in place of chlorine alone to disinfect the filtered water; and, to reduce corrosion, "the final pH is now being corrected by a combination of aeration and chemical treatment" (7).

What wonder, in view of all these agencies used in 1934, that Stein's small settling reservoir and small upward-flow filters of 1832 proved utterly inadequate! Or that, having put in a second unsuccessful filter, Richmond went on with muddy water for 75 years and rounded out a full century before it had an adequate purification plant!

Nineteenth Century American Literature on Filtration.

Before detailing American progress in water filtration, it will be illuminating to see what native guides to the art of filtration were available to the American engineer during the nineteenth century. These were few and inadequate until about 1870.

Loammi Baldwin II, called the Father of American Engineering, in his report of 1834 on a new water supply for Boston, embodied excerpts from foreign descriptions of the Greenock upward-flow, backwash filters of Robert Thom and the somewhat similar filters of Albert Stein at Richmond.* He barely mentioned the Quai des Celestins filters opened in 1806 at Paris and the filters and filter galleries at Glasgow, dating from 1808–10. Baldwin visited European engineering works in 1807 and again in 1823–24 and was the first American engineer on record to make such a tour. So far as appears in the Boston report, he saw no filters while abroad. The foreign data in the Boston report dealt chiefly with the flow of water in conduits, as was natural, since the report recommended a gravity supply for Boston, with no proposal for filtration (4).

Charles S. Storrow read diligently in the extensive civil engineering library of Loammi Baldwin at the time of his graduation from Harvard in 1829. In December of that year he went to Paris where he attended the Ecole des Ponts et Chaussées and lectures at the Ecole Polytechnique. In 1835 he published the first American treatise on water works (8), but the book was written in 1832 while he was in Paris. It was chiefly concerned with hydraulics but included a few lines advising combined settling and storage reservoirs, plus filters, and six paragraphs on Thom's and Simpson's filters (see Chap. V). Apparently Storrow did not see the Quai des Celestins filters while in Paris.

Dr. Robley Dunglinson, in his American book, *Human Health*, also published in 1835 (9), treated briefly filtration, boiling, softening, distillation and aeration—the last to restore deficiency of air caused by distillation. He gave a concise summary of Lowitz's studies on the use of charcoal in water treatment, announced in 1790. His most significant statements, in view of the date, pertained to chlorination, which will be discussed later (see Chap. XIV). He mentioned no specific water purification plant, not even the filter put in use at Richmond in 1832, near the University of Virginia, where he was professor of chemistry.

* The first Loammi Baldwin—cabinet maker, surveyor, soldier of the Revolution, canal builder, "man of learning" and originator of the Baldwin apple—was made an Honorary Graduate of Harvard in 1785. His civil engineering library and those of his sons, Loammi and George, were presented to the town of Woburn, Mass., home of at least three generations of Baldwins [*Dictionary of American Biography*, which cites "Sketch of the Life and Works of Loammi Baldwin, C.E.," by Prof. George L. Vose (1885)]. William D. Goddard, Librarian, Woburn Public Library, states that the Baldwin Library of 2,110 volumes was given by a Baldwin descendant to the Woburn Library in 1899 and transferred to the Massachusetts Institute of Technology in 1914.

The Journal of the Franklin Institute, founded in 1826, made available in 1838 a translation of Arago's report on the Fonvielle pressure filter, which worked at a high rate and was cleaned by backwashing (10) (see Chap. IV). It is not likely that many American engineers read Arago's report or even saw Dr. Dunglinson's book. For many years there were no American additions to literature on water treatment.

Kirkwood on Filtration in Europe.—When the American Civil War was over, James P. Kirkwood, an eminent water works engineer of Brooklyn, N. Y., was engaged by the city of St. Louis, Mo., to recommend improvements to its water supply. He advised filtration and was sent abroad to gather information on filtration in Great Britain and on the Continent. While he was abroad, the city decided not to build filters. On his return he submitted a report which was finally published in 1869 (11).

Kirkwood's St. Louis Report described and illustrated the filters and filter galleries of nineteen European cities. Until the report of 1895 by Allen Hazen, another American engineer who went to Europe, Kirkwood's was the only book in any language devoted to the filtration of municipal water supplies. He described the filters of Leicester, Liverpool, seven of the eight London metropolitan water companies, Wakefield and York, England; Edinburgh, Scotland; Dublin, Ireland; Marseilles and Nantes, France; Altona and Berlin, Germany; and Leghorn, Italy. Filter galleries described were those at Perth, Scotland; Angers, Lyons and Toulouse, France; and Genoa, Italy.

The object of filtration through sand underlaid by coarse material, says Kirkwood in his general summary, is to remove suspended matter, including not only earthy materials but also "fine vegetable fibers and the minute organisms, vegetable or animal, which in all river waters prevail more or less during certain of the summer months."

Not for a long time, says Kirkwood, would our rivers carry as much organic matter in suspension as did European streams. He therefore turns his attention to the "clayey discoloration" of American rivers. According to him, this renders their water very objectionable to sight and for industrial uses and is no contribution to health and cleanliness. Custom, as on western rivers, might reconcile persons to the use of muddy water, especially where clearness is associated with the hard and unpalatable waters of limestone springs. Such of the sediment of muddy waters as will fall by its own weight in 24 hours could

FIG. 31. JAMES P. KIRKWOOD (1807–1877)
Engineer who designed first successful slow sand filters in America for Poughkeepsie, N.Y., 1872; author of early treatise on European filtration practice; second president of American Society of Civil Engineers
(From "Early Presidents of the Society," *Civil Engineering*, 6:338 (1936))

be more economically removed by sedimentation than by filtration. In fact, successful filtration presupposes sedimentation.

Abroad, wherever sole reliance had been placed on filtration it had failed, as in France, or had but partly succeeded, as at one of the London works. Settling reservoirs, writes Kirkwood, were being used more and more as a relief to filters at London, although recognition of their economy had come slowly. Besides reducing the load on filters they had become valuable expedients for water storage, especially on the Lea, or New River works of London. At Liverpool, Leicester, Edinburgh and Dublin the large valley reservoirs required to store compensation water for mill owners served also for presedimentation.

Of the five "natural filters" or filter galleries seen by Kirkwood, three were in France. To him these seemed to be as satisfactory as the filter plants of that country were unsatisfactory. Any deficiencies in the galleries were in quantity rather than quality of product.

Kirkwood virtually sums up the best practical points, of the filter plants he had visited by presenting in a short text and two plates a design for filters at St. Louis. He assumed, on the basis of experiments, that 24 hours' detention in settling basins were enough and that four basins should be provided: one for filling; one for settling; one for decanting; and one being cleaned. He assumed a rate of filtration of nearly 3.4 mgd. per acre, allowing for the area out of service for cleaning. Having remarked previously "that the English filters are all deficient as regards any arrangement for measuring the precise flow from each filter or the precise head on each filter while it is in action," he provided in his St. Louis design a small well at the end of the main drain of each filter, a sluice gate working downward, with the top of the sluice acting as a weir, thus indicating the head on the filter and its yield.

On the quality of water in relation to health, Kirkwood says little. He presents the prevalent theory that, by filtration, water cannot be "dispossessed" of any "noxious gases which may have . . . [been] absorbed" from sewage pollution, "nor of some of the very minute organisms due" to pollution. Besides the inadequate knowledge on these subjects at that time, it should be remembered that Kirkwood's instructions in December 1865 were "to proceed at once to Europe and inform himself in regard to the best process in use for clarifying river waters used for the supply of cities, whether by deposition alone, or by deposition and filtration combined." His commission was well executed, but St. Louis, as has been pointed out, had rejected filtration before Kirkwood's return.

Poughkeepsie and Hudson, N. Y., employed Kirkwood as filtration engineer in the early seventies and he was consulting engineer for the water works at Lowell and Lawrence, Mass., both of which included filter galleries, but with these contacts his filtration work ended.

An American Pioneer on Quality of Water.—Professor William Ripley Nichols of the Massachusetts Institute of Technology was for some years the leading American authority on the quality of water supplies. In 1878 he contributed a long review of filtration to the annual report of the Massachusetts State Board of Health (12). In it he set forth clearly and simply the theory and practice of "natural" and "artificial" filtration (infiltration galleries and slow sand filtration) at home and abroad. It contains a notable section on algae in reservoirs and the resultant tastes and odors. His conclusions were:

Sand is the only practical medium for large-scale filtration.

There is as yet no evidence that sand filtration will efficiently purify polluted water, although, properly carried out, it will "lessen the liability of ill effects."

All visible suspended and an appreciable part of dissolved organic matter may be removed by sand filtration.

For the present, artificial filtration should be regarded as a means of removing suspended matters only, "although under the management of a person of intelligence, education and experience, the simple sand filter is capable" of reducing the organic matter. Such management cannot be expected in ordinary practice. Removal of color and taste, in the light of experience, should be regarded as incidental, varying much with the condition of the filter.

It would not be worth while for a town to build sand filters, he continues, unless it were willing to spend enough money in construction and operation to make the scheme efficient. Requisites for efficiency include ample settling basins; at least duplicate filters, which should be covered; frequent cleaning and renewal of filtering material; and covered clear-water reservoirs, which should be emptied and cleaned if required. Finally, says Nichols, no town should undertake artificial filtration unless it is willing to face the possibility of spending $2.50 per mil.gal. for operation alone (12).

Much of the information and opinion in Nichols' essay of 1878 is repeated in his book of 1883 (13). This is the first American book devoted entirely to the sanitary aspects and the chemistry of water supply.

In 1884, Nichols was even more skeptical than in 1878 regarding the advisability of filtration for American cities. But he was chiefly concerned with the reduction of color, tastes and odors. His extensive experience as consultant had been chiefly confined to such waters as were found in Massachusetts and adjacent cities in New York. In the United States, he said, large-scale filtration was almost unknown, while in Europe filtration of surface supplies was general. The reason for this difference was usually attributed to the cost of filtration and the difficulties due to hot summers, and in Northern states, cold winters. When consulted, he had been deterred from advising filtration neither by cost nor climate, but by the fact that experiments, by himself and others, and experience at existing works had convinced him that sand filtration would not remove the color generally affecting American surface supplies nor the disagreeable tastes and odors to which they are liable. Although carefully conducted sand filtration can improve such supplies, Nichols doubted whether it was worth the cost, especially if it were unsatisfactory at the season when most necessary (14).

Fanning's "Water Supply Engineering."—Colonel John T. Fanning, who in the last third of the nineteenth century was dean of American hydraulic engineers, completed a large treatise on water works in 1876 and published it in 1877 (15). It was the first American treatise on water works to appear since Storrow's little book of 1835. Fanning reviewed the quality of water and noted that it was a vehicle for the spread of diarrhea, dysentery and typhoid. He summarized the chief means of water purification devised abroad, including plain sedimentation, coagulation by iron salts, the use of charcoal in filters, filters and infiltration galleries. He mentioned recent filter basins and galleries in America, and cited the filters recently completed at Poughkeepsie as the first of the kind in America. He emphasized the need of roofing filters against the effects of both low and high temperatures.

Croes on Filtration in America.—The possibilities and limitations of filtration as seen in 1883 by an engineer of large experience and wide observation were briefly stated in the introduction to a paper by J. J. R. Croes (16). The object of filtration then was to remove visible impurities. With clarification there was a limited reduction of chemical impurity. Processes of filtration applicable on a large scale, it was generally believed, "do not make a polluted water fit for use."

FIG. 32. ALLEN HAZEN (1869–1930)
Author of first treatise on art and science of water filtration; Chief Chemist of Lawrence Experiment Station of Massachusetts State Board of Health during filtration experiments
(From photograph made during Hazen's period at the Lawrence Experiment Station (probably 1892); made available for use here by his son Richard Hazen)

The principle of filtration, Croes says, is that, by the slow passage of water through very small orifices, matters in suspension are deposited on the orifices. If the water is passed through rapidly or under great pressure, the sides of the orifices are washed clean. These words were written on the eve of rapid mechanical filtration. The paper is highly valuable for its brief descriptions of filters built in the United States and Canada up to its date. The descriptions were drawn from the author's *History and Statistics of American Water Works,* a series of articles that had been run in *Engineering News* from early 1881 until 1886 (17).

Lawrence Experiment Station.—Charged with the duty of giving advice on water supply and sewerage, the reorganized State Board of Health of Massachusetts established, late in 1887, the Lawrence Experiment Station. Its function was to study water and sewage treatment. This it is still doing in the 1940's. In 1890, the board published a special report (18). One volume dealt with the *Examination of Water Supplies;* the other, *Purification of Sewage and Water,* reviewed the experimental work at Lawrence. The experiments on water treatment centered largely on nitrification of organic matter by intermittent sand filtration, but much attention was also given to the reduction of bacteria. The early chemical results, including reduction of matter in solution, were better than the bacterial. At the outset of the experiments, the filtration rates were so low that they were worthless, but soon the rates were increased to a practical point and the bacterial results were considered satisfactory. On the basis of the results, a water filtration plant for the city of Lawrence was designed and built under the direction of Hiram F. Mills, engineer-member of the State Board of Health. The filter was put into use in September 1893. Here, as well as in the preceding experiments, nitrification of organic matter, particularly to reduce available food for bacteria, was considered of great importance, so intermittent filtration was used.

Time, it may be interjected, soon showed that, for water, intermittent filtration was a fallacious practice taken over from sewage treatment where oxidation of the large amount of organic matter to prevent a putrefaction nuisance was often demanded. True, the Merrimac River supply at Lawrence was heavily polluted. How this pollution compared with that of the Thames at London, Mills and his colleagues do not appear to have considered. At London, a high bacterial removal by continuous slow sand filtration had been demonstrated by Percy Frankland, shortly before the Lawrence experiments were begun (see Chap. V) but Mills did not know this or else disregarded it.

The opening of the Lawrence filter plant was followed by a marked reduction of typhoid in that city. This, and the prestige of the Massachusetts State Board of Health, established American confidence in filtration at a time when water-borne typhoid, endemic and epidemic, was taking a heavy toll; at a time, also, when American cities and water companies were at last willing to pay the cost of efficient purification.

Epochal confidence in water filtration was created in America by the Lawrence experiments and the city filter, and by the studies by Allen Hazen, Chief Chemist at Lawrence, of effective size and uniformity coefficients of sand grains and of frictional resistance to the passage of water through sand and gravel (19).

Hazen's Pioneer Treatise on Filtration.—Hazen's work at the Lawrence Experiment Station was soon followed by a trip to Europe where he studied the workings of many slow sand filters, including some of those visited by James Kirkwood 30 years earlier. Hazen's observations were embodied in his book of 1895. Kirkwood had gone abroad with little knowledge of the art of filtration at a time when the relation between water and public health was not understood. Hazen had the advantage of the greatly improved knowledge of his day. Kirkwood made a valuable report on his foreign observations but Hazen wrote the first treatise on the art and science of water filtration (20). The scanty treatment of rapid or mechanical filtration in Hazen's book is understandable in view of his previous connection with the ultra-conservative Massachusetts State Board of Health and the newness of mechanical filtration in 1895. This deficiency was partly remedied in the edition of 1900.

Reports and Journals.—In the last decade of the century came: the report of Edmund B. Weston, Assistant Engineer at the Water Department, on experiments with rapid and modified slow sand filters at Providence, R.I., but centering on one make of rapid filter (21); the classic report by George W. Fuller, Chief Chemist and Bacteriologist, on tests of rapid filters at Louisville, Ky. (22); Hazen's report on tests of slow and rapid filters at Pittsburgh (23); and a second report by Fuller, this one describing tests at Cincinnati of modified English slow sand filters and of American or rapid filters (24). These were forerunners of many later reports on municipal filtration experiments conducted by skilled engineers, chemists and bacteriologists, unrivaled in scope and importance elsewhere in the world.

New means for disseminating information on water purification among water works engineers and superintendents were provided in the seventies and eighties by the organization of the American Water Works Association (1881) and the New England Water Works Association (1882), each with its published proceedings, and by the establishment of engineering journals which devoted much of their space to water works problems and progress. These, with the proceedings of national engineering, chemical and bacteriological societies, made available the latest ideas and accomplishments in the art and science of water purification, in striking contrast to the paucity of information in the first half of the nineteenth century.

Early Makeshifts and Failures

Makeshift Filters or Granular Strainers.—After the failure of the upward-flow filter at Richmond in 1832, no further attempt at large-scale filtration was made until about 1870. In fact, diligent search disclosed no filter installation of any kind or size until 1849. Then came various devices that at best were only rapid, granular strainers. Structurally, these were cribs, boxes, chambers, trenches and banks. Media employed, singly or in combination, were sand, gravel, charcoal and sponge. More pretentious and more efficacious were the filter galleries and some of the later upward-flow filters.

Altogether 51 filters of various types were built in the United States and Canada in the period 1849-93. Roughly these were divided as follows (16, 17, 25, 26):

Filter Cribs: Eleven, built from 1849 to 1882, beginning at Kingston, Ont.

Charcoal, Sand and Gravel Filters: Sixteen, built between 1855 and 1893, the first at Elizabeth, N.J., and followed at Elmira, N.Y., in 1857, and at Stockbridge, Mass., in or about 1862. The different media at Stockbridge were separated by perforated tile. (Not counted here are some of the upward-flow filters.)

Sponge, Charcoal and Sand Filters: Eight, in the period 1875-82, the first at South Norwalk, Conn. All but one of these were designed by William B. Rider of that town. Raw water passed laterally through the sponge and charcoal into a small chamber. The eighth filter in this class was built at Hannibal, Mo., in 1882. Water passed through

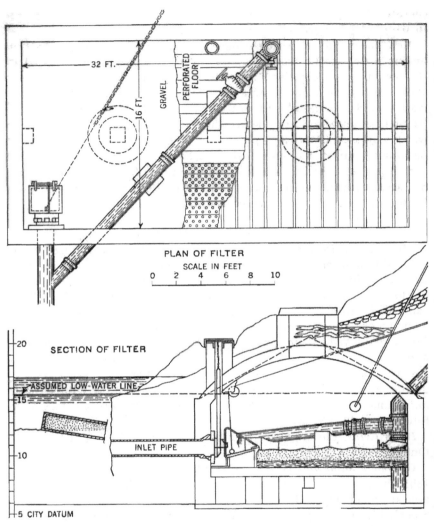

Fig. 33. Miniature Slow Sand Filter at Marshalltown, Iowa, 1876
One of the earliest filters west of the Atlantic seaboard states; designed by
T. N. Boutelle, engineer
(From original drawing supplied by H. V. Pedersen, General Manager,
Marshalltown Water Works)

3 ft. of sponge and 2 ft. of sand and gravel before reaching the pump well.

Filter of Sponge Only: One, at Alton, Ill., in 1882.

Sand or Sand and Gravel Filters: Fifteen, built between 1866 and 1883. The first of these was built at Annapolis, Md. Water was passed through 3 ft. of sand and gravel into a "brick trough" located 3 ft. beneath the bottom of the reservoir. The trough was covered by wooden slats ½ in. apart. At Carthage, Mo., in 1882, a relatively large filter was provided. It was 25 x 50 ft. and 12 ft. deep. At Clinton, Iowa, in 1874, sand and gravel were placed in boxes which could be raised for cleaning. At Marshalltown, Iowa, in 1876, a miniature slow sand filter was built. It was 16 x 32 ft. in plan, roofed by a masonry arch. Its engineer was Thomas N. Boutelle, who two years later designed a covered upward-flow filter for Burlington, Iowa.

Abortive Attempts at Filtration.—Nearly all the filters of charcoal, sponge and sand just reviewed were small. However useful or useless they may have been, they served for a time and doubtless were pointed to with pride. In striking contrast were a half dozen large-scale abortive filter projects dating from 1871 to 1893.

The first of these, completed late in 1871 at Providence, R.I., was a large infiltration basin (see below). Also in 1871, Columbus, Ohio, completed a filter. It had an area of 8,742 sq.ft. and consisted of 7 in. of sand on 47 in. of broken stone. In 1873, at Springfield, Mass., an elaborate set of lateral-flow excelsior filters, placed in tiers of crates which could be lifted for renewal, came to grief as soon as put into use (see below). At Toledo, Ohio, in 1875, a single filter, instead of the three (with presettling reservoirs) planned by the engineer to treat water of the turbid Scioto River, soon clogged up. In 1880, at Brockton, Mass., 55,000 sq.ft. of the bottom of a reservoir on Salisbury Brook were prepared to act as a filter by laying 4-in. drain tiles 8 ft. apart, 30 in. below its bottom. The yield was so impregnated with iron that it was never utilized. In 1887, at Easton, Mass., a similar reservoir was completed but never used. In 1893, the water company at Wilkes-Barre, Pa., built a filter in a recently constructed impounding reservoir to alleviate complaints of taste and odor. The unit had an area of 12,000 sq.ft. It was supported on planks 2 ft. above the bottom of the reservoir. On the planks was a 10-in. layer of 2-in. broken stone, covered with 6 in. of river gravel and topped by only 4 in. of Long Island sand. From this, the water passed to and through a chamber 6 x 6 ft. in plan, containing two trays of animal charcoal 1.5 ft. deep above 3 ft. of coke. In 1895 the company installed a 10-mgd. rapid filtration plant (27).

Filter Projects Never Executed

Eleven cities in the United States and one in Canada considered filtration between 1836 and 1876 without a single adoption. In some cases the process was only mentioned as an undesirable possibility.

Boston, June 16, 1836.—Robert H. Eddy, a civil engineer, in a report on possible sources of water supply (28), advised going far enough afield to get water that would not need filtration—a doctrine followed by Boston from that day to this.

Pittsburgh, Pa., October 1847.—A communication from J. H. Laning, of Cincinnati, "on the subject of a smoke consumer and water filterer" was laid before the common council. The portion on the smoke consumer was referred to the special committee on that subject and the balance to the water committee. The select committee concurred. That seems to have ended the matter.

Albany, N.Y., 1849 and Later.—George W. Carpenter, a civil engineer, reported on May 14, 1839 (29), on a possible supply from the Hudson River and assumed that it would be passed "through a coarse filtering bed to free it from such materials as might injure the pumps." W. J. McAlpine, a civil engineer, in a report August 3, 1850, advising a supply from Patroon Creek, answered objections to its hardness and frequent turbidity by saying that there was no remedy for hardness, but that turbidity "can be easily corrected either by filtering the water or by extending the pipes to the upper pond," enlarging it and building a division wall to form a settling basin (30).

Philadelphia, 1853–54.—On October 27, 1853, the city council directed its watering committee "to inquire into the practicability of erecting at Fairmount a filter of sufficient capacity to filter all the water (from the Schuylkill River) before it enters the distributing system." This action resulted in a notable engineering report by Frederic Graff, superintendent of water works, supplemented by a report by two chemists (31). The studies resulted in the conclusion by all hands, including the watering committee, that filtration was unnecessary. After mentioning some of the London filters and their chemical results and comparing analyses of the water being supplied to Philadelphia with that of a number of other American cities, Graff declared: "I am fully convinced that no adequate result can be obtained by the enormous expense which it would be necessary to incur in building and keeping in order such large filter beds [717,392 sq.ft.] as we should require, and probably the certainty of constant supply and efficiency of the works might be impaired by such troublesome and expensive and, I think, needless apparatus." The Watering Committee reported to the city council on May 3, 1854, that, in the light of the analyses presented, it was "perfectly satisfied with the extraordinary quality of the Schuylkill water" and that there was "no necessity whatever for its filtration."

Trenton, N.J., About 1856.—The Tenth Census of the United States, in data gathered about 1881 regarding the water works of Trenton, listed a filter "50 x 60 ft. in area; tile at bottom, 4 layers of gravel" (25). A report made on May 3, 1856, to the directors of the water works stated that construction of a filter, apparently in a reservoir then nearly completed, was proposed. No record of such a filter has been located in city offices and libraries in Trenton.

Chicago, 1860–63.—On March 7, 1860, the water commissioners passed a resolution requesting E. S. Chesbrough, Chief Engineer of the Sewerage Board, "to submit a project and estimate for extending the inlet pipe so far out into the lake that the water obtained shall be free from the wash of the lake shore and the flow of the [Chicago] River." On February 25, 1861, Chesbrough reported: "In order to obtain pure and clear water at all times it is proposed to construct a filter bed at the east end of the lot on which the present pumping works stand." He proposed to inclose his filter by a cofferdam 1,400 ft. in circumference. He envisaged winter difficulties in cleaning the filter. As an alternative that would serve for a time he considered a 40-mil.gal. settling reservoir, inclosed by cribwork. In conclusion, he advised postponement of construction until after further studies with the aid of chemical analyses (32).

In 1862 or 1863, consideration was given to building a "filter trench" or gallery 3,000 ft. long, carried 10 ft. below lake level. Skepticism as to its success prevailed. Decision to build a new lake-intake tunnel was made (33). Subsequently longer and larger lake tunnels were built; a huge pumping station was erected to divert a part of the sewage-laden Chicago River into the Illinois & Michigan Canal; the

Chicago Drainage District was created and the Chicago Drainage Canal was built for more extensive diversion and dilution of the sewage; then several immense sewage treatment works were constructed to lighten the burden of disposal by dilution. At long last a program of water purification was adopted (34). A 320-mgd. unit was under construction in 1942, but when the entire one-billion-gallon filtration program was finished nearly a century had elapsed since water filtration was proposed but rejected.

Cincinnati, Ohio, July 3, 1865.—Filtration was proposed by James P. Kirkwood in 1865 but apparently his advice was not given serious consideration (35). Rapid filters were completed in 1907.

St. Louis, Mo., 1865–66.—James P. Kirkwood's relations with the St. Louis water commissioners and their rejection of filtration has been described above. Subsequent events are reviewed below.

Oswego, N.Y., September 1866.—Filtration was considered but dismissed on the ground that sedimentation would suffice, in a report by William J. McAlpine (36). Water from the Oswego River, he said, was sometimes too turbid for domestic use but, as the quantity required for drinking, cooking and washing would not be 10 per cent of the water pumped, it would be cheaper to have a filter in each house. Or, if desired, water after standing a few days in the storage reservoirs "might be run through a large filter for a few specified hours each day," after which each family could draw off and store enough water for a day. Such a filter would not cost much but was not included in his estimate. Filters of gravel and sand, with a capacity of 1 mgd., were "in constant use at Kingsford's Starch Factory," * wrote McAlpine, yielding "beautifully clear" water (37).

Schenectady, N.Y., February 1, 1868.—In a report advising a supply from Sand Creek, William J. McAlpine says that the Bonnie Kill plan would require "filter beds; and even then the works will not [give] water as pure as in the accepted plan" (38). A supply from a filter gallery was introduced in 1871 (see below).

Manchester, N.H., November 23, 1869.—"A set of expensive filtering apparatus" comprising three units of ¼-acre each would be required

---

* The Kingsford Starch Works were established in 1848 by Thomas Kingsford, who came from England shortly before then. Filters were in use in 1850, and probably in 1848. Thomas Kingsford III stated in 1940 that he could find no data on the filters except that they used a large quantity of charcoal and sand (37).

for a supply from the Merrimac River, reported J. B. Sawyer, a civil engineer, to the City Aqueduct Co. He advised Lake Massabesic instead (39). Later, the city built works drawing on that source.

Lowell, Mass., Summer of 1869.—After reviewing the various possible sources of supply, including the Merrimac River, either settled or filtered, J. Herbert Shedd, a civil engineer from Boston, recommended Beaver Brook, without treatment (40). Experiments with settling and with filtering Merrimac River water led to the conclusion that if that source were adopted filtration would be preferable to sedimentation because of the necessity for large and costly settling reservoirs. The authorities proposed to build a settling basin and filters, then a settling basin only, but soon built a filter gallery, instead. In 1876, a filter, feeding into the gallery, was built beside the river (25). Mechanical filters were considered at length by the water board in 1888–90 but were not adopted (41). A decarbonation, iron-and-manganese removal plant, including rapid filters, to treat water from driven wells was completed in 1915.

Detroit, August 15, 1874.—George S. Greene of New York City and G. Weitzel of Detroit recommended the construction of two filters, at an estimated cost of $370,000, with others added for each 100,000 increase in population. The size and character of the filters was not stated. Instead, a settling basin was built. In 1923 rapid filters were completed.

Montreal, March 23, 1876.—At the suggestion of some of the members of the water committee, Louis Lesage, Superintendent of Water Works, "was directed to study the question of filtering basins and to prepare plans, with an estimate of cost." This he did but he concluded that it would be sufficient to provide a settling reservoir with a capacity for a few days' detention to clarify the water of the St. Lawrence when turbid in the spring (42).

Laggard Growth of Slow Sand Filtration

No successful slow sand filters for city supply were built on this side of the Atlantic until 1872, 40 years after the earliest ones had been constructed in England and fifteen years after they had been put into use in Germany. At the close of 1900 there were approximately twenty slow sand filters in the United States and five in Canada. Even then they were far outnumbered by rapid filters. In 1940 there were about

100 slow sand filtration plants in the United States as compared to about 2,275 rapid filters. All in each class had hygienic purification as their primary objective. In addition there were a few slow sand filters, many rapid filters and still others variously classified or unidentified in purpose (43).

Canada, in 1940, had about twelve slow sand and 120 rapid filters.

Poughkeepsie, N.Y.—Poughkeepsie, a city then having a population of 20,000, built the first slow sand filter in America. The filters were part of the first complete water works system of Poughkeepsie, put into use late in 1872. Creditable as the plant was for the time and place, it was but a simple beginning. Most notable of the changes which have taken place since are the adoption of coagulation, chlorination and prefiltration.

Poughkeepsie's filter plant is the more notable because through all these years it has treated one of the most polluted and potentially dangerous water supplies in the world; and because the Hudson River was adopted as a source of supply by the water commissioners of the day against the advice of their engineers, chief of whom was James P. Kirkwood, the father of slow sand filtration in America.

In 1855 a report on four possible sources of supply was submitted to a special water committee by William McCannon (44). He recommended near-by Morgan Pond. Its only rival, in his opinion, was the Hudson River, but it was objectionable because of turbidity during spring freshets and "brackishness during a long-continued south wind," especially at low stages of the river. Turbidity could be overcome by a settling reservoir but there was no cure for brackishness. No positive action was taken until 1869, when under legislative authority "an overwhelming" popular vote was cast in favor of building works and of a water commission created for the purpose (45).

In its first report (February 7, 1870), the commission said that before appointing a construction engineer they had engaged James P. Kirkwood as consulting engineer and W. Davis as resident engineer (46). In company with these engineers, the commissioners had visited the available sources of supply. The waters of the Hudson River, Fallkill, Crum Elbow and Wappinger's Creek had been chemically analyzed.

Strangely, although reports by both Kirkwood and Davis had been received and the latter was printed at length, the commission passed over Kirkwood's report with the statement:

AMERICAN SLOW SAND FILTRATION

Mr. Kirkwood says in his report now on file: "My opinion at present is in favor of the Fallkill as on the whole the best. . . . The water appears excellent. The larger part of the water, and probably all of it, could be delivered by gravity" (46).

Not a word on what their consulting engineer said about the Hudson! Thomas Lawlor, Director of Public Works, wrote in 1935 that the Kirkwood report has never been found although a very diligent search was made over a period of years (47).

Davis's report of January 31, 1870, was chiefly a categorical account of his surveys of the sources of supply considered but it contains this significant remark on the Hudson: "The water, if taken from this source, to free it from mechanical impurities will have to be passed through filters." Nothing is said about pollution (46).

In marked contrast with this weak remark by Davis, was the strong condemnation of the Hudson in the first report made by the chief engineer, J. B. G. Rand, on April 12, 1870 (48), two months after the report of the commission. Rand submitted cost estimates for a water supply from the Fallkill, the Hudson River and Wappinger's Creek. The purest water of the three was the Fallkill. Wappinger's Creek was hard and polluted by chemical refuse. As to the Hudson River, Rand declared that:

———regard for the recent taste of the people would prevent us from taking water from any source which has received sewage, and might therefore contain the living germs of cholera, typhoid fever, dysentery, tape worms, etc., without using every reasonable means to guard against them—the question is not so much—is the water wholesome now, as what will be its condition in the future. . . . I know no means of getting rid of this noxious matter on a large scale. By careful filtration large quantities may be greatly improved, but not entirely freed from the poison (48).

Settling basins, but not filters, were included in Rand's estimates for a supply from the Hudson River. His intake location (subsequently adopted) was presumably chosen to avoid direct pollution by the sewage of the city, but was subject to tidal influence and to the discharge from a projected State Hospital outlet sewer.

A stand against taking a supply from the Hudson was made by the Poughkeepsie *Daily Eagle* in the summer of 1870, as extracts from editorials of the period indicate:

July 30, 1870: The Water We Are Expected to Drink.—The opinions of all the engineers who had examined the subject—Mr. Kirkwood, Mr. Davis and Mr. Rand—gave decided preference to the Fallkill.

August 27, 1870: THEY HAVE DECIDED. . . . Previous to their decision, the matter had been discussed pro and con in the newspapers but we had no official word from the commissioners or their agents, except the reports of the engineers, Messrs. Davis and Rand. Both these showed plainly that the Fallkill was in their opinion the best source of supply. . . . Mr. Rand [held that the river was] almost out of the question. Mr. Kirkwood, the best authority in this country on hydraulics, agreed with Mr. Rand on all points and was in fact the advising and consulting engineer in the whole investigation. . . .

At a subsequent meeting of the Board a member moved that Mr. Kirkwood be requested to give his opinion in writing as to the sources of supply, but the majority promptly voted down the motion, thus in effect declaring that they were determined on river water or nothing, right or wrong, no matter who or how many advised another source. . . .

There must have been some reason why the opinions of all the engineers and chemists should have been disregarded and a conclusion in direct opposition to their advice should have been adopted (49).

Despite the black picture of the Hudson, Kirkwood's recommendation of the Fallkill, and the opposition of the *Daily Eagle,* the water commission reported on February 1, 1871, that "after mature deliberation" it had chosen the Hudson as the source of supply (46). It was maintained that, with filters, the cost would be considerably less than with a supply from any other possible source. Moreover, "analyses and practical tests prove the Hudson to be of superior quality."

Kirkwood resigned as consulting engineer December 31, 1872. How much of a part he took in designing the works and how much of it was done by Rand as chief engineer the reports of the water commission do not disclose. Fowler, who wrote in 1888, after having been in charge of the works for seventeen years, gave full credit to Kirkwood.

The treatment plant, as described in the water commissioner's report for 1872, consisted of a small inlet basin intended for the deposit of "heavier particles of mud"; two filters, having an area of $\frac{1}{3}$ acre each; and two small clear-water basins, operated in series. The depth of the filters was 72 in., with 24 in. of sand; 18 in. of gravel graded $\frac{1}{4}$ to 1 in. in size; 6 in. of 2-in. broken stone; and 24 in. of 4- to 8-in. stone "fragments." The cost of the treatment works, including land, was $76,915. The works were put into use about December 1, 1872.

The filters were used intermittently for several years, according to the condition of the river water. In 1875, they were used a total of about six months, except when being cleaned; in 1876, almost constantly. In the next two years the entire supply was filtered, except when stopped by ice or algae. The report of the commission for 1878 states: "The consumers, accustomed to drink filtered water, will accept nothing else; nor will they accept any . . . complication of circumstances" for the non-use of the filters (46).

Tastes and odors due to organic growths seem to have given trouble from the start. To obviate these, Davis, in the summer of 1875, began the practice of bypassing the filters when the temperature of the river water rose to 70°F. and tastes and odors could be detected in the water in the distribution system.

Davis, who was resident engineer in constructing the works and became superintendent in 1871, was succeeded in 1881 by Charles E. Fowler, who continued in office until his death January 4, 1908.

When Fowler took office in January 1881, he found the filters not in use because a heavy coating of ice made cleaning difficult. The ice was removed and the units restored to service, after which all the water supplied to the city was filtered up to April 1892 (50). Three times in the early winter of 1895, 500 tons of ice were removed from the filters. On the third occasion, the sand froze and the cleaning took 24 men two days (51). In 1903, an elevator was installed in the new filters to help remove ice.

In several of his annual reports (46), Fowler noted trouble from algae growths on the filters. From August to September 1889, these required constant attention and the labor of two men. From July to October 1891, algae growths stopped filtration within ten or twelve days after they appeared and in another period within seven days. In 1891, the two clear-water basins were covered with wooden roofs to exclude sunlight and dirt.

On December 17, 1896, after the filters had served 24 years, a large unit in two equal compartments was added, bringing the area up to $1\frac{1}{4}$ acre. The top layer of the new filter was 31 in. of Long Island sand. This rested on gravel, below which was broken stone. At the bottom, 6-in. tile underdrains were laid. Until June 1, 1897, both old and new filters were used, but the new one did nearly all the work. The old ones had become so compacted that they passed but little more water after than before cleaning. The 15 in. of sand had become clogged throughout. Sand had been carried down to the bottom of the 4 ft. of gravel and stone. The sand was removed down to the gravel and 30 in. of the new sand put in. Although Fowler had advised roofing the filters years before, construction was not authorized until 1904. The new units were covered and in use in November of that year and the old ones in August 1906. Groined concrete arches were used to cover the filters and to replace the wooden roof of the clear-water basins.

In a paper read in 1898, Fowler said (51) that when he took charge of the filters in 1881, the dirty sand was cleaned by passing it through two 12-ft. troughs containing running water. In 1886, a trough 200 ft. long was substituted, reducing cost of washing from $2.50 to $0.61 per cu.yd. In 1892 he installed a "hollow tank," with motor-driven perforated revolving arms, whereupon cost was $0.64 per cu.yd. In 1895–96 cost was reduced to $0.54 by introducing a jet washer. In March 1897, use of a double jet reduced cost to $0.24 per cu.yd. In 1914, the primitive method of loading sand into barrels and wheeling it out gave way to removal by ejectors. The following year an ejector system of washing the sand thus removed was adopted (52).

The first chemical analyses of raw and filtered water at Poughkeepsie were made in November 1887 by Professor William Ripley Nichols. Similar analyses were made by Professor Thomas M. Drown in 1889 and 1891. These analyses, said Fowler (51), showed reductions in albuminoid and free ammonia and thus a "material increase" in filter efficiency between 1887 and 1891. Two bacterial counts reported by Professor Drown in 1891 showed a reduction from 1,160 in the raw to 62 in the filtered water in one case and from 1,576 to 34 in the other, or 95 and 98 per cent. But in January 1892, a reduction of only 82 per cent was noted. Four sets of bacterial counts made by D. B. Ward, M.D., before the filters were enlarged, were given in the annual report for 1896 (46) thus:

	Raw-Water Inlet Basin	Clear-Water Basin	Percentage Removed
February 14	1,064	736	31
February 26	80	50	37
May 22	102	32	68.6
June 5	4,016	85	97.8

In 1898, after two filters had been added, Dr. Ward made twelve bacterial counts. The range shown was: inlet basin, 27,000 to 4,200 per ml. on November 10; percentage removal by old filter, 99.7 on January 6 and 73.33 on November 12; new filter, 99.36 on February 4 and 88.76 on December 12, the respective counts before and after filtration being 13,950 reduced to 88 and 2,240 reduced to 272.

A coagulating and settling basin, with in-and-out baffles, was put into use December 27, 1907, a few days before Fowler died. A coagulant was used intermittently until 1929; since then continuously. Another important event of 1907 was cutting out the connection through which the filters had sometimes been bypassed.

A rapid or mechanical filter of the high-pressure type was given a trial at some date before Fowler wrote his paper of 1892 (50). The result during a period of turbidity, using alum as a coagulant, "was clear, bright water, but the quantity of alum required was not only very great but an appreciable amount was left in the effluent." (This may have been because of low alkalinity of the water.)

In the same article Fowler noted that the "Anderson process" would be given a trial in connection with the filters in the hopes that it would produce a colorless water and remove the clay turbidity that sometimes clogged the beds. Annual reports of the water commission for 1892–94 (46) show that mechanical difficulties and illness of the engineer in charge of the "revolving purifier" (to produce comminuted metallic iron coagulant) prevented a test of the process.

A drought beginning in the summer of 1908 reduced flow in the Hudson to such an extent that the salt water line had reached upstream to Poughkeepsie by early fall, resulting in a perceptible salty taste in the drinking water. When winter cured that ill, it brought another, for the river froze over and began to yield a raw water "concentrated with sewage, but with no turbidity," rendering both sedimentation and coagulation processes ineffective. And when the filters also failed to respond to any treatment, significant bacterial counts were soon encountered in the filter effluent.

To meet this situation, George C. Whipple, as consultant, recommended the application of chloride of lime in place of alum. Thus, on February 1, 1909, by means of the coagulant apparatus, chloride of lime was introduced into the low-lift pump suction line. By February 12, a temporary dosing appliance, consisting of two barrels and a "regulating box," permitted transfer of the application point to the inlet of the sedimentation basin. And on March 17, 1909, in view of the notable success of the process, regular chlorination was begun with a permanent apparatus (53).

Abandonment of the Hudson River for an upland gravity source of supply was considered by Allen Hazen in a report made in 1913. He thought the works capable of treating 5 mgd. for several years.

He outlined a plan for extensions during the next ten years, including additional coagulation basins and filters. Advisory control of the filters by Hazen's firm was begun in 1913 and continued until his death in 1934. Chester M. Everett and then Malcolm Pirnie succeeded Hazen.

Prefilters designed by Hazen, Whipple & Fuller were put into use early in 1920. They were of the rapid or mechanical type, in four units having a total area of 1,645 sq.ft., and a combined rated capacity of 4 mgd. The prefilters were duplicated early in 1928.

A simple aerator was installed in 1920 and a more elaborate one in 1926. The first aerator consisted of 1,000 $\frac{1}{4}$-in. holes drilled in the top of a pipe which discharged settled water into a basin before it passed to the slow sand filters. The second aerator, as described in the 1926 report of the chief engineer, Walter E. Walker, included 94 conically arranged nozzles throwing a fine, widely distributed spray. It received water from the prefilters under a low head before it went to the final filters. It was designed by Hazen & Whipple (46). Pirnie, then with that firm, described aerators of this type at Poughkeepsie, Providence and elsewhere (54). Removal of 40 to 70 per cent of carbonic acid gas (CO_2) was effected by the spray nozzles in 1931. A single spray nozzle, designed by Chester M. Everett, was used in the winter of 1932–33. It eliminated frazil ice but was only about 75 per cent as effective as the regular nozzles, wrote Cole, in the annual report of the water works for 1932–33 (46).

Later changes at Poughkeepsie included pre-ammoniation, tried experimentally in 1931 to keep down chlorine taste and odor in winter and algae in summer and soon adopted for regular use; use of lime to prevent the red-water plague in hot-water systems, begun in 1933; and black alum, tried in 1933 and put into use on a large scale November 30, 1934.

Raw sewage from the State Hospital was discharged into the Hudson River 2,000 ft. above the water intake for 60 years, but the state began to treat the sewage August 12, 1933. Although the year was half gone the count of bacteria in the river water for 1933 was the lowest in years, averaging 2,742 per ml., but ranging from 25,000 to 300. The range for coagulated and settled water in 1933 was 170 to 0, with an average of 3.4; prefiltered, 150 to 0, with an average of 3.8; laboratory tap, 4 to 0, with an 0.2 per ml. average. All *Esch. coli* samples from the laboratory tap were negative in 10-ml. and 1-ml.

samples (46). Turbidity and color reductions by means of coagulation, sedimentation and double filtration were notable. In July 1942, all 1-10-ml. raw-water samples were positive; settled-water samples ranged from negative for 1-ml. to 50 per cent positive for 100-ml. samples; double-filtered water showed less than 0.02 per cent positive in 100-ml. samples (52).

Typhoid deaths from 1881 to 1890 and cases and deaths from 1891 to 1935 have been compiled from various sources with the aid of Dr. W. H. Conger, Health Officer at Poughkeepsie (55), and Mr. Cole (52). The data are too voluminous for presentation here. Broadly, they show a heavy but erratic typhoid toll up to 1910 and a rapid decline to zero for the half-decade 1931–35. Improvement in the character of the water supply contributed largely to the reduction but data are not available for evaluating the factors. Without question, the rate would still be high if the water were not adequately treated.

This is the history of the first municipal slow sand filtration plant in America. Water treatment at Poughkeepsie has been from the start a struggle against heavy odds in which the engineer, the chemist and the bacteriologist have cooperated and triumphed.

Hudson, N.Y.—Two years after Poughkeepsie completed the first slow sand filter in America a second one was put in use by its up-river neighbor, Hudson, N.Y. That was late in 1874. Like Poughkeepsie, Hudson disregarded the advice of its engineers and pumped water from the Hudson River instead of taking an upland gravity supply.

In 1876 a joint stock association began delivering spring water to houses and to sidewalk cisterns equipped with pumps (56). After operating for 30 years as "proprietors of the aqueduct," the owners, under the name "Hudson Aqueduct Co.," by which they were commonly known, were granted a charter by the legislature. The company continued to supply spring water until 1908, when it was dissolved.

Water commissioners, appointed by the city council in 1872, engaged William J. McAlpine as engineer. In a report dated September 19 (57), McAlpine favored a gravity supply from Lake Charlotte rather than pumped and filtered water from the Hudson, which had been proposed by the city council. The lake, he believed, would give purer water than the river, with less likelihood of interruptions to service, but the capital and operating costs would be larger.

Contamination and defilement of the Hudson from the upriver towns and Erie Canal water were mentioned by McAlpine but he refrained from "any stronger expression of opinion" on the "river water, in deference to its practical use by your shipping for so many years without apparent injury."

In the event that a supply were to be taken from the river, McAlpine advocated a filter in the distributing reservoir, consisting of a "pyramidal mound" of stone covered with sand and gravel. A longitudinal passage at the bottom of the mound would contain supply, distribution and waste pipes and also a pipe for collecting the filtrate. Gates in a chamber outside the reservoir would control the pipe connections.

The water commission transmitted McAlpine's report to the city council in September 1872, with an endorsement of his plan for a gravity supply from Lake Charlotte. A referendum vote on May 7, 1873, on sources of supply stood: Hudson River, 184; Hudson Aqueduct sources (springs), 108; Lake Charlotte, only 67; aqueduct and river, 1; total, 361 (58).

On September 1, 1873, J. B. G. Rand was appointed engineer for the proposed works. He had been chief engineer of the new water works and filters at Poughkeepsie, completed late in 1872, and afterwards had visited filters in England and on the Continent. Water was admitted to the distribution system on November 1, 1874.

Water drawn from the Hudson River at a depth of about $8\frac{1}{2}$ ft., at a point where the river was 35 ft. deep, was forced up the hill to a filter with an area of a little over 0.2 acre, then passed into an adjoining clear-water reservoir with a capacity of 3.2 mil.gal. The filter was 72 in. deep, as at Poughkeepsie, with practically the same depths and character of the various layers of media. In his report for 1874–75, Rand said that he had made plans for covering both basins, to prevent trouble and expense due to ice in winter and "aquatic plants" in summer (59). These were not adopted.

At the close of his report for 1874–75 Rand stated that a 10-in. gravity supply main from Lake Charlotte would cost less than $70,000, or $12,000 less than the capitalized cost of coal alone for pumping from the Hudson. The pumping plant at the river could be held in reserve for emergencies.

A second filter, with an area of 0.53 acre, was designed by Professor John Emigh, of the Rensselaer Polytechnic Institute, Troy, N.Y., and put into use in September 1888. Although provided with only central underdrains on each axis, Emigh proposed to clean the filter by reverse-flow wash from a central inlet, supplemented occasionally with surface scraping (59; report for 1887–88). The method proved unsuccessful.

H. K. Bishop, who became superintendent of public works in 1899, immediately pointed out the need for renovating the filters of 1874 from top to bottom, supplementing the central underdrains by laterals, and the need for more boilers and pumps as well. But why, he asked, should money be spent for improving, extending and operating the works at the river to force to an elevation of 310 ft. water polluted with the sewage of the city and of the towns and cities on the Hudson and Mohawk rivers when plenty of pure wholesome water could be found which would run downhill to the reservoirs? All things considered, a gravity supply of pure water would cost less per year than a pumped and filtered supply from the polluted Hudson (59).

A typhoid epidemic in 1899–1900 emphasized the need for safer water. In his report for 1901, Bishop, in urging action for a new supply, wrote: "The season is now approaching when we may expect to have the dreadful typhoid with us again."

But until 1905, Hudson continued to pay heavily in sickness and deaths for choosing and continuing the river as a source of water supply against the advice of successive engineers. For years the filters were not under technical control and no attempt was made by either the water or the health department to correlate the water supply with the abnormally high typhoid cases and deaths. Not until 1900 was such a study made, and then not by a city official. A young physician, who had graduated from a civil engineering course at Cornell, delved into the vital statistics of the city and presented before the local University Club a remarkable study, including figures for 1885–1900, of water supply and typhoid at Hudson (60). This was Clark G. Rossman, M.D., who became Commissioner of Public Works of Hudson in 1935 and still filled the position in 1941.

The primary reason for the heavy typhoid toll in Hudson, Dr. Rossman showed convincingly, was the pollution of the Hudson River by the sewage of up-river cities and towns, strongly reinforced by the sewage of the city itself. The filters were unable to cope with this pollution. Typhoid was higher in winter than in summer and autumn, contrary to general experience. The curve was highest when the winters were coldest and the river covered with ice the longest. The punishment was all the more severe because the filters were not covered and heavy ice prevented cleaning through the long winters. Thus the sewage-polluted water was least exposed to air and light when the weather was coldest. This coincidence Dr. Rossman established by records showing periods during which the river was closed to navigation.

His paper and the reports of Superintendent Bishop that followed, coupled with a steep rise in the typhoid curve, led to the introduction, in 1905, of a new water supply. C. C. Vermeule was the consulting engineer and Bishop was the chief engineer of construction. The Hudson River pumping station was shut down late in the year. The new supply was delivered into one or both of the existing filters. Eventually the old one was converted into a reservoir.

Late in 1941, J. McClure Wardle, Superintendent of Public Works (61), wrote that since 1934 the water had been pretreated at the storage reservoir with chlorine and ammonia to extend the filter runs from six weeks to six months, thus permitting filtration throughout the winter. Postchlorination and postammoniation gave what for a time were considered good results, but, when it was found that during the summer months bacterial counts were sometimes high and gas formers were present, free residual chlorination was adopted. That has practically eliminated gas formers and brought the total bacterial count down to reasonable proportions.

The typhoid rate reached its worst in 1904 with 152 cases, 17 fatalities and a death rate of 167.1 per 100,000. After the abandonment of the Hudson, the rate dropped rapidly. Doubtless other causes than the quality of the water were responsible for much of the typhoid before and after the date of Rossman's paper, but what the typhoid rate would have been if the river water had been used unfiltered is appalling to contemplate.

St. Johnsbury, Vt.—The third slow sand filter in America was completed late in 1882 by the St. Johnsbury, Vt., Aqueduct Co. After functioning more than a dozen years it was replaced in the nineties by the earliest of the picturesque group of circular filters with steeply pitched roofs which catch the eye of tourists passing from St. Johnsbury to Littleton in the White Mountains of New Hampshire. All these filters treated water from Stiles Pond, introduced in 1827 to supplement a still earlier supply from springs. At the start, the pond water had been passed through a "filter" in a bulkhead. A filter of coarse sand only 400 sq.ft. in area and 15 in. deep was provided. In 1882 a much larger unit was built alongside the small one and the latter was reconstructed. The two units had an area of 2,101 sq.ft., and a depth of 30 in.; 18 in. "finest sand" on top, then 6 in. of fine and 6 in. of coarse gravel (62).

In 1895 the company built a slow sand filter 50 ft. in diameter. To this there were added two of the same size in 1897 and a fourth in 1912 (63). These were built after plans by E. H. Gowing of Boston

FIG. 34. COVERED SLOW SAND FILTERS AT ST. JOHNSBURY, VT.
Three of four filters built by the village in 1895–1912 to supersede private company filters of 1882
(From photograph supplied by R. C. Wheeler, Barker & Wheeler, Engineers)

(64). Their combined area was 7,850 sq.ft., or nearly 0.18 acre. These filters were still being used in 1942. The filters of 1882 were abandoned about the time the first circular filter was completed.

The round filters were inclosed in a twelve-sided building having a pyramidal roof (65). The inside of each building—sides, roof down to the ceiling, and ceiling—was sheathed with 1-in. unmatched boards. Doors and windows were double. In 1934 during excavations for a new pipeline, a 13-in. brick wall was found. This probably inclosed the filter of 1882 (66), which was rectangular and was also inclosed in a wooden building (65).

The need for inclosing filters at St. Johnsbury will be apparent to those familiar with winter temperatures in the Passumpsic River Valley. United States Weather Bureau records show average monthly temperatures at St. Johnsbury for the 27 years 1894–1930 for the four winter months as: December, 20.7°F.; January, 15.7°; February, 16.8°; March, 22.8°. By single days the lowest temperatures were: November, — 13°; December, — 43°; January and February, each — 38°; March, — 22°; and, for good measure, May, — 1°.

C. H. Bowman, resident attendant at the filters, stated in 1935 that "a film of ice" sometimes formed on the water above the filter but did not "bother much" when the filter was being cleaned. One or two of the four filters were cleaned each week. A total of 12 in. of the 20-in. layer of sand was taken out on an overhead trolley and wasted. New sand was taken from a pit, screened but not washed, and used to bring the layer back to its 20-in. depth. In the autumn, lily seeds in the pond water necessitated more frequent cleaning. In 1935, a Venturi recording meter was installed on the main outlet from the filters.

The drainage area of Stiles Pond is sparsely populated. A sample of pond water taken in May 1932, and examined by James M. Caird of Troy, N.Y., had a color of 20, a slight turbidity and "a number of organisms, including *Dinobryon* and *Asterionella*."

What person was responsible for the slow sand filters of 1882 and why they were built at that early date is now only a matter for conjecture. A descendant of one of the Fairbanks brothers, chief owners of the Aqueduct Co., states that members of the family had traveled abroad, read widely, were interested in improvements, and "probably the idea was theirs" (67).

Competing water works were built by the village of St. Johnsbury in 1876. Water was pumped from the Passumpsic River, within the village. As the river was subject to some local pollution and to turbidity, sawdust and shavings when the stream was in flood, the works included a small upward-flow filter, which appears to have been a failure. Late in 1892, a Jewell rapid filter was built. It was operated without a coagulant and was abandoned in 1894 (68).

In 1905, a project for filtering the Passumpsic supply was rejected by popular vote. In 1906, the State Board of Health condemned the village supply, leaving the Aqueduct Co. without competition. In

1924, the village bought out the company. Rapid filtration was recommended by Barker & Wheeler in 1932, but not adopted (66). In 1942, the consumption of water from the slow sand filters was about 1.5 mgd. (65).

Examination by the State Board of Health of numerous samples of filtered water from Stiles Pond for the period 1925–41 showed a few coliform organisms in only six instances (69).

These annals of the quest for pure water in a small New England village * supplement the meager and scattered data that have previously been published regarding the third American slow sand filter plant. They also bring into the St. Johnsbury picture the village-owned upward-flow filter antedating the first filter of the company, followed by a hitherto almost unknown rapid filter installation.

A Few Subsequent Slow Sand Filters.—Ilion, N.Y., put the fourth slow sand filter in operation September 20, 1892. It was still being used 50 years later (70). Nantucket, Mass., built a filter and aerator for removing tastes and odors in 1892 but since algae gave no trouble that year it was not used until 1893 (71). It was not a success (see Chap. XVI).

The intermittent filter at Lawrence, Mass., which began service in the latter part of 1893, stands sixth chronologically among American slow sand filters but for a time it was ranked first in importance. Its design was based on the treatment of both sewage and water at the Lawrence Experiment Station of the Massachusetts State Board of Health. It was the first practical demonstration in America of the bacterial efficiency of filtration (72). Intermittent filters were built at Mt. Vernon, N.Y. (73), and at Grand Forks, N.D. (74), in 1894, but never again for municipal supplies in America except for use at Springfield, Mass., in 1906, to cope with algae growths in the notorious Ludlow Reservoir pending introduction of a new supply (Chap. XVII).

Albany, N.Y.—Filters at Albany, N.Y., were completed in 1899 to treat the polluted water of the Hudson. In their design Allen Hazen profited by his experience at the Lawrence Experiment Station and the observations abroad and at home embodied in his book on filtration (20). He employed pre-aeration and presedimentation but not intermittent filtration.

* The population of St. Johnsbury village was 3,360 in 1880; 5,660 in 1900; and 7,437 in 1940.

Albany began pumping water from the Hudson in 1875 to supplement near-by gravity sources of supply, some of which had been introduced at the beginning of the century. Strong opposition to utilizing the river developed and was continual despite filtration and the subsequent elaboration of the plant to include coagulation, chlorination and prefiltration. In 1932 the river was abandoned for a distant gravity supply, treated by rapid filtration and various accessories.

The quest for pure water at Albany began in the seventeenth century. Some of the events at that time and up to the Hudson River Tragedy, as the adoption of that source became, will be noted briefly.

On April 30, 1680, Dankers and Sluyter, in a journal of their American tour, stated that the inhabitants of Albany, a town of 80 or 90 good houses, "have brought down a spring of water, under the fort, and underground into the town, where they have in several places always fountains of clear, cool water" (75). These houses were located between the old Dutch stockade of Fort Orange and the new English fort. An entry of August 1686 in Reynolds' *Albany Chronicles* (77), says that water was then furnished from a pond or "Fountain" created by a dam at the head of Yonkers (State) St., from which it was delivered through 2-in. bored logs to "a city well in each of the three wards." It may be inferred that this supply was introduced some years before 1680, for on Sept. 14, 1686, the city council ordered the pipes repaired because in some places they were decayed or at least had become "so leaky that the wells are quite useless" (77).

Pehr Kalm, an eminent naturalist from "Swedish Finland," noted on June 20, 1749 (79), that the "water of the several wells" in Albany was very cool but had an "acid taste, which was not very agreeable." He found "little insects in it, which were probably monoculi." They were pale in color, very narrow, and ranged in length from one-half to four "geometrical lines." Their heads were about the size of a pin. Their tails were in two branches, each ending in a black globule. They swam "in crooked or undulated lines almost like tadpoles." Water containing monoculi, Kalm said, did not seem to harm the inhabitants of Albany but he thought it not wholesome to those unaccustomed to it. He had been obliged to drink water containing monoculi several times, after which his throat felt as though there were a pea or a swelling in it.

In the first edition (1789) of his *American Geography*, Jedediah Morse (80) said the well water of Albany was "extremely bad, scarcely drinkable by those who are not accustomed to it. Indeed all the water for cooking is brought from the river and many families use it to drink." In a revised edition of his book (1793) Morse said that the inhabitants were about to construct works to bring in good water.

Minutes of the Albany Water Works Co., from March 1800 to 1851, when the property of the company was acquired by the city, show that in 1800 the company was then supplying or ready to supply water (81). Theodore Horton states (82) that the earliest works included dams and reservoirs on two small streams north of the city from which "bored logs with circular wrought-iron straps" led to "masonry cisterns located at various points in the lower part of the city." The cisterns, he adds, "were first used for fire purposes and to pump from but later the pipes were extended into residences." The wooden pipe was soon replaced by cast-iron mains, costing $150 a ton.

Cholera, which caused hundreds of deaths in the summer of 1830, was attributed by some "to the impurity or peculiarity of the water in city wells." The health board had the waters of fourteen wells examined by Drs. Romeyn Beck and Philip Ten Eyck, who pronounced them "free from any impurities which could be injurious to health."

In the years 1841–49, six engineers reported on pumped supplies from the Hudson and Mohawk, and gravity supplies from Patroon's Creek and Norman and Hunger Kills (83, 84, 85, 86). By far the most noted of these engineers was Major David Bates Douglass, who had planned the first Croton Aqueduct. In 1846 he recommended Patroon's Creek as "decidedly the softest and purest of all" sources considered, the others being the Hudson and the Mohawk. A committee endorsed his recommendation and transmitted the report to the city council, which took no action (84). George W. Carpenter proposed a supply from the Hudson River, passed through "a coarse filtering bed to free it from such materials as might injure the pumps," but again no action was taken (87). The great fire of August 17, 1848, started by "a washerwoman's bonnet" in the Albin Hotel, spurred the city authorities to action. Although it was not so disastrous as the Chicago fire of 1871, attributed to Widow O'Leary's cow and an overturned lantern, the Albany conflagration swept over 37 acres and destroyed 600 buildings. On November 7, 1848, a popular vote of

4,405 to 6 was cast in favor of city-owned works (76). On April 9, 1850, the legislature created a special commission with authority to build works to supply the city with "a sufficient quantity of pure and wholesome water." The act required the city to buy the works of the Albany Water Works Co. and to employ a civil engineer as superintendent of water works (87).

William J. McAlpine was appointed chief engineer of the projected works on May 1, 1850. After considering the reports of the last decade, he recommended Patroon's Creek. Its frequent turbidity, he said, could be easily corrected either by filtration or by constructing a division wall in the pond and allowing the incoming water to settle in one basin before it passed to the other, but a filter, he thought, would save the cost of a division wall (88). The water commissioners adopted McAlpine's plan, which called for a gravity supply from an impounding reservoir on Patroon's Creek. Until the new works should be completed "the waters of Maezlandt Kill, the old fountain head," would be continued in use. On May 1, 1851, the city began operating the works bought from the old company. On November 4, water was let into the new aqueduct from Rensselaer Lake (87). George W. Carpenter became superintendent on April 7, 1853 (87), and served for 39 years. During that period he made some of the earliest and most noteworthy reports on tastes and odors due to organic growths in the reservoirs.

Need for more water led the commissioners to engage James P. Kirkwood early in 1872 to submit a plan and estimate for pumping water from the Hudson to the existing Bleecker Street Reservoir. Apparently his opinion on the quality of the water was not requested (89). That subject was left to Professor Charles F. Chandler of Columbia University (90). His report of May 15, 1872, introduces one of the most tragic chapters in the history of American water works. He approved the quality of a source of supply already dangerously polluted and bound to become more so as the cities above Albany on the Hudson and Mohawk rivers increased in population and industry —a supply destined to cause thousands of cases and hundreds of deaths from typhoid. Based on only one sample, collected and sent to New York by Carpenter on March 14, Chandler pronounced the Hudson water safe. His method of analysis is interesting if not amusing. "The suspended impurities which rendered the water turbid being temporary in character, were allowed to subside." The clear water was examined for various salts, organic and volatile matter and hardness.

"The organic matter in water," he wrote, "really demands the closest scrutiny." That derived "from sewage, though highly dangerous in certain stages of decomposition, [is] speedily changed by the oxygen held in solution in running water, to forms which are innocuous." The Hudson water, Chandler said, was satisfactory in organic content.

Natural aeration was given much weight by Chandler. Glens Falls, the Falls of the Mohawk and the State Dam at Troy, he said, "are the most effective means contrived by nature and art for preparing the water for the use of your citizens." Except at Troy, no sewage of consequence, Chandler wrote, is discharged into the river. Its volume "is so small in comparison with that of the river" that "the most careful examination of the water has failed to reveal anything to sight, taste, smell or analysis, which can be considered as throwing the slightest suspicion upon the purity of the Hudson, or its fitness for supplying a perfectly wholesome beverage for the city of Albany."

In transmitting, with approval, the Kirkwood and Chandler reports of 1872 to the council the water commissioners said that, while some of the citizens favored the Hudson, "a large number have so strong prejudice against its use that, no matter how clearly its purity may be demonstrated, they are unwilling to have it selected. . . . It is extremely difficult to educate the public mind up to the belief that the impurities flowing into a stream, no matter what its volume, its velocity, or its length, do not retain their deleterious and objectionable forms" (89). The commissioners adopted the Hudson, thinking to educate opponents by forcing the water down their throats. But apparently they had misgivings for they said in their report that although Kirkwood's estimates included nothing for filters they believed their cost could be met with the means available. The special committee of the city council to which the commissioners' report was referred, endorsed "the plan recommended by the Water Commissioners . . . including the necessary filteration [sic] . . . or a subsiding reservoir." The resolution to adopt the plan was rejected by the council, 11 to 4, then reconsidered and laid on the table. There it lay for a year. On June 9, 1873, it was taken up and adopted, 12 to 4. Filtration did not come until 1899.

Opposition to the Hudson had been voiced by the Albany Institute directly after the date of Chandler's report. It declared that within eight miles of Albany a tremendous load of sewage and industrial wastes was being poured into the river by upstream cities and towns, and pointed out that if the people of Albany supinely submitted to the pending water supply plan, the results would be disastrous to the health of the city. The committee advocated complete utilization of Patroon's Creek, as originally intended, and reforestation of its drainage area. The report was adopted by the Institute on June 12, 1872 (91). The Institute, at a meeting attended by three civil engineers and five physicians, again condemned the Hudson (92). The Albany Homeopathic Medical Society, whose opinion had been requested by the city council, disapproved the river in a report submitted on June 9, 1873. "Irrefragable evidence," it said, "holds that the organic material most active in the production of disease is of an animal nature," not removable by "any known means of filtration." Notwithstanding condemnation by engineers, physicians and others, the council approved the Hudson River project on the very day the adverse report of the Homeopathic Society was received. I. C. Chesborough took engineering charge of the project in August 1873. Water was first pumped to the Bleecker Reservoir on September 14, 1875, and repumped to the Prospect Hill Reservoir on February 3, 1878. Thus ended the first chapter in the Hudson River Tragedy.

The filters, assumed by the water commission in 1872 to be necessary, and, with a settling reservoir as an alternative, approved by the city council, had not been built. Advocacy of a gravity supply not needing filtration continued. For some years the commission and council were at odds, the latter opposing the Hudson River supply. The council refused funds for another pump to lift water from the river; asserted that the commission, to bolster up its pet Hudson River supply, had let the upland storage reservoirs fill with debris and sediment. The council opposed a bill before the legislature to grant power to the commission to go over the council's head and borrow money, and nearly voted to ask the legislature to oust the commissioners and substitute men of its choosing. But the legislature sided with the water commissioners and, on May 12, 1884, granted them full power to extend and improve the existing works and made it mandatory for the council to issue $400,000 of bonds for improvements.

Grover Cleveland, then governor of the state of New York, was petitioned by river-pier owners and boating clubs to direct the State Board of Health to investigate the pollution of the Albany Basin. The petitions were referred to the board and an investigation was made by its committee on sewerage, assisted by Horace Andrews, a civil engineer. In a report of November 22, 1884, Andrews said that the lower end of Patroon's Creek, carrying considerable sewage, emptied into the Hudson about 3,000 ft. above the water works intake. All the sewage of Albany discharged either into or below the basin. Float experiments showed that at flood tide the entire volume of water flowed upstream, sometimes 3,400 ft. or more. "Manifestly," he said, "it is most improper" to pump water to the reservoirs for the last three and one-half hours of each flood tide and the first one and one-half hours of each ebb tide. The committee of the state board recommended that the governor declare the basin "a nuisance dangerous to health" and prohibit the discharge of sewage into it. It also advised an investigation of possible new sources of water supply for Albany (93).

The city council adopted a resolution December 15, 1884, requesting the water commission to desist from letting a contract for a new pumping engine "until the people may be heard from" on a question affecting "the health and lives of a hundred thousand citizens." On the same day the council appointed a committee to consider a proposal that had been made to supply Albany with pure spring water for $1,000,000. As in the case of schemes to bring down a gravity supply from the Adirondacks, on which reports were made by Col. John T. Fanning, a civil engineer (94), nothing came of the project.

Attacks on the Hudson by the council and others led the water commissioners to call on Professor Chandler a second time to report on the quality of the Hudson. His report of January 31, 1885, was even more disastrous than his endorsement of 1872, because meanwhile much had been learned by others, but overlooked or ignored by him, about water-borne typhoid. Notwithstanding this and the increasing pollution of the Hudson, the professor persisted in declaring the supply safe and did not even suggest filtration (96).

Chandler said he had been asked if his opinions of 1872 had not been changed by subsequent events, and specifically whether new methods of analysis, "especially microscopic and culture experiments, may not record the presence of dangerous organisms which would escape every method of chemical analysis," and whether the "knowledge of zymotic diseases has not advanced to such a degree as to compel different conclusions; and finally, whether the test of experience in the city of Albany has not demonstrated the danger of making use of this source of water supply."

An emphatic "No" was the answer to all these questions. He did not claim "that chemical tests will detect the specific poisons of zymotic diseases in water" but, he added, no method of investigation yet proposed will do this "except the actual production of the diseases, and no one has ever found in a river-water the specific poison of any zymotic disease." He said that the only reliable authority who has claimed to have found disease germs in water was Dr. Koch, who "thinks he observed the cholera bacillus in a water tank [reservoir] at Calcutta in which persons suffering from cholera bathed and washed their clothes." In the course of a long discussion of bacterial examinations of water, Chandler "waxed sarcastic," thus:

> Under these circumstances it would appear that counting the number of bacteria that will develop in gelatin, or in the culture media, on the addition of a sample of water, is not a very reliable method of determining the danger of water for domestic purposes, although some enthusiastic microscopists, carried away by their skill in raising bacteria in their microscopic gardens, have said that the days of chemical analysis of water supply are numbered. . . .
>
> When the biologist learns to detect in water 'the specific poisons of zymotic diseases,' and to distinguish them from harmless organisms that we eat, drink, and breathe with impunity all our lives, then we may set up biological analysis as superior to chemical analysis, for selecting drinking water.
>
> Up to the present time, however, biological analysis will not tell us anything with regard to the Hudson River water that we do not already know. The river receives a small amount of drainage, and thanks to the oxygen and the micro-organisms, it becomes so thoroughly purified that, when it reaches the Bleecker Reservoir for distribution to Albany, it may be drunk without danger to health (96).

A very different opinion on the quality of the Hudson River water from that expressed by Professor Chandler was given a little later by Professor William P. Mason of Rensselaer Polytechnic Institute at Troy. The Albany Board of Health, which does not seem to have concerned itself with the water supply before, engaged Mason to determine "whether the influence of Troy sewage is felt in the Albany water supply." In a masterly report dated April 23, 1885, and based on analyses of many samples of water taken from above Troy down to Albany, Mason declared that "the influence of the Troy sewage is felt just below Troy," and "there is no material change for the better by the time the water reaches Albany." Slight "evidence of self-purification" was due to dilution. On the evidence obtained, and because no other adequate source of supply was apparent, Mason advised sedimentation and filtration of the Hudson. He urged construction of a proposed intercepting sewer to prevent the return of Albany sewage to the water intake (97).

On December 19, 1883, the council, by 19 to 0 vote, denied a requisition of the water commission for $400,000 to enlarge the Hudson River plant. The council refused it on the ground that the water was "detrimental to the health of the citizens," and that the commission had failed to submit plans for the work. A blanket report on various subjects referred to the council's water committee during the previous three months was submitted on April 6, 1885. Seven hearings had been held at which many citizens and the water commissioners had been heard. "A very large proportion of the citizens," the committee reported, "are anxious to be relieved from the necessity of using water contaminated with sewage." Professor Chandler's report was characterized as face saving, merely backing up his report of thirteen years before. After declaring that the Hudson supply contained sewage from Albany on the up-tide and Troy on the down-tide, the committee recommended "a different source of supply or proper filtration of the present supply."

Of "many valuable suggestions" for a new supply the committee was impressed with the "Drive Well System" of William D. Andrews & Bro., New York, then "in successful operation in Brooklyn." Forty wells already driven on lowlands north of the city had convinced the owners that adequate supplies could be obtained from that source.

Finally, the law committee of the city council reported that "little or no relief" could be expected from the water board and proposed a "new and impartial commission" to investigate "various improvements and sources of supply." By a vote of 15 to 0 the council adopted the report. Six weeks later (May 22, 1885) the legislature created a special commission to tackle the water problem. The commission was given six months to submit to the council a general plan and estimate for a new supply or the purification of the old one or else retire. If the council approved the plan, the commission was to proceed with its execution at a cost of not over $1,200,000. This looked like business, but twelve years of planning and counterplanning passed before construction of filters was started and two more before filtered water was turned into the mains.

William E. Worthen was engaged as engineer and Professor Albert R. Leeds as consulting chemist. On November 20, 1885, the commission transmitted to the council reports by Worthen and Leeds, a proposal from the Newark Filtering Co. to build mechanical filters and offers from Andrews & Bro. to supply water from driven wells. Worthen submitted estimates for a new gravity supply from the Hudson above Glens Falls, from the Mohawk River, from Norman Kill, and from Wyant's Kill, east of the Hudson. He also outlined a plan for supplementing the existing gravity system. Leeds (98) proposed a dual supply, consisting of driven wells, aerated, and Hudson River water, both aerated and filtered.

Leeds' report is as notable for being up-to-date as was Chandler's, ten months earlier, for being antiquated. But unfortunately, as later events showed, he was influenced by a commercial bias. This lay in his advocacy of aeration and filtration, in both of which he had or was soon to have financial interest (99).

He had been engaged, Leeds said, to inquire into the available sources of pure and wholesome water and what, if any, method could be adopted for its purification. He had also been authorized to make experiments on purifying the existing Hudson River supply. He had considered the relative advantages of "(a) artificial oxidation—so-called aeration methods; (b) natural and artificial filtration; (c) combinations of (a) and (b)." His "experimental method" consisted of an analysis of the raw water followed by treating the water with air under pressure and then by filtration. The difference between analyses of the raw and treated water indicated the "benefit possible through artificial methods of purification." From the "gratifying results" of his laboratory experiments with oxidation and filtration, Leeds concluded that there would be no practicable obstacle to the application of these methods on a large scale to city water supplies.

After reviewing his data on various surface sources of supply he rejected all except the Hudson at Albany. This, he assumed, could be so purified [oxidized?] and filtered as to eliminate "organic matters dangerous to health" and "organized particles in the form of germs capable of injuring health." The only water analyzed that met Leeds' approval was from a test well recently put down north of the city— probably a part of the driven-well project then brewing.

The citizens of Albany, Leeds declared, were "drinking a residual portion of the sewage of Troy and a part of their own sewage," and

Troy was drinking unoxidized sewage from points above it on the Hudson. Legislative action and "a humane and wise public opinion" should compel these and all other communities "to reclaim their sewage before emptying the purified effluent into a flowing stream." He added that "simple, economical and completely effectual methods of doing this are known and practiced by sanitary engineers." Even so, no one can be guaranteed a safe water supply "unless it can be thoroughly purified immediately before use. This can be done artificially and naturally" (98). Strong doctrine far ahead of the times!

In 1885, there was no united public opinion in Albany or elsewhere demanding the legislation proposed. No state had adopted such legislation. No American city was treating its sewage effectively, if at all. Only three cities had slow sand water filtration plants. Rapid or mechanical filtration was on the threshold with only one or two little-known plants operating on municipal supply.

Leeds' conclusions on the quality of the various sources were accepted by the special commission in its report of November 30, 1885, but instead of endorsing his recommendations for a joint supply from wells and the Hudson, each treated, the commission advised, as its final choice, driven wells for the main supply and improvements to the existing near-by gravity sources. But if the council did not approve driven wells, aeration and filtration of the Hudson was advised.

The estimated cost "for aeration and filtration complete" was "say $200,000." The estimate was based on an informal proposal by the Newark Filtering Co., under date of August 26, 1885. The company offered to install a plant to supply 20 mgd. of "bright, clean and wholesome water," produced by means of seven pressure filters 30 ft. in diameter. The intention was, the proposal said, to employ "our new method of purification by the use of metallic iron, which method has been perfected by our Mr. [John W.] Hyatt and obviates the use of alum."

This is the first known offer to install mechanical filters on so large a scale. The special water commission said that the "system will involve the aeration of the water by forced air." As for "microbes or bacterial life," the commission added, no method had yet been suggested which would "destroy such life in water in large quantities except at enormous expense." But Mr. Hyatt had shown to the commission "plans for a system by which he claims such destruction may be effected at reasonable expense."

For some years the water supply question at Albany continued to be in a muddle unparalleled in the history of American water works. The driven-well fiasco resulted in a large expenditure by the special water commission with nothing to show for it except a thousand useless wells and pumping machinery held in storage. The city sued the contractor on its guarantee but so far as can be learned recovered nothing. In the course of this fiasco, the commission obtained legislation authorizing it to let contracts without competitive bidding.

Double filtration was suggested October 7, 1889, by a Committee of Thirteen that had been in existence four years. After characterizing the driven-well project as a "great and costly experiment" that had failed, the committee advocated diversion of the sewage below the intake and double filtration of the Hudson River through sand to free it from floating impurities before it was pumped to the upper reservoir, then through charcoal to remove "impurities detrimental to health." If Albany would build sewage treatment works on the island below the city (which was done years later) there was reason to think it would "be able to enjoin" Troy and other cities on the Hudson from pouring their sewage into its water.

To release the stored machinery, the special water commission, the old water commission and the city council agreed to let the machinery be used to reinforce the Hudson River pumps. Finally, the special commission threw up the sponge and the old water board regained power for a time. It advised filtration of the Hudson but was never given power to build a plant.

Another special water commission was authorized in 1892. The mayor then appointed seven water commissioners whose duty it was either to improve the existing supply or procure "pure and wholesome water" from another source. Subject to approval by a two-thirds vote of the council, the new commission might adopt and execute plans at a cost of not over $500,000. About this time, George W. Carpenter, who had been superintendent of water works for 39 years, was succeeded by George I. Bailey. He, too, as the act of 1850 required, was an engineer. The new commission reported to the council on December 5, 1892, in favor of a Kinderhook Creek supply. Accompanying this recommendation were reports from Clemens Herschel and J. J. R. Croes, consulting engineers, and reports by others on biological and chemical analyses. The Herschel-Croes report was brief and said nothing on the quality of the Hudson. The council immediately approved the proposal for a supply from the Kinderhook by a vote of 11 to 6.

The Kinderhook project hung fire during 1893, but on December 18, Frederick P. Stearns, chief engineer of the Massachusetts State Board of Health, reported on the comparative merits of the Kinderhook, unfiltered, and the Hudson, filtered. He declared against continuing the use of the Hudson, adding that a water once polluted, even though frequently filtered, "cannot be compared favorably with unpolluted water, such as can be obtained from Kinderhook Creek." Death came to the Kinderhook project, however, in 1894 by means of pressure brought on the legislature. A contract was let for a 5-mgd. pump to supplement the Hudson supply.

No progress toward a better supply was made for two years. Allen Hazen, then of Boston, recommended slow sand filtration of the Hudson in a report approved and transmitted by the water commissioners to the council February 13, 1897 (100). After various delays, an act abolishing the old water board was passed by the legislature, rejected by the city on a close vote (unanimous vote required), then repassed by the legislature and signed by the governor. On June 1, 1897, Hazen was appointed chief engineer of filtration works to be built by the new commission. The plant designed by him was put in use by stages from July 27 to September 6, 1899. Hazen's construction report, dated December 29, 1899, appeared in the report for that year (100). (For other descriptions, see References 101, 102, 103.)

The abnormally high general death rate, the incidence of typhoid and presence of diarrheal disease at Albany for many years past, said Hazen in his report of 1897, reflected the pollution of the Hudson at Albany. A comparison of the typhoid death rates of Albany, without filtration, with Poughkeepsie and Lawrence, and with various foreign cities having filtered supplies from polluted rivers, was decidedly unfavorable to Albany, Hazen said. Comments on mechanical filtration occupied eight pages of Hazen's report.

Estimates showed that mechanical filters would cost much less than slow sand filters and that capitalized construction and operation costs would be somewhat less. But, said Hazen, no city had yet used rapid filtration for water so highly polluted as the Hudson. Slow sand filtration was recommended as "the only system which has been demonstrated to be capable of purifying such a source of supply."

On March 1, the council received a long communication from the New York Filter Manufacturing Co. designed to show that the "American" would be less costly than the "European" system of filtration. Although the communication was merely "received" by the council, it led the Medical Society of Albany County to obtain a public hearing at which its committee expressed the belief that physicians of Albany were practically unanimously in favor of slow sand filtration.

The purification plant designed by Hazen received water from a new intake two miles above the old one. Eleven fountain aerators discharged into a settling basin from which water passed to eight filters of an area of 0.7 acre each. At a filtration rate of 3 mgd. per acre, and one filter out of use for cleaning, the plant had a capacity of 14.7 mgd. The walls, floors and roofs of the filters and clear-water basin were of plain concrete, Hazen not yet being fully sold on reinforced concrete. Bailey was credited with the general plan. A chemical and bacteriological laboratory was provided.

A bacterial efficiency of 99 per cent, said Hazen, in a paper read January 3, 1900 (103), had already been attained. It was expected that a great reduction in the city's death rate for water-borne diseases would follow. The plant removed "part of the color and all of the suspended matter and turbidity."

An extension of the water intake from the back to the main channel of the river was put into use August 24, 1907. Experiments extending through 21 months showed, said Wallace Greenalch, then commissioner of public works, that with both presedimentation and prefiltration, a slow sand filter operating at 6 mgd. could produce as good water as with water presettled only at half the rate, and that, considering capital cost, double filtration would be cheaper than single (104). Sixteen sand and gravel prefilters or scrubbers, having a total area of 0.3 acre, went into operation on October 29, 1908.

Hypochlorite of lime was applied to the water at various points beginning on June 26, 1909. Coagulation in the settling basin was used 111 days in the year 1913–14. The next year the prefilters received coagulated water 259 days, partly due to repairs to the final or slow sand filters. In 1915–16, the water applied to the slow sand filter was coagulated only 67 days, because of the increased cost of sulfate of alumina due to the war demand for sulfuric acid.

Conditions at Albany early in 1920 led Theodore Horton, the chief engineer of the state department of health, to advise conversion of the prefilters or scrubbers into "straight mechanical filters," changing one of the basins containing the slow sand filters into settling reservoirs, use of coagulation throughout the year, and substitution of cast iron for the deteriorated steel conduit leading from the purification plant to the pumping station (105).

The *Esch. coli* load carried by the Albany purification plant, declared George E. Willcomb, chemist and sanitary engineer of the works in his report for 1920–21, "is one of the greatest borne by any works in this country." Cohoes, Troy and Watervliet were the chief contributors of sewage to this load. Willcomb suggested joint sewage works, "embodying oxidation and disinfection of the effluent." The conduit from the filters to the pumping station, said the water works report for 1921–22, was "in a dangerous condition." In the report for the following year the prefilters were reported to be in bad shape with half of their area clogged and cracked.

It was evident that a new supply was imperative. Pending that, improvements of the filters were to be made. To design these, Hazen & Whipple were engaged. The work was carried out in stages and completed late in 1925. A account of the improvements was published by Hazen in 1926 (106), stating that when he

———undertook reconstruction of the treatment works in 1923 he believed, as he did in 1897, and does now, that an unpolluted gravity supply from upland sources is best for Albany. The pollution of the present source has doubled since 1899. Very few American cities take water from sources so badly polluted. . . . [The old settling reservoir not having been built for coagulation], pumping the coagulated water to the prefilters broke the floc into finely divided turbid matter which passed into and sometimes through the filters.

Chief among the many improvements designed under the direct charge of Chester M. Everett, of Hazen & Whipple, were: a new 10-mgd. coagulation and sedimentation basin; two sets of aerators, one in advance of and one after the prefilters; conversion of the prefilters or scrubbers into mechanical filters; inspection and repair of the 48-in. conduit from the reservoirs to the pumps, primarily to maintain excess pressure on the conduit during floods, thus preventing leakage into the conduit of polluted water from the canal in which the conduit was laid; changing the old settling reservoir to provide storage between the prefilters and final filters. Construction costs during 1923–25 were almost $700,000—for improvements to serve only until a new gravity supply was introduced.

A typhoid outbreak in the spring of 1924 brought the cases for 1923–24 up to 169 and the deaths to 14, compared with 61 cases and 2 deaths the previous year. After internal inspection of the steel filtered-water conduit, Theodore Horton attributed the outbreak to leaks in the pipe. Numerous damage claims were brought against the city by or in behalf of sufferers from the typhoid outbreak. A test case led to an award of $3,000 damages. This was sustained by the appellate division of the Supreme Court on June 28, 1928, and by the Court of Appeals February 13, 1929 (107). It was held that the water supply became polluted by an overflow of the Hudson River into the Erie Canal and thus into the steel conduit and that the city delayed for at least ten days putting into use a chlorination plant at the lower end of the conduit. The other claims involved lesser accounts.

Convinced at last that the Hudson River pumping and filtration works must be abandoned for a gravity supply from sparsely populated gathering grounds, the city began taking action to that end in 1926. Nicholas S. Hill Jr., New York City, after investigating 24 possible sources of supply within a radius of 50 miles of Albany, advised taking Kinderhook Creek, east of the Hudson, as had been recommended several times before (108). On July 6, Robert E. Horton, of Albany and Schuylerville, was appointed as consulting engineer and directed to investigate sources west of the Hudson, lying chiefly in Albany County. In a report on October 11, 1926, he recommended Hannacrois and Basic Creeks, to be supplemented later by Catskill Creek (109). This source was adopted. The main elements of the project were an 11-bil.gal. impounding reservoir, a 4-ft. conduit 20 mi. long and a distribution reservoir at Loudonville. No filters were proposed but they were required by the State Water and Power Commission. Horton had not thought filtration necessary, as there were only 315 residences on the drainage area of 49 sq.mi., and the two reservoirs would provide 400 days' storage (110), nor had Hill proposed filters.

The project was carried out with Whitman, Requardt & Smith as engineers and Robert E. Horton as consulting engineer. The purification plant had a capacity of 32 mgd. It included coagulation, sedimentation, chlorination and provision for the use of lime to control acidity and alkalinity. The plant went into partial use November 10 and full use December 3, 1932. The old Hudson River filters were then abandoned, after having been used 33 years.

Algae troubles in the impounding reservoir were first met by applying copper sulfate from a boat. In February 1941, *Synura* appeared after the lake froze over. An attempt was made to control taste and odors by applying activated carbon to the water ahead of the filters. This being ineffective, free residual chlorination was used with fair success. It produced a woody taste but complaints diminished considerably. Latterly, the entire reservoir has been successfully treated with copper sulfate just before it freezes.

Manganese has been high at times, particularly in July to September. This is attributed to drawing water from close to the bottom of the reservoir throughout the year to avoid surface growths of algae and supply cool water in summer. Before the summer of 1942, iron and lime were substituted for the usual aluminum sulfate coagulations at such time to provide the pH value of 9.0 required to precipitate the manganese. In the summer of 1942, H. C. Chandler, the supervising chemist (111), adopted a plan used by him on other supplies. Lime only was added to the settled water just ahead of the filters, raising the pH to 9.0 and precipitating the manganese on the filters. During these two months the manganese reached a high of 3.47 and an average of 1.87 ppm. and was reduced to a minimum of 0.10 and an average of 0.20. Meanwhile the use of aluminum sulfate was continued.

During the period from 1871 to 1935 the typhoid death rate per 100,000 ranged from 171 in 1888 to 0 in 1926. Omitting the earlier years, the half-decade averages were 78.6 for 1888–90; rose to 91 for 1891–95; fell to 83.8 for 1896–1900; dropped to 21.8 for 1901–05 and then, except for one slight rise, fell steadily to an average of 1.1 for 1931–35. For 1936–40 the rate was 1.2. The single year 1939 showed no death; for 1940 the rate was 0.8, all reported as non-residents; in 1941 there were no deaths from typhoid.*

Introduction of unfiltered water from the Hudson to supply only a part of the city was followed by an increase in the typhoid death rate. In the nine years after filtration typhoid cases decreased 66.8 and typhoid deaths 70 per cent compared with the previous nine years. Sea-

---

* These and other figures to follow, together with the comment here presented, are based mainly on statistics and other information supplied for use here by Theodore Horton and George E. Willcomb. The more recent figures are from the annual review of "Typhoid in Large Cities of the United States," published in the *Journal of the American Medical Assn*. It should be understood that the figures for the 1870's and in lesser degree for some years later were probably incomplete and otherwise deficient, as in most American cities of the period.—M. N. B.

sonally, the decrease was 84.6 per cent in cases and 76 per cent in deaths for November–April and 61.2 and 60 per cent for May–October. This indicated that typhoid in the cold months came largely from the water supply and in the warm months from the typhoid fly and returning vacationists.

In the nine years before filtration, typhoid cases totaled 3,854; in the nine years after filtration they fell to 854. But in 1909–17 the cases rose to 993. This rise was largely, if not wholly, due to a typhoid epidemic in April and May 1913, attributed by Theodore Horton to a freshet that flooded the filters (112). The epidemic of May–June 1924, already noted, with 104 cases, attributed to the break in the steel filtered-water conduit, brought the typhoid rate for that year to 12.6 per 100,000, but the average for 1918–26 was only 4.6, compared with 16 in the preceding nine years. The first full year after the near-by unfiltered sources of water supply had been shut off and the entire city supplied with water from the revamped Hudson River purification works was also the first year during which no typhoid cases were reported. In the previous eighteen years these improvements had been made: prefilters in 1908; hypochlorite used intermittently from 1909 on; liquid chlorine in 1915; coagulation at intervals from 1913 to about 1925 and continuously thereafter (114).

Albany suffered from typhoid fever for many decades because of blind persistence in drinking untreated water from one of the most heavily polluted rivers in America. The ravages of typhoid, although lessened, continued after the city built one of the best filter plants ever designed. This plant was of the slow sand type, long used abroad but never employed here for so large a city. Addition of prefilters, coagulation and chlorination, combined with skillful operation, worked marvels but after many years the city decided to go further afield and obtain a gravity supply from a sparsely populated area. It was assumed that this would not require filtration. To the amazement of its expert engineers and the city authorities, the state agency, which by that time had sanitary control of new water projects, insisted that the new supply be filtered. Meanwhile, slow sand filters had so far given way to rapid filters that the latter were built for the new supply. In epitome, Albany is typical of American cities that persisted in using polluted water at their door, despite the necessity for purifying it sooner or later, instead of going to a more distant and purer source—and even so being compelled finally to use some measure of treatment.

CHAPTER VII

Inception and Widespread Adoption of Rapid Filtration in America

While slow sand filtration was gaining a foothold here and abroad, in America rapid filtration appeared as a rival. The new process was soon well in the lead, then forged ahead until adoptions of the older process virtually ceased. Meanwhile old plants of the slow sand type either had their burdens lessened by the addition of rapid or other types of filters or else were abandoned.

The rapid filter was an American product, although its basic principles and the elements of the apparatus used were anticipated abroad both in patents and practice. In America, too, patents anticipating some elements of the rapid filter were granted and in one or two instances put in use in a small way before the American rapid filter began its remarkable career.

For many years the American rapid filter was called "mechanical" because it was cleaned by mechanical means in contrast to manual labor used in cleaning slow sand filters. Then came the contrasting terms, "American" *vs.* "English," then "rapid" *vs.* "slow," with or without the adjectives "American" and "English." The chief mechanical devices employed were jets of water applied on or just below the surface; reverse-flow wash, which operated through the whole depth of the filter; and revolving sand agitators or stirrers, which loosened the media from top to bottom.

Most of the early American rapid filters operated under considerable pressure. Often this was provided by pumping from the source of supply through a closed filter tank into the distributing mains of the water works system.

Direct pumping and negative head were covered in many British and a few French patents long before the advent of the American rapid filter. An English patent issued in 1819 to Henry Tritton claimed "rarefaction or exhaustion of air" by an air pump applied to a chamber receiving filtrate from an open gravity filter. Darcy's French and English patents of 1856 included a negative head of 5 m.

FIG. 35. BIRDSILL HOLLY (1820–1894)
Patented direct pumping system, 1869, and reverse-flow wash filter, 1871. Although no filter of his design is known, in the 1880's and later, the principle of pumping water through rapid filters direct to consumers was applied in many instances
(From portrait in *District Heating*, New York, 1915)

or 16 ft., provided by carrying the filter outlet pipe to the bottom of the clear-water well.

In England, in 1791, James Peacock patented a filter with a false perforated bottom and reverse-flow wash. Robert Thom completed a filter with these elements at Greenock, Scotland, in 1827. In France, in 1835–38, Henry de Fonvielle patented a closed filter which was cleaned by opposing countercurrents of water under pressure. One was installed at a Paris hospital in 1836. Henry Darcy obtained French and English patents in 1856 on a filter similar to those of Peacock and Thom, plus horizontally directed jets of water just above the surface of the filter to prevent or lessen mud deposits and also a revolving broom to help clean the top of the filter.

Three engineers were granted noteworthy American patents in this field before 1880: Henry Flad of St. Louis, in 1867, patented a filter closely resembling Peacock's design of 1791. Several small Flad filters were installed in St. Louis, including one for a hotel. Birdsill Holly, of Lockport, N.Y., well known for his direct pumping engines, was granted a patent in 1871 for a filter provided with perforated underdrains which served also to distribute water for reverse-flow wash. J. D. Cook, of Toledo, Ohio, in 1877 patented a series of filters washed by reverse flow, with pipe connections permitting any filter to be cut out for washing while the others remained in service. This was a modification of a plant he had designed for Toledo but which was not constructed.

Mechanical sand agitators to aid in filter cleaning seem to have made their first appearance in patent claims in 1858, in a communication to the British patent office from William Clark describing a filter working under pump pressure, the filter having within it a vertical revolving shaft with prongs which "operate on the layers of sand." An American patent was granted in 1872 to Andrew J. Robinson, of Troy, N.Y., the single claim of which was for "an agitator arranged within an air-tight water filter to remove the accumulated dirt from the same by agitating the filtering material while the pressure is on." Eight years later, Daniel C. Otis, of New York City, obtained a patent on an apparatus for an upward-flow filter with reverse-flow wash aided by a sand agitator. He did not limit himself to a specific type of agitator, but showed a drawing of one composed of a series of horizontal blades attached to a rotatable central vertical shaft. The Otis patent

FIG. 36. JOHN WESLEY HYATT (1837–1920)
President, Newark Filtering Co. and Hyatt Pure Water Co., 1880–1892; took out scores of patents on filtration procedure and apparatus
(From portrait in *Cyclopedia of National Biography*)

was granted July 6, 1880, just before the advent of mechanical or rapid filtration.

The Hyatt Mechanical Filter

Four men whose names deserve to be linked at this point are Patrick Clark, John Wesley Hyatt, Col. L. H. Gardner and Isaiah Smith Hyatt.

Clark's contribution to the art of mechanical filtration was the application of vertical jets of water to aid in cleaning filter media supported on a false bottom. He built a filter at Rahway, N.J., in or just before 1880. In October 1880, he applied for a patent. In December, he, John W. Hyatt and Albert C. Westervelt incorporated the Newark Filtering Co.

John Hyatt, an inventor and manufacturer of Newark, N.J., applied for a patent February 11, 1881, on what was virtually a stack of Clark's filters, placed in a closed tank and operated each independently of the others by means of common supply, delivery and wash pipes. His application, like Clark's, was granted on June 21, 1881, and assigned to the Newark Filtering Co. On the same day, Hyatt obtained a patent in England.

Col. L. H. Gardner, Superintendent of the New Orleans Water Co., after making small-scale experiments on coagulation at New Orleans, was convinced that it was more efficacious than filtration for the clarification of muddy water (1).

Isaiah Smith Hyatt, older brother of John, while acting as sales agent for the Newark Filtering Co., was baffled in attempts to clarify Mississippi River water for a New Orleans industrial plant. Colonel Gardner suggested using a coagulant. This was a success (1). Isaiah Hyatt obtained on February 19, 1884, a patent on simultaneous coagulation-filtration. Although unsound in principle, it largely dominated mechanical filtration for many years.

The first Clark-Hyatt filters are reported as having been "introduced" to municipal supplies in the latter part of 1881 on a new water supply for Frankfurt am Main, Germany, but they may have been temporary (2). The two earliest, and for about three years the only, Hyatt plants in the United States were installed in 1882 at Somerville, N.J., and Newport, R.I. Priority is given to Somerville because the plant there is the better authenticated of the two and because the Clark-Hyatt filters of 1882 were displaced in 1885 by John Hyatt's

Fig. 37. Isaiah Smith Hyatt (1835–1885)
Patented simultaneous coagulation and filtration in 1884; method first applied to municipal water supply in 1885 by Somerville & Raritan Water Co., New Jersey
(From photograph supplied by Ralph W. Hyatt, son of John W. Hyatt)

sand-transfer wash filters combined with Isaiah Hyatt's coagulation-filtration system. Before the close of 1882 a Hyatt filter had been installed for an urban water works in France (3).

Thus in 1880–85 did four men join in the evolution of mechanical or rapid filtration. Clark soon faded out of the picture. Gardner entered it only by suggesting to Isaiah Hyatt the use of a coagulant, and Isaiah Hyatt, still a young man, died in March 1885. John Hyatt was then alone. Already he had taken out 20 filter patents while only two were granted to his older brother. By the close of 1889, John had obtained about 50 patents. Scattered grants in the 1890's brought his record above 60. Most notable of all these were three on washing systems, including sectional wash; several on strainers for underdrain systems; and two on aeration, primarily in connection with filtration. The Hyatt aeration patents, like those granted to Professor Albert R. Leeds a little earlier, were of little practical importance, but they marked an era in water purification during which stress was laid on the removal of organic matter.

The advent of the mechanical filter in the field of municipal water supply has long been dated 1885 and placed at Somerville, N.J., both the earliest filters at Somerville and their forerunner at Rahway being ignored. This has been partly due * to regarding coagulation as an element rather than an adjunct of mechanical filtration.

Coagulation and Mechanical Filtration.—No coagulant was used in the various antecedents of the American rapid filter; nor by the Hyatts in their first years; nor at the start by Warren, one of the most important early rivals of the Hyatts; nor in England with many if not most of the early mechanical filters; nor with many of the later ones up to the present day. Here in America, Montreal affords a notable instance of mechanical filtration without coagulation.

Alum was not even named by Isaiah Hyatt in his coagulation-filtration patent. He had "successfully purified the water of the Mississippi River at New Orleans," he said, by using "perchloride of iron," but claimed use of "any other suitable agent which is capable of coagulating the impurities of the liquid and preventing their passage through

---

* A considerable degree of responsibility for this rests on me. In compiling the various editions of *The Manual of American Water Works* (1888–1897), on which various writers have drawn, I overlooked a brief description of Clark's Rahway filter in a United States Census report (4). I also overlooked a brief contemporary mention of the Somerville filter of 1882 (5).—*M. N. B.*

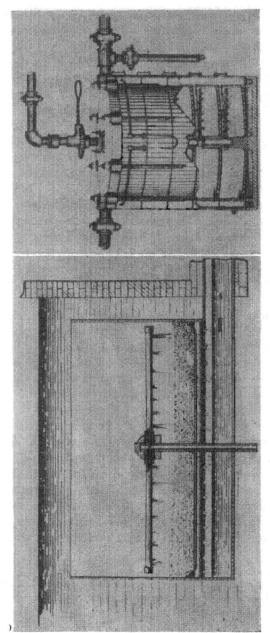

Fig. 38. Evolution of Rapid Filter by Clark and Hyatt

Patrick Clark suspended shallow filter in river and provided surface-jet wash from perforated revolving arm; loosened dirt was swept downstream; filter installed about 1880 at Rahway, N.J., Water Works
(From U.S. patent drawing, June 21, 1881)

John W. Hyatt superimposed several Clark filters in closed tank and serviced them with common pipe system; method first applied in 1882 by Somerville & Raritan Water Co., New Jersey
(From "The Multifold Water Filter," *Eng. News*, January 1, 1882)

the filter bed." A patent granted to Isaiah Hyatt on the same day as the one under consideration claimed a filter composed of some "inert material and metallic iron in comminuted form thoroughly commingled" and stated that he had "used with good results 1 part to from 15 to 20 parts sand or analogous material." In place of sand either quartz or sulfate of baryta might be used. For liquids con-

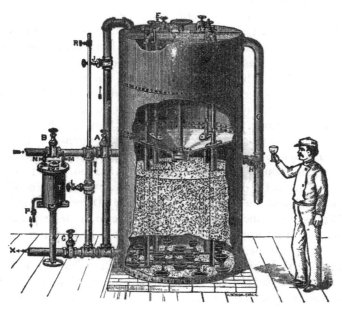

FIG. 39. HYATT SAND-TRANSFER WASH FILTER

One or more pipes led from near bottom of filter to top of closed upper chamber; by opening valves at upper end of pipes and introducing water under pressure through the underdrains, the sand and its load of dirt were transferred to the top of the upper chamber, from which dirty water was wasted and clean sand returned to the lower chamber

(From Geyelin's "Analysis of Mechanical Filter Actions," *Proc. A.W.W.A.*, 1889)

taining a high percentage of impurities an increased ratio of iron would be necessary. So far as is known, this combination of filtering material was never employed.

Evidence that John Hyatt was not committed to alum in the latter part of 1885 is afforded by his promotion efforts at Albany, N.Y. In an informal proposal August 26, 1885, for installing a 20-mgd. filter plant, the Newark Filtering Co. said its intention was to use "our new method of purification by the use of metallic iron, which method has just been perfected by our Mr. Hyatt, and obviates the necessity of using alum." A demonstration plant at the company's factory in Newark would be ready for inspection "next week." The filters would be 30 ft. in diameter and work under pressure. How the comminuted iron was to be used was not stated (7). Commingling it with filter sand may or may not have been proposed, as it was in a second Isaiah Hyatt patent of February 19, 1884. A little later in the same year, John Hyatt exhibited to the Albany special water commission a system of filtration in which iron and lime were substituted for alum and the water was to be aerated by forced air (8). This he was willing to install at Albany. As to bacteria, the commission had not found in practical operation any system which would "satisfactorily destroy such life in large quantities except at enormous expense." But Hyatt had "shown to the commission plans by which he claims such destruction may be effected at reasonable expense" (7).

Hyatt's European Campaign.—German and French sources afford information not found elsewhere regarding not only Hyatt's promotional campaign in Europe but also his earliest American installations, both industrial and municipal. Henry Gill, Director of the Berlin Water Works, published late in 1881 a German description of an American filter that appeared before the Clark-Hyatt filter had been described in any American journal (2).* His report mentioned a notice published November 24, 1881, to the effect that the mayor and council of Berlin had asked a commission charged with building a sand filter at the projected Lake Tegel works to consider whether it "could recommend that the new American filter could advantageously be substituted for the customary wall-inclosed sand filter which is being proposed." The American filter, he said, had been brought to Europe by Amassa Mason. After an unsuccessful attempt to introduce it in England it "was introduced . . . for the first time in Frankfurt am Main" to purify the water of the river when the spring water supply of the city became inadequate. The manufacture of the filter, after a German patent had been obtained, was given over to a

* Simultaneous articles appeared in *Engineering News* and in *The Scientific American* early in January 1882 (9, 10). Each contains little if anything not given in John Hyatt's patent of June 21, 1881, except that the *Scientific American* article credits Patrick Clark, of Rahway, N.J., with the invention of the filter in its "original form," adding that it "has been brought to its present state of perfection by Mr. John W. Hyatt, a prominent inventor of Newark, N.J."

Karlsrühe manufacturer, Herr Ruhl, who formerly lived in Berlin and had his apparatus on exhibit there. The article describes in detail the original Clark-Hyatt multifold filter with surface-jet wash. The Frankfurt am Main wash, however, was modified to force the revolving perforated arms down through the sand—a possibility suggested by Hyatt in one of his early patents.

Gill damns the filter with faint praise by saying that for factory use, where muddy river or sea water must be freed from gross impurities, the American filter had "a not too inconsiderable worth." The chief superiority claimed for the filter was speed. For "a city population," Gill declares, "water cannot be sufficiently freed of suspended particles when it is filtered quickly." He cites the typical English slow sand filter as the model to go by, including its slow rate of filtration, assumes that the American filter in question would have to be operated at the same rate, rules out stacking the filters because that would require extra pumping lift, and concludes that for the proposed capacity of 17.4 mgd. at the projected works, fourteen buildings, each 478 ft. square, all heated in winter, would be required. Because of their design and the necessity for heated housing Gill puts the life of the American filter at ten to fifteen years, against 100 years for "the common walled sand filter." To operate them, he says, 56 men would be required against ten for slow sand filters. The only data presented on actual operating experience tells that to wash the sand, in place, 20 per cent of filtered water was found necessary. Considering all the drawbacks, Gill concludes that the American filter had no advantage in either capital or operating cost. Therefore it was not recommended for the Berlin water works (2).

Professor William Ripley Nichols, in an abstract of a later German version of Gill's report, states his belief that the report was on a Hyatt filter (11). This was confirmed soon afterwards in a letter from the Newark Filtering Co. (12). The letter also states that 20 per cent of wash water was not required, American practice showing that 2 per cent or less was enough and that the depth of sand in the Berlin filter was 10 in. instead of 6 in., as stated by Gill. More significant was the company's statement that it was in receipt of an order for a filter with a capacity of 1.5 mgd. to be delivered at the Berlin water works. It may be added that in 1883 slow sand filters were built for the Lake Tegel supply of Berlin (13).

Early Hyatt Filters in Belgium and France.—An article published October 1, 1882, in *Le Génie Civil* (3) stated that Messrs. Hyatt and Rapp, "American engineers," had taken out a French patent on a filter and confided its construction to M. Rikkers of St. Denis. Clark-Hyatt filters had been established in a distillery at Antwerp, Belgium, and on an urban water works in France. Hyatt, instead of Clark, was credited for the filter placed in a stream for the water supply of Rahway, N.J. Hyatt filters, the article said, had been in use a year on the municipal water supplies of Newport, R.I., and Somerville, N.J., and on industrial supplies for a paper mill at Rochester, N.Y., aniline works at Albany, N.Y., and at the Warmouth Sugar Refinery in New Orleans. By lumping all these separate installations together, the time during which the Newport and Somerville plants had been in use was exaggerated.

Hyatt Filters at Newport, R.I.—Verification of the mention of Hyatt filters at Newport in *Le Génie Civil* of October 1, 1882, has been made through the recollections of Bradford Norman, son of George H. Norman, founder of the Newport water works (14). The younger Norman stated that in 1882 a filter was in use consisting of [rectangular] cast-iron sections bolted together and provided with sight ports. The filtering media were sand and mica. The underdrains were formed by winding wire around Maltese-cross-shaped iron castings. Reverse-flow wash was used, without separate sand agitators. After about six years the filter was abandoned because it was so corroded and clogged that it would not function.

Before the Hyatt filter of 1882 was installed, George H. Norman, then owner of the Newport water works, built in succession filters of his own design, consisting of one size of sand, confined in timber cribwork with brick coping, the latter extending slightly above the ground surface. Difficulty in cleaning led to their abandonment, Harold E. Watson, the present commissioner, believes.

Clark's Rahway Filter.—The only known description of Clark's filter at Rahway reads: "Clark filter, 16 ft. sq.; sand 6 in. deep, on fine wire cloth; cleaned once in 24 hours." The supply was the north branch of the Rahway River. The water was characterized as "generally good; in spring polluted by refuse of felt factory above, also dead vegetable matter" (4). The water works were built in 1871–72 with George H. Bailey of Newark as engineer.

In 1876, Patrick Clark, then city engineer of Rahway, reported to the water board that he visited the pumping station on March 27 under directions to "suggest some remedy for the almost constant turbid condition of the water delivered to the consumers." The original plan of the works, he stated, was unfortunate because "the filtering basin built along the river bank," on the assumption that it would yield spring water, actually supplied, in large part, water from the river, rendered turbid by every rain. He recommended construction of "a large settling basin and filtering apparatus." The following summer, Bailey, engineer for the original works, was about to make a report and estimates on filtration when the board decided to give up the project. On the succeeding May 7, the board ordered on file a communication from H. R. Worthington "relating to J. D. Cook's plan of filtration" (see below). On May 8, 1876, the board appointed Clark as its Chief Engineer. On October 21, the minutes of the board noted his resignation, with permission to let his filter remain at the works. C. W. Ludlow, superintendent in 1938 (15), could not learn when the filter was installed and abandoned but gave as the recollection of an assistant engineer, who was at the pumping station shortly after the filter was built, that "the filter was inadequate, did not serve its purpose and a short time after installation was discontinued."

From the scanty data available, it is assumed that the Clark filter was put in use early in 1880 or late in 1879 and used from 12 to 18 months. As its area was 256 sq.ft. and the water consumption in 1880 was 0.394 mgd. (16), it worked at the rate of about 64 mgd. an acre. Its importance lies in the fact that the Clark and Hyatt patents of June 21, 1881, were the basis of the American mechanical filter.

Hyatt Filters With Coagulating Apparatus for Somerville and Raritan, N.J.—The Somerville Water Co., with George H. Pierson as engineer, built water works in 1881–82 to supply Raritan and Somerville, N.J. The supply was taken from the Rahway River, a stream carrying suspended matter from a red-shale drainage area (5, 6).

Company minutes show tentative acceptance, in August 1881, of an offer from the Newark Filtering Co. to furnish four pressure filters (19). Although the water company began to supply consumers by May 1, it was not until May 8 that it contracted for filter connections. A description of the filters published in August 1882, from data supplied by the designing engineer, stated that there were four filters of

Fig. 40. George F. Hodkinson (1868–   )
With Newark Filtering Co., 1886–88; Secy., Hyatt Pure Water Co., 1888–92; in charge of Chicago office of New York Filter Co., 1892–93; with Western Filter Co., St. Louis, 1893–95; with O. H. Jewell Filter Co., Chicago, 1896–1900; Engr. & Mgr., Roberts Filter Mfg. Co., Philadelphia, 1904–08; Mgr., Filtration Dept., American Water Co., Philadelphia, 1908–34; head of G. F. Hodkinson Co., Philadelphia, 1934–
(From photograph supplied by Hodkinson)

the Hyatt multifold type, each 5 ft. in diameter and 8 ft. high, and that the water consumption was 0.2 mgd. (6). From this and other data it is inferred that each filter tank contained a stack of Clark-Hyatt rectangular filters, washed by surface jets and reverse flow, combined.

On April 7, 1885, the water company minutes state, a communication was received from the Newark Filtering Co., proposing to furnish new filters for $7,000. The executive committee was directed to obtain other proposals. The only one received was from the Crocker Filtering Co., of New York, dated April 29. It offered to put in two of its largest "filtering machines for $5,000." On May 23, 1885, the water company ordered "four Hyatt filters, $6\frac{1}{2}$ ft. in diameter by 15 ft. high, capable of withstanding a pressure of 150 psi. and *accompanied by suitable coagulating apparatus.* [Author's italics.] They must be guaranteed to deliver 0.5 mgd. of bright, clear and wholesome water, if washed once a day." The filters were to be set up by the filter company on foundations provided by the water company. If the filters were satisfactory to the water company on a test made "after the river had been muddy ten days" the water company would pay $6,500 for them, less $2,500 for the old filters, which were to be removed at the expense of the filtering company. The new filters appear to have been put in use in July 1885. The net result of later negotiations was that the water company paid $3,950 for the new filters and retained the old ones (19).

A report made by a representative of the New Jersey State Board of Health after a visit of inspection in 1911 states that there were then in use four Hyatt filters of "a very old type," washed by ejecting the sand by means of a water jet. The process took a half hour per filter. The Hyatt filters were replaced in 1913 by two Jewell filters. Since then three additional Jewell filters have been installed and also one filter has been provided by the American Water Softener Co. of Philadelphia (19).

Subsequent Hyatt Installations.—As far as can be determined over a half century later, the only Hyatt filters put into permanent use on municipal supplies in 1885 were at Somerville, N.J., in May, and a very small one at Tunkhannock, Pa., in August.

An advertisement, dated November 1886, listed five municipal plants as "in operation." Besides the two plants completed in 1885 these were located at Charleston, W.Va., Belleville, Ill., and Rich Hill, Mo.

All have been authenticated. An advertisement probably referring to the close of 1888 stated that Hyatt filters had been "adopted" by 30 cities and towns and thousands of manufacturers and private consumers. This seems to have been the peak year for the Hyatt filters. A compilation from all available sources shows a few plants put in use in 1889–91 and only one in January 1892. The decline after 1888 was apparently due to competition and protracted infringement litigation.

Pre-aeration was used at three of the Hyatt plants: Belleville, Ill., and Greenwich, Conn., put in use in 1887, and Long Branch, N.J., completed in 1888. Scanty data in local newspapers establish this fact for the first two places, indicating that air was sucked into the water flowing through a vertical tube placed above ground. At Long Branch a similar device was sunk deep into the earth. These aerators were in line with patents granted to John Hyatt in 1885 and 1887 (see Chap. XVI).

At Long Branch the filtering media were sand and prepared coke, 3 to 1. Wash water, admitted to the bottom of the filter through a perforated pipe, found its way to the top, from which it ran to waste (20).*

The Atlanta filters had the next to the largest capacity (3 mgd.) of any Hyatt filters installed before consolidation with the New York Filter Co. There were 12 units, 8 ft. in diameter by 13½ ft. high, in two compartments arranged for washing by the sand-transfer system. They were completed in 1887. Originally the filters were 60 in. deep, composed of 3 parts of sharp screened sand and 1 part of ¼-in. screened "coke" from locomotive boxes. In 1893, 30 in. of uniform-sized sand was substituted (when the filters were moved). "Crystal alum," applied in the influent pipes about 30 ft. ahead of the filters, was used until 1903, when a coagulating basin was completed at the new site. The strainers were copper saucers or cones, filled with 0.1-in. copper "shot" and covered with perforated plates. These filters originally treated water from a 52-acre impounding reservoir on South River, a stream draining red clay land. In 1893 they were moved to the Hemphill station on the Chattahoochee River and put alongside a battery

* The Long Branch filters were still in use (summers only) a half century later at the West End Station of the Consolidated Water Co., Long Branch, but they had been changed from pressure to gravity type and new underdrains installed, probably in 1897 when clear-water basins were added (21).

of eight horizontal filters, 8 ft. in diameter and 20 ft. long, just completed. The latter were of the Hyatt "sectional wash type," patented by John Hyatt, August 27, 1889. All these filters worked under a 17-psi. head, given by settling basins.

The first, or sand-transfer wash, filters were put into use about the middle of November 1887. They were shut down in or about 1912 but were not removed until 1933.

The horizontal sectional wash filters of 1893 were added to at intervals until the number reached 36 in 1910 and their combined capacity 18 mgd. These filters were washed by passing water at high velocity up through the filter. Three inlet valves, separately operated, theoretically divided the filter into as many sections (22). About 1935, the practice of opening all the valves at once was adopted. Renovation of this battery of 36 filters was scheduled in 1942, but only to the extent of repairing valves, rustproofing all metal surfaces and cleaning and reclassifying the 28-in. sand and 19-in. gravel strata. The filters were still turning out water of an excellent quality, bacterially and chemically. "Modern" rectangular filters of 36-mgd. capacity were put into use in 1932, giving a total capacity of 54 mgd.

The flow diagram in 1942 showed: storage reservoir giving three days' detention; dosing with sulfate of alumina; around-the-end baffled mixing chamber, 1,800-ft. travel in 45 min.; addition of ammonia and chlorine for disinfection; ten hours' sedimentation; addition of activated carbon for taste-and-odor control; filtration; lime applied to filtrate to control corrosion and red water; clear-water reservoir.

A 6-mgd. plant at Oakland, Calif., was the largest and latest known Hyatt installation before the consolidation. Twelve 0.5-mgd. units went into service in May 1891. They were still used occasionally up to July 29, 1942, wrote J. D. DeCosta, Engineer of Distribution of the Bay Cities Utilities District, but "many years ago new underdrains and egg crates * were installed. The tank shells had been in use about 50 years.

The Warren Filter

John E. Warren, agent of the paper mills of S. D. Warren & Co., Cumberland Mills, Me., planned and constructed a filter plant for the

* Egg crates is the name given to compartments in the gravel layer of the filter, formed by vertical partitions extending from the filter floor to the base of the sand. Their object is to prevent the wash water from taking lateral paths through the gravel, disrupting it and turning it over (26).

Fig. 41. JOHN E. WARREN (1840–1915)
Designed and installed twelve large rapid filters for paper mill of
Cumberland Manufacturing Co. in Maine
(From portrait supplied by Joseph A. Warren, son of the inventor)

mills in 1884. It was for some years the largest mechanical filter plant in existence, with a rated capacity of 12 mgd. It was put in use after the Clark filter at Rahway and the first Hyatt filter for the Somerville Water Co., but before the second Hyatt plant for that company. A diary (27) kept by Warren notes the inception and completion of the filters in 1884 but no other mention of them appears until 1889:

May 12, 1884: We are making no additions to the mills this summer except to provide a general water supply and filtering arrangement which the fouling of the water by the Saccarappa manufacturers renders necessary. This will be accomplished by a device of my own which I am confident will be most effective.

Sept. 9, 1884: My scheme for filtering which has been the principal job of outside work is complete and bids fair to be a complete success although we have hardly given it a fair test as yet.

Jan. 2, 1889: In my record of 1884 I mentioned a system of filters which I had then constructed here and which proved an eminent success, so much so that patents were secured and plants have been constructed by us elsewhere. . . . the Cumberland Manufacturing Co. has been organized. . . . , a Mr. Nye as general manager (27).*

The filters of 1884 at the Cumberland paper mills were of the gravity open wooden tank type, in contrast with the more commonly used closed metal tanks. They were 20 in number, each 8 ft. in diameter. In a description published in 1887 (29) their capacity was given as 12 mgd., but some years later measurements showed 7 to 8 mgd. was being treated. These filters, wrote Joseph A. Warren in 1942, were used until about 1896 when enlargements of the mill required their removal. They were succeeded by 30 filters, 10½ ft. in diameter, which were still being used in July 1942. They are of the early type, with revolving rakes driven by a shaft and raised and lowered by a vertical shaft having a screw cut on an extension, with a nut that can be run

* Joseph A. Warren (28), of the Research Laboratory, S. D. Warren & Co., and son of John E., states that his father was connected with the paper mill from 1868 to his death in 1915. John Warren was the nephew of S. D. Warren, who began manufacturing paper in 1854 at Cumberland Mills, Me. The "Saccarappa manufacturers," mentioned in the diary, owned a cotton mill located on the Presumpscot River, about a mile above the paper mill. The fouling of the water was partly caused by fiber waste from the cotton mill. Joseph A. Warren was 14 years old when his father built the filters. He recalls seeing a pressure sand filter, probably supplied by Hyatt, at the paper mill, and thinks that Hyatt filters were tried on special water services in the mill and may have suggested to his father the building of gravity filters. Walter B. Nye was manager of the Cumberland Manufacturing Co., builder of Warren filters, until the company sold out to the New York Filter Manufacturing Co. in 1898.

in either direction. A 0.35-mil.gal. precoagulation tank with Carmichael dosing apparatus was provided for these filters. "Alum" was used for only a short time as the improvement it effected in the filter effluent was not sufficient to warrant its cost. The Warren filters have been supplemented by rectangular Norwood filters (28).

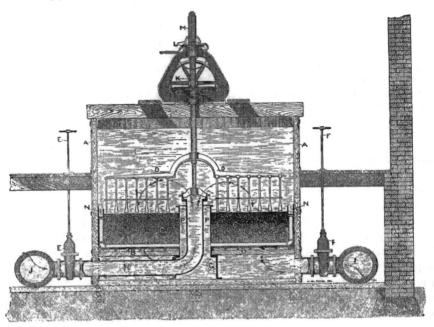

FIG. 42. SECTION OF WARREN GRAVITY FILTER
Revolving rakes forced down into filter media to aid reverse-flow wash
(From Geyelin's "Analysis of Mechanical Filter Actions," *Proc. A.W.W.A.*, 1889)

John E. Warren's first filter patent, dated December 22, 1885, was for the combination of a filter with a false bottom, a toothed sand agitator, vertically adjustable, and a plunger for rotating and lowering the agitator. A second patent was granted on March 12, 1888, and a third on May 28, 1889, both covering details of the typical Warren filter. The third patent claimed the combination of a filter, a central well extending up through and also below the filter, the filter being closed at the bottom to receive sediment and provided with a discharge outlet and an inlet extending into the well. There were a peripheral gutter connected with the well and a revolving rake agitator. A later adjunct to the Warren filter was an alum dosing apparatus, patented in July 1890, by the inventor Professor Henry Carmichael, of Malden, Mass., who assigned it to the company. It consisted of a six-arm pump made up of curved tubes attached to a hollow hub, driven by a propeller located in the raw-water supply main. The speed of the pump varied with the velocity and therefore the volume of the raw water. The dosing could be further regulated by varying the level of the alum solution in the dosing tank, so that the pump would pick up more or less alum as desired.

Little publicity was given to the Warren filter until 1887 (29). Early in 1894, descriptions of nine Warren plants built for municipal service were published (30). The first two of these were put into use at Augusta and Brunswick, Me., in 1887, the third in 1890, at Oshkosh, Wis. Not until the fourth plant was installed in December 1892, at Macon, Ga., was a coagulant used. It was employed on the four latest plants described in the article. Four of the five plants using a coagulant, and possibly the fifth, included small precoagulation tanks. A 12-mgd. settling reservoir was placed ahead of the coagulation tank at Athens, Ga. The rights in the Warren filter were acquired by the New York Filter Manufacturing Co. on or about April 1, 1898.

The National Filter

The National Water Purifying Co. of New York City was incorporated August 20, 1886, to promote a filter patented by William M. Deutsch, who had been a salesman for the Newark Filtering Co., builder of Hyatt filters. Albert R. Leeds, Professor of Chemistry, Stevens Institute of Technology, Hoboken, N.J., transferred rights in his water aeration patents to the company, became its professional adviser, and invested and lost more and more of his professional earnings in the stock of the company (31).

All told, Deutsch was granted ten American patents in 1886–90 and was assignee of two by Claude Deutsch. British patents of December 28, 1886, and October 25, 1892, were virtually duplicates of American patents of the same dates.

Deutsch obtained his first patent (No. 355,004, December 28, 1886) very soon after quitting as salesman of Hyatt filters. Its most significant claims were for horizontal perforated wash pipes at two levels, one just beneath the surface and another at the bottom of the filter.

the latter serving also as underdrains; and for sectional wash made possible by diametrical partitions carried from the bottom to the surface of the filter. The underdrains consisted of two eccentric perforated pipes, the space between which was filled by coarse granular material. As far as is known, such underdrains were never used.

A patent application on September 4, 1886, was held up nearly six years (granted June 7, 1892; No. 476,737). Its only claim was for a filter divided into a number of compartments, with a separate discharge pipe for each, to provide for separate or sectional washing.*

Deutsch's patent of March 29, 1892 (applied for October 8, 1887) was for "a method of and apparatus for cleaning and purifying filter

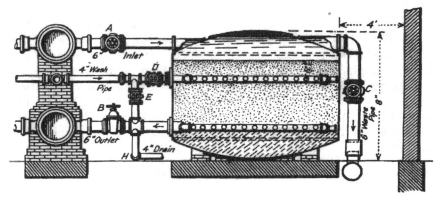

FIG. 43. NATIONAL FILTER AT TERRE HAUTE, IND.
Double reverse-flow wash: *A*, for top layer; *B*, for entire filter through underdrains
(From *Eng. News*, February 7, 1891)

beds" by aeration. The preferable method was to admit air to the underdrains from an air chamber placed above the filter tank, thus filling with air the space above the filter. Then, by turning on water above the filter the air would be compressed and forced down through the filter. A part of the air would "remain in the interstices, aerating

* On November 30, 1888, an examiner in the U.S. Patent Office reporting on the interference claim by Deutsch, ruled that John W. Hyatt was first in the field with sectional wash (32). The examiner stated that while Deutsch was a salesman for the Hyatt filter he had access to Hyatt's drawings for a sectional wash filter. Hyatt filed an application for a sectional wash filter patent September 7, 1885. His application was renewed May 15, 1886, but the patent was not granted until August 27, 1889 (No. 409,970).

and purifying the bed." As an alternative, air might be supplied from a "source independent of the bed, as from a compressor." No evidence that this scheme was ever used has been found.

Air-and-water wash for cleaning filters was the subject of an apparatus patent granted to Deutsch January 26, 1896 (No. 553,641; application October 21, 1891). The claims were for cleaning a filter by compressing air within the filter and strainer by the pressure of a body of water, then allowing the compressed air and water to escape suddenly. An apparatus patent dated May 29, 1900, covered the combination of a large number of water pipes, a superimposed filtering bed and a multiplicity of air-supply pipes having air-distributing nozzles.

Sedimentation, coagulation and filtration were the subject of three Deutsch patents, all dated June 19, 1900. Application for the third of these was filed by W. M. Deutsch on October 27, 1897, soon after Fuller completed his Louisville experiments, demonstrating, as experience had shown, the necessity of clarifying highly turbid water before filtration. The patent was for a combination of settling basins, filters, means for forcing a coagulant into settled water before its admission to the bottom of the filters, connections for washing the filters, in sections, by reverse flow. The other two patents were granted to Claude Deutsch. One was similar to the W. M. Deutsch patent; the other was on a process for passing water to be filtered and a coagulant into a settling tank, introducing carbonate of lime and allowing the water to settle.

The earliest publicity for the National filter that has been found is a small advertisement in Croes' *Statistical Tables of American Water Works* for 1887. It claimed surface washing as "a new principle in filtration." Its declaration, "No Infection From Cholera and Typhoid," seems to have been the first such claim for the American mechanical filter.

A short article published early in 1887 stated that both the National filter and Professor Albert R. Leeds' system of aeration were controlled by the National Water Purifying Co. (33). Charles B. Brush, of Hoboken, N.J., in a paper read before the New England Water Works Association on June 16, 1887, said: "Mr. Deutsch and Dr. Leeds are the inventors of the National System." Brush mentioned litigation between the Newark Filtering Co. (Hyatt) and the National Water Purifying Co. This and other infringement litigation, apparently on coagulation, was also noted in an advertisement of the Hyatt Pure Water Co. appearing in *The Manual of American Water Works* for 1888 thus:

A decree has been obtained against John E. Johnson, representing the Continental Filter Co. manufacturing the Roeske filter. We are now in the United States Courts prosecuting 'The National Co.' of New York, Wm. M. Deutsch, Albert R. Leeds, the 'Jewell Pure Water Co.' of Chicago, Ill., and others.

More spicy was the following:

We are the pioneers in rapid and perfect chemico-mechanical water purification and maintain supremacy in spite of illicit *imitators* who brazenly pirate our property, copying *our processes and mechanical devices,* attempting to evade our patents, adopting and advertising as advantageous, features which we have superseded by improvements—surface washing in particular.

Thus began the litigation between the early filter companies that contributed toward the various consolidations.

Four National filter plants were put in use in 1887: Chattanooga, Tenn., and Champaign, Ill., in June; Exeter, N. H., in August; and Winnipeg, Manitoba, in December.

Champaign, Ill.—At Champaign no record or tradition of a National filter plant could be found although inquiries were sent in 1940–41 to the water companies, University of Illinois professors and Champaign and Urbana libraries. Evidence that such a plant was in use is found though, in a trade catalog of the New York Filter Co., published in 1893. The catalog contains a testimonial from S. L. Nelson, Superintendent, Union Water Co., dated January 14, 1888. It states that a National filter was put in operation in June 1887, and had been working satisfactorily for six months. It combined aeration, lime precipitation and filtration. Besides rendering the water "clear and bright, free from odor and vegetable matter" it also "removes hardness." The source of supply was an abandoned coal prospecting shaft, but it was soon given up for deep wells.

Chattanooga, Tenn.—At Chattanooga, ten vertical pressure filters were completed in June 1887. Two smaller ones were added late in the year, presumably to meet the original contract guarantee of 3 mgd. Additions in 1887–91 brought the total to 20, with a nominal capacity of 5 mgd. Twenty-four Jewell pressure filters were added in 1892 and two in 1894, with a combined capacity of about 3 mgd. Water from the Tennessee River, highly turbid at times, was pumped through the filters into the distribution system. In 1897 a settling basin and a clear-water basin were constructed after which the National and Jewell filters, although still in closed tanks, were operated under gravity head. Twenty 1-mgd. open concrete gravity filters were built in 1914, 1917, 1925, and 1930. In 1911 a second and in 1925 a third settling basin were added. In 1940 the old National and Jewell filters were replaced, after a service of 50 years, by six open gravity filters with a capacity of 2 mgd. each, bringing to 32 mgd. the capacity of the existing filters. A steel mixing tank, supplying basins No. 1 and 2, was erected early in 1940.*

Terre Haute, Ind.—At Terre Haute, an early installation of National pressure filters is notable because of the high head under which it worked for 24 years and because the filter tanks were still in use after over 50 years of service. The Terre Haute Water Works Co. completed works in 1873 under a charter or franchise requiring "filtration." The supply was taken from the Wabash River. A filter crib was provided but was submerged a large part of the year, sometimes to a depth of 25 ft., and in general was unsatisfactory. In September 1889, after a change in ownership, a contract was let for 12 National high-pressure filters, 10 ft. in diameter and 7 ft. high, including the slightly dished heads. They were put into use in July 1890, but because they did not come up to the contract requirement of 3 mgd., a thirteenth filter was added in 1891. It was horizontal, 8 ft. in diameter and 20 ft. long.

Fortunately, a detailed description of the earliest filters, written just after they were put in use, is available (35). Some of the tanks, the superintendent, L. L. Williamson, wrote (36), had "withstood a test pressure as high as 208 psi., without leakage . . . borne entirely by the flange of the heads." The plates, wrote the manager, W. H. Durbin, in 1940 (37), were $\frac{11}{16}$ in. thick for the heads and $\frac{3}{8}$ in. for the sides, all of open-hearth steel. Because under fire pressure "several of the dished heads were ruptured at various times," the new settling basins were changed to a few feet of gravity head in 1924. W. E. Taylor, Chief Engineer of the works, 1908–40, states (38) that the working pressure ranged from 60 psi. domestic to 120 psi. fire. Major and minor ruptures combined probably numbered several a year. Bracings were added at unknown dates, but did not prevent ruptures,

---

* Chattanooga data are based on a contemporary description of the earliest filters (33) and on slightly later descriptions (34), supplemented and brought up to 1940 by A. F. Pozelius, Manager, City Water Co. of Chattanooga, and L. E. Wickersham, Sanitary Engineer, American Water Works & Electric Co., New York City.

which usually occurred under fire pressure. Deformation by "breathing" about ¼ in. at the center of the heads probably started incipient cracks at the inner part of the flanging radius. This also applied, Taylor adds, to the earlier of the horizontal 8 x 20-ft. Jewell filters installed in 1900. The same type and size of Jewell filters installed later in 1900, but having "a much easier and longer flanging radius," did not rupture. The thickness of the plates of the Jewell tanks, states Durbin, was ¾ in. for the heads and ⅝ in. for the sides.

A double upward-flow washing system with water under pump pressure, wrote Williamson, was used for the National filters: (a) from perforated pipes located 6 in. below the surface of the filters, "thus breaking up the surface layer of the bed where most of the sediment accumulated"; (b) from the underdrains, "lifting the whole mass of sand and washing out the finer particles of sediment which have lodged in the lower part of the bed." The time required for washing was in the first case 3 to 5 min. and in the latter 5 to 7 min. Dirty wash water was discharged from above the filter through a single pipe. The amount of water filtered averaged 2.75 mgd., against a rated capacity of 4 mgd. Loss of head in the filters ranged from 10 to 30 psi. The filters were washed from one to six times a day using filtered water for the pressure filters and unfiltered water for the gravity filters, the total wash water required amounting to 3 to 30 per cent of the pumpage. The filtering material was sea sand and coke. "Alum" was used as a coagulant.

The underdrainage system of the National filters at Terre Haute consisted of semicircular pipe, embedded in the concrete bottom of the filters, with strainers tapped in the pipe. After six or seven years these were replaced by a center manifold and 1¼-in. laterals, with Jewell strainers tapped in every 6 in. In 1924, mushroom strainers were substituted. The wash-water outlet at the side of each tank, originally used, has been replaced by a central funnel. The filtering media of sand and coke, mixed, have been replaced by 30 in. of sand on 9 in. of gravel.

Two Jewell open gravity filters with a capacity of 0.5 mgd. each were installed in 1891, delivering their filtrate to a 0.775-mil.gal. clearwater basin for fire use. In 1900, nine high-pressure horizontal Jewell filters were added. Each was 8 x 20 ft. in plan. Until 1902, there was no presedimentation. Even then, only a 0.72-mil.gal. settling basin was provided.

The thirteen National and nine Jewell pressure filters and the small settling basin served until 1924, when they were supplemented by four concrete open gravity filters with a total rated capacity of 4 mgd. and a 2.5-mil.gal. settling reservoir. At that time all the old pressure filters were changed to operate under the low head of the new settling reservoir, thus becoming gravity filters, although the tanks remained closed. The rated capacity of the plant in 1942 was 12 mgd. The population of Terre Haute increased from 30,217 when the earliest filters were installed in 1890, to 62,693 in 1940.

Chlorination with hypochlorite of lime was introduced in 1911. In 1916 [should be 1913?], says Taylor (38), an "electrolytic chlorine double cell was installed by the Chloride Process Co., under the supervision of Omar H. Jewell of Chicago. Difficulty of control and dosage led to the abandonment of the apparatus after a few years." In 1916 liquid chlorine was introduced.

New Orleans.—New Orleans undertook, a half century ago, one of the boldest and most disastrous attempts ever made to filter the water supply of a city. Cocksure of the efficiency of its filters, disregarding a pending lawsuit for infringement of Isaiah Smith Hyatt's patent on simultaneous coagulation and filtration without presedimentation, and despite the advice of its noted chemist and heavy stockholder, Albert R. Leeds (31), the National Water Purifying Co. contracted to build a larger mechanical filtration plant than had yet been constructed. It went so far as to guarantee a constant supply of clear water at a stipulated cost for coagulant.*

The contract was signed in 1891. The plant was put in operation in March 1893. After tests the New Orleans Water Works Co. refused to accept the plant on the ground of non-fulfillment of guarantee. Its refusal was upheld by the state courts. This left the company with a 14-mgd. filter plant to dispose of piecemeal elsewhere if and when it could, together with a heavy loss of money and prestige.

Due very likely to their dismal failure, data on these filters are scanty. The plant was featured in the first trade catalog of the New York Filter Co., issued in 1893, a few months after the consolidation of the National, Hyatt and American filter companies. The cata-

---

* The officials of the water company, stated General Superintendent George G. Earl, in a report to the New Orleans Sewerage and Water Board, dated July 6, 1900, "advised the filter people to put up one filter instead of thirty and find out first what could be accomplished" (39).

FIG. 44. ALBERT R. LEEDS (1843–1902)
Chemist and stockholder, National Water Purifying Co.; patentee of apparatus for treating water by electrolysis and by forced aeration
(From portrait in Leeds obituary, *Stevens Indicator*, April 1902)

log contained an imposing view of the 30 filters at New Orleans, each 8 ft. in diameter and 30 ft. long, beneath which appeared: "Filters the Entire Supply of the City." In the technical press the first mention of these filters was a brief notice that the company had been defeated in its suit to collect $134,500 claimed under the contract, followed by a statement that the decision had been appealed (40).*

When George G. Earl became Chief Engineer of the Sewerage and Water Board in 1892, the water company

> . . . had its pressure filter plant under contract, with guarantees as to performance which proved utterly impossible to meet. . . . The water as delivered was not fit for any use except lawn sprinkling. The water mains were so clogged with mud that the water delivered was often muddier than the river itself. The average turbidity of the river water was around 625 ppm. and the attempt to purify this water without preliminary sedimentation, by the use of pressure filters, in line between the pumps and the distribution system, was predoomed to failure (42).

F. W. Capellen, City Engineer of Minneapolis, has reported filter operating conditions found by him on a visit to New Orleans in 1894 (43). The filter contract called for the delivery of 14 mgd. of "crystal clear water, free from opalescent hue. Loss of head in the filters was not to exceed 15 lb. when the water was dirty nor 6 lb. when clean; and the filter company was permitted to use not to exceed $8 worth of coagulant per million gallons per 24 hours." During his stay in New Orleans, "the filters had to be washed every four hours, due to a large amount of fibrous sediment in the river water." Summing up, Capellen said: "The filter company has . . . undertaken a task that has no parallel anywhere, either here or in Europe, and it would appear that the water should be settled . . . and then filtered. I understand that the plant has been put in at a loss."

Professor Albert R. Leeds, in his now-it-can-be-told confession before the American Water Works Association in 1896, soon after the filter contract was voided by the Louisiana courts, drew a picture of the New Orleans tragedy (31). After reviewing events leading up to the forma-

---

* Nothing further regarding this lawsuit has been found in the technical press. G. A. Llambias, Special Counsel, Sewerage and Water Board of New Orleans, at the request of A. B. Wood, General Superintendent, states that the suit was brought to obtain payment for the amount claimed under a contract to furnish a filter plant. The defense was that two tests proved the plant to be entirely insufficient and unfit for delivering the guaranteed quantity and quality of water. A plea for a part of the contract price was denied (41).

tion of the National Water Purifying Co., including the taking over by it of his aeration patents and his professional savings, and saying that he was the "largest holder" of its $1,500,000 stock, he continued:

. . . the company, against professional advice, against everything I could say, undertook to purify the waters of the Mississippi at New Orleans. . . . That company put in a very great plant there, that cost $130,000. It undertook by mechanical filtration to take the water directly from the Mississippi River at all seasons of the year, at all times of flood; with the Red River and the Arkansas River pouring different sorts of constituents into it; it undertook to . . . [pump water through the filters to the water mains under a varying domestic and fire demand] and to supply—what do you suppose?

I think you will say that they ought to have been satisfied if they could have supplied wholesome water. But, gentlemen, the contract signed by the filter company against the professional protest of a person who had repeatedly analyzed the Mississippi water, was, under all these extravagant engineering conditions, to supply at all times, clear and wholesome water, free from opalescence; in other words, the water of the Mississippi river, containing sometimes 2,000 parts [per million] of solid matter and full of filth at all times, . . . was to be made as pure as the most crystal spring water, the penalty for failure to do which was the loss of their money.

. . . an engineer spent a year there testing [the plant] under these conditions . . . finally, the filter company brought the water company into court. The case was carried to the Court of Appeals and decided against the filter company, for the reason that, singularly enough, [although] a great deal of filtered water was clear and free from opalescence, there was also some that was not. That adverse decision came during a financial and commercial crisis, and the company was broken to pieces, . . . (31).

Stunned by the failure of the filter plant, the water company made no further attempt to filter its water supply. In 1891 the number of service connections not in use exceeded those in service. In 1896 there were only 4,800 connections on 118 miles of pipe. Meters could not be used. For decades rainwater cisterns were depended on for domestic supply. The population of the city was 242,000 in 1890 and 287,000 in 1900.

After the city took over the water works from the company, an experimental water purification plant was installed and operated under the immediate charge of Robert Spurr Weston. In 1909 a coagulation, sedimentation and filtration plant, with a normal capacity of 40 mgd., was put in operation. The filters were of the Jewell type, using perforated plates between ridge blocks. Equipment was supplied by the Roberts Filter Manufacturing Co., of Philadelphia. The chemicals used were lime and sulfate of iron, thus clarifying and softening the water as well as removing bacteria. Chlorination with hypochlorite of lime was used from May 1915 to July 1916, when it was supplanted by liquid chlorine. Enlargements in the plant completed in 1932 brought its normal capacity to 112 mgd., which it still was in 1942. Until 1936, the lime and sulfate of iron were applied simultaneously. Since then split treatment has been practiced. Recarbonation was used during most of the year 1941–42.*

The American Filter Company

The American Filter Co. of Chicago was the third, chronologically, of the concerns that united to form the New York Filter Manufacturing Co. early in 1892. A former Hyatt representative was a patentee of the American company's filter and a stockholder in the company. This was Ernest H. Riddell, once of Riddell & Kerrick, agents of the Newark Filtering Co., at Chicago and Cincinnati.† The other patentee, also a stockholder, was Chester B. Davis, of Chicago.

In early advertisements, the American Filter Co. offered to supply filter plants for any purpose and to furnish plans and specifications and to undertake construction of the whole or any part of water works systems. From other sources of information it appears that of ten water works for which Chester B. Davis was engineer up to 1889 the only three having filters had American Filter Co. installations. These were at Elgin and Rogers Park, Ill., and Mount Clemens, Mich. When the Elgin filters were under criticism in 1900, the mayor stated that the engineer for the original water works was financially interested in the filters then installed.

* The information in this paragraph was brought up to September 28, 1942, by A. B. Wood, General Superintendent, New Orleans Sewerage and Water Board. For details of changes completed in 1936, see article by Carl C. Friedrichs Jr., in the April 1936 *Journal American Water Works Association* (44).

† G. F. Hodkinson, of Philadelphia, who joined the Hyatt staff in 1886 and was secretary of the Hyatt Pure Water Co. in 1892 when the New York Filter Co. merger occurred, says the American Filter Co. was organized January 25, 1888, with a capital stock of $100,000, of which Riddell and Chester B. Davis received $25,000 each for their patent rights in the United States and Canada. Riddell went to England about 1894 to promote the sale of the patent rights there and became associated with Leonard H. Bristowe (53). Testimony to Riddell's success in England was supplied by C. L. Simpson, grandson of James Simpson of London, who stated in 1909 that "about 20 years ago he used an American rapid filter called the Riddell . . . a pressure or pump-through filter"; and by Managing Director Spence, of Peter Spence & Sons, Chemical Manufacturers, Manchester, England, who wrote in 1937 that in 1895–96 "Messrs. Bell Bros. installed a number of Riddell filters."

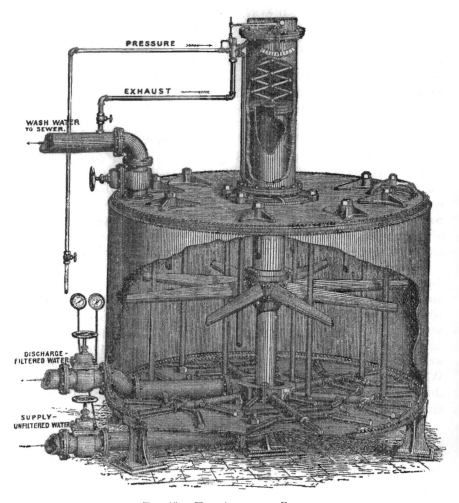

Fig. 45. The American Filter

Raw water admitted above filter through central vertical pipe to and through radial perforated arm; washed by reversing flow through filtered-water discharge pipe, vertical feed pipe and radial arms, latter being forced into media by hydraulic piston and then oscillated; when long horizontally placed cylindrical tanks used, two or more sets of washing arms employed

(From Geyelin's "Analysis of Mechanical Filter Actions," *Proc. A.W.W.A.*, 1889)

Three United States patents were granted to Davis and Riddell. Application for the first was filed December 12, 1887. The patent was granted June 11, 1889. It covered a filter in a closed vertical tank, underdrains of perforated pipe having wire closely wound around them, and a washing system composed of horizontal oscillating perforated radial arms moving up and down through the filter by means of a hydraulically operated piston working in a central vertical cylinder. Wash water was distributed through the perforated arms. A British patent, also dated June 11, 1889, was similar to the first American patent. Two United States patents dated April 14, 1891, were much the same as the first one. The washing system was the same, in principle, as one described in an early Hyatt patent.

Nine installations by the American Filter Co. have been authenticated. By dates put in use, these were: Elgin, Ill., June 1888; Sidney, Ohio, September 1888; Greenville, Texas, November 1888; Mount Clemens, Mich., and Rogers Park, Ill., March 1889; Cairo, Ill., June 1889; Streator, Ill., August 1889; Davenport, Iowa, and Little Rock, Ark., June 1891.

Elgin, Ill.—At Elgin, the first six filters were 10 ft. in diameter and height, had a total rated capacity of 1.5 mgd., were supplied by the Chicago Steam Boiler & Engine Works for about $17,000, and treated water from the Fox River. In 1893, the New York Filter Co., successor to the American Filter Co., added two 8 x 10-ft. filters, bringing the capacity to 2 mgd. The earliest bills for a coagulant, dated June 1888, were for alum. Peculiar tastes and odors in the filtrate in 1893 led to the removal of all the filter material in one of the original filters. The wooden supports of the underdrains were found to be "slightly soured." To this the tastes and odors were attributed. One by one in each filter a concrete bed was placed to support the underdrains and the depth of sand was increased. Thereafter the quality of the filtrate was satisfactory and the coagulant was cut in half. The filters were shut down in 1903 on the introduction of artesian well water. Twenty years later their use was resumed to supplement the artesian supply but shortly afterward the river suction pipe was broken by order of the State Board of Health. In 1938 a softening plant was installed.

Rogers Park, Ill.—The Rogers Park filters are of interest because they supplied water for 25 years, 1889–1914, to an area annexed to Chicago in 1893. The city took over the property of the Rogers

Park Water Co. in 1907 and operated the filters until 1914 for the benefit of such consumers as wished filtered water instead of water from the city's Lake View intake which extended 6,000 ft. into Lake Michigan, compared with 3,200 ft. for the company intake.

Streator, Ill.—At Streator, four American filters were still in use in 1940, after over a half century of service. The shells were 10 ft. in diameter and 7 ft. high, of $\frac{5}{8}$-in. plates. Twelve $2\frac{1}{2}$-in. stay rods tied the top and bottom together. Six Western pressure filters were also in use in 1940. These were horizontal, 8 ft. in diameter and 21 ft. long, with "bumped" ends and center walls dividing each filter into two compartments. Three of these were installed in about 1896 and have $\frac{1}{2}$-in. shells. Until about 1907 the American and the three earlier Western filters operated on the discharge end of the high-pressure pump and considerable sand was carried into the distribution system. Settling and clear-water basins were then installed. Since then all nine filters have worked under an 11-ft. head, but as closed low-pressure filters.

The underdrain system of both makes of filters consisted of Cook well strainers. During the last two years, wrote H. J. Adams, Superintendent of the Northern Illinois Water Corporation, on September 16, 1940, the underdrainage system in the American and three earlier Western filters has been removed, owing to corrosion and wear. It has been replaced by $1 \times \frac{3}{16}$-in. steel grating, supported on 4-in. channel irons. The 2 ft. of concrete below the filter has been removed, so the channel irons rest on the steel bottoms of the tanks. On the grating was placed 18 in. of 2-in. gravel and above that 26 in. of standard filter sand. These changes give, in effect, a false bottom to the filters and provide a chamber which equalizes the distribution of wash water and reduces friction loss, and increases the yield of the filters. All the American and Western filters at Streator were still in use in 1940 but it was hoped that within a few years they could be replaced by open-type gravity filters.

Davenport, Iowa.—American filters at Davenport, put in use in June 1891, were still being operated in November 1941, but only the shells and piping remained. As originally installed, there were ten horizontal cylindrical units, $7\frac{1}{2} \times 30$ ft. in dimension, divided by a vertical bulkhead into two compartments, each with a sand area of 100 sq.ft. The nominal capacity of the plant, with one unit out of service for cleaning, was 5 mgd.

Notes taken by Alvord, Burdick & Howson during a valuation of the property in 1902 showed that the filters then contained 4½ ft. of sand, but no gravel. The strainer system was a cast-iron manifold into which 2-in. brass pipe with No. 7 slots was screwed. The filters were backwashed in the "usual way," but "each 30-ft. shell," says Mr. Burdick, "was equipped with four plunger cylinders on top, 12 in. in diameter, 3½ ft. long, with a piston and a plunger and a hollow eight-legged spider inside the filter which normally was above the sand line, but when the filter was being washed could be forced down nearly to the bottom of the sand. This device did not work well mechanically. . . . The filters were being backwashed without agitation in 1902." In a paper read before the Illinois Society of Engineers in 1936, Mr. Burdick said, "Good success was only attained after the addition of sedimentation basins about 1900."

In 1908, the filters were equipped with Jewell strainers and air-and-water wash. Subsequently air agitation was given up, not only for the American filters but also for Jewell filters installed in 1908. About 1920, the inside surfaces of the filter tanks were cleaned by air blast and lined with cement brushed on. The dished heads of some of the early filters gave way before the practice of subjecting them to 80–100 psi. fire pressure was stopped. Repairs were made by machine flanging, hand flanging having been used in the original fabrication.

For many years Charles B. Henderson was in charge of the Davenport water works. He was succeeded in the late 1930's by J. H. Wells as manager. Wells wrote November 14, 1940, "These filters are still in use and in good condition . . . [but] we expect this winter to change them from pressure to gravity."

The Blessing Filter

James H. Blessing, of the Albany (N.Y.) Steam Trap Co., entered the field of mechanical filtration a few years after the Hyatts. He applied for four pressure filter patents in 1886. By 1895 he had been granted seven patents in the United States and two in England. His strainer system was unique, his sand agitators less so but interesting. He supplied only two municipal water works with filters but installed many for houses, offices and industrial plants. His business was absorbed by the New York Filter Manufacturing Co. in 1896 for which he received only $2,400 of the $600,000 stock of the company. He

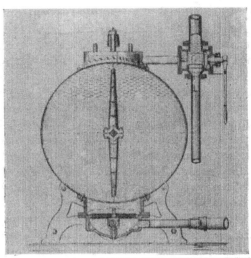

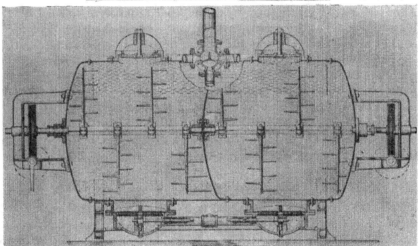

Fig. 46. Blessing Horizontal Duplex Filter

First permanent rapid filter in municipal supply in New England put in use at Athol, Mass., 1889; lower section contains sand screens; agitation by horizontal revolving arms

(From undated trade catalog of Albany Steam Trap Co.)

seems to have been equalled by none of his rivals in studying the art of water purification—as disclosed by patents.*

A "sand screen" of gravel in a chamber below the bottom of a filter and a revolving agitator to clean the gravel were covered in Blessing's first patent (March 20, 1886). In his second (November 23, 1886) a new sand screen was substituted. This was a perforated plate in the bottom of the filter tank into the holes of which pins or plungers mounted on a reciprocating head were inserted from below. A slight difference in the diameters of the holes and pins allowed the filtrate but not the filter sand to pass through. Reciprocation of the pins, effected by a cam mounted on a horizontal revolving shaft, removed any sand lodged in the holes. This patent also covers a filter-sand agitator, consisting of a revolving central shaft combined with which were either "chain sheaves" or three revolving horizontal arms provided with vertical teeth on their upper and lower sides.

A duplex filter patent, also of November 23, 1886, covered two filters similar to those of the second patent. British patents essentially like the second and third American, also dated November 23, 1886, were granted to Blessing. All the filters thus far described were placed in vertical or cylindrical tanks.

A "charcoal purifier," to receive effluent from the duplex filters, was included in Blessing's fourth patent (April 20, 1889). In a Blessing trade catalog of 1890 it was stated that "Mr. Blessing has devised a Purifier" which may be attached "to the filter of his invention," and used or not used as desired, "the animal charcoal in the Purifier being finally relied upon to reduce to a minimum the micro-organisms which may possibly remain in the water passing through the filter." Testimonials in the catalog show that the purifier had been supplied with filters for house use.

A "chemical chamber," for use in treating effluent from a duplex filter, was the subject of an application filed by Blessing August 25, 1887, but held up until September 13, 1892, then assigned to the New York Filter Co., which had recently acquired the Hyatt coagulation patent.

* Evidence of this is afforded by a bound volume of American and British patents dated from 1812 into the 1880's. This volume was given by Blessing to Wallace Greenalch, one-time Water Commissioner of Albany, who presented it to the author, together with a Blessing trade catalog, in 1937. Both are in the author's collection of source material deposited in the Library of the United Engineering Societies, New York City.—*M. N. B.*

In a trade circular regarding the duplex filter in 1887, Blessing boldly declared:

> As to the matter of treating water chemically, we will say we are prepared to so treat water by any of the well-known methods, such as practiced by the Messrs. Clark, Graham, Darcet, Miller, Hofman, Le Tellier, Holden, Demalley, Maingay, Jaminet, Frost, Spence and many others, depending entirely on the circumstances and the requirements of the water to be treated.

A number of these processes were covered by patents, copies of which are in the bound volume already mentioned. All differed from the claims of the Hyatt patent on coagulation.

Horizontal filters with sand agitators differing somewhat from those already noted were the subject of two patents granted to Blessing on December 9, 1890.

Athol, Mass.—At Athol, Mass., Blessing's filters were installed by the Athol Water Co. in 1887 and added to in 1890.* Algae growths in an impounding reservoir seem to have led to the installation of the Blessing filter. It was 8 ft. in diameter, 16 ft. long, divided by a vertical diaphragm into equal and independent units, each provided with a "sand screen" in its bottom and sand agitators revolving on a horizontal arm. The filter medium was sand.

Percy M. Blake, of Hyde Park, Mass., reported on the Athol filters in May 1896 (45). A solution of crude sulfate of alumina was added to the water at the gate house, 4,781 ft. above the filter plant, which was on the 10-in. gravity supply main. During the summer of 1895 it was "necessary to wash and scour each filter twice daily." The effluent after washing was not "thoroughly freed from sediment and products of decay, although the process is probably as effective for water of the kind obtained from any mechanical process."

Robert Spurr Weston saw the Athol filters in 1903. He says they were then used as strainers only, treating water badly infected with algae. His firm made plans for a coagulation basin in or about 1905. The town bought the works from the company in 1906. Slow sand filters were built in 1912, after plans by James L. Tighe of Holyoke, but they treated water from another reservoir.

* The Athol plant has been called the earliest mechanical filter installation on a New England municipal water supply, but hitherto unpublished data show that a Hyatt aeration and filter plant was put into use in 1887 or early in 1888 on the works of the Greenwich Water Co., supplying Greenwich, Conn., Rye and Port Chester, N.Y. The first Athol filter was used for half a century.

A central rapid filtration plant superseded both the rapid and slow filters June 7, 1937. On November 27, 1937, Frank Hall, Superintendent, wrote that the Blessing filters were being scrapped. During the last three years of their use a coagulant was employed the year around. Latterly, the sand in the Blessing filters had been taken out and washed once a year. Black and lumpy sand in the ends of the filters where there was no agitation by the revolving arms, was replaced by new sand.

Thus came to an end the earliest mechanical filters in Massachusetts and the earliest permanent installation in New England, after a half century of service.

Ottumwa, Iowa.—At Ottumwa, Iowa, a thousand miles west of Athol, the only other known Blessing filter on a municipal supply was installed late in 1890 or early in 1891. It was preceded by one of the largest known American sets of charcoal filters and followed in 1895 by Jewell gravity filters (34).

Three Jewells and Their Filters

As the Hyatt, National and American filter companies were about to merge and the Warren filter was coming to the front, the Jewell filter entered the municipal field. It had been under promotion for several years, both at home and abroad, but chiefly in the industrial field. Once established in the municipal field it attained prominence. The three Jewells—Omar H., and his sons, Ira H. and William M.—took out about 50 patents in 1888–1900. The earliest were granted to Omar H. Jewell; later ones to him and one or both sons, or to one son alone.

Omar H. Jewell was born at Wheaton, Ill., June 1, 1842. As master mechanic for grain elevators he became interested in improving the quality of boiler feed water from the notoriously foul Chicago River. The first of his filters seen by his son William was located in Elevator L, of Armour, Dole & Co., on the South Branch of the Chicago River. Probably it was built in or about 1885. The Jewell Pure Water Co. was organized and largely financed by James B. Clow & Sons, well known Chicago dealers in water works supplies. W. E. Clow, chairman of the Clow concern, stated in 1937 that in the earlier years he had charge of sales and sold filters in England, France, Germany and Italy. The elder Jewell devoted himself to manufacturing the filters.

FIG. 47. THE THREE JEWELLS—FATHER AND SONS
Top: OMAR H. JEWELL (1842–1920)
Left: IRA H. JEWELL (1869–1940)   *Right:* WILLIAM M. JEWELL (1870–1940)
(From photographs supplied by Ira and William Jewell)

William M. Jewell became chemist of the Jewell Pure Water Co. in 1887, following graduation from the College of Pharmacy, University of Illinois, at the early age of 17. With his father and brother the firm of O. H. Jewell & Sons was formed in 1890. It was soon incorporated as the O. H. Jewell Filter Co., which continued in business until it merged with two other companies in 1900. William Jewell was in Europe from October 1888 to December 1889, assisting Jewell agents in erecting "steel tank pressure filters," made in Chicago and shipped abroad complete. These seem to have been for industrial use. In the late 1890's while George F. Hodkinson was manager of the O. H. Jewell Filter Co. at Chicago, Ira and Ariel * Jewell were sent to Moscow to supervise the operation of an experimental filter plant for a few months. Nicholas Simin had previously visited the United States and arranged for the installation. The Morison-Allen Co. and then the Morison-Jewell Filtration Co. represented the O. H. Jewell Co. in New York and Philadelphia from about 1888 to 1898. William B. Bull of Quincy, Ill., was Vice-President of the O. H. Jewell Filter Co. in the late 1890's.

The first Jewell filter patent was granted to Omar H. Jewell February 7, 1888, and nine more followed. Five of the ten were for a feed-water purifying apparatus.

Electrodes placed in a dome located on the top of a filter tank were an element of one of Omar H. Jewell's early applications for a patent filed December 17, 1887, and granted July 10, 1888 (No. 386,073). The electrodes were connected with a battery or other source of electricity. This was one of the earliest patents on the use of electrolysis in water treatment. Among the many other patents granted to Omar H. Jewell were several on strainers or screens for underdrain washing systems; revolving sand agitators; a settling chamber below the filter; and means for maintaining a partial vacuum in filters (negative head), both process and apparatus.

Ira H. Jewell's earlier patents included one on apparatus for continuous cleaning of filters of large area by lifting, successively, portions of the sand and supernatant water by a pump on a truck on a movable platform, above which was a screen to separate the water and its load of impurities from the filtering material as the water went to waste.

* Ariel Clyde Jewell, wrote Ira H. Jewell on Aug. 8, 1936 (46), "built up a large business in water distilling, operating under the trade name 'Polarstill.' . . . He died several years ago."

Perforated pipes were placed beneath the screen to convey steam or other sterilizing agents. The sand thus cleaned and sterilized was returned to place. (Patent dated August 10, 1897; filed November 7, 1892.) A later patent granted to Ira H. Jewell (May 8, 1900) covered a traveling filter washer similar to the one just described, combined with a horizontal-flow surface-jet wash, revolving sand stirrer arms and a multiplicity of air pipes adapted to discharge air under pressure upward through the filter.

An electrolytic process for producing hydrate of iron for use as a coagulant to be applied to a filter was patented by William M. Jewell July 17, 1900. Two other patents of the same date were on a method and apparatus for producing "a purifying reagent" (coagulant) by "subjecting water to the action of sulfurous-acid gas, passing the solution so formed over iron and converting the resulting solution into ferric sulfate by oxidation." A filter rate-of-flow controller was patented by William Jewell February 23, 1897. A similar device was used on the Jewell filter during the Louisville filtration experiments, 1895–96. William Jewell believed this to have been the first use of such a controller on a mechanical filter.

Brockton, Mass.—The first known attempt to introduce Jewell filters into an American municipal supply was made at Brockton, Mass., in 1888. The report of the Brockton Water Board for that year states that the Morison-Allen Co., New York City, petitioned the mayor, council and water board to allow a demonstration of one of its small filters on the Brockton works. Apparently Brockton would have installed Jewell filters had it not been for the strong objections of the Massachusetts State Board of Health.

Rock Island, Ill.—The first Jewell filter for a municipal supply was put into use at Rock Island, Ill. Installed in 1891, its object was "to clarify the water." Because it was too small it was shut down in two months, but after enlargement, operation was resumed. In 1899, settling basins and slow sand filters were built. In 1911 a return was made to rapid filtration, using open rectangular filters.

Subsequent Plants.—Five other Jewell plants for city water supply were put in use in 1891. By May 1896, a total of 21 plants had been completed and one was under construction. The largest of these was a 10-mgd. plant for the Wilkes-Barre, Pa., Water Co., put in use in 1895. Next in size was a 4.5-mgd. plant completed by the Niagara Water Works Co. in March 1896. Gravity filters were used in eighteen of the 21 plants. Of the three pressure plants, the one at Chattanooga was installed to work with existing filters of the pressure type and those at Terre Haute and Lake Forest were put in on direct pumping systems. Steam for cleaning and sterilizing filters was used at all the plants (34).

A multiplicity of strainers appears to have been the rule in the early Jewell filters. The strainer described and illustrated for the Wilkes-Barre filters, and apparently the type generally used, was a perforated aluminum-bronze plate, placed across a cup screwed into the underdrain and wash pipe. Between the plate and the bottom of the cup was a deflector to spread the wash water and steam. The filters were washed by reverse flow, aided, in the typical gravity filters, by revolving-rake sand agitators. These consisted of horizontal arms from which numerous rods extended to the bottom of the filter. Subsidence or coagulation chambers were provided in most of the plants described in the article of 1896. These were small compared to the rated daily capacity of the filters, except at Creston, Iowa, where the chamber capacity was 0.5 mil.gal. compared with a filter capacity of 0.7 mgd. This was really a presettling reservoir.

After protracted litigation brought by holders of the Hyatt coagulation-filtration patent of February 19, 1884, against users of Jewell filters employing a coagulant, the U.S. Circuit Court of Appeals upheld preliminary injunctions against the Elmira Water Works Co. and the Niagara Falls Water Works Co. in 1897 (47). The settling chambers below the filters, it was held, were too small to avoid infringement of the patent.* Nor did they provide the independent settling reservoirs for coagulation which Hyatt claimed his process made unnecessary.

Early in 1898, the O. H. Jewell Filter Co. settled with the New York Filter Manufacturing Co., holder of the Hyatt coagulation patent, for infringement, and took out a license for the use of the Hyatt patent in the central area, including Tennessee and Kentucky, and west of the Mississippi River. The New York Filter Manufacturing Co. agreed to confine itself to the eastern area, within which it was to supply the Jewell filter. The Morison-Jewell Filtration Co. of New York and Philadelphia settled for infringements and retired from business,

* Coagulation basins with a holding capacity of 2.5 mil.gal., or nearly half the daily capacity of the filters, were constructed at Elmira in 1937. The old chambers beneath each Jewell filter had a detention period of about 20 min. (48, 49).

but its Vice-President and General Manager, Samuel L. Morison, of New York, became general manager of the New York Filter Manufacturing Co. (50). In 1900, the New York Filter Manufacturing Co., the O. H. Jewell Filter Co. and the Continental Filter Co. consolidated as the New York Continental Jewell Filtration Co. Omar and William Jewell made a five-year contract with the consolidated company. Subsequently, William Jewell began private practice, largely as a consultant. Ira Jewell continued to operate independently, as he had been doing for some time past. He was engaged in much litigation, notably an unsuccessful suit against the city of Minneapolis for alleged infringement of down-draft or negative-head and central operating-control patents.

A subsurface filtering and washing system introduced by Ira H. Jewell about 1935 has been applied to a number of plants (51). Wire mesh screens set in large castings near the top of the filter are supplied from below with water for filtration or for washing. The main portion of the filter is washed by the usual reverse-flow method from the bottom. In principle, this was similar to the subsurface washing system of the National filter, forty years earlier.

Omar Jewell died in 1930 at the ripe old age of 88. Ira and William died in 1940.

The Jewell Export Filter Co. was organized about 1900 to exploit filters overseas. In a catalog published in 1903 it was stated that the company had offices at York, England; Berlin, Germany; Trieste, Austria; Moscow, Russia; Alexandria, Egypt; and Arnheim, Holland. In 1912 it had representatives also in Johannesburg, South Africa; Japan; and China. It then reported fifteen installations in Russia and ten in India, besides plants in many other countries. It had only three plants in England in 1912. On the Continent it had few installations in 1912, outside Russia, but some years later Jewell filters were installed at Warsaw, Poland, and the plant was said to be the largest in Europe up to that time. It was reported to have been destroyed or at least badly damaged during the Nazi blitzkrieg of 1939.

In 1903, S. L. Morison of New York, who had been connected with various companies promoting Jewell filters for many years, was Vice-President and General Manager of the Jewell Export Filter Co., and R. W. Lawton was Engineer of Construction. Edmund B. Weston, who as Assistant City Engineer in charge of water works at Providence, R.I., conducted the Providence filtration experiments in 1893–94, was

President and General Manager of the company for many years before his death late in 1916. His headquarters were at Providence, but he traveled widely, going abroad annually for many years before the outbreak of World War I.*

The Continental Filter Company

The Continental Filter Co. came to the front in the early 1890's and obtained enough importance by 1900 to have its name included in the third great consolidation of filter companies, the New York Continental Jewell Filtration Co. It was incorporated in West Virginia, November 20, 1891, with offices in New York. At the time of its consolidation it had built filter plants for six municipal supplies. It was not dissolved until June 1927.

Three Williamsons—David, David Charles and James E.—were Continental filter patentees and engineers. David was the pioneer. He was the Chief Engineer of the Continental Filter Co. during the nineties. David C. was successively draftsman, erecting engineer and assistant engineer of the company from 1897 until it entered the consolidation of 1900. After being with the new company six years he became chief engineer of its filter department, which position he held at least until the close of 1912. Charles L. Parmelee was Chief Engineer of the Continental Filter Co. from March 1899 until the consolidation in 1900. Apparently he was the engineer for the New York Continental Jewell Filtration Co. for a short time in 1900 and held the position until a few years before the company was sold by receivers to the American Water Softener Co. in 1925. He was in private practice at the time of his sudden death in March 1937.†

Nine American patents were granted to the Williamsons between 1892 and 1900, and one to James E. Williamson in 1908.‡ The first patent was for a centrifugal filter. It was issued to David Williamson February 16, 1892. David Williamson's second patent (June 28, 1892)

* The Jewell Export Filter Co. still had offices at Providence in September 1942, but the nature and extent of its recent operations could not be ascertained.

† These notes on the Williamsons and Parmelee have been drawn from testimony by the latter and by David C. Williamson in Defendant's Record, Ira H. Jewell vs. City of Minneapolis, December 1912; from correspondence with George F. Hodkinson, of Philadelphia; and from records of the American Water Works Association.

‡ One of the earliest of the nine patents was assigned to the Continental Filter Co. Several of the others were assigned to H. B. Anderson, New York City, who, in 1900, was one of the incorporators of the New York Continental Jewell Filtration Co.

covered double filtration, under pressure, in a horizontal cylindrical tank, divided by a vertical partition into two compartments, which had downward filtration in the first and upward in the second. The filtering material in each unit was supported on a series of perforated plates, forming a false bottom, with a chamber below extending the whole length of the tank. Perforations in the bottom of this chamber afforded passage into a mud drum, also extending the whole length of the cylindrical tank. Raw water under pressure was admitted to the top of the first filter through holes in a curved plate attached to the top of the tank shell. It passed down through the unit and false bottom, then into a chamber beneath the second chamber, then upward through the second unit and into filtrate collecting pipes. Wash water took the same course as the water that had been filtered, except that it was admitted through a single perforated pipe above the first filter and withdrawn through perforated pipes above the second filter. Downward filtration through one or more filters or compartments under either pump or gravity pressure was covered by David Williamson's third and fourth patents, granted early in 1894. The object stated in the first of these was to concentrate the supply and delivery pipes and points of control and facilitate cleaning.

A patent granted to David C. and James E. Williamson late in 1894 claimed concentration in action of wash water to cause the mass of filtering material to be effectually "stirred, ground and thereby cleaned." This was to be effected by discharging wash water into the filter from a central pipe, having an enlarged bottom, with slitted outlets.

Of three patents granted to J. E. Williamson, one issued November 22, 1894, was for aerating the filter material while being washed. Air under pressure was applied from a chamber a little above the filter. The second patent (July 3, 1900) was on an apparatus for cleansing filtering material by water-and-air wash, applied through a single set of nozzles at the bottom of the unit. After reverse-flow wash, the water was shut off, pipes partly drained but with enough water left to form a water seal, and air supplied. The third patent was on a multiplicity of automatically controlled settling tanks and means of supplying a coagulant thereto (September 4, 1900).

D. C. Williamson was granted a patent on December 29, 1908, on a traveling suction pipe and pump for removing sludge from sedimentation basins and reservoirs. The pump could be moved horizontally in either direction along the bottom of the basin; could be provided with an agitator; and could be operated intermittently or continuously.

Six Continental filter plants were installed on municipal supplies: Two of an early type in 1893 and 1895 at Atlantic Highlands and Asbury Park, N.J.; and four of a late type in 1899, at Middletown and Stamford, N.Y., Vincennes, Ind., and Louisiana, Mo.

The New Jersey plants used double filtration, under pressure, for removing iron from deep-well water. They were cleaned by reverse-flow wash and compressed air, used alternately. Each of the two filter tanks used at Atlantic Highlands was 6 ft. in diameter and 11 ft. long. The four tanks at Asbury Park were 6 x 28 ft. The media were coarse sand in the primary and animal charcoal in the secondary filter. No chemicals were used.*

Middletown, N.Y.—Of the four plants of 1899, the one at Middletown, N.Y., included two wooden settling tanks, four gravity filters in wooden tanks and four steel horizontal pressure filters (53). No description of filters at Stamford, N. Y., has been obtained but Superintendent Charles R. Mattice wrote on March 31, 1941, that two Continental filters were "in service and functioning perfectly." The "grids" were changed in 1927 and remodeled in 1937.

Louisiana, Mo.—At Louisiana, Mo., two coagulation-sedimentation tanks and a single gravity filter were put into use in 1899. The settling tanks were used alternately on the fill-and-draw plan. They gave a nominal subsidence period of 108 min. The tank effluent was applied to the filter through 6-in. overhead pipes, perforated with 1-in. holes, 6 in. apart. Air-and-water wash was used, with air applied first. Sulfate of alumina was used as a coagulant. The supply was from the Mississippi River. A week's test of the plant in late September 1899 showed an average bacterial reduction of 97.3 per cent, ranging from 99.9 to 96.5 per cent. The raw-water count was 57,000 to 400 per ml. (54).

Vincennes, Ind.—At Vincennes, Ind., a 2-mgd. plant was put in use late in October 1899, to treat water from the Wabash River which carried considerable loam. It included coagulation tanks and six

---

* Data on both plants were gathered by me on inspection trips in December 1905. General data were supplied by David Williamson, Chief Engineer, Continental Filter Co. (52). Unfortunately my article of 1896 (34) did not describe the piping connections and flow sequence for the New Jersey plants. Presumably the filters were in general accordance with David Williamson's patent of June 28, 1892.—*M. N. B.*

open gravity filters, all built of wooden staves. The settling tanks were operated in pairs, on the fill-and-draw system, with 21 min. of quiescence. The filtering material was 36 in. of sand on 6 in. of gravel. Water from the settling tank flowed over circumferential weirs onto the filters and was collected by brass strainers. Cleaning was by reverse-flow wash and compressed air. An account of the operation of the coagulation-sedimentation basins by Charles L. Parmelee, then Chief Engineer of the Continental Filter Co., was given in a contemporary article (55). A special feature of the Vincennes filters was central control of hydraulically operated valves, designed by David C. Williamson (56).

New York Continental Jewell Filtration Company

In the twenty years after Patrick Clark and John W. Hyatt joined forces and incorporated the Newark Filtering Co. late in 1880, many filter companies sprang up, competition and patent litigation were rampant, consolidation after consolidation occurred until July 25, 1900, when the New York Continental Jewell Filtration Co. stood almost alone in the field. It was heir to scores of patents, including those granted to Leeds, Deutsch, Warren, Blessing, the Hyatts, two of the three Jewells and the three Williamsons. Some of these had run out; others were about to expire. The Hyatt coagulation-filtration patent, upheld by the highest court after costly and bitter litigation, had about a year of life. Moreover, both practical experience and the Louisville experiments had shown that it was based on a fundamental misconception and that as a rule precoagulation was desirable if not essential. In addition to all this, the underlying principles of water purification had been or were being established on a firm basis of engineering, chemistry and sanitary biology, so that filter design was open to any engineer and filter construction to any contractor. The new company held some unexpired patents on equipment, but so did other companies.

Following its organization in 1900 the New York Continental Jewell Filtration Co. built filter plants after either its own plans and controlled patents or, more and more frequently, the plans of independent engineers. A notable early contract was the equipment for the large filter plant of the East Jersey Water Co., at Little Falls, N.J., completed in 1902 (57). In advertisements in the *Journal of the American Water Works Association* the company announced that the number of its plants had reached 220 in 1906 and 360 in 1909. The New York Continental Jewell Filtration Co. was sold at a receiver's sale held at Nutley, N.J., March 26, 1925. The purchaser was the American Water Softener Co., of Philadelphia, through George F. Hodkinson, who for many years was manager of the latter company's filtration department.

Little Falls Marks New Era *

The Little Falls filters of the East Jersey Water Co. inaugurated a new era in the design of rapid filters. In shape they were rectangular instead of round; in structure they were of reinforced concrete rather than of wood or iron. In them, application of the coagulant had been transferred from the point where the raw water flowed to the filters to the point where the treated water entered a detention basin in which it was held for a period to permit chemical reaction and flocculation. And as an aid in cleansing the filter sand it was agitated by means of compressed air introduced underneath the filter before reverse-flow wash rather than by the old method of stirring with revolving rakes operated from above.

Except for the shape of the filter tanks, these new features were merely adoptions and enlargements of designs already in use at the Continental plants mentioned above. Reverse-flow wash had been patented in England by Peacock in 1791 and had been put to use by Thom at Greenock in 1827. Thom's filters there and at other localities in Scotland had been placed in rectangular masonry tanks larger than those at Little Falls.†

The nominal capacity of the Little Falls filters when placed in service early in September 1902 was 34 mgd. Main elements of the plant were chemical mixing tanks, a coagulation basin, 32 filters—each 15 x 24 ft. in plan and 8 ft. deep—and a clear-water basin beneath the filters. To protect the strainers, which were of the Continental type,

* Although in the design of the Little Falls filters, Fuller and his associates made use of what he had learned in the Louisville and Cincinnati experiments of the late 1890's, described below, the filters for those cities were not placed in service until a few years after those at Little Falls. For that reason, and because, chronologically, they belong after the several installations of Continental filters, the Little Falls installation is described before the experiments and plants at Louisville and Cincinnati.

† See above for a brief review of British and American anticipations of elements of the design of the Little Falls filters and see Chapter V for a resume of Peacock's patents and theories and for descriptions of Thom's installations.

2 in. of broken quartz and then 5 in. of coarse quartz were placed above them. The filtering medium was 30 in. of sand.

In the design of the plant, George W. Fuller was in full charge on behalf of the East Jersey Water Co. and Charles L. Parmelee represented the New York Continental Jewell Filtration Co. J. Waldo Smith, Chief Engineer of the East Jersey Water Co., had general direction over the design and construction. A detailed description of the design of the plant and its early operation was presented before the American Society of Civil Engineers by Fuller in 1903 (57).

Louisville Rapid Filtration Experiments

Many know of the Louisville experiments on water purification in 1895–97 and their contribution to the art and science of rapid filtration. Few ever heard that for several years in the eighties experiments with slow sand filtration showed that it could not cope with the highly turbid water of the Ohio at Louisville. This turbidity had been a source of concern to the directors of the water company and to Chief Engineer Charles A. Hermany since the water works were completed.

Although the later experiments under Hermany, in direct charge of George W. Fuller, were completed in 1897 and described in Fuller's classic report of that year, it was not until July 1909, that the Louisville rapid filters were put into successful operation. Changes in the plant were completed in 1909 and others were made which extended through several years. Finally, the original plant was reconstructed. New features of design and operation were adopted. Some of these were merely in keeping with the ordinary progress of the time; others were innovations. Valuable records have been kept which show, year by year, the work accomplished, classified under sedimentation, coagulation, filtration and chlorination (58).

When the Louisville Water Co., nominal owner of the water works, incorporated October 6, 1854, was unable to sell its stock, the city bought a majority of it in 1856 and all of it later, but the works have always been operated as a company. Conditions before the Civil War delayed construction so that the works were not put into use until October 16, 1860. Consumers were few until the war was over.

The source of supply chosen was the Ohio River, above "town drainage." Theodore R. Scowden was chief engineer during construction. In his report of November 1, 1859 (58), he said that the quality of the water "was attested" by Dr. Locke of Cincinnati and other eminent chemists but he stated their findings in general terms only. "Frequent freshets discolor [!] the water but do not vitiate its ... qualities." Suspended matters "can be easily removed by filtration or subsidence"—which did not prove to be so easy.

Apprehension that the citizens would be loath to change from the clear water of wells to the turbid water of the Ohio was expressed by Scowden. This he sought to quiet by stating that those accustomed to the use of the Ohio River where works had been established "appear to pay little attention" to turbidity "as the taste is agreeable and its qualities are healthful."

Charles A. Hermany, who had become chief engineer of the works on their completion late in 1860, noted in his annual report of 1864 a lack of "reservoir capacity to obtain a clear water at all times by subsidence."

On March 29, 1876, a committee on water works extension, of which Hermany was a member, reported that surveys had been made for settling basins and filters on the river bottom near the pumping station. It recommended experiments to determine the proper ratio between capacity of settling basins and area of filters and to ascertain "the composition of the filtering media, for the reason that there is no experience in our country by which the problem has been solved." The experimental basins and filters should be of small area but of full working depth (59).

No experiments were made. A contract was let in 1876 for a 100-mil.gal. reservoir, in two compartments, located on Crescent Hill. Water was piped to it on December 15, 1879. It was still in use in 1942.

For 20 years the president of the company, Charles R. Long, led a campaign for filters, during which time two sets of filtration experiments were made: one on slow sand, which was found impracticable; and the other on rapid filtration, which was found efficient *if* changes were made in the design of the commercial filters tested and if adequate precoagulation and presedimentation, also lacking in the filters tested, were provided.

Few data on the slow sand filtration tests are available. President Long noted them in five successive annual reports: 1880, proposed; 1881, plan ready; 1882, under way, good results hoped for; 1883, "working with a reasonable degree of satisfaction"; 1884, "an unsettled problem as to practical feasibility and reasonable cost of maintenance, however desirable and necessary of accomplishment for the comfort, health and welfare of our citizens." Here Long's remarks on the slow sand filtration tests end.

Chief Engineer Hermany's reports contain nothing on the tests except expense items, running from 1882 to 1887 and totaling only $1,941, including labor and material. The largest annual totals were $795 in 1882, for two 12-ft. tanks, including masonry foundation, and $889 in 1883, for "labor in setting tanks" and numerous sundries. Later items may have been for dismantling the plant.

Fortunately, Fuller, in his Louisville report on the rapid filtration experiments of 1895–97 gives more specific information, doubtless obtained, in part, directly from Hermany. Fuller says the tests were conducted under Hermany's direction for eight months in the fall of 1884 and spring of 1885. There were two filter tanks 12 ft. in diameter, "after the English plan described and recommended [for St. Louis] by Mr. Kirkwood. The sand had an effective size of 0.36 mm., agreeing very closely with the size employed in the best English filters in Europe." Summarizing, Fuller said:

> As a result of the tests of 1884–85 it was learned that the clay could be removed and an effluent free from turbidity secured by the English filters at a net rate of about 1.5 mil.gal. per acre daily. But the principal point of practical significance was the marked indication of the clay passing into the sand layer, and the necessity for cleaning and reconditioning the sand layer at periods of comparatively short duration (60).

Soon after President Long gave up hope of clarifying the water by slow sand filtration his attention was directed to rapid filtration where it remained centered until the process was adopted. In his report for 1887 he expressed the hope of being "able to test in a practical way some of the new methods of filtering and purifying water . . . notably the Hyatt Pure Water System, which is now in use in twenty odd cities and towns and . . . appears to meet all essential requirements necessary to a pure and healthful water supply." Less optimistic in 1889, he said no system of filtration "has been practically tested, used or adopted for filtering successfully the water supply of any city of the magnitude of Louisville, to say nothing of the peculiar characteristics of the Ohio River water" that would justify the company in building "a filtering plant of any of the systems attracting attention." In 1891 and again in 1893 he declared that the problem was still unsolved. In 1894 Long said that Lawrence, Mass., was trying a system in a small way somewhat similar to slow sand filtration, but it would not be practicable for Louisville. Mechanical filtration was in use in a number of small cities but none approximating the size of Louisville, except in New Orleans, where the plant had failed to meet the requirements and was not accepted. Satisfactory results in smaller cities and many recent improvements led the Louisville Water Co. to hope that it could find a way for testing "some one or more methods in a practical manner."

Meanwhile, for a year in 1893–94 the company employed a chemist and bacteriologist who made tests and monthly reports. The tests showed the water to be "reasonably free from every species of zymotic or disease-producing germs"—marvelous to relate!

After years of investigating, said President Long in his report for 1895, the company finally adopted a plan for "a series of experimental tests of filtration, embracing agreements with four companies." In 1896, Long reported that these tests were about closed. They had been made under the direction of George W. Fuller, Chief Chemist and Bacteriologist. With pardonable pride, Long characterized the tests as "the first practical step taken towards solving the vexing question of filtration and purification of water supplies for large cities upon modern scientific principles. . . . We have reason to hope that the plans, specifications, and contract for such a system can be made, and the work gotten well under way, [in] the ensuing year."

Instead of such speedy action, further experiments were made. The cost of the tests to the end of 1896 was $26,776. During 1897 the water company continued, "on its own hook," further tests at an additional cost of $15,418. These tests were concluded in January 1898.

The Jewell Filter Co. of Chicago, the Cumberland Manufacturing Co. (Warren filter) of Boston, the Western Filter Co. of St. Louis, and the Harris Magneto-Electric Filter Co. of New York entered competing bids under which each supplied, installed and operated at its own expense a 0.25-mgd. unit, except that there were both a gravity and a pressure Western filter.

Besides Charles Hermany as Chief Engineer and George W. Fuller as Chief Chemist and Bacteriologist, the Louisville Water Co. was staffed as follows: Charles L. Parmelee, Assistant Engineer; Robert Spurr Weston, Assistant Chemist; Dr. Hibbert Hill, Assistant Bacteriologist; Joseph W. Ellms, Assistant Chemist; George A. Johnson, Clerk and Assistant Bacteriologist; Reuben E. Bakenhus and Harold

C. Stevens, assistants. Mechanical analyses of the filter sands used were made by Harry Clark of the Lawrence Experiment Station.

The Jewell filter was represented by the two brothers, William M. and Ira H. Jewell. The Warren filter was in charge of George A. Soper, who made a special report on its operation for its proprietor (61). Charles T. Whittier was in charge of the two Western filters.

FIG. 48. THE STAFF AT THE LOUISVILLE EXPERIMENT STATION IN 1896

Back row: Robert Spurr Weston, Asst. Chemist; William, the Janitor; George A. Johnson, Clerk & Asst. Bacteriologist; Harold C. Stevens, Asst.
Front row: Reuben E. Bakenhus, Asst.; Joseph W. Ellms, Asst. Chemist; Charles L. Parmelee, Asst. Engr.; George W. Fuller, Chief Chemist & Bacteriologist; Hibbert Hill, Asst. Bacteriologist; [?] Benton, Asst. Bacteriologist
(From A.W.W.A. *Manual of Water Quality and Treatment*, 1940)

Fuller began his duties October 1, 1895; completed his main report on the apparatus of the four contestants at the end of November 1896. Studies were then made of the corrosive action of filtered water on pipes and boilers and on boiler incrustation. The water company also tested electrolytic apparatus, devised by Palmer and Brownell of the Louisville Manual Training School, and "the MacDougal Polarite System" (settling tank, clay extractor and polarite filter). Finally "devices of the company" were tested. Fuller's final report was signed October 7, 1897. Fittingly it was brought out in a thick volume (60) in a style conforming with Kirkwood's classic St. Louis report on *Filtration of River Waters in Europe* and by the same publisher.

The water experimented with at Louisville was taken from the Ohio River at the pumping station. The turbidity ranged from 1 to 5,311 ppm. The population above the pumping station was estimated at 4,500,000, of which 1,575,000 lived in 220 cities and towns. The nearest city discharging sewage into the river above the pumping station was Madison, Ind., 50 miles above Louisville, with a population of about 12,000. Although attention was given to the reduction of bacteria, interest centered in clarification—so much so that bacterial efficiency is not mentioned in the final summary and conclusions.

The apparatus and processes other than the three makes of rapid filters are dismissed briefly in the conclusions. Of two of these the report says: "The Harris Magneto-Electric System was a complete failure. The MacDougal Polarite System, as it was tested by this Company, was not applicable to the purification of the Ohio river water." Bare mention is made, in the final conclusions, of the application to the Jewell filter for a brief period of electrolytically produced chlorine.

The tests of all three makes of filters showed that the combination of sedimentation, coagulation and filtration was "correct in principle," but as used at Louisville had "several weaknesses," the most important being "the totally inadequate facilities, in all cases, for the employment of subsidence to its proper economical limits." Fuller also states:

> In addition to plain subsidence and to coagulation prior to filtration, there are times when coagulation in conjunction with subsidence can be employed to advantage in keeping clay and other suspended matter from passing on to the sand layer.

This seems to imply coagulation-aided subsidence prior to the coagulation and sedimentation practiced with the rapid filters of 1896–97. Further emphasis on the necessity for lightening the burden on the filters and for giving time for coagulation is thus stated:

> The evidence is very decisive that so far as practicable the suspended matter should be removed before reaching the sand layer, and that, at that point, the water should be thoroughly coagulated. Further, it is clear that subsidence

should be employed with waters of this character to a degree where the amount of coagulant to be applied *just before the entrance to the filter* [author's italics] should not frequently exceed 2 grains per gallon.

The full significance of this emphasis on what is now understood as presedimentation and precoagulation cannot be grasped without knowing the state of the art of rapid filtration when the Louisville experiments were being made. The Hyatt process patent of 1884 was based on the combination of coagulation and filtration, without sedimentation. In the next dozen years this was modified to give a short time for the coagulant to act and a little opportunity for sedimentation. (This appeared in some of the rival filters, partly to avoid infringement of the Hyatt patent.) In the filters tested at Louisville, there was no presedimentation and practically no time for the coagulant to act before the water passed to the filters—this, too, notwithstanding the high turbidity at times, and the fact that only a few years earlier a filter company had been brought to ruin by guaranteeing to clarify Mississippi River water by rapid filtration alone.

Three coagulants were tried at Louisville: alum, or basic sulfate of alumina; potash alum; and lime. Of these, alum was found most suitable. Independently of the filter tests, the water company made laboratory studies, the report states, of "the Anderson process" for preparing "iron hydrate directly from metallic iron"—scrap iron in a revolving horizontally placed cylinder through which water is passed. The tests indicated that the process "is not applicable for the economic and efficient purification of Ohio River water."

In addition to the inadequacy of the Jewell, Warren and Western experimental plants for presedimentation and precoagulation, Fuller describes another weakness as follows:

The several filters represent the prevailing size in practice [sand surfaces 9.5 to 12.15 ft. in diameter], but for economy in operation the individual filters should be much larger, the limit to be determined by the successful operation of mechanical appliances to stir the sand layer effectively while it is being washed by a reverse flow of water.

Unfortunately, when the filters were built, some years later, their area was so great and the sand-stirring apparatus so cumbersome that, combined with an unsatisfactory strainer system, the filters had to be reconstructed. It should be said that in their original design, Fuller took no part. He embodied the lessons of the Louisville tests in the Little Falls plant of the East Jersey Water Co., completed in 1902.

The final conclusion of the report follows:

The general method of subsidence, coagulation and filtration is applicable to the satisfactory purification of the Ohio water at Louisville; but, as practiced by the Warren, Jewell, and Western systems during these tests, its practicality is very questionable if not inadmissable. By removing the bulk of the suspended matters from the water, large reductions could be made in the size of the filter plant, amount of coagulant, and cost of operation. On the basis of 25 mgd., these reductions when capitalized at 5 per cent would represent about $700,000. There is no room for doubt but that for a less sum than this satisfactory provisions for subsidence as outlined herein could be provided, which would not only aid in furnishing a filtered water of better quality, but would also give the water consumers a better service in other regards (60).

Convinced of the practicability of rapid filtration the directors of the company determined, August 23, 1897, "to construct a system of filtration and water purification . . . plans for which are being prepared." The estimated cost of a 25-mgd. plant, including pumping station and water tower, was $500,000. Postponements covering two years occurred. At the close of 1900, said the annual report, $254,000 had been paid on clear-water basin and filter house construction.

The directors of the water company had given Hermany a free hand to plan "such a system of filtration as he thinks will fully meet the requirements and the best interests of the company." To protect his novel features he filed a blanket caveat in the patent office "upon all designs, members, devices, combinations, arrangements and mechanisms, with their functions pertaining to the art of water filtration and purification."

Four functions of the plant designed by Hermany were removal of turbidity, color, organic matter and bacteria. Water for the existing 100-mil.gal. Crescent Hill subsiding, storage and distribution reservoir was to go to 12-mil.gal. coagulation chambers and thence to filters with a capacity of 37.5 mgd.

The chief distinguishing features of his scheme for mechanical filters, according to Hermany, were:

. . . the relatively large size and the rectangular shape of the filter tanks, each one having an area of 0.1 acre; the sand support and strainer system, composed of layers of wire netting and wire cloth; and the sand agitator system, by means of which a set of 72 agitators can be used to stir the sand in one whole bed at a time and moved bodily to any other bed, either transversely or longitudinally. In addition to these important features of the filters proper there are some new ideas in the coagulating section of the plant, which include an immense steel coagulating tank, with three concentric tanks above for preparing the coagulant; and a new coagulant feed pump, which is designed to be more exact and reliable than anything heretofore developed. [Eventually] there will be nine filters in groups of three, but only one group is to be bid upon now. Each tank is 146 ft. 11⅜ in. long, 30 ft. 3 in. wide and 8 ft. deep in the clear, giving a sand surface of 0.1 acre for each, or a unit capacity of 12½ mgd. when working at the rate of 125 mgd. an acre (62).

Construction progress was slow for several years. Whatever the cause, on March 26, 1906, the water works was put into the hands of a board of works (Trustees of the Louisville Water Co.), which included the mayor. The only director of the company on the new board was Charles R. Long, who as president of the company for many years had worked persistently for filtration. Before the end of the year he retired. The new board stated in its report for 1907 that it had found the water works in bad condition, except for the filter plant, then under construction.

Charles Hermany's death, January 18, 1908, terminated his 51 years of service on the Louisville water works, during 47 of which he had been Chief Engineer and Superintendent. The post was filled by S. Bent Russell who for many years had been on the engineering staff of the St. Louis water works, during which time he had made important studies of filtration.

Conditions encountered by Russell at Louisville and reasons for his early resignation were thus related by him 26 years later:

> When I became chief engineer of the Louisville Water Co., with which I had had no previous connection, one of the most pressing problems was to give the city clear water without further delay. The new filter plant was presumably complete. It had some new features meant for cleaning the filter tanks which proved inadequate as built although previous experiments on a small scale were reported to have given promise of success.
>
> After a while a change in the small group controlling the company forced me into a false position. I knew that radical changes would have to be made in plant and methods and also that I would not be allowed to make them. I would have great responsibility without proportionate authority. I decided I ought to resign. I made a public statement to the effect that I thought the chief engineer should have more authority. After I left, the filter plant was altered and made useful (63).

Theodore A. Leisen became chief engineer of the Louisville Water Co. August 1, 1908. He left a similar position at Wilmington, Del., where he had found an upward-flow filter plant in a useless condition and had promptly shut it down. His 1908 Louisville report stated:

In February the Crescent Hill filters, as then constructed, were completed, and arrangements made to start them in operation. It was demonstrated in the trial operation that the filtering process was satisfactory, but the devices for cleaning proved defective for the reason that a uniform distribution of the wash water could not be obtained, and in consequence continuous operation was temporarily discontinued. Investigations were started with a view to remedying the defects and after careful study and submission of the designs for the proposed operations to eminent authorities, it was decided to reconstruct the strainer system to provide means for securing an adequate and uniform supply of wash water, without otherwise materially altering the original design of the filters. Plans for the contemplated changes were immediately commenced, and the work of reconstruction carried out as rapidly as circumstances would permit. The inevitable delays consequent to such work have retarded its progress beyond expectations, but . . . it is confidently hoped that the filters will be in successful operation in May 1909.

In addition . . . plans are being prepared for a 12-mgd. coagulation basin, which will be built the coming season [1909], and, pending the completion of this, Crescent Hill Reservoir will be utilized as a coagulating basin . . . (58).

The "eminent authorities" mentioned by Leisen as approving his proposed changes were Rudolph Hering and George W. Fuller. Their report (64) indicated that, owing to imperfect working of the filters, the sand layers became clogged by the accumulation of mud removed from the water. They advised that each of the three 0.1-acre filter units be divided into two; that the sand agitating machine be abandoned; that the strainer system be remodeled in general accordance with the one designed for the Cincinnati filter plant, to give a vertical rise of wash water of 20 to 24 in. per minute; that better means be provided to remove the wash water; that in place of taking wash water from the elevated tank and distribution system a separate washwater tank be provided; and that a 12-mgd. coagulation basin be constructed. The estimated cost of these improvements was $150,000.

The filter plant was put into "successful operation" July 13, 1909, the board of works stated, but important additional changes were to follow. Outlay to the end of 1909 had totaled $1,960,000. Leisen, in his report for that year, noted that, in addition to the changes already outlined, lateral wash-water waste troughs had been installed and means provided to supply the coagulant directly to one compartment of the Crescent Hill reservoir.

All elements aiding in the work of purification were working together in 1912, wrote Leisen in his report for that year. Until then the "full benefits from both the sedimentation reservoirs and the coagulation basins were never attainable at the same time."

Chief Engineer Leisen was succeeded in 1914 by James B. Wilson. In his report for that year Wilson states that a supplementary plant of 36-mgd. capacity, built by the Pittsburgh Manufacturing Co., was put in use on November 1. In his report for 1915 Wilson noted that the new filters required an "abnormal amount of wash water." Experiments showed that rather than change a part of the filter material it would be better to lower the wash-water troughs. This had been done for three of twelve filters and had decreased the wash 20 per cent. The same thing was done for the other nine filters in 1916, saving 30 per cent of wash water. Use of the new filters, except for emergencies, was temporarily discontinued.

Reconstruction of the new filters was carried out early in 1924. Reasons for their almost complete abandonment for many years were stated by Superintendent of Filtration W. H. Lovejoy in 1925:

> In 1914 a second battery of twelve 3-mgd. units was built in the north half of the filter house, which had been left for that purpose in the original construction. Soon after starting these latter beds certain weaknesses began to show up that were due to features overlooked in both the design and the equipment. After running these units for two or three years under the constant burden of frequent breakdowns and repairs, we finally were forced to let them lie idle and keep only a part of the units ready for service to help out the old filters in case of peak loads. This was the situation up to about a year ago, when our normal peak loads had increased to such a point that we were forced to start preparations to put these north filters in shape for dependable use (65).

Summarized, the troubles with the filters in question were: (1) filter gallery crowded; (2) operating floor too narrow; (3) troughs too high above sand; (4) all valves too light, and operating motors also too light; (5) effluent controllers stuck and so located that they couldn't be taken apart for repairs.

Lovejoy credited the Hermany filters of 1909 (which were soon revamped, as already described) with having been in continuous and successful operation up to 1925. By 1927, the 72-mgd. combined capacity of the old and new filters was sometimes exceeded due to "operating troubles from algae." Moreover, there was "some doubt as to the condition of the steel filter shells and steel false bottoms in the Hermany filters." Accordingly, Alvord, Burdick & Howson were instructed to plan a 48-mgd. addition to the plant, thus permitting an entire rebuilding of the Hermany filters if this should prove necessary" (58). The third set of filters, combined with the second, gave a capacity of

84 mgd. Removal of the Hermany filters by company forces was begun October 28, 1930. Six new 6-mgd. filters built by contract were put into use in June 1931. They occupied the same space and had the same capacity as the old Hermany filters and restored the capacity of the plant to 120 mgd.

Another important addition to the physical equipment of the plant was made in 1933 when Dorr flocculators were installed in the coagulation basins.

Algae troubles were encountered immediately after the original plant was put into use and again the next year. The three sieges of 1909–10 were described by Lovejoy, who was serving as chemist and bacteriologist of the plant (66). About August 1, the growths cut the filter runs from twelve hours to five. From August 5 to 21, the runs were about six hours; then the filters slowly recovered.

A worse siege of algae occurred from September 1 to October 1, with filter runs reduced to an average of two and one-half hours. The turbidity averaged 50 as against 150 during the August siege. The third siege was worse than the first two. It began August 12, 1910. In those days average length of filter runs was cut from fourteen to one and one-half hours. Use of copper sulfate was decided on but it took a week to get it. On August 20, application of the agent was begun. In three days the filter runs were increased to thirteen hours.

In 1913, Stover, then chemist and bacteriologist of the plant, noted that it would naturally be supposed that the operation of a filter plant would be easiest when bacteria and suspended matters in the water were lowest but many filter superintendents had found "that warm weather and clean water bring troubles peculiarly their own." When the Ohio River has a turbidity below 30 ppm., he said, the length of runs almost invariably decreases, and "if such turbidities are accompanied by micro-organisms, still greater decreases follow. Copper sulfate increased the length of the runs during the recent algae troubles but hypochlorite did not (67).

Returning to the subject of his earlier paper, Lovejoy wrote in 1928: "After experiencing recurring sieges of algae regularly for about eighteen years with their serious and sometimes disastrous effects on plant operation a new scheme of treatment was adopted" (68).

A double-purpose dredge boat was built, tried experimentally in 1927 and put into use in 1928. It was 20 x 40 ft. and 4 ft. deep. On it was mounted an 8-in. electric-driven dredge pump. To this was attached a flexible hose with a suction hood at its lower end. An 8-in. pipeline, on pontoons, was provided to carry the sucked-up mud to a drain when the apparatus was used to clean the basin. But when used to create artificial turbidity the mud was discharged into the raw water at the inlet end of the settling basin. Some of the suspended matter is re-deposited in the settling basin. Some is carried forward to the coagulating basin. "The residual turbidity of 100 ppm. entering the basin," says Lovejoy, "is sufficient to provide a nucleus for an excellent floc which carries down an additional quantity of organisms . . . before the water goes to the filters." He knew artificial turbidity was used for laboratory work but not "on a large scale" (68).

Lovejoy's article prompted Lewis V. Carpenter to report two earlier instances of this method of algae control—one in 1925 at Evanston, Ill., and the other in 1924 at Huntington, W.Va. (69).

In his report for 1928, Chief Engineer Chambers said that the dredge was used for mud diffusion 110 hours in August–September. Besides removing "practically all the mud" remaining in the north basin, it held the filter runs "up to an average well above 20 hours." In 1929, the dredge was operated 28 days for a total of 143 hours in the south basin solely to create artificial turbidity for algae control. This also cleaned the basin.

Comparative tests of copper sulfate, prechlorination and artificial turbidity for algae removal in August–September 1929, indicated that the latter is the most effective and cheapest of the three. The two basins had not been emptied for cleaning since 1924. Under the old methods they would have been due for cleaning in 1927 and 1929. Dredging had kept them "virtually cleaned without interruption in the service" (58).

The worst drought in a hundred years, with only 8.47 in. of rainfall in April–November against a normal of 23.47, occurred in 1930. Suspended matter in the raw water averaged only 60 ppm. This produced the worst algae trouble in 22 years, said the 1930 report, but

. . . by the use of artificial turbidity and a high washing rate in the filters we were able to hold up filter runs better than ever before. . . . Difficulties from high algae content, high hardness and bad tastes and odors complicated our operation very seriously. Phenol taste developed late in December, and this was handled fairly successfully with pre-ammoniation. However, following this a mouldy organic taste appeared which has so far withstood all the varied treatments which we have tried. This was prevalent all along the Ohio River from Pittsburgh to Cairo and apparently none of the plants on the river were able to remove it or combat it perceptibly.

Besides this calamity and a spring cleaning, a fall cleaning was necessitated by the "accumulation of a heavy mat of organisms and organic matter which caused septic action to take place in the basins . . . the first time this has happened in the history of the plant operation. [It] was caused by the long period of drought *and clear water in the reservoir* [Author's italics]."

Prechlorination by hypochlorite of lime was mentioned as under consideration in Leisen's report of 1911 (70) but was not put in use until May 1, 1914. Liquid chlorine was substituted in 1915 and became fixed practice. Postchlorination was added in 1924. Hydrated lime for pH or acidity control has been used continuously since August 1931. Copper sulfate to keep down algae has been used from time to time since August 20, 1909, only 38 days after the plant was put into operation, but from 1927 onward chief reliance has been put on artificial turbidity when the river water has been relatively clear. Pre-ammoniation to prevent the formation of phenolic tastes and activated carbon to control tastes and odors due to micro-organisms and organic matter were adopted in 1930 and 1931 respectively. Since 1936, wrote Filtration Superintendent Lovejoy on August 26, 1942 (71), "pre- and superchlorination have supplanted artificial turbidity for combating algae and for the lengthening of filter runs. Prechlorination at rates of 20–30 lb. per million gallons has been highly successful in solving this problem."

In common with southern cities generally, typhoid was high in Louisville until the twentieth century and persisted at a higher rate than in many other parts of the country until quite recently. For the years 1906–10 the typhoid death rate was 52.7 per 100,000. It fell to 19.7 for 1911–15, to 9.7 for 1916–20, 4.9 for 1921–25 and on down to 0.9 for 1936–40 (72). The filters had gone into use in 1910 but not until 1912 was there complete functioning together of all three purification elements—filtration, sedimentation and coagulation. In 1914–15 prechlorination was begun and in 1924 postchlorination was added. Doubtless these improvements in the water treatment contributed to the decline of typhoid, following the law of diminishing returns and supplemented by other causes for typhoid reduction.

A memorial gateway to Hermany was dedicated in 1930. On a bronze tablet stating that he served as Chief Engineer of the Louisville Water Co. from 1861 to 1908, these words appear: A PIONEER IN WATER FILTRATION.

A pioneer Hermany was. He conducted experiments with slow sand filters for several years in the 1880's. Finding that such filters could not clarify the muddy water of the Ohio he watched the early use of rapid or mechanical filters. Conflicting claims of rival filter companies led Hermany and the directors of the Louisville Water Co. to give the filter companies a free hand for each to show what his make of filter could do with standard working units. Each company was to operate its own filter, with its own staff, at its own expense, but subject to the technical supervision of Hermany and an able laboratory staff headed by George W. Fuller.

To Fuller belongs the credit for drawing revolutionary conclusions from the Louisville experiments, but he had no chance to apply them at Louisville. Hermany designed the plant. Although Fuller understood that he would be consulted, he was not.

The greatest lesson of the Louisville experiments was the important role of presedimentation and precoagulation in the operation of mechanical filters treating highly turbid waters. Fuller's own words on this subject, written to the author nearly forty years after the Louisville experiments and only a few months before Fuller's death, follow:

[The work at Louisville] was genuine pioneering in dealing with coagulation and preparation of turbid waters for application to rapid filters comparable to practice of that date [with filters] operated for the most part under guidance of filter companies without technical advice (73).

How little work filters perform at Louisville compared with the previous treatment of the water is amazing. Figures supplied in 1934 by Lovejoy show that for the years 1910–33 the average percentage of reductions by the three original elements of the plant were:

Turbidity: Filters, 10 per cent, with yearly range from 16.5 in 1923 to 5.5 in 1912.

Bacteria: Filters, 9.6 per cent, with two years showing increases due to aftergrowths caused by prechlorination. For the three years the bacterial removal was reported for filtration and chlorination in combination. For the other years the range was high—0.02 in 1919 to 30.9 for 1930. Chlorination alone during the 19 years it was practiced removed an average of 0.9 per cent of the bacteria in the raw water, with a range from 1.8 to 11.4 per cent.

As will be understood by most readers, the percentage distribution of work done by the main elements of the process depends largely upon the turbidities and bacterial counts of the raw water. This is illustrated by the year of the great drought, 1930, when with a raw-water turbidity of only 60 ppm. sedimentation did only 20 per cent of the work. Coagulation came to the rescue with 67 per cent (both unprecedented figures). Filtration did 13 per cent. In this same drought year the bacteria in the raw water averaged 3,492 and were reduced 14.7 per cent by sedimentation. Here also coagulation did a good job (47 per cent reduction) but not so good as in other years nor as good as sedimentation the same year. The 30.9 reduction by filtration in 1930 was three times the average for the whole period and was approached no nearer than 13.4 per cent. In this dry year chlorination did valiant work with 7.4 per cent of the bacterial reduction to its credit.

Cincinnati's Rapid Filters

Shortly after George W. Fuller completed his Louisville experiments he became Chief Chemist and Bacteriologist for a filter testing station at Cincinnati. The object of the test, wrote George H. Benzenburg, who succeeded George F. Bauscaren as Chief Engineer of the filter plant finally built, was to obtain "reliable data as to the relative merits of the so-called English and [the] mechanical filtering systems in the purification of the Ohio River water" (74). The main tests extended from March 28, 1898, to January 25, 1899. The final conclusion of Fuller's report was that either modified English slow sand filters or American rapid filters would purify the Ohio River but that the latter would be preferable (75). After many delays a 112-mgd. plant, using lime and sulfate of iron as coagulants, was put into use October 22, 1907.

The crude beginnings of a public water supply through private enterprise at Cincinnati date from 1797, when water from wells was hauled through the streets and sold from door to door. Competitors soon began to sell Ohio River water, delivered first by sled, then by cart. Running water was first distributed through pipes in 1821. In 1824 there began a movement for municipal ownership that persisted despite three defeats until the city bought the works in 1839. From that time until the rapid filtration plant was completed in 1907 there was an almost continuous movement for more and better water than the Ohio River in its natural state afforded. Settling reservoirs were recommended in 1861; both these and filters by James P. Kirkwood in 1865; settling reservoirs again in 1871; these and filters in 1880 or 1881. In 1888 sedimentation experiments were made. In 1896 settling reservoirs and filters were recommended for the third time but in 1897 the commission to which the matter was referred advised that further investigations be made before deciding on any method of treatment. These resulted in the testing station and then in the large plant already mentioned.

Major changes in the Cincinnati plant since it was put in use are typical of the problems and progress in the art of water purification during the past thirty-five years. These changes were summarized for use here by Clarence Bahlman (76), Superintendent of Filtration:

Chlorination of the filtrate was begun in 1911; first, occasionally with hypochlorite of lime; then intermittently with liquid chlorine (1915–17); and continuously with liquid chlorine from 1918.

The brass screen which originally separated the 8-in. layer of gravel from the 30-in. sand bed having failed, it was removed in 1915; the depth of the gravel was increased to 14 in. and the sand placed directly on the gravel.

Alum as a coagulant, applied in the presedimentation reservoirs at times of low runoff in the river, was introduced in 1926. It was used from 5 to 20 per cent of the time up to 1934. Iron sulfate and lime are applied constantly to the coagulation basins.

Growth of sand from incrustations caused removal of the excess in 1929 and the placing of 12 in. of new sand above the old sand left in the filter. This was the only renewal of sand from the opening of the plant in 1907 to late in 1934.

Algal growths in the river, following the completion of additional dams in the Ohio River, have often resulted in very short filter runs. The wash-water reservoir of 182,000 gal. was increased to 843,000 gal in 1930.

Intense phenolic tastes which first appeared in 1918, but sometimes do not occur for several years, were completely eliminated between 1927 and 1931 by suspending chlorination and resorting to double coagulation and high lime treatment, up to 1 to 3 parts of caustic alkalinity in the effluent.

Ammonia-chlorine treatment to combat occasional phenolic and algal tastes was provided in 1931. It has been very successful but in one case of a heavy "slug" of phenol did not completely prevent taste formation.

Activated carbon treatment at the entrance to the coagulation basins was provided for in 1931 but not used until 1933 and 1934 when intense algal tastes occurred. Although not always completely successful, it was sufficiently so to stop all complaints.

Bloodworms or Chironomous in the larval state have caused growths which have broken loose from the side walls and slopes of the 19-mil.gal. open clearwell. Contrary to experience elsewhere, the higher residual chlorine made possible by ammoniation has not eliminated these growths.

The drought of 1934 resulted in maximum capacity rates of filtration on several days and led to a study for probable enlargement of the plant.

Enlargement of the plant to a capacity of 160 mgd. was made in 1936–38. Since then it has sometimes been run at a rate of 200 mgd. New features introduced were pneumatic unloading of chemicals, dry chemical feed, mixing by hydraulic jump, flocculators, sludge collectors and continuous removal of sludge.

In the earlier years of my quarter-century experiences at this plant, it seemed to us that a sparkling clear water of uniform bacterial quality was all that we were striving for. In the last decade or so, however, we note an ever increasing consumers' demand for a water free from all objectionable and unnatural tastes, and an increasing interest on the part of industry as to the chemical constitution of water. Although the average annual total hardness is only 100 ppm., the hardness during summer months often approaches 170 ppm., of which 75 per cent is calcium sulfate. Softening the water is, therefore, receiving some attention.

Typhoid was high at Cincinnati until the filtration plant was put in operation. The epidemic of 1887 caused a typhoid death rate of 142 per 100,000, according to Johnson's "Typhoid Toll" (77), compared with a range of 42 to 56 for the five preceding years, and 40 to 71 after that, the 71 being for the year immediately following the epidemic. Filtered water came into use late in November 1907, thus having little effect on typhoid that year. The rate for the years 1902–06 was 59 and for 1908–13 only 11 per 100,000 (77). For 1916–20 the rate was 3.4, for 1936–40 was 1.1; for 1941, 0.2 and for 1942 it was 0.4 per 100,000 (72).

Cincinnati's quest for pure water through a century and a half centered almost wholly on quantity until 1860. Then it became a quest for purer and ever purer water, culminating in rapid sand filtration to which have been added such accessories as chlorine disinfection and taste and odor control by ammonia and activated carbon.

St. Louis's Rapid Filters

Need for removing the heavy burden of sediment from the water supply of St. Louis was recognized at the inception of the first water works in 1829, but the little that was attempted by sedimentation during the next thirty years was nullified by deposits in the small reservoirs provided. In 1865, James P. Kirkwood made plans for settling basins and filters (78). They were approved by the special water commission and he was sent to Europe to get data on practice there. Strong opposition to filtration resulted in the recall of Kirkwood before his report (79) was completed and to building new works with settling basins only.

It is now obvious that the slow sand filter proposed by Kirkwood in 1865 would have been a sad failure at St. Louis, even with more presedimentation than he proposed. Coagulation was necessary. L. H. Gardner showed its possibilities in 1884 by his large-scale demonstration in a St. Louis reservoir, but that seems to have made little or no impression on the city officials (1). Tests of upward-flow filters with reverse-flow wash soon followed (78). Pioneer studies of sedimentation were made by James A. Seddons in 1888-89, preparatory to an extension of the settling reservoir (80). Tests of the "Anderson process" of precoagulation and slow sand filtration, the coagulant being comminuted metallic iron produced in revolving cylinders, were made in the early nineties (78) (see Chap. XVIII). In view of the scientific direction given to these tests it seems a pity that they were not concerned with rapid filters, then widely used but on a small scale. The coming World's Fair at St. Louis led or drove the city to adopt coagulation as an aid to sedimentation and prepared the way for rapid filtration, achieved in 1915, fifty years after the water commission had been forced to resign for its approval of Kirkwood's plan for sedimentation and filtration. The plant of 1915 was the largest installation of rapid filters yet made, with a capacity of 120 mgd. It was located at Chain of Rocks, on the Mississippi River, some miles above the old settling reservoirs and pumping station. In 1929, an 80-mgd. plant was completed at Howard Bend, on the Missouri River, 13 miles west of the city limits, thus providing two independent sources of supply.

Operating methods at both plants, as described February 28, 1934, for use here, by Chief Chemist August V. Graf, were: At both plants, double coagulation, first with milk of lime and sulfate of iron, then with sulfate of alumina, is practiced. Secondary coagulation is necessary to provide floc for successful filtration. Occasionally more sulfate of alumina than is necessary for flocculation is provided to aid in reducing the normal carbonate alkalinity below 30 ppm. The average amount of lime used at Chain of Rocks in the ten years 1924-33 was 5.2 gpg., resulting in a reduction in hardness of 74 ppm. (81).

At the Howard Bend plant, in the language of Mr. Graf, the flow pattern is as follows:

The water enters a shore intake, is pumped into a rising well from whence it flows through a number of presedimentation basins, thence through a rapid mixing conduit where milk of lime and ferrous sulfate are added to the water, through five tangential conditioning basins, through the primary coagulation basins, thence through two large sedimentation basins operated in parallel, through another rapid mixing conduit where aluminum sulfate is added, through a third large basin and thence to the carbonation chamber where carbon dioxide is added, through a conditioning basin and thence to the filters.

Sludge removal at the two plants is a large undertaking, as the following data from Mr. Graf's notes of 1934 show:

At the Chain of Rocks the amount of mud removed averaged 350,000 tons a year for the past 10 years. About half of this was removed by opening the sewer gates for one-half hour at varying intervals and the remainder by labor and teams. The teams were used to draw scrapers which cut off portions of the mass of mud and dragged them to a central gutter, through which water was flowing. The men were provided with scrapers which were used as such and also as braces to keep small A-shaped boxes in place as the mud drawn by the horses and the water used to aid in removing the mud were carried along by the boxes. Since 1928, small tractors, equipped with bulldozers, have been used to push the mud to the central gutter. A few men, provided with scrapers, are used to remove the mud from the sloping sides and to clean up after the tractors.

At Howard Bend the sewer gates on the presedimentation basins are opened as often as necessary to keep the sludge depth below a certain height. When enough solid material, which is not removed by opening the sewer gates, has accumulated, the basins are taken out of service and are cleaned by means of streams from fire hoses. The basins are comparatively small and the bottoms have decided slopes to the sewer outlets. The primary coagulation basins are equipped with mechanical sludge removers which makes draining and cleaning of these basins unnecessary. The sludge is pumped from these basins, a small part being added to the raw water on its way to the presedimentation basins and the greater part being used to fill the low places around the plant. Sludge disposal is not a problem at either plant (81).

In epitome, St. Louis depended on water from wells or the Missouri, hauled to customers from about 1800 to 1829; then, for decades, on water pumped from the river to settling reservoirs of inadequate capacity. After a century of such service, sedimentation was aided by coagulation. Rapid filtration was provided in 1915 and has been supplemented in later years by various accessory innovations, as already outlined.

CHAPTER VIII

Upward Filtration in Europe and America

Utilization of the force of gravity to throw down suspended matter was the primary motive of most of the many designers of upward-flow filters from the seventeenth to the twentieth century. "Upside-down" would be an apt term to apply to the upward-flow filter, for as a rule the media were in layers graded, from bottom to top, from coarse to fine. It was assumed that gravity, acting on the suspended solids in the ascending water, would draw them to the bottom of the raw-water chamber beneath the filter, not only from the water in the chamber but also from the interstices in the filtering media, particularly from the coarse layers of the filter. Likewise, in washing, the larger interstices at the bottom of the filter were supposed to facilitate carrying the dirt below. It was not even suspected that the major force in filtration was adhesion to the media rather than straining. This is made clear by repeated assertions that filters were merely strainers, the small interstices barring the passage of the suspended particles of larger size. This theory was pressed home with supposedly telling force when removal of disease germs, as well as mud, was claimed for filtration.

Such claims, however, did not arise until long after Porzio described multiple filtration by successive pairs of upward-downward filters in 1685 (see Chaps. II and IX). Amy, aware or not of Porzio's plan, brought forward the same general scheme in the 1750's (see Chap. IV). To meet an urgent need during the Siege of Belgrade in 1790, Mederer von Wuthwehr, Surgeon General of the Austrian Army, improvised upward filters in the holds of old transport ships lying in the river.

British and American Patents

The very next year, James Peacock, in the first British patent on filters, claimed as a new and useful departure from current practice, filtration by ascension and cleaning by reverse flow. Whether his patent was ever put into effect, except for a trial about 1800 on three ships of the British Navy, is unknown (see Chap. V). In 1798, Joseph Collier went Peacock one better by patenting a combination of upward sedimentation and double filtration (see Chap. IX). During the next century scores of British and a few American patents on upward filters were taken out. Most of them came to nothing.

Of the few American patents on upward-flow filters, the most notable were four granted in the last half of the century. Henry Flad, an eminent engineer of St. Louis, Mo., was granted a patent March 5, 1867, on a filter much like the one patented by Peacock in 1791 but with better provision for regulating its operation. A few were installed in local buildings. J. D. Cook, of Toledo, Ohio, water works engineer of note in his day, obtained a patent November 6, 1877, on a series of pairs of filters, which alternated upward and downward flow (see Chap. IX). John W. Hyatt, prolific patentee of filters, was granted a patent December 10, 1889, on "Apparatus for Upward Filtration." Raw water was admitted to a filter in a closed tank through strainers at the bottom of the unit. The filtrate was drawn from strainers below the filter surface. No installations have been found recorded. George H. Sellers, a Philadelphia engineer, included upward filters in three American and a British patent taken out in 1896.

Nineteenth and Twentieth Century Installations

At Greenock, Scotland, filter beds designed to work by either upward or downward flow, with reverse-flow wash, were completed in 1827. They were designed by Robert Thom, who built similar filters at Paisley and Ayr, Scotland, about ten years later. At Glasgow, in or about 1830, the Cranston Hill Water Co. built an upward-flow filter (see Chap. V).

Albert Stein, a German-American engineer, completed upward-flow filters at Richmond, Va., in 1832. They failed immediately. This was the first American attempt to filter a municipal water supply (see Chap. VI).

At Leghorn, Italy, three pairs of upward-or-downward-flow filters were installed about the middle of the nineteenth century.

First of the American upward-flow filters after the ill-fated ones at Richmond, Va., was a small one built in 1874 at New Milford, Conn., after designs by B. H. Hull, of Bridgeport. Gravel, sand and charcoal were placed in a brick chamber in an impounding reservoir. Apparently its use was soon discontinued. Croes, in his paper of 1883 on early American attempts at filtration (1), describes briefly, besides the Richmond and New Milford filters, five other upward-flow installations. The locations of these, in chronological order * were: St. Johnsbury, Vt., 1876 or 1877; Burlington and Keokuk, Iowa, 1878; Lewiston, Me., and Stillwater, Minn., 1880; and Golden, Colo., 1882. Filtering media were: charcoal, sand and gravel at Keokuk; gravel and charcoal at Stillwater; sand or gravel at the other places. False bottoms, so far as can be learned, were generally planks or boards on edge, set ¼ to ½ in. apart. The Stillwater filter was 50 × 100 ft. in plan. Next in size was the Burlington filter, 120 × 20 ft. It was built by the Citizens

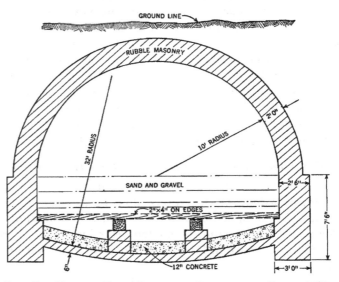

FIG. 49. UPWARD-FLOW FILTER AT BURLINGTON, IOWA, 1878
Designed and built by T. N. Boutelle, Engineer; 20 × 130 ft. in plan
(From blueprint supplied by Frank Lawlor, Supt., Burlington Water Works)

Water Co., of which T. N. Boutelle was Chief Engineer. The Lewiston filter had an area of 400 sq.ft. and was designed by M. M. Tidd, a well-known Boston engineer.

At Pawtucket, R.I., a small upward-flow filter was built in 1883 and superseded in 1888 by a larger one at a new pumping station. Both were designed by Edwin Darling, Superintendent of Water Works, who got the idea for the first installation from the filter at Lewiston. The false bottom of the earlier filter was boards set on edge; of the later

*Some of the dates given by Croes have been slightly changed to agree with later information.

one, an iron grating. The earlier filter consisted of 24 in. of stone, egg sized at the bottom, decreasing to the size of peas at the top. The second filter, from the bottom up, was: stone, 2 in. and smaller in size; 1 ft. of birch and maple charcoal; and 18 in. of stone, graded to the size of pea gravel at the top. When the writer visited the new filter in 1889, soon after its completion, Darling stated that the object of the earlier filter was to remove fish, spawn and inorganic matter of comparatively large size. The new filter, he said, removed nearly all suspended matter and, "it is claimed," also "a large percentage of microorganisms" (2). It should be remembered that even the filter of 1888 was designed when there were only three typical slow sand filters in the United States. The water consumption at Pawtucket in the first full year after the completion of each was 2.3 mgd. and 3.35 mgd. This would give working rates of 200 mgd. an acre for the first and 75 mgd. for the second filter. To clean the larger filter, stated Darling in 1889, the supply was shut off, the filter drained, and water under 40 psi. pressure applied to its top through hose attached at point after point in a pipe running the length of the filter. The larger filter was used for 42 years, or until 1930, when it was given up to make room for a 54-in. supply main. Fine-screening and chlorination were then adopted (3).

Storm Lake, Iowa, put an upward-flow filter in use in 1891 (4), as did Bartlesville, Okla., in 1894 (5). At Bartlesville highly turbid creek water was coagulated and settled before filtration. Superintendent C. E. Perkins stated in 1933 that the scanty information available indicated that the filter was soon abandoned because it could not be cleaned successfully (6).

Coagulation with comminuted metallic iron, produced at the plant, followed by aeration and upward filtration was put in use at Tacoma, Wash., in 1892 and at Wilmington, Del., in 1894. The Tacoma plant was hastily improvised by A. McL. Hawks, under the general direction of George H. Sellers, who had in mind utilization of the so-called Anderson process of coagulation, aeration and filtration—but not by upward flow. Sellers elaborated the Tacoma plant in one built for Wilmington. Susequently he patented his combination. The Wilmington plant worked at a high rate, probably with little benefit either from coagulation or aeration. It was abandoned in 1903 when Theodore Liesen became chief engineer (see Chap. XIII).

Revival of Upward Filtration in England

In the 1920's an upward-flow filter, similar in general principle to Peacock's patent of 1791 and Thom's filters of 1827 and later at Greenock and elsewhere in Scotland, was patented in England by Pennell. It has since been utilized at two municipal water works in England and at many industrial plants in that and other countries. It is a rapid filter, washed by reverse flow. Either open or closed tanks may be used, but the filtering head is low and the washing head only a few feet greater. The municipal filters are small. One at Blackburn was put into use in 1925. At Grange-Over-Sands a filter was completed in 1927 and duplicated in 1930. All these are in open-topped rectangular masonry tanks. The Grange-Over-Sands filters, writes Thomas Huddleston, engineer of the district (7), "are based on the Pennell-Wylie patent," controlled by F. W. Brackett & Co. The makers (8) state that the chief function of the carefully graded sand and gravel used as media is to remove peaty matter from moorland water.

The Brackett Upward-Flow Filter, as the manufacturers have named the Pennell or Pennell-Wylie apparatus,* has a hopper-bottomed settling chamber below the perforated false bottom of the filter. In the open-topped filter, raw water enters the top of a large central vertical tube, passes down into the settling chamber, rises through the false bottom and the sand. The filtrate is drawn off from just above the filter tank. To wash the filter a lever above the filter tank is pushed to one side. This lifts a wash-out valve in the bottom of the hopper. The downward rush of water from above the sand automatically closes the raw-water inlet and carries down suspended matter from the filtering unit and sediment from the hopper. On closing the wash-out valve, the raw-water inflow is resumed automatically. In pressure filters, the raw water is admitted directly into the hopper, just below the perforated false bottom. The filtrate is collected in two concentric pipe rings, just above the unit; one at the circumference, the other a third of the way across. The air between the water surface above the sand and the tank dome is said to be compressed and to reinforce the down-draft of water when the wash-out valve in the hopper is opened. A coagulant is admitted directly to the incoming raw water if needed to help remove either color or suspended matter.

* A British patent was granted to Reginald Humphrey Lee Pennell September 11, 1922, for improvements in the filtration of turbid water or liquid. This patent seems to be the basis of the Brackett filter.

CHAPTER IX

Multiple Filtration: Seventeenth to Twentieth Centuries

Multiple filters were installed in a few places abroad in the nineteenth century. Since then they have come into extensive use in France and have been used less commonly in the United States, Great Britain and other countries. The number of successive filtrations has ranged from two to six. A roughing and a final filter is the general rule, outside of France. Long before there were any permanent installations, small-scale experiments with multiple filters were made and plans for such filters were published.

Francis Bacon, in his *Sylva Sylvarum* of 1627 (1), mentions two sets of experiments on freshening sea water by passing it through ten vessels in one case and twenty in another, the first being a failure, the second a success (see Chap. II). In 1711, Marsigli (2) reported that sea water was not made drinkable by filtering it fifteen times through superimposed vessels filled with clay and sand (see Chap. II).

Porzio, in his book on military camp sanitation, published in 1685 (3), described at length and illustrated a series of multiple filters, the first of each pair working downwards, the second upward. One scheme was for three pairs of filters in a boat, floating in the water to be purified, the other for treating well water on land (Chap. II).

Patents of Three Centuries

As in other fields, so in this one—few of the many patents granted got beyond the patent stage. For the first time, their number and variety is here exhibited.

Joseph Amy, leader of the procession, outdid most of those who followed, in that he made and sold hundreds of filters. In 1745 he applied for a French patent on passing water through sponges inserted in the side of a boat, supplemented, if need be, by filtration through sponges inserted in cross-partitions in a second boat. Sponge having been disapproved, a patent was granted to Amy in 1749 for a filter of sponge and sand placed in a lead, pewter or earthenware container. Among various types of filters described and illustrated by

FIG. 50. J. D. COOK (1830–1902)

Chief Engineer, Toledo, Ohio, Water Works during construction in 1873–74; designed settling reservoirs and filters for original works, but city authorities eliminated reservoirs and reduced filter area, thereby rendering filtration unsuccessful; Cook then designed multiple filters which city did not build; latter design patented in 1879

(From portrait in *Cyclopedia of National Biography*)

Amy in a book published in 1754 was a large one designed for a military garrison. It included three pairs of filters working in series. In each pair the water passed downward, then upward, through 3 ft. of sand, making 18 ft. in all (see Chap. IV). The basic design closely resembles that of Porzio.

The first of a long series of British patents on multiple filters was granted in 1798 to Joseph Collier for an upward-flow filter working under hydrostatic pressure. After passing upward through a settling chamber and a leather diaphragm for clarification the water continued upward through charcoal to "sweeten" it or free it from any "putrid or noxious particles."

Triple filtration, preferably in a watertight chest, was patented in 1810 by Joseph Stephenson. The sequence was downward through sponge compressed by the weight of the water, then downward through sand and charcoal and sand, and then upward through charcoal topped with sand.

Henry, Count de Crouy, took out a patent December 12, 1839, for a "filtering machine" consisting of two or more sets of filters, working under head provided by a cistern.

Double lateral filtration through (a) sand and (b) artificial slabs of coke or coal cinders and coal tar was one of several elements in a patent granted February 14, 1873, to F. H. Atkins.

Caesar Gerson, M.D., of Hamburg, Germany, took out a British patent August 27, 1877, for double filtration: (a) upward through layers of sponge and powdered pumice stone or gravel, each impregnated with insoluble iron salts; and (b) downward through a mass of Swedish iron ore and layers of sand, powdered glass and wool shearings, all impregnated with iron salts. (An American patent was granted to Gerson on March 19, 1878.) A modified form of Gerson's filter was described in 1887 (4), with a statement that it was designed for industrial plants and town water works, requiring for the latter only a ninth of the space needed for sand filters. Several industrial installations in England and on the Continent were mentioned.

Elaborate multiple pressure filters were proposed for construction at Toledo, Ohio, by J. D. Cook, engineer and superintendent of the water works of that city, but the plan was not adopted. He was granted an American patent on a slightly different system of multiple filters on March 6, 1877. The design called for several pairs of narrow settling chambers and filters, as many as might be thought necessary. The flow was downward in the first and upward in the other filters. Any filter could be washed while the others were in use. Broken stone, gravel, sand, charcoal, sponge or other filtering material could be used. No filters of this kind were built.

To remove micro-organisms from liquid which preferably had been passed through an ordinary filter to free it from mechanical impurities, K. Möller proposed, in a British patent of January 14, 1885, secondary filtration through compressed asbestos, amianth, stagwool or other fibrous material.

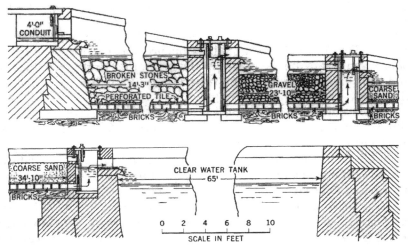

FIG. 51. MULTIPLE FILTERS AND AERATORS AT GLASGOW
Designed and built in 1846–48 by William Gale Sr., Chief Engr., Gorbal's Gravitation Water Co.
(From copy of Gale's drawing supplied by John Cochrane, Chief Engr., Glasgow Corp. Water Works)

Andrew James Bell, of Manchester, England, was granted a British patent March 2, 1885, for a combination of a downward-flow filter to remove grosser impurities and, below it, lateral-flow filters in vertical tubes for additional purification. The filters were cleaned by a reverse-flow current of water or steam. Double filtration following the underlying principle of the Möller and Bell patents was patented in England October 24, 1888 by G. D. Gerson. His primary filter, to remove suspended matter, was composed of sponge, placed at the bottom, and pumice stone at the top, both treated with tannate of iron.

His secondary filter was of pumice stone. Filtration was upward in both filters.

Alexander T. Walker, water works engineer of Reading, England, was granted a British patent on multiple filters February 8, 1893. His design called for a filter of coke or other material followed by a series of polarite filters, all working by upward flow. He had already built multiple filters on the same general principle at Reading, as noted below.

The Reeves Patent Filters Co. and W. Reeves were granted a British patent November 4, 1897, on a combination of presedimentation, upward-flow filtration through coarse material placed above, and downward-flow filtration through a rapid filter cleaned by reverse flow aided by a horizontal revolving rake working in the surface of the filter.

Armand Puech, of Mazemet, France, took out a British patent on May 14, 1898, on multiple filters. The filters were a result of experiments made at his cloth factory in southern France, where he put in a working filter, apparently in or before 1897. With later improvements designed by H. Chabal, this became the Puech-Chabal system, mentioned below.

Installations in the Eighteenth and Nineteenth Centuries

Europe.—The first working multiple filters, which were also the first known filters for municipal supply, were completed in 1804, at Paisley, Scotland. Water was taken from a river through a filter trench of stone to a settling basin, then passed laterally through two filters, the first of gravel, the second of sand. Triple filtration was begun in 1848 by a company supplying a suburb of Glasgow, Scotland.

At Glasgow, Scotland, in the ill-fated filters designed by Thomas Telford and put into use in 1807, water was passed in succession "through a series of cells, filled with sand" (see Chap. V).

In 1812, "Paul found in Geneva" a series of twelve, fifteen or twenty filters working in tandem. The containers had a diameter of 6 in., were 2 ft. high, filled nearly to the top with sand and hermetically sealed. If ten or twelve of these filters were in use, says Delbrück (5), the water would pass through 16 to 20 ft. of sand. No further information on these filters has been found.

John Williams, in a book designed to promote his patented scheme for "Sub-ways," published in London in 1828 (6), suggested that water taken from the Thames above London might be passed downward through a succession of filters, located at lower and lower levels, "until sufficiently pure to enter the pipes."

While at Leghorn, Italy, in 1866, James P. Kirkwood (7) saw a sextuple filtration plant consisting of three pairs of filters. The first filter of each pair worked by downward and the second by upward flow. The filter media in the first five filters were, from top to bottom, fine gravel, coarse gravel, wood charcoal, coarse gravel. In the sixth filter, gravel only was used. As the source of supply was springs, normally clear and bypassed when turbid, the filters and monumental filter house seem to have been more ornamental than needful. This may have been due to the fact that Paschal Poccianti, who designed the water works, was a Florentine architect of "high reputation." Kirkwood does not say when the works were built but as the architect had been "some time dead" it may be assumed that these multiple filters were installed in the 1850's. It may also be assumed that Poccianti got his ideas for multiple filters from Porzio's book of 1685.

Two of the London water companies employed double filtration in the latter part of the nineteenth century, beginning in 1866 (7) and 1891 (12), respectively.

In Belgium, in 1881, after experiments made in 1879–80 by Professor Gustav Bischof, the Antwerp Water Works Co. put in prefilters to treat settled water. They were composed, from the top down, of 2 ft. of river sand, 3 in. of gravel and a 3-ft. mixture of spongy iron and gravel. The final filters were of sand. Provision was made for aeration between the prefilters and final filters. The layer of spongy iron and gravel matted together and clogged. As there were 27 in. of sand and gravel above it, renovation was no easy task (see Chap. XIII).

At Stamford, England, double filtration, first upward then downward, was adopted early in 1882. The media were spongy iron, sand and gravel. A 6-in. layer of spongy iron was separated from the other material by perforated brick, laid dry. Both units were washed by reverse flow. Whether the spongy iron was at the bottom of the filter, as at Antwerp, is not stated in a contemporary description of it (8).

In Holland, double filtration, first through coarse sand and gravel, then through fine sand, was put into use at Zutphen, in 1889, and at Schiedam in 1890. At Zutphen the object was to remove iron and manganese from well water. Before prefiltration, the water was aerated by passing it over weirs. In 1923 a new but similar plant was built for Schiedam, except that sprays were used for aeration.

The double filters at Schiedam were also of coarse and fine material. They treated river water after sedimentation. These filters were used from 1890 to 1921, when a supply, also twice filtered, was obtained from Rotterdam.

Eugen Goetz (9), engineer of the water works of Bremen, Germany, introduced there an unusual system of double filtration in 1895 or earlier and at Altona, Germany, in or about 1896. The water of a newly cleaned filter was siphoned onto an adjacent filter that was "ripe" or in good working order. Siphonage made it possible to have all the filters on the same level. According to Carriére (10), Jewell rapid or mechanical filters, supplied with water settled for ten hours, were installed in 1913 and 1917 to treat water before its delivery to slow sand filters.

Amazing as it will appear to those who have seen the clear water of Lake Zurich, the supply of Zurich, Switzerland, taken from that source, has been twice filtered for four decades. This is largely on account of plankton in the water, two-thirds of which are removed by the secondary filters. Bacteria are reduced one-half by the prefilters and up to 50 per ml. by the final filters. Slow sand filters had been used for many years before prefiltration was introduced. In 1898–99, Water Engineer Peters built prefilters of coarse sand under the roof of the filter house and above the slow sand filters, and substituted finer sand than that previously used. In 1912–14, a new double-filtration plant, higher up the lake, was built. When the author visited the plant in 1904 the prefilters were cleaned by reverse-flow wash, aided by compressed air once in 3 to 4 days; once in five years, the entire bed was removed, washed and replaced. Four prefilters, of fine sand, were added in 1928. [The preceding statements were revised in November 1939 and March 1940, by O. Lüscher, Director of the Zurich water works. He stated that the final filters were cleaned once a year and the sand renewed once in ten to twenty years (11).]

At Reading, England, Borough Water Engineer Alexander T. Walker experimented with multiple filtration in 1889, installed a plant in 1890 and in 1893 took out the British patent already mentioned. Each unit of the plant of 1890 included five prefilters with media progressively diminishing in size, and a final slow sand filter. Somewhat remodeled by Leslie G. Walker, son and official successor of Alexander, these filters were still in use late in 1941. Complete "modernization," planned by the second Walker, was deferred by the

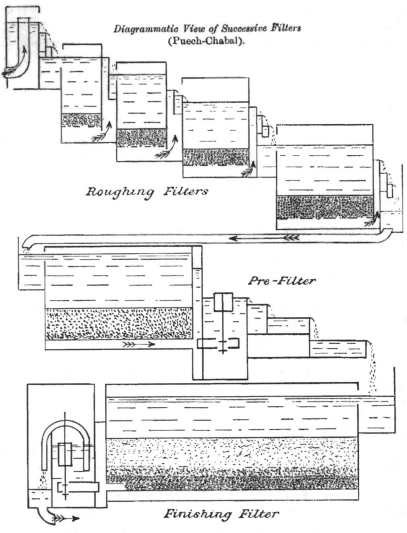

Fig. 52. Puech-Chabal Multiple Filters and Aerators

Flow is through four *degrossisseurs,* a prefilter and a final filter with five cascade aerators

(From John Don's "Purification of Water for Public Supply," *Proc. Inst. M.E.*, January 1909)

second world war. Walker prefilters on another site of the Reading water works deliver water, strange as it may seem, to rapid pressure filters. Postchlorination was begun at Reading in 1910.

Under the leadership of Armand Puech, whose patent of 1898 has been mentioned, and his engineer-lieutenant and successor, H. Chabal, many Puech and Puech-Chabal filters have been constructed. The first of these for municipal supply was completed in 1899 to treat a part of the water supply of Paris. Information supplied in 1935 by H. Chabal et Cie, Paris, showed that about 125 municipal Puech-Chabal plants had been built in France, nearly 20 in Italy, a number in other countries to the south or east and two in England. Each unit includes three or four *dégrossisseurs* or roughing filters, a so-called prefilter and a final filter, the whole working in series or tandem. The filtering material decreases successively in size and the unit-area rate of filtration also decreases from filter to filter. As a rule, there was in the early days no presedimentation, no coagulation, no disinfection. In the passage of time, the roughing filters and the intermediate prefilters have been virtually converted into rapid or mechanical filters, but not so called by the promoters.

The Paris filters of 1899 were installed by Puech before he was joined by Chabal. They were located at Ivry on the Seine and treated water from a settling reservoir. There were three *dégrossisseurs* of gravel and a final slow sand filter. The rated capacity of the plant was 20,000 cu.m. or 5.25 mgd. By 1935, the capacity of the plant had been increased to 300,000 cu.m. or 82 mgd. Meanwhile, the settling reservoir had been converted into roughing filters; prefilters had been inserted ahead of slow sand filters; and air-and-water wash had been provided—at least for all but the final filter.

Of the two Puech-Chabal installations in England, one was completed in 1902 by the East London Water Co. shortly before the Metropolitan Water Board acquired the works of all the water companies in the Metropolitan District. Water taken directly from the Thames at Hamworth was passed through three roughing and one slow sand filter. These filters were abandoned in May 1914, when use of the intake supplying them was discontinued (12). William B. Bryan was engineer of the East London Water Co. when it adopted the Puech system on his recommendation. He journeyed to Paris to see the Puech system there when, by an hour's journey to Reading, he could have seen the Walker sextuple filters that had been in use for a decade.

The second and only other installation of Puech-Chabal filters in England was completed in 1912 for part of the water supplied in bulk by the Derwent Valley Water Board to Derby, Leicester, Nottingham and Sheffield. There are three sets of roughing filters and one of slow sand filters. Before the second world war, R. W. S. Thompson, engineer of the board, began changing the slow sand to rapid gravity filters. Up to December 1941, the work had progressed slowly, due mostly to labor shortage. The Puech-Chabal filters were still in use. Ultimately, it is expected that the roughing filters, which have limestone media, will be used to harden the water before rapid filtration. Another part of the water supplied by the board is treated by rapid pressure filters.

Filters at each end of its gravity aqueduct from Wales were put into use by Birmingham, England, on July 21, 1904. They are notable for being the first large and apparently still the largest installation of double filters in Great Britain. The primary filter is unique because its object, states Chief Engineer A. A. Barnes (13), is to remove "fine solids of a peaty nature, together with *Crenothrix* and kindred iron bacteria" which, says a booklet of 1908, would be deposited in the aqueduct and reduce its carrying capacity (14). Following is Barnes' description of the filters as they were on May 18, 1942:

The primary or roughing filters are of coarse sand. Their area is 13,500 sq.yd. and their maximum capacity, at a rate of 4,500 gpd. [Imp.] per sq.yd., is 60 mgd. [Imp.]. The filtrate is hardened by lime treatment at the rate of 60 lb. per mil.gal. [Imp.].*

The water thus filtered and hardened passes for 73.5 miles through cast-iron and steel aqueducts to open storage reservoirs with a total capacity of 700 mil.gal. [Imp.]. The final filters are of sand, 2.75 ft. deep and 68,500 sq.yd. in area. Their capacity at a maximum filtration rate of 560 gpd. [Imp.] per sq.yd., with a final head of 3.75 ft., is 38 mgd. [Imp.]. Each filter is hand scraped once in ten weeks.

[In U.S. units the roughing filters, working at a rate of 26.1 mgd. per acre, have a capacity of 75 mgd., and the final filters, at a rate of 3.25 mgd. an acre, a 45.6-mgd. capacity.]

Two rapid filters, originally intended for primary filtration, were put into use for Birmingham August 2; 1938. Each unit is 15 x 60 ft. in plan, with sand 2 ft. deep. Working at a rate of 20,000 gal. [Imp.] per sq.yd. per day [about 106 mgd. (U.S.)] per acre, their joint capacity is 4 mgd. [Imp.]. They are backwashed for 10 min. once in 48 hr., with filtered water under pressure

* This was advised, at the start, by the medical officer of health, to prevent action of the soft water on lead service pipes of the distribution system (13).

at three times the rate of filtration. The normal filtering head after backwashing is 2.33 ft., running in 48 hr. to 4.25 and 4.75 ft. The water is so clarified by these filters that no secondary filtration is found necessary. The filtrate is chlorinated at a rate of 0.3 ppm. [Imp.].

To sum up Great Britain's story, where multiple filtration has been adopted the general custom has been to put rapid or mechanical filters ahead of slow sand filters, and not to use a coagulant. This was done at York, in 1900, when an American make of filter was installed under the direction of W. E. Humphrey, Engineer of the York Waterworks Co., who became British agent for the Jewell Export Filter Co.

More than two score years later, after a number of similar adoptions in Great Britain, the London Metropolitan Water Board cautiously began tests of rapid prefilters. In 1925 a small plant was installed. Since then rapid prefilters have been constructed at several of the board's filtration works. The main object has been to increase the capacity of slow sand filters. It should be noted that ample storage and presedimentation of river water has long been provided at London, with pump shutdowns when the water is highly turbid; also that during the first World War, coagulation was introduced to reduce the cost of pumping to storage reservoirs.

United States.—In the United States, multiple filtration came late. Atlantic Highlands and Asbury Park, N.J., included double filtration in iron-removal plants completed in 1893 and 1895 (see Chap. XXI). At Wayne, Pa., double filtration was used in 1895–96, then abandoned.

Maignen scrubbers or roughing filters, generally consisting of sponge above stone, were installed at five Pennsylvania water works in 1900–09. In partly modified form, they were installed at Wilmington, Del., in 1910. All have since been abandoned. The first of these plants was put into use by the Huntingdon Water Supply Co. in 1900. Maignen scrubbers were completed at two of the Philadelphia filtration plants: Lower Roxborough in 1902 and Belmont in 1909. The scrubbers were introduced at South Bethlehem in 1904; Kittaning in 1905; and Lancaster in 1906.

Philadelphia has the largest double filtration plant in the world, at Torresdale, on the Delaware. To lessen the burden on slow sand filters having a nominal capacity of 220 mgd., rapid filters using air-and-water wash, were completed in 1909. In 1920, a presedimentation reservoir, with a detention capacity of twelve hours, was added to supply the rapid filters. Subsequently, coagulation by means of sulfate of alumina was introduced for use at times of high turbidity. At the Queen Lane reservoir, Philadelphia, a 60-mgd. double filtration plant was completed in 1911. In 1922, half of the prefilters were converted into rapid sand filters, with air-and-water wash, raising the capacity of the plant to 100 mgd. (15). Continuation of double filtration at Philadelphia, with probable extension of the use of rapid prefilters, was projected in 1940. Plans were being made in 1942 (15).

At only a few other cities in the United States has double filtration been employed. At Steelton, Pa., prefilters of the rapid type, but using anthracite coal screenings, were completed in 1908, to treat water laden with coal dust and dirt, plus the sewage of Harrisburg.

At South Norwalk, Conn., also in 1908, a double filtration plant with double aeration was put into use to cope with algae causing tastes and odors. Slow sand filters were used, with fine-grained primary filters working at a low rate and a single coarse-grained bed working at a high rate, thus reversing the usual practice. This plan was adopted on the advice of Harry W. Clark. Associated with him in its design was William S. Johnson (16). In 1936, states Engineer Elmer F. Bracken (17), the secondary filter was converted into a clear-water basin (see Chap. XVI).

The only other known prefilters in the United States were installed to reduce the burdens on existing slow sand filters treating highly polluted water. There were three of these: Albany and Poughkeepsie, N.Y., and Lawrence, Mass. At Albany, scrubbers were installed in 1908. They were converted into rapid filters, put in operation January 1, 1928, and used until the abandonment of the Hudson River supply in 1932. At Poughkeepsie, after the slow sand filters of 1872 had been used for nearly 40 years, rapid prefilters were completed in 1920. At Lawrence, rapid filters were completed late in 1938 and put in use ahead of the slow sand filters (see Chap. VI).

Canada.—In Canada, double filtration was put into use at Montreal in 1918. It was practiced intermittently until given up in 1935. Since that time single filtration, through mechanical filters, has been employed. Coagulation was never used at Montreal (18).

The almost complete absence of multiple filtration in the United States today and its complete absence in Canada is due to confidence in single filtration, generally of the mechanical type, supplemented, in nearly all cases, by disinfection with chlorine. The reliability of single filtration and chlorination is backed by public supervision of the hygienic qualities of public water supplies.

CHAPTER X

Drifting-Sand Rapid Filters

A type of rapid filter called "drifting-sand" was put into use at Merthyr Tydfil, Wales, in 1913, and soon afterwards in a half dozen cities of the Western Hemisphere, notably at Toronto and at Recife, State of Pernambuco, Brazil. It was exploited as novel but its basic principle—ejecting filter sand from bottom to top of a filter—had been applied by Hyatt in the United States 30 years earlier and by Bollmann, on the continent of Europe, ten years earlier. The same general principle was adopted by a filter manufacturer at Belfast, Northern Ireland, about 1925, and by a French concern at Paris in the early 1930's. Outlines of these five systems of washing rapid filters, used before or after reverse-flow wash, are here brought together for the first time.

Hyatt Sand-Transfer Wash

In the second stage of the development of the American mechanical filter by John W. Hyatt, the revolving-jet and reverse-flow wash gave place to the sand-transfer system. In this the entire body of sand was transferred by reverse flow through vertical pipes extending from the bottom of the filter unit to the top of an upper compartment and dropped there. Attrition of the sand grains against each other during the transfer dislodged the dirt which was discharged with the wash water from the top of the compartment. On completion of the washing, the cleaned sand was dropped through valves back into the filter compartment. This washing system was first used for a municipal supply in 1885 in the second installation of Hyatt filters for the Somerville and Raritan Water Co., in New Jersey. Subsequently it was used in the much larger Hyatt plant for Atlanta, Ga., and apparently in some other municipal installations.

Hyatt took out two United States patents on sand-transfer wash apparatus, covering many variations. The first of these was dated March 6, 1883 (application filed September 15, 1882). The bulk of the specifications was devoted to describing a quadrangular group of filters interconnected to operate in series, with one empty to receive sand being washed in transfer from one being cleaned. The sand-transfer pipe rose at an angle of 45 degrees from the bottom of the side of one filter to the top of the next. Provision was made for introducing a jet of water in the inclined riser pipe to accelerate the movement of the sand.

The second patent (February 19, 1884; application August 11, 1883) was on a washing process and apparatus by means of which the filter was continuously washed without interfering with filtration. This was effected by "moving a current of the filtering substance, together with a stream of water, from the lower to the upper part of the bed without checking the flow of liquid to be purified, whereby the particles are separated and washed in transit and the impurities are carried off by the escaping water without interfering with the filtering process." This patent covered (a) reverse-flow wash through the underdrainage system, and (b) a "jet pipe placed directly below" the lower end of a central sand-transfer pipe (an ejector). The cleaned sand brought up through the central transfer pipe forms a mound on the top of the bed.

Bollmann Sand-Transfer Wash

A sand-ejector washing system used after the common reverse-flow wash is the leading feature of a rapid sand or mechanical filter made by the Bollmann Filter-Gesellschaft of Hamburg, Germany. The manufacture of this type of filter was begun in 1903. In the early 1930's a more elaborate form was put into use. The first form is called *"einem Lenkkörpern"* and the second *"mehreren Lenkkörpern,"* or "one guide" and "many guides" to direct the sand to the ejector by a "spray wash."

In both forms the upper part of the filter tank is cylindrical, the lower part is shaped like an inverted truncated cone, and the ejector is placed at the base. The ejector discharges into a chamber formed by two cones, the lower cone being inverted. The lower or inverted cone is concentric with the inverted conical bottom of the tank, with a narrow space between the two, down which the sand flows to the ejector when it is in action. Between the upper cone and the conical bottom of the filter tank is a V-shaped space, filled with sand. A "rinse wash," supplied by a separate pipe, aids the flow of the sand down the upper cone of the ejector chamber. This sand joins with that flowing down the inside of the conical bottom of the filter tank to the ejector.

DRIFTING-SAND FILTERS 267

In the simpler type of filter, the combined underdrain and reverse-flow wash apparatus is located just above the bottom of the cylindrical portion of the filter unit and also just above the top of the upper cone of the ejector chamber. This apparatus, as shown in the trade catalog, consists of closely spaced perforated pipe extending in both directions from a central pipe or manifold.

In the latter type of filter, the combined underdrain and reverse-flow wash system is located at a lower level than in the earlier type,

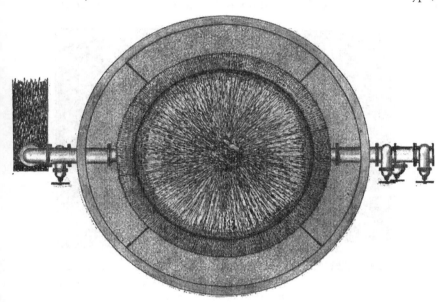

FIG. 53. BOLLMANN COMBINED REVERSE-FLOW AND SAND-EJECTOR-WASH FILTER
View from top of "flat dome-shaped roof or spreader"
(From Bollmann Trade Catalog, Hamburg, Germany, received 1934)

well down in the space between the upper cone of the ejector chamber and the conical bottom of the filter tank. There are two concentric troughs, with perforated bottoms. The filtrate passes up through perforations into the troughs and out through a filtered-water pipe. These troughs are covered by steeply pitched roofs. The roofs are the *"mehreren Lenkkörpern"* or "many guides" which direct the sand each way when being sucked down to the ejector, that is, toward the sloping side of the bottom and upper cone of the ejector chamber.

When water is turned into the ejector of either type of filter, it passes up, laden with dirty sand, through the ejector chamber and then up through a central pipe to the top of the filter. There it overflows upon a flat dome-shaped roof or spreader, then runs down the gentle slope into a circular gutter, from which the dirty water is wasted. The washed sand is discharged from the dome-shaped roof upon the top of the filter unit, forming two concentric craters. When the ejector or secondary wash is completed and the supply valve closed, the reverse-flow wash is turned on and restores the level of the top sand.

FIG. 54. SAND TRAPS, EXTRACTOR PIPES AND WASHERS IN TORONTO DRIFTING-SAND FILTERS
Small inclined pipes lead from sand traps to washer, which extends up through bottom of tank; there are 264 traps and sand pipes and 30 washers to each of the ten filters
(From *Eng. News*, September 21, 1916)

The sand is 1 to 2 m. (39 to 78 in.) deep, varying with the character of the raw water and desired quality of the effluent. The sand is of uniform size. Water is applied to the top of the unit. Filtered water from the bottom of the unit passes up through the bottom of the collecting channels. These, as already stated, also serve to distribute the water for reverse-flow wash.

The maker of the Bollmann filter claims that the whole of the unusually deep unit is effective; "that catalytic and biochemical processes are not confined to the surface layers; that catalysis, dialysis, absorption and agglutination also take place in the lower layers during filtration." The company also claims that reverse-flow wash, even when reinforced by compressed air, does not thoroughly clean the filter, but the cleaning process is completed by the attrition of the sand grains on each other as they are ejected. Mud balls, the maker states, cannot form in the filter.

The largest plant listed in their catalog of 1930 was installed in 1914 and extended in 1925 at one of the stations of the water works of Berlin. Its enlarged capacity was 70,000 cu.m. or 18.5 mgd. (U.S.) (1).

Ransome or Gore-Ransome Drifting-Sand Filter

A decade after the Bollmann sand-transfer wash filter first came into use, the "drifting-sand filter," also employing an ejector washing system, made its appearance in Great Britain, was patented there and was introduced at a few places in Canada and elsewhere. So far as can be learned, the first drifting-sand filter plant ever built, and the only one in Great Britain, was the small one put into use for Merthyr Tydfil, Wales, in 1913. On December 4, 1913, a British patent on the filter was granted to William Gore and Martin Deacon, engineers of Westminster, England. Gore, it is said, was the inventor. The Ransome ver Mehr Co., maker of the Ransome concrete mixer, took over the manufacture of the filter. After the operation of a demonstration plant at Toronto, the city contracted with the John ver Mehr Engineering Co. of Canada for drifting-sand filters with a total capacity of 72 mgd. (U.S.). The plant was put into use in 1918.

The Toronto plant consists of ten main cylindrical filters, each divided into 30 units having a hopper bottom. At the bottom of each hopper there is a sand ejector into which the water to be filtered is discharged and passes upward through a central pipe which serves the double purpose of inlet for a part of the raw-water supply to the filter and sand-transfer pipe. Sand to be cleaned is brought to each ejector by many small tubes, called sand extractors, leading from points on the outer edge of the unit near its bottom. Washing is effected in the ejector, from the top of which dirty water is drawn to waste. The cleaned sand passes up through the sand-transfer pipe and drifts out in every direction, forming a mound shaped like a volcanic crater in each segmented unit of the filter. Sand is constantly drawn from the

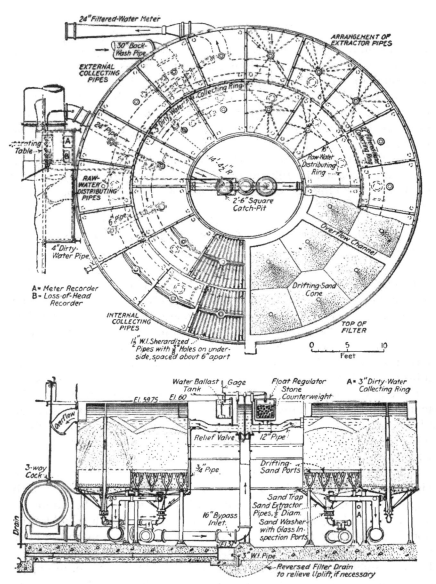

Fig. 55. Gore-Ransome Drifting-Sand Filter at Toronto
Horizontal and vertical sections; continuous washing effected by multiple ejectors which carry dirty sand to top of filter from which it drifts down to a flat conical surface; each filter made of many units, each with one ejector supplied by many extractor pipes
(From *Eng. News*, September 21, 1916)

bottom of the bed, washed and deposited at the top without interfering with filtration. When needed, reverse-flow wash is used (2, 3).

The Toronto drifting-sand filters were built to supplement slow sand filters whose capacity was outgrown. In 1941, a large installation of the latest type of rapid filters was put into use, giving the city three filter plants. Trouble with clogging at the Toronto drifting-sand filters was stopped by probing, and wear on pipes was reduced by replacing $\frac{3}{8}$-in. with $\frac{1}{2}$-in. pipes.

Drifting-sand filters were installed before 1920 at Brampton, Oshawa and Rockland, Ont. (2, 3); at Kingston, Jamaica, B. W. I., in 1915 (4); and at Pernambuco (Recife), Brazil, in 1918—added to in 1921 and in or about 1930 (5). Except at Pernambuco, all these plants have given way to the more common type of rapid filter, either by conversion or new installation. At Pernambuco, a battery of Bollmann filters was added in 1938. Such evidence as can be obtained indicates that the sand ejector pipes gave trouble by clogging.

The drifting-sand filters at Pernambuco were a part of an extensive addition to the water works designed by the late distinguished Brazilian sanitary engineer, Saturnino de Brito. They treated impounded water. Later in 1943, H. C. Baity (6), after interviewing the younger Brito, wrote from Rio de Janeiro as follows:

Sr. Brito tells me that all of the drifting-sand filters which were installed at Recife, including the eight units of Ransome filters constructed in 1918 and the eight units of Bollmann filters added in 1937 or 1938, are in continuous and satisfactory operation. Coagulation and sedimentation facilities were added to the Ransome units after their original installation. During times of heavy raw water turbidity, I am informed by Sr. Brito, the filters often have to be washed as much as three times a day. Whether this can be called satisfactory operation depends upon the standard of comparison.

The drifting-sand filters installed for Merthyr Tydfil were still in use in 1942, being operated by the Taf Fechan Water Supply Board which was also supplying water passed through regular British makes of rapid filters (7).

Turn-Over Filter Company, Belfast

The Turn-Over Filter Co., Belfast, Northern Ireland, in 1925 added to the product from which it had taken its name filters of the sand-transfer-wash type. Up to January 13, 1925, wrote Charles B. Bramwell, director of the company, a number of its "Uneek" filters had been installed on municipal and other water supplies (8). In 1940 the filter was still being advertised. The filter medium is a single-sized fine sand, confined in cylindrical closed tanks, placed vertically or horizontally. An ejector is used to move the sand through a pipe leading up outside the filter and to discharge it onto the top of the same or an adjacent filter—substantially as in the Hyatt patents of 40 years earlier. Reverse-flow wash is also used.

Trailigaz Filters in France

Seventeen water works in France installed rapid sand filters washed by sand transfer and reverse flow during the five years before the outbreak of war in 1939. These filters had the trade name "Trailigaz." They were provided by Societé Traitement des Liquides et des Gaz, Paris.* An installation seen at Rennes, France, in 1938, had a daily capacity of 10,000 cu.m. (2.64 mgd.). It treated highly colored water, containing clay, organic matter and iron salts. Before filtration the water received sulfate of alumina and was settled. After filtration it was chlorinated. The filter layer consisted of 5.75 ft. of sand, placed in a reinforced-concrete tank nearly 10 ft. high, having a hopper bottom. The underdrains had very fine slits. Besides being cleaned by reverse-flow wash, the filter sand was also cleaned by forcing it to the top of the bed by an ejector, placed in the bottom of the hopper. From the top of the discharge tube the washed sand fell back in *gerbes* (showers, as in pyrotechnics) (9).

The basic principles of all the sand-transfer methods of washing filter sand here described are identical: Eject dirty sand, with water, from the bottom to the top of the filter, thus detaching and eliminating material retained by filtration. In effect this is what is done by ordinary reverse-flow wash, which, it should be noted, is used in conjunction with sand-transfer wash. Although John W. Hyatt seems to have put into use only his simpler form of sand-transfer, his patent specifications described apparatus like those of the later inventors: notably, washing of sand continuously and drifting of sand down on top of the filter. As a final tribute to this versatile inventor, it may be noted that his patents anticipated various other devices, including a porous false filter bottom for reverse-flow wash.

* In an advertisement appearing in *L'Eau*, of Paris, it was stated that, after a competition, Paris had decided to put in a Trailigaz plant of large capacity.

CHAPTER XI

Natural Filters: Basins and Galleries

"Natural filters" originated in Europe early in the nineteenth century. Haltingly at first, then rapidly, their use was introduced in America during the last half of the century. Supposed by many to be more efficient and less costly than "artificial filters," they soon gave way to slow sand filters in Great Britain, but continued in favor a long time in France, while in America they gave way to both slow and rapid filters after the late 1880's, or to wells or gravity sources of upland water.

"Natural filters" were built alongside rivers or lakes in permeable alluvial deposits. Usually they were open-jointed conduits built in trenches which were then backfilled. Rarely they were shallow open basins with unpaved bottoms. In theory, they yielded water from the adjacent turbid river or lake, clarified by lateral filtration through 20 to 50 or more feet of sand or gravel. In fact, their yield was largely from the underground flow. To that extent, filter galleries were horizontal instead of vertical wells.

The first filter gallery on record was built in 1810 by the Glasgow, Scotland, Water Works Co. immediately after the crude filters of Thomas Telford had failed (see Chap. V). In addition to being the first, the gallery is notable for three reasons: Boulton and Watt designed a submerged pipeline, with flexible joints, to bring the water from the gallery beneath the Clyde to the existing pumping station; these pump makers disclaimed responsibility for the success of the gallery and urged another attempt with filters instead; about 1830, adjacent lands were flooded to increase the yield of the gallery.

About 1818 the Cranston Hill Water Co., at Glasgow, forced by public sentiment to abandon an intake in a downriver stretch of the polluted Clyde, built a filter gallery near that of its rival company. To the consternation of the second company, the yield of its gallery was so loaded with iron that it was unusable. Consequently, it built a second and then a third set of filters—which were only partly successful (see Chap. V).

Filter Gallery at Toulouse

The second city to be provided with water by "natural filtration" was Toulouse, France, in the early 1820's. The development of this project indicates a mistrust of the "artificial filters" of the period; a choice of "natural filtration" based on analogy drawn from ordinary wells; and the conversion of an open filter basin into a closed filter gallery due to the earliest recorded instance of algae troubles in a municipal water supply.

After the destruction of a Roman aqueduct in the Middle Ages, says Imbeaux (1), Toulouse depended on a comparatively tiny flow of water piped from a spring to a single fountain; on spring water sold by merchants; on water "good to drink" brought to the city by boat from the Garonne; and on public and private wells. At some unstated date, filters of unspecified nature were established at the Samaritain and by licensed starchmakers. From other sources it appears that as early as 1771, Charancourt, when seeking a French patent, presented a certificate from the authorities of Toulouse stating that his process of water treatment was in successful use in that city, but what the process was he did not disclose (see Chap. IV).

Monsieur Lagane, impressed with the inadequacy of the water supply of his city, and wishing to provide a memorial to his wife, willed 50,000 livres to the city in 1789 to be used for the introduction of water from the Garonne. He stipulated that the water must be "pure, clear and pleasing to the taste." The legacy was contingent upon the completion of a "conduit" within six years after the decease of Madame Lagane. Not until April 2, 1817, did she die. On that very day a commission on fountains was created. Its leading member was the engineer, Jean Francois D'Aubuisson, who has been described as the "soul" of the enterprise. He became historian of the new water works (2).

Some method of purifying the water of the Garonne was necessary to meet the terms of the legacy. Four plans for artificial filters were considered and rejected in 1817–20. The commission then invited further proposals.

Abadie, whose plans for pumping equipment were finally executed, proposed that masonry compartments filled with sand and gravel be constructed on the bank of the Tounis Canal. Water would be passed through the filtering material laterally to a pump in the center. Reverse-flow wash, made feasible by local conditions, could be used to clean the filters. Viribent, a local architect, "proposed as a substitute a method used for some years in the establishments that furnish all the potable water to the inhabitants. Here it is purified in traversing, not horizontally, a mass of sand, as proposed by Abadie, nor from above, as in the ordinary filter fountains, but from below and with many repetitions." These two plans were submitted to the Toulouse Academy of Sciences and referred to a committee. Magues, who reported for the committee, said that "when the water of the Garonne was very dirty it was not completely purified when passed in succession through four beds of gravel and sand, each 4 ft. deep, and that 1 sq.m. of these beds, placed one above another, would not clarify 20 cu.m. or one *pouce* in 24 hours." (This would be at the rate of about 20 mgd. per acre of ground space.) Following the "inclination" of Magues and influenced by the fact that local wells always yielded water which was "abundant and limpid," the city decided to dig a test pit or infiltration basin.

Dependability of the proposed filter pit was questioned by D'Aubuisson. Prudence, he said in his *Memoir*, led the water commission to investigate "artificial clarification." Thereupon it called to its aid an architect named Raymond, who had built artificial filters in the city. Raymond submitted plans for a filter to be located near the site chosen for a pumphouse; the nature of this plan is not stated. He assumed that, to produce one *pouce* (20 cu.m. or 5,284 gal.) of water per second, 5 sq.m. of filter surface would be required at a cost of 500 francs per unit, or 100,000 francs for the 200 *pouces* required. This seems to have been considered prohibitive (2). (The yield of Raymond's filter would have been about 4 mgd. per acre.)

Decision to build a test pit or infiltration basin was made in 1821. As executed, this was an excavation in sand and gravel near the Garonne. It was elliptical, 8×14 m. at the bottom and 3.1 m. deep ($26 \times 46 \times 10$ ft.). Three pumping tests by means of a screw of Archimedes led to the belief that by an enlargement of the basin the desired 200 *pouces* (about 1 mgd.) of water could be obtained. The excavation was enlarged to a bottom area of 10×108 m. (approximate), or some 11,622 sq.ft. Many gagings showed a yield of 88 to 98 *pouces* or less than half of what was needed. At first the basin yielded very good water but in following years "aquatic plants" appeared due to the "sun's rays" and "strong heat." Various means employed to remedy the evil "were without effect; the reptiles gained; and these

Fig. 4.
Coupe en travers de la Galerie filtrante N° 3 de Toulouse.

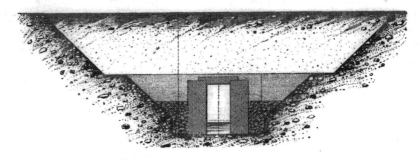

Fig. 5.
Galeries filtrantes de Toulouse.

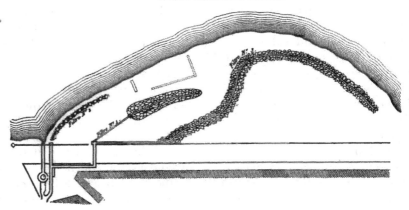

FIG. 56. FILTER BASIN AND GALLERIES AT TOULOUSE, FRANCE
Small open basin of early 1820's gave rise to algae nuisance; open-jointed conduit built on bottom of basin and basin refilled with gravel and sand; other filter galleries subsequently added
(From Darcy's *Fontaines Publiques de la Ville de Dijon*, Paris, 1856)

plants, these animals, died and putrefied in a water lukewarm, making it very bad. Finally it became absolutely intolerable." A committee of the Toulouse Academy of Sciences reported that the water entering the basin was good, but that that leaving it was corrupt.

Covering so large an area as the basin was considered to be impracticable. Instead, at the suggestion of D'Aubuisson, the bottom of the basin was cleaned, "a little aqueduct of brick simply superimposed" was built the length of the basin, well-washed pebbles were placed to the height of the mean level of the river, then smaller pebbles, then sand and other material until the top of the dike was reached. By this means good water was obtained. Its yearly range of temperature was from 17 to 8°C., or 63 to 38°F. "Precious advantage," declared D'Aubuisson. "Cool in summer, it presents an agreeable drink coming from the fountain; warm in winter, it guarantees our fountains from freezing" (2).

Thus did Toulouse afford the first large-scale example of tastes and odors from algae growths in a public water supply and show that they could be prevented by excluding light and air. Thus, also, did it provide the first example of an infiltration basin and then become the second-known city to have a water supply from a filter gallery.

The dates of the various stages of the basin and its conversion into a gallery are uncertain. It is known that, although the city decided to try a "natural" filter in 1821, various delays occurred, involving particularly the design and construction of the pumps. It may be assumed that water from the enlarged basin was delivered in 1825 and from the gallery in 1827, possibly later. A second gallery was soon built, and in 1864 a third, says Kirkwood (3). These galleries, supplemented by springs, supplied Toulouse for about a century (1).

Natural Filtration at Nottingham and Perth

Soon after the conversion of the Toulouse filter basin into a filter gallery, "natural" filters were also built at Nottingham, England, and at Perth, Scotland. At Nottingham a filter basin was built alongside the River Trent, with a filter gallery extending upstream. These, and the entire water works, wrote Water Engineer B. W. Davies (4), "were designed at the very early age of 25, by the father of modern water engineering, the late Thomas Hawksley, who was then engineer to the Trent Waterworks Co." The filter basin and gallery were put into use in August 1831.

Fig. 57. Adam Anderson (Early 19th Century)
Designer of filter gallery for Perth, Scotland, placed in service in 1832 and used for a full century
(From reproduction of a painting by Thomas Duncan supplied by Cyril Walmesley, Engr. & Mgr., Perth Water Dept.)

Testimony of Hawksley before a British commission a few years after the Nottingham system was completed shows that the filter basin there, like the one at Toulouse, was subject to growths of *Conferva*. These were removed at three-week intervals in the summer and six-week intervals in the winter by pumping out the water and sweeping the bottom with a broom (5). Hawksley assumed that the water from the river percolated through 150 ft. of sand and gravel into the basin and gallery. The current of the stream, he said, cleaned the bed of the river opposite the basin and gallery. This doubtless explains why both remained in service for many decades.

At Perth, the filter gallery was designed by Dr. Adam Anderson, Rector of Perth Academy, who investigated various possible sources of water supply from 1814 to 1829 and was engineer for water works put into operation by the town in 1832. From a copy of a report made by Dr. Anderson in 1834 (6), it is evident that he studied various sites for the filter gallery before choosing the one adopted and that to these studies may be attributed a century of service by the gallery. These studies included driven-well tests to show underflow water levels and hardness determinations. They are the most scientific and extensive found among the descriptions of filter galleries built in the nineteenth century.

Records of only a few filter galleries built in Europe after the one at Toulouse have been found. One was included in water works built at Vienna in 1836–41. In the fifties, galleries were constructed at Derby and Newark, England; Lyons and Angers, France; and perhaps at Genoa, Italy.

Kirkwood's Report

This brings our story down to James P. Kirkwood's European observations of 1866. His classic *Report on the Filtration of River Waters, for the Supply of Cities as Practised in Europe* (3) included illustrated descriptions of filter galleries at Perth, Scotland; Angers, Lyons and Toulouse, France; and Genoa, Italy. At Lyons there were two galleries and two filter basins. All the galleries were built of masonry and had unpaved bottoms. The latest of the two galleries at Lyons had the abnormal width of 33 ft.

When Kirkwood's report was published in 1869, there were no filter galleries in America but a filter basin had been built at Hamilton, Canada, in 1859, and one was under construction at Newark,

N.J. Although Kirkwood described artificial filters seen by him in a dozen European cities and natural filters in only five, it was the latter that impressed American water works engineers.

Immediately following the appearance of the report, a number of filter galleries and two basins were built in America: in 1870 at Whitinsville, Mass.; in 1871, at Schenectady, N.Y., Columbus, Ohio, Indianapolis, Ind., and Des Moines, Iowa; in 1872 at Lowell, Mass., and (a filter basin) at Waltham, Mass.; in 1874 at Decatur, Ill.; in 1875 at Brookline and Lawrence, Mass.; in 1878 at Rutland, Vt.; in 1880 at Nashville, Tenn., and Ft. Wayne, Ind.; in 1888 at Green Island and Hoosick Falls, N.Y., and at Springfield, Ill.; in 1891 at Reading, Mass. This is not a complete list. After that, few natural filters were built in the United States. In Canada, Toronto built a filter basin in 1875.

Massachusetts, owing partly to geological conditions, was far ahead of the other states in the number of natural filters built. According to a comprehensive paper by Kingsbury (7), 23 filter galleries and five basins were built in the Bay State in the period 1870–1934, but only five galleries and one basin (Waltham) were in use in 1939, of which none furnished the entire supply for its town. By dates of inauguration, the galleries were: Brookline, 1876; Taunton, 1876; Wellesley, 1884; Newton, 1890; Shirley, 1903; and Great Barrington, 1904.

Des Moines Filter Gallery

Most noteworthy of all the natural filters in the United States is one completed by the Des Moines Water Works Co. in 1871 and still in use by the city, much extended, more than 70 years later. From its length of about $3\frac{1}{2}$ miles in 1938 the city drew its entire supply of about 14 mgd. In dry periods, adjacent land is flooded with water from the Raccoon River, which the gallery parallels. In 1907 Alvord, Burdick & Howson (8) concluded that, with further extensions, the gallery would be capable of supplying the city indefinitely. These conclusions were based on studies of ground water levels in the valley of the Raccoon and of water levels in the river and in the pump well. These and other data indicated that the water in the gallery came largely from the river. So far as appears from other sources of information these were the most complete if not the only scientific studies of the origin of the yield of a filter gallery made since those by Dr. Anderson at Perth about 80 years earlier.

Following the early French filter galleries already mentioned, two notable ones were built in that country: at Nimes in 1871 (9), and at Nancy in 1875 (10, 11). Among still other galleries in France are those described by Debauve and Imbeaux (10) as existing in 1905 at Albi, Carcassonne and Vichy. These authors also mention galleries at Dresden and Essen in Germany; at Bologna, Italy; and at Budapest, Hungary.

A unique development of an additional water supply for Moose Jaw, Saskatchewan, Canada, was reported in 1940 (12). Water is collected in a filter gallery along the South Saskatchewan River and pumped to an open canal leading 68 miles to a 50-acre saturation storage reservoir, formed by removing 2 ft. of top soil above a stratum of sand from 20 to 30 ft. deep. From this the water is collected by 192 well points spaced 15 to 30 ft. apart.

Theory and Fact

Since Kirkwood's report (3) was doubtless the fountainhead of American ideas on natural filters, it is unfortunate that he was not more clear and consistent on the origin of their yield. Apparently he had in mind that some but not all of the water came from the underflow down and across the valley in which they were built. In one place he said that the natural filters at Lyons "are technically filtering galleries but in reality they are collectors or conduits for gathering the water already filtered by natural processes." Whether the water thus collected came from the river or the underflow he did not specify but the context gives the idea that some of it came from the underflow. In his foreword Kirkwood declared that the yield of a filter gallery is "mainly derived from the adjacent stream."

Professor William Ripley Nichols of Boston, writing on natural filtration nine years later, but after there had been some experience with it in the United States, aptly said:

The most favorable situation for a gathering well, basin or gallery is . . . in the neighborhood of a lake or river . . . because: (1) at such a place there is almost certain to be a decided movement of the ground water towards the stream; (2) the water from the river can make up any deficiency caused by the removal of the ground water.

It was formerly supposed, and is so even now by many persons who have not made a study of the subject, that in such cases the water is entirely from the river, and filtered by passing through the intervening sand and gravel. While I would not deny that in some cases a considerable amount of water comes from the stream, I believe that, as a rule, the smaller proportion of the water is thus derived, and in many cases none at all. (13)

Nichols then elucidates his theories at some length, citing the much greater hardness of water from wells and filter galleries in Massachusetts, France and Germany than the water of adjacent rivers and ponds. He summarized his own studies of the yield of the Waltham filter basin, showing that it came from the underflow. He urged that careful studies be made before adopting "natural filtration," including the sinking of perforated well points, regularly spaced, at right angles to an experimental dug well.

Shortly after the date of Nichols' paper, John T. Fanning brought out his *Water Supply Engineering,* which for years was the bible of American water works engineers (14). He noted that wells, filter basins and galleries were sometimes constructed "in the porous margin of a lake or stream, down to a level below the water surface, where the water supply will be maintained by infiltration." In some cases, he said, filter basins are intended to intercept the flow from the land side quite as much as the flow from the water side, this idea being supported by analyses and temperature observations.

From Nichols and Fanning American water works engineers might have learned that the origin of the yield of natural filters could be determined by observations of temperature, hardness (or other analyses) and water elevations. Few if any appear to have done so until the Des Moines studies of 1907.

In view of the early abandonment of most of the natural filters on the one hand and the long service of a few on the other it is pertinent to ask for an explanation. Aside from general obsolescence, such as changes to new and more productive sources of supply, the chief reason for abandonment or material decline in the yield of filter galleries and basins was lack of proper engineering study to choose sites or determine whether local conditions warranted their adoption as sources of supply. Such studies would have included the probable division of yield between the adjacent body of water and the underflow, and particularly the liability of clogging of the material between the river or lake and the gallery or basin. On the subject of clogging Kirkwood wrote after visiting in 1866 five European cities having natural filters:

> The river water which finds its way into the deposit of sand or gravel where the galleries are placed, must have deposited somewhere the sediment held in suspension while in the river channel. I could not learn, however,

that the filtering galleries became unserviceable from any such causes. The deposit which takes place upon the river bottom in the ordinary and in the low stage of its water is removed, it is asserted, in times of floods, when the bottom is scoured of all its light matter, and the coarser earths composing it become in this way periodically exposed. (3)

Kirkwood laid stress on the alternate clogging and clearing of the river bottom. In the Des Moines report of 1907, mention is made of the fouling of shoals in the river. It is important to point out that consideration should also be given to the sealing of the river bank and to the chances that a gallery located parallel to a convex bank is more likely to be kept clean by scour than one alongside a straight or concave bank. The effect of the width of channel in relation to scour should also be considered.

Little has been said regarding the design of filter galleries. All have a family resemblance but like members of other families there are notable individual differences. Structurally, all are long conduits but of varying cross-sectional sizes and shapes. As a rule, they have had arched tops, with tight joints; open-jointed side walls, but sometimes tight-jointed except for openings near the bottom; and either unpaved or open-jointed bottoms. The material usually has been brick, stone or concrete, sometimes vitrified pipe, rarely wood.

Furthest from the family type was the original filter gallery built in 1880 on an island in the Cumberland River at Nashville, Tenn. It was a stone-filled iron cage, 152 ft. long, 6 ft. high and 10 ft. wide, inside. The framework of the cage resembled that of a bridge. Each side wall had an upper and lower chord of 18-in. channel beams. Vertical iron rods spaced 2 in. apart in the clear, extending from chord to chord, served as braces. Two-inch horizontal tubes extending between the top and bottom chords afforded cross bracing. These tubes butted against the beams and were held in place by rods passing through the tubes and channel beams and secured by nuts. Diagonal tie rods were provided. Both ends of the cage seem to have been similar to the sides except that near the bottom of the outlet end there was a long rectangular opening, connected by a reducer to a 36-in. pipe leading to the pumping station. In constructing the gallery a trench was dug on a level with the river bottom. The cage was then erected and the gravel removed from beneath it until the cage rested on bedrock. A cover of "railroad iron" was then added and the trench backfilled to a depth of 20 ft. (15, 16).

Elm planks were used to form all four sides of a filter gallery built at Springfield, Ill., in 1888 and of a second built in 1889. Each gallery was $3\frac{1}{2} \times 4$ ft. in cross-section. The planks were 3×8 in. laid to leave $\frac{1}{8}$-in. spaces between them. Coarse gravel and then sand were placed around the structures (17). Porous rings of reinforced concrete, 36×36 in., placed $1\frac{1}{4}$ in. apart, were used in 1911 to extend one of the galleries (17).

Hemlock "filter boxes" with sides of galvanized wire of 2-in. mesh were laid in a trench at the edge of Lake Erie at Painesville, Ohio, in 1893 or 1894. They were 14 in. wide, 20 in. high, 8 ft. long, and

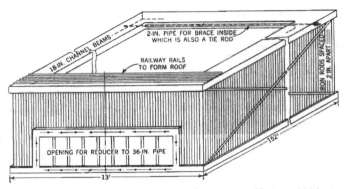

FIG. 58. FILTER GALLERY AT NASHVILLE, TENN., 1880

Originally iron crib, $13 \times 152 \times 6$ ft. high, filled with stone; later additions: 1. sidewalls of stone roofed with stone slabs, the whole surrounded by quarry spalls, around and above which was screened gravel; 2. new gravel (in use in 1889) substituted; 3. cylindrical brick conduit penetrated by 3-in. drains on sides, with broken stone beneath and on sides (1898)

(From sketch by Robert L. Lawrence Jr., Supt. & Chief Engr., Nashville Water Works, based on description in *Tenth Census of United States*)

placed end to end. The trench was over 4 ft. wide and was carried $6\frac{1}{2}$ ft. below the lake level. Two feet of gravel, coal and charcoal were placed on each side and above the top of the boxes, and sand was laid above that. Wave action was to keep the sand clean. (18, 19).

Perforated vitrified pipes were laid at Stonehaven, Scotland, in 1931, to collect water from a gravel stratum beneath and adjacent to a stream. The pipe discharged into a well 4×40 ft. in plan and 9 ft. deep. The pipe and well were located above a submerged dam built across the valley and carried down to impervious material (20, 21).

Possible Revival of the Filter Gallery

The possibility of reviving the filter gallery seems worth considering where local conditions appear to be favorable. Comparative cost estimates might be made after studying all the factors controlling the yield: character of the stratum of permeable material; depth and slope of the water table, both down and across the river valley; probable rate and volume of flow into the gallery, from both the land and the water side; comparative temperature, hardness, iron and manganese contents of the river water and of the underflow, the latter as observed in test wells essential to an investigation. Most of these factors have been used as studies preliminary to developing underground water for city and irrigation supplies. Some of them have been employed, though but rarely, in studying the feasibility of "natural filters."

In considering the possibility of constructing a "natural filter" (or a gallery rather than wells for underflow collection) there would naturally be taken into account trenching and backfilling by motor power instead of the manual labor employed on most of the old galleries; also to be considered in connection with the gallery structure would be various improved means for producing the units, such as large blocks of terra cotta or concrete and precast concrete rings, plain or reinforced, porous or perforated.

CHAPTER XII

Plain Sedimentation

From time immemorial clarification of turbid water by subsidence has been practiced: first, as a household art; then, though rarely, on gravity water supplies brought down by masonry aqueducts; finally, since about 1800, on near-by turbid river supplies, lifted by pumping machinery. Rain water stored in cisterns has benefited by sedimentation, either incidentally or by design. Subsidence by repose has been aided by coagulation from early times, but until well into the twentieth century only on a small scale; recently its use has been on an increasingly larger scale and now is a concomitant of rapid filtration in thousands of water works of all sizes. During the progress of the art of water treatment, removal of color and bacteria, as well as turbidity, has been the object of sedimentation. This chapter, however, will be devoted chiefly to plain sedimentation.

Laodicea may have been the earliest city to have had settling reservoirs to clarify a turbid water supply brought to a city by an aqueduct. In or soon after 260 B.C., Antiochus Theos built an aqueduct some four miles long from the River Caprus to the city, crossing two shallow valleys on arches and a deeper one by twin inverted siphons of bored stone which at one point were under a pressure of about 60 psi. At the terminus of the aqueduct there were two chambers, one 46 × 46 ft. and the other 15 × 15 ft. in plan. A study, made about 1888, of the ruins led to the conclusion that the chambers were settling reservoirs (1).

Remains at Carthage of two many-chambered vaulted reservoirs have been repeatedly described by travelers who visited the site of ancient Carthage from the twelfth to the twentieth centuries. The date of these structures has been the subject of frequent archaeological discussions, particularly in regard to the "lesser cisterns," built for public use to supplement the innumerable private rain water cisterns. For present purposes it will be assumed that the public rain water cisterns were built by the free Carthaginians, before the Romans carried out their declaration *Carthago delenda est*, consummated in seventeen days of carnage in 146 B.C. With reasonable certainty it may be said that

PLAIN SEDIMENTATION

the "greater cisterns" at the lower end of the famous aqueduct were built by the Emperor Hadrian (ruled 117–138 A.D.) to supply the new Carthage built by the Romans in and after 122 A.D. Interest in both reservoirs is heightened by assertions that one or the other was used either for sedimentation or filtration.

The Arabian geographer Idrisi visited the ruins of Carthage in the twelfth century. He reported 24 rain water cisterns, built upon a single line, "the whole disposed geometrically, with great art" (2).

Five hundred years later Thomas Shaw, an English traveler, visited the site of Carthage (3). Besides the cisterns "appertaining to individual houses," he noted "two sets Belonging to the Publick." The smaller of these was at a higher elevation than the larger, "near the Cothon; having been cuntrived to collect the Rain Water which fell upon the Top of It, and upon some adjacent Pavements, made for the Purpose." He added that it "might be repaired with little Expence; the small earthen Pipes through which the Rain Water was conducted, wanting only to be cleaned." Fortunately, Shaw included in his published *Travels* a plan of the rain water cisterns, inserted in a map of ancient Carthage and environs. His plate, with the map curtailed and simplified, is here reproduced, as is also a view of the remains of the cisterns as they stood more than a century later. The general agreement of the plan and view is striking.

Dr. N. Davis, who explored the ruins of Carthage in 1860 and measured and described both sets of cisterns, states in his *Carthage and Her Remains* (4) that the lesser cisterns were beneath the outer walls of the later temple of Aesculapius, were eighteen in number, each $93 \times 19\frac{1}{2}$ ft. in plan, with their water line 17 ft. above their bottom and $10\frac{1}{2}$ ft. below the inner summit of their arched roof. They extended from northwest to southeast. Quoting now:

> On each side there is an arched gallery upward of six feet wide, which communicates with the cisterns, and was probably intended for the convenience of the public in drawing water. Originally there were six circular chambers with cupolas, one at each of the angles, and two in the centre. These may have contained statues, and served, at the same time, as a shelter for those who had charge of the cisterns. Our photographic sketch embraces the ruins of the only cupola still remaining, toward the western angle, as well as that portion of the cisterns upon which the decaying effects of time and the merciless grasp of the barbarian have told least. Near this cupola we obtain a good view of the central division of the whole range of cisterns, the first of which are either partly or entirely filled up, but the remainder are still in

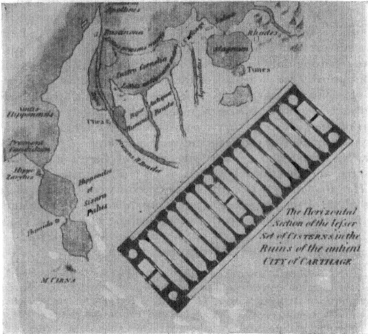

FIG. 59. RAIN WATER STORAGE AND SETTLING RESERVOIRS AT ANCIENT CARTHAGE

Top: Ruins of rain water reservoirs; orientation agrees with plan below
Bottom: Plan, with map of Carthage and vicinity; chief objects on map relettered; end of later aqueduct shown
(Top view redrawn by Whymper from photograph in Davis's *Carthage and Her Remains*, New York, 1861; bottom plan from Shaw's *Travels*, Edinburgh, 1738)

excellent preservation, contain water to this day, and might be restored at a small cost. . . .

These cisterns, which appear to have been surrounded by a colonnade, were supplied by a vast terrace above, which collected the rain-water; and, no doubt, some of the edifices within the precincts of Aesculapius, so close by, also contributed toward filling them. (4)

Near by Davis found a subterranean aqueduct running toward the reservoir at the lower end of the aqueduct.

As seen by Thomas Shaw in 1738, "the grand Reservoir for the Aqueduct lay near the western wall of the City and consisted of more than twenty contiguous cisterns," each about 100 ft. long and 30 ft. broad (see more exact figures by Davis below). "Adjoyning these we see the first ruins [lower end] of the ancient and celebrated Aqueduct, which may be traced as far as Zow-wan and Zung-gar, to the distance of at least fifty miles [about 60] . . . two leagues to the Northwest of Tunis several of the arches are entire, which I found to be seventy feet high" (3).

Instead of the twenty cisterns seen by Shaw in 1738 Davis found only fourteen in 1860. Each was about 400 × 20 ft. in plan. Their depth he did not ascertain as they were filled to the level of the imposts of the arches with an accumulation of earth. Six or seven were entirely available for Arab dwellings, forming the village of Malkah. A fifteenth structure, which may have been a gallery rather than a cistern, abutted the group of cisterns at one end. It was about 18 ft. wide and had an arched roof a little higher than those of the cisterns. Remains of a conduit that once delivered water from the aqueduct to the cisterns were seen. Near by were the "immense ruins" of the aqueduct. Work was in progress for utilizing portions of the aqueduct (filling in gaps with iron pipe) to convey "the delicious water from Zoghwan spring . . . not to magnificent Carthage, but to the filthy, miserable and wretched city of Tunis" (4: pp. 362–63 and 269).

J. J. R. Croes, eminent American water works engineer, during a visit to Tunis in 1880 visited the lower end of the aqueduct and the site of one or both of the many-chambered cisterns or reservoirs. After describing the rain water cisterns he states that

———there is another great reservoir . . . divided into eighteen compartments, two evidently intended for filters or settling basins, and the rest for storage of the settled water. These reservoirs have been restored, and are now used for the water supply of the towns of Goletta and Marsa, water being brought to them from the hills of Zaghouan . . . utilizing for a portion of the way the magnificent aqueduct built 1,670 years ago by the Emperor Hadrian. (5).

Unfortunately Croes did not state his reasons for believing that two of the compartments of the reservoir were used for water clarification, nor whether this dated from their construction or was a modern innovation.

Audollant, writing on Roman Carthage twenty years later, declared, in general terms rather than by naming specific structures, that the Carthaginians built their cisterns in compartments so as to give "progressive filtration" (sedimentation?). He reviewed the statements and opinions of many writers on whether the rain water cisterns were built by pre-Roman Carthaginians or by Romans. Unlike Davis, who argued the subject at length, Audollent concluded that the rain water cisterns were of Roman origin (6). Careful study of all the data the author has seen, including rainfall and topography records and reports from other parts of North Africa, leads to the belief that the rain water cisterns were constructed by the Carthaginians.

Finally, Imbeaux's *Annuaire* of January 1, 1930, states that the aqueduct was constructed in the second century A.D. by the Emperor Hadrian to make up for the deficiency in the ancient cisterns in times of drought, and the report adds that it was cut by the Vandals in the fifth century, re-established after the Arab conquest, and finally plundered anew after the Turkish conquest of 1374. In 1859–64, says the *Annuaire,* the aqueduct was restored by Collin, a French engineer, and in 1886 by the Tunis water administration. Various portions of the aqueduct have been replaced or supplemented by modern types of conduits, including *"ciment armé centrifugé système Hume."* Another up-to-date practice of the Tunis water administration is the sterilization by "javellization" [chlorination] of its water supply (7).

At Rome, a settling reservoir at the head of one of the aqueducts, apparently put into use in 52 B.C., and *piscanae* located on the lower part of each of six of the nine aqueducts, are mentioned by Frontinus (8). His descriptions are the only contemporary engineering records of works built to improve the quality of a public water supply until long after the beginning of the Christian Era. Of the settling reservoirs, Frontinus says:

> The intake of New Anio is on the Sublacensian Way, at the forty-second mile-stone . . . from the river; which flows muddy and discolored even without the effect of rainstorms, because it has rich and cultivated lands

adjoining, and, as a result, loose banks; for this reason a settling reservoir was built upstream from the intake, so that in it and between the river and the conduit the water might come to rest and clarify itself. But in spite of this construction the water reaches the city in a discolored condition, whenever there are heavy rains. The Herculanean Brook, which has its source on the same way, at the thirty-eighth mile-stone, opposite the springs of Claudia and beyond the river and highway, joins it, being of itself exceeding clear, but losing the charm of its purity by admixture. (8: p. 19)

Fig. 60. Piscana on Roman Aqueduct Virgo

Two-story, four-chambered catch-basin; water enters upper left chamber *B* at *A*, drops through *C* to chamber *D*, passes through *E* to chamber *F*, rises through *G* to chamber *H*, returns through *I* to aqueduct; mud or slime passes through flood gate *K* to cloaca built by Tarquinus Piscus
(From Fabretti's *De Aquis et Aquaeductibus Veteris Romae*, Rome, 1680)

Further on Frontinus uses the following graphic language, in speaking of structures on some of the aqueducts:

Of these waters six are drawn into covered catch-basins [*piscanae*] . . . in which, resting as it were from their run and taking a new breath, they deposit their sediment. Their volume is also determined by gauges set up in these basins." (8: p. 21)

Clemens Herschel calls the *piscanae* "covered catch-basins," designed to intercept pebbles that came rattling down the aqueducts which, if not removed, would have plugged the lead service pipes of water consumers. He bases this conclusion, in part, on the fact that in a garden which he visited in Rome, the paths were paved with pebbles said to have been removed from a *piscana*. A model of a *piscana*, in a museum at Rome, was photographed by Herschel and is shown in his book (8: p. 199). It closely resembles the sketch here reproduced from Fabretti.

Vitruvius, writing about 15 B.C., after telling how to wall up wells and cisterns, says:

If such constructions [cisterns] are in two compartments or in three so as to insure clearing by changing from one to another, they will make the water much more wholesome and sweeter to use. For it will become more limpid, and keep its taste without any smell, if the mud has somewhere to settle; otherwise it will be necessary to *clear it by adding salt* [author's italics]. (10)

Vitruvius seems to have had in mind clarification on the premises of the consumer. It would be interesting to know just what salt he had in mind and what specific examples of its use he could have cited in the first century before Christ.

When Caesar began his successful attack on Alexandria, in 47 B.C., he found the royal area of the city supplied with water brought from the Nile through aqueducts to private cisterns which afforded both storage and clarification. In these cisterns, says Hirtius (11), the water "settles by degrees, and becomes perfectly clear . . . the water of the Nile being exceedingly thick and muddy, is apt to breed many distempers."

At Lugdunum (now Lyons, France), the aqueduct system built by the Romans included three reservoirs with vaulted roofs. These were called settling basins in a lecture delivered by William Corfield, Professor of Hygiene and Public Health, University College, London, in 1873 (12). He states that they were on the hill of Fourvieres. One was 48 ft. long, 44 ft. broad and 20 ft. high. Water was admitted through two conduits. It is said that water was drawn through "several round holes in the roof." A second reservoir was 110 ft. long, 12 ft. broad and 15 ft. high, divided by a wall into two chambers. A third was described by Corfield as "large," and as having five of the supporting arches still in place. There was also a discharge conduit 1½ ft. wide from which lead pipes distributed water to the palaces and gardens. He does not give the date these reservoirs were constructed. It may have been 13 A.D. or later.

Modern Times

No more such settling reservoirs appear to have been constructed until modern times. The first one then recorded is notable because it is the earliest baffled settling reservoir to be described. It was apparently built late in the seventeenth century. Henry C. Engelfield reported in 1804:

> A most excellent arrangement for the purification of river water on a large scale is mentioned in the writings of De Luc or De Saussure, but I cannot turn to the passage in their works.* It was applied with complete success to the stream which supplied a large town in Switzerland (13).

FIG. 61. BAFFLED SETTLING RESERVOIR IN SWISS STREAM
(From *Journal of Philosophy, Chemistry and the Arts*, October 1804)

A sketch of the device, here reproduced, shows a series of under- and-over baffles. By means of these, wrote Engelfield, "all floating

---

* Engelfield's forgotten source may have been one of the works of either Jean Andre Deluc, F.R.S., a Swiss geologist and meteorologist (1727–1817), or Horace Benedict de Saussure, a Swiss physicist and geologist (1740–99). Such of their works as I have located lack indexes. I have not seen de Saussure's *De Aqua* (1771).— M. N. B.

impurities will be left at the top, and the heavier mixtures will subside." To clean the reservoir, the partitions might be removed or a man sent down between the baffles.

The nineteenth century opened with three notable instances of sedimentation in Europe, all to lessen the burden on the earliest-known filters for public water supplies. At Paisley, Scotland, a ring-shaped settling basin surrounded the circular filters—the first known filters for general city supply—completed in 1804 by John Gibb. At Paris, two years later, a series of small settling chambers preceded the filters built by Happey on the Quai des Celestins. At Glasgow, in 1807, Telford built very shallow settling basins ahead of his ill-fated filters for a water company. About this time two proposals for settling reservoirs in Great Britain were made: the earliest, in 1804, by Ralph Dodd, a civil engineer of note, for two projected London supplies, another one, by Dr. Hope, in a report on an additional water supply for Edinburgh. These settling reservoirs were in lieu of filters, for which favorable sites were not available. (See Chap. IV for Paris reservoirs and Chap. V for the other reservoirs mentioned here.)

In the 1820's and 1830's, in Great Britain, sedimentation and filtration each had their supporters, with the former leading for a time and the two soon being joined. James Simpson, in the early stages of his investigations, held that settling reservoirs were inadequate and too costly for water clarification. He included a small one in his experimental plant at Chelsea but none in his pioneer slow sand filter plant in 1829 (see Chap. V). Some of the other London water companies put sole dependence on sedimentation for a time but either voluntarily or by Parliamentary mandate built filters before 1860.

In America, until late in the nineteenth century, reliance was placed on settling reservoirs rather than on filters of proper design and capacity, but most of the reservoirs were too small to do the work needed. The first American settling reservoir on record was a part of water works completed in 1829 for Lynchburg, Va. (14), but it should be remembered that up to 1825 there were only 32 water works in the United States, mostly in small places and supplied from springs or clear upland streams (15). The Lynchburg water works was notable both for its settling reservoir and for the pumping machinery that forced James River water to the reservoir under a 245-ft. head. The works were designed by Albert Stein. He it was who completed water works for Richmond, Va., in 1832, which included the unsuc-

PLAIN SEDIMENTATION

cessful first filter in America. Inadequate sedimentation preceded filtration at Richmond (see Chap. VI).

Turning briefly from America to France, we review some little-known experiments on sedimentation and the notable opinions of a French savant. Arago, reporting in 1837 to the French Academy of Sciences on Fonvielle's rapid pressure filter, mentioned experiments on sedimentation * made at an unstated date by Leupold at Bordeaux. The little that Arago said about the experiments is followed by remarks on the limitations of sedimentation which are as true today as they were when made a century ago. Quoting:

> From the very interesting experiments and calculations made at Bordeaux, by M. Leupold, we learn that after ten days of absolute repose, the water of the Garonne, taken at the time of a freshet, had not returned to its natural limpidity. At the commencement, it is true, the larger particles subside very fast, but the finer go down with a slowness which would put all patience at a stand.
>
> Simple repose then cannot be resorted to as a means of clarifying the water destined for the supply of a large city. Who does not perceive that eight or ten separate basins would be necessary for a day's consumption? Add to this that in certain places and at certain seasons, water exposed in a stagnant condition to the open air during ten consecutive days, would become foul and taste badly, either on account of the putrefaction of innumerable insects which would fall into it from the atmosphere, or in the consequence of the vegetation which would begin to take place on its surface.
>
> Repose, however, may be considered as one valuable means of getting clear of the grosser particles which are held in suspension. It is under this point of view *only,* that basins and reservoirs have been contrived and established in England and France. (16) (See also Chap. XIII.)

At Quebec, Canada, in 1848, a settling reservoir for a proposed water supply from either the St. Charles or the Montmorency River was considered. George R. Baldwin † reported that from what he had seen "waters more highly charged with sediment than either of these streams are very often allowed to enter the pipes of distribution in some of the large cities in England and Scotland. The character of the sediment here is probably a sand, coarser or finer according to the violence of the flood" (17).

---

* The date and further details of these experiments have not been found. They must, of course, have been prior to Arago's report of 1837.—*M. N. B.*

† Brother of Loammi Baldwin, better-known early American engineer. George had inspected water works in Great Britain prior to 1834 and perhaps subsequently. —*M. N. B.*

A settling reservoir for use only when the stream was muddy was built in 1854 by the borough of West Chester, Pa. It was 125 ft. square and was located at the pumping station on Chester Creek. This was perhaps the first American instance of a settling reservoir designed for occasional use only and one of few anywhere (18).

Two settling basins and a clear-water basin working in series were put into use at Augusta, Ga., on March 6, 1861. They received water from the local power canal leading from the Savannah River several miles above the city. In rainy seasons the water carried much clay in suspension. Water from the canal was admitted to a settling basin 10 ft. deep, passed to a 200 × 200-ft. basin, 15 ft. deep, then through $\frac{1}{4}$-in. joints in brick paving to a clear-water basin, also 15 ft. deep. The second basin was intended to contain a filter but the Civil War prevented its completion as such (19).

At Kansas City, Mo., a settling reservoir in two long, parallel compartments, with water from the Kaw River flowing lengthwise through one and back through the other, formed a part of water works built in 1874. Each compartment was 350 × 100 ft. in plan and 13 ft. deep. The combined holding capacity was 7 mil.gal. In a description of the reservoir published in 1887, Galen W. Pearson said that it differed from any he knew when he designed it, "in allowing the maintenance of a uniform water level in the subsiding reservoirs"—evidently meaning that it was operated on the flow-through rather than the fill-and-draw plan. He stated that sedimentation gave good results while the consumption did not exceed about 2 mgd. ($3\frac{1}{2}$ days' detention or less) with the important exception that in times of high turbidity "the water had really human aversion to settling" (20).

In or about 1887, a group of settling basins of 60-mgd. capacity was completed at Quindaro, on the Missouri River. The group was horseshoe-shaped, with three basins adjoining a smaller basin of similar shape. Water was admitted near the center of the bottom of the smaller basin. It was withdrawn over a weir at the farther end of the basin into a conduit and passed to the bottom of basin No. 2. From this it was withdrawn at the top and introduced into the bottom of No. 3. The process was repeated in No. 4 (21). Pearson was chief engineer of the National City Water Works Co., which built the works and operated them until they were bought by the city in 1895. Under Wynkoop Kiersted, as consulting engineer, the mode of operating the Quindaro settling basins was changed in

1898–99 by cutting long notches in the dividing walls so that the water flowed from basin to basin at the surface. A large additional basin was put into use in 1910, after designs by Kiersted. It was 600 ft. long by 480 ft. wide and had its bottom laid out as six rectangular flat inverted pyramids. A central pocket, with mud valve, was connected with a pipe leading to a main mud discharge pipe. The new basin was fed from the group of old ones through a 48-in. pipe, its center 5.65 ft. below the water surface. Besides, three 30-in. pipes were laid down the slope to permit feeding from the bottom. If desired, it could be used as a clear-water reservoir (22).

Before the additional basin was built, the use of a coagulant was begun. The first record of such use is in the annual report for 1902, which states that 670,000 lb. of alum were added to 5,467 mil.gal. of water during the year.

After having depended on sedimentation, either alone or aided by coagulation, for 54 years, the city put into use in May 1928 a rapid filtration plant, preceded by sedimentation and coagulation. Chlorination has been practiced since January 1911, beginning with hypochlorite of lime (23).

Other large American cities that were supplied from settling reservoirs before the days of coagulation and filtration included Cincinnati and St. Louis (see Chap. VI) and Omaha (see Chap. XIII).

The natural laws governing sedimentation were first studied in 1888 or 1889 by James A. Seddons, at St. Louis, Mo., preliminary to designing a new system of settling basins for that city. With these studies as a basis it was decided to adopt the fill-and-draw system. The general conclusions drawn by Seddons were:

(A) The basins should not be covered. (B) In the time that could be allowed for settlement there would be no material difference in clearness from the top downward. (C) The time to be occupied in filling the basin could not be counted in the period required for ultimate clearing; for though a considerable quantity of the heavier sediment was deposited during the operation of filling, this filling must stop before the internal currents set up commence to die away. These internal currents were the controlling element in the final clearing of the water. (D) The size and shape of the basin would have little effect in retarding the internal currents. This retardation really depended upon the rate at which the initial motion was consumed by friction between the several interlacing internal currents; the effect of the confining surface upon these currents is simply a factor in the general results. (24)

Fifteen years later Allen Hazen wrote a notable paper on sedimentation. His introduction and final summary follow:

> Since Seddons published his paper on "Cleaning Water by Settlement" (*Jour. Assoc. Eng. Soc.*, 1889, p. 477) there has been but little published discussion on the theory of this subject, but the practice of building and operating sedimentation basins has advanced materially. For example, it has been found in St. Louis that continuous operation, that is to say, a continuous flow of water into, through and out of the basin, gives quite as good results as the intermittent operation which was studied by Seddons, and the new arrangement allows the effluent to be delivered at a higher level, the economic advantage of which is evident. The use of baffles has also been learned, and it has been shown clearly that a well-baffled basin will do as much work as a much larger basin without baffles. A discussion of the subject from a theoretical standpoint, in view of these developments, may lead to a better understanding of it, to the collection of better data, and to improvements in design.
>
> The processes which take place in sedimentation are extremely complex; to discuss them in their entirety seems hopeless. . . .
>
> The fundamental propositions may be very concisely expressed. They are: first, that the results obtained are dependent upon the area of bottom surface exposed to receive sedimentation, and that they are entirely independent of the depth of the basin; and second, that the best results are obtained when the basins are arranged so that the incoming water containing the maximum quantity of sediment is kept from mixing with water which is partially clarified. In other words . . . practically accomplished by dividing the basins into consecutive apartments by baffling or otherwise. (25)

A few other studies of sedimentation, more or less theoretical in character, together with hundreds of references to descriptions of plants, are listed in a bibliography published in 1938 (26). Many references to papers in the journals of the American and New England Water Works Associations may be found in their general indexes.

CHAPTER XIII

Coagulation: Ancient and Modern

Coagulation as an aid to sedimentation for the clarification of individual household and small industrial water supplies has been practiced from ancient times. At large industrial plants its use seems to have begun in the first third of the nineteenth century. It was seldom if ever used for municipal supplies until the fourth quarter of that century. In most cases, coagulation was not adopted as a separate process but was used in conjunction with filtration; it was rarely employed with slow sand filtration but was widely used to precede rapid filtration from 1885 onward. Before this combination came into general acceptance in the early 1900's, coagulation as an aid to sedimentation only was adopted by a few American cities, most of them large ones, and all using water highly turbid at all times and excessively so at intervals. Sooner or later, these cities built rapid filters, with pre-coagulation followed by sedimentation. As time went on, coagulation was introduced where prejudices against it had led to the adoption of slow sand rather than mechanical filtration. Finally, slow sand filtration was almost wholly replaced by rapid filtration, of which coagulation is an essential part. Thus coagulation and filtration, each used in a small way through the ages, were wedded. The practice spread over the world but less rapidly and completely in Great Britain and France than elsewhere.

The agent most generally used as a coagulant is aluminum sulfate—commonly known as alum, filter alum or sulfate of alumina. Lime and iron are frequently used together as coagulants; lime is sometimes employed alone; iron has been used in comminuted metallic form but is more generally employed as an iron salt. Other materials, improvised for time and place, have included almonds, beans and nuts, toasted biscuits and Indian meal.

The underlying principles of coagulation were scientifically defined earlier than those upon which filtration is based and might earlier have been utilized for the improvement of public water supplies had the public demand for clear water been more insistent and had it not been for the influence of ill-informed or prejudiced persons whose word was respected—notably Arago of France in 1837, the Massachusetts State Board of Health a half century later, as well as medical men who protested frantically against both coagulation and rapid filtration in the late nineteenth and early twentieth centuries.

Almonds, Beans, Nuts and Alum

Egypt.—Although coagulation must have been practiced from ancient times in Egypt, no record of it has been found until the close of the sixteenth century. Prospero Alpino, Italian botanist and physician, in a late sixteenth century book on Egyptian medicine, gives an eyewitness account of the use of almonds to aid sedimentation. After noting that "Galenus said in Book I, 'simp. med. facult.,' " that Egyptians once used water filtered through earthen jars, Alpino added:

I saw another method, too, employed by them and frequently made use of, too, by which they made water clear and pure. For as soon as they had brought home water from the stream in camel skin bags they put it away in oblong vessels having a rather wide and round belly and holding two amphoras of water each. When the water had been placed therein and immediately settled they smeared the edges of the vessel with five sweet almonds properly crushed and grasping the almonds in the hand suddenly plunged hand and arm into the water up to the elbow and moved elbow and hand vigorously and violently this way and that through the water stirring it up until they had made it far more turbid than before. Then they withdrew the arm from the vessel, leaving the almonds in the water. They let it clarify and it was properly clear in three hours' time. They put the water from the great vessel into little earthen jars in which it became more clear and cooled. (1)

Felix D'Arcet, French chemist, industrialist and writer on scientific subjects, after a "sojourn in Egypt," wrote a notable paper describing filtration through porous jars in that country, and coagulation with almonds in northern Egypt and with beans in the Soudan. He also described a method of coagulation by alum that he had devised (2). The water of the Nile, he said, was always muddy but during the inundation contained eight grams per liter of suspended matter. Clarification was always employed but rather to avoid mud than as a hygienic measure. The poor, who could not afford the large, costly filter jars used by the rich, used almonds, ground and made into cakes. After placing the turbid water in jars, they rubbed by hand a cake of almonds on the inside of the jar, circularly from top to bottom. Next, the water in the jar was shaken vigorously in all directions.

The jar was then covered and allowed to stand four or five hours during which time it became clear. The operation was performed by the "sacer" or water porter who brought the Nile water to the houses daily. In all the bazaars of Egypt small cakes of almonds could be bought for five parats or about four French centimes. Their average weight was about 64 grams and one served for a month. In the process of clarification the particles of almonds formed a sort of emulsion with the water; the oil united with the earth and the latter settled to the bottom of the jar. At Sennar and Dongola in the Soudan, instead of almond cakes, beans were used—broad, kidney or castor oil—but they did not give full limpidity.

Seeing the imperfection of water clarification by almonds, D'Arcet tried alum which both he and his father had used with success in France for clarifying the water of the Seine. With 0.1 gram of alum per liter of muddy water he obtained complete coagulation within an hour.

Alum, says D'Arcet, gave no cause for alarm on account of health. In its decomposition the excess of acid was taken up by the carbonate or bicarbonate of lime in the water, while the alum itself was entrained in the earthy particles thrown down. Large pieces of crystals of alum were preferable. These were attached to a thread and moved about in all directions. Formation of voluminous flocs and precipitate were signs that the alum had been dissolved. Powdered alum might be used more easily than lump. It was sprinkled on the surface of the water, avoiding pronounced agitation. Or a solution of alum might be employed, in which case the surface of the water was agitated lightly. In Egypt alum with a potassium base was used but an ammonia base would give the same results (2).

India.—The nut of the *Strychnos potatorum* was one of the seven substances used to purify water named in the *Sus'ruta Samhita*, according to a manuscript of 400 A.D. which summarized ancient Aryan and Indic lore (see Chap. I). This nut, wrote James F. W. Johnston in 1854, was at that time used to clarify the marsh waters of India. A supply of the nuts, he said, was often carried by travelers in that country. One or two of them "rubbed to a powder," on the sides of the earthen vessel containing the water, soon caused it to subside (3).

China.—Despite many references to the use of alum for small-scale water clarification in China, almost no writers go beyond mentioning that it was an ancient practice. The earliest eyewitness account of this custom is given by Navarette, a Spanish missionary of the latter part of the seventeenth century. He also mentions the use of coca for the same purpose and refutes theories of Chinese philosophers as to why the Yellow or Red River keeps its color from its source to the sea, attributing the color to the earth over which it runs. The Spaniard wrote:

> The yellow or red river is a remarkable thing, and is therefore called *Hoang Ho*. It springs in the west, runs many leagues without the wall, fetches a great compass about it, and returning again courses through *China* until it comes into the province of *Nan King*, where it falls into the sea. Its course is about eight hundred leagues, it is very rapid, and from its sources keeps a bloody hew [sic], without changing, or altering its colour in any place. When we went to court, we sailed on it two days and a half, and were surprised and astonished to see its whirlpools, waves and colour; its water is not to be drunk, and therefore we laid in our provision before-hand. Afterwards we observed a secret in nature, till then unknown to us, which was that the waterman and servants filled a jar of this water, putting in a little allum [sic], they shaked about the jar; then letting it settle two hours, it became as clear and fair as could be wished, and was so delicate, that it far exceeded the other we had provided, though it was extraordinarily good. In *Canton* I learned another easier and wholesomer cure for it, and it is only putting some small grains which make fish drunk (and in Spanish are called *coca*) into a jar, and the water will become clear in a very short time. (4)

General William Sibert, U. S. Army, of Panama Canal fame, saw water being coagulated in China in or about 1914. He stated:

> When first I boarded our houseboat I saw a Chinaman moving a cane back and forth in a large earthen vessel of water. I saw that the water gradually was being cleared and I asked the Chinaman what he was doing. I found that the cane had been pierced with small holes and that it was full of powdered alum. This alum, in dissolving, clarified the water. . . . This means of clarifying water I found had been used in China for centuries. (5)

England.—The earliest-known reference to the use of alum for water clarification in England is found in Dr. John Rutty's treatise of 1757 on mineral waters (6). Rutty states that "in Kent, where they have little but muddy pond water, it is cleared by throwing into it a little alum." Rutty also says: "I have found several spring waters to become more limpid upon the admixture of alum."

In London, wrote Dr. John Bostock (7), the water served by the New River Company in December 1827 was "very turbid and dark colored." After some hours of sedimentation the water was nearly transparent "but the dark colours still continued." Neither boiling nor filtration through sand and charcoal removed the color but "alum and certain metallic salts (sulfate of iron the most effective), especially when heated, threw down a precipitate and left the water without colour." Three years later, Abraham Booth (8) wrote that alum "clears the foulness of water readily. The salt is decomposed by the carbonate of lime and the alumina carries down all impurities."

The horror of an early nineteenth century Englishman over putting alum or any other "foreign matter" into water to clarify it at a time when most of the London water companies were serving turbid and heavily polluted water from the Thames was cited by Arago in 1837 as follows:

——speaking one day of the aluming of water to an English engineer, whose extensive experience had given him much practical acquaintance with the habits and feelings of the public, and who was lamenting the imperfection of the means now in use for purifying water, . . . "what are you proposing," said he immediately, "Water, like Caesar's wife, ought to be beyond all suspicion." * (9)

In 1843, James Simpson stated that "he had in some instances accelerated precipitation by previous mixture of alumina or pipe clay and other materials and had succeeded in throwing down the colouring matter so that the filters produced perfectly pellucid water" (10). No earlier use of coagulation before filtration has been found. Apparently these "instances" were temporary experiences only.

The possibility of clarifying the water of the Thames at London was noted in 1851 by a chemical commission in a report to the General Board of Health (11). The pronouncement was notable because it stated the cause and character of the turbidity concerned, noted the complete disappearance of the "alum" employed and also the nature of chemical and physical reactions. At that time, it may be added, all of the London water companies serving water from the Thames had or were about to have slow sand filters. Quoting now from the report:

Like rivers generally, the Thames is liable to turbidity from floods. It then acquires a yellow colour, well known as the flood tinge, which is of an unusually persistent colour, and only very partially removed by sand-filtration. . . . This clay tinge, which resists the action of acids and does not even fall down with carbonate of lime precipitated in the water, is known to be

* The shades of Caesar's wife have often been invoked to emphasize the importance of water from an unpolluted source but I have never seen the invocation credited to Arago's English engineer.—M. N. B.

removable by alum. We were informed by an officer of one of the companies, that seven grains of alum per gallon [Imp.] would be sufficient in general to precipitate the clay completely, and to produce a perfect discolouration. The alumina is itself entirely removed, but sulphuric acid is introduced which, by converting lime into sulphate, would induce a hardness permanent on boiling. In floods also the water often tastes disagreeably of vegetable matter. (11)

England's earliest adoptions of alum as a coagulant have been summarized in substance as follows by Peter Spence, Managing Director of Peter Spence and Sons, Ltd., chemical manufacturers in Manchester:

As far back as 1881 a reservoir containing the water supply of Bolton was so turbid that it was objectionable. It was treated at the intake at the rate of 1½ grains per gal. [Imp.] and this gave perfectly clear and colourless water. [At Bacup, early in the twentieth century,] a reservoir was purified by applying aluminoferric and chalk from a raft sailing from bank to bank. (12)

Despite the clear understanding of the nature and possibilities of coagulation in England, as shown above by quotations from seventeenth to nineteenth century writers, and by its successful application in at least two cases, it has been practiced there in comparatively few instances. Not even since the tardy and limited rise of rapid filtration there has it been widely used. Under the emergency conditions of the first World War it was introduced on a large scale by Sir Alexander Houston, Director of Water Examinations, Metropolitan Water Board, and his engineering coadjutors, but the purpose was primarily to reduce the cost of pumping to settling reservoirs.

French Experiments and Opinions

Some industrial works in Paris used alum to clarify the waters of the Seine as early as the 1820's. Evidence of this is found in two paragraphs in a memorial on water purification by Raymond Genieys (13). His clear but cautious remarks on coagulation, freely translated, follow:

Clarification by the use of reagents, notably alum.—Divers means have been employed to hasten the separation of suspended matters. In some establishments at Paris they have tried the use of salts that, by a double decomposition with those contained in the water, form salts of a high enough specific gravity to deposit promptly and carry down with them the matters in suspension. But this means has to be modified according to the almost continual and unexpected changes of dissolved salts in the water, thus requiring many precautions. The only application that we cite is its use in many manufactories and hospitals that, for their service, cannot employ the water of the Seine as pumped directly from the river when the latter is suddenly charged with muddy particles. The means used consists of alum, or sulfate acid of alumina, and of potassium or of ammonia; this salt acts with much efficiency for separating foreign matters suspended in the waters. They do not yet clearly explain the mode of action in this operation; they know only by experience, that if to a hectoliter of very muddy water they add about 5 grams of alum, the water becomes very limpid in a short time.

They think that the elements which this process introduces in the water are in too small proportion to be harmful in the ordinary uses. But this means is not yet in common use, and filtration, which in other respects has all the advantages, without presenting the inconveniences, is today the process more generally prevalent.

A specific instance of an experiment on the use of alum at Paris, in 1828, follows (14).

"Tainted animal matter" was added to water to give it a bad taste and smell. A drachm of powdered alum was added to 2 gal. (Imp.) of this water. After stirring the water, then allowing it to stand for 24 hours, a deposit of 1 in. was found, the water above which was "perfectly pure in taste and smell" and "more clear and sparkling" than a sample that had passed through a thin layer of animal charcoal. A test showed that at least a third of the alum had been neutralized and that the remainder had not caused "astringency which could at all interfere with the valuable properties" of the water or injure its consumers. An equal weight of carbonate of soda was subsequently introduced to neutralize any acidity remaining in the water.

The earliest disapproval of the use of a coagulant in water treatment was expressed by Arago (9). In a report written in 1837, approving the filter of Fonvielle, there is a long passage, heretofore overlooked by historians of coagulation. It says in substance:

Science, or rather chance, brought to light the means of hastening considerably, or rendering almost instantaneous, the precipitation of earthy matters held in suspension by water. This means consists in adding powdered alum to the turbid water. . . . If it is true that water after having been alum'd, still requires filtration, we can easily conceive why the employment of alum, as a means of clarification, has not become general. Besides, in the large way, the price of the salt in addition to other means, might be objectionable. Another more serious objection is that it affects the chemical purity of river water, that it introduces a salt which it did not before contain,—that in supposing this salt wholly inactive in certain proportions, consumers might fear that at times these proportions might be very materially exceeded, and that this might easily happen through the negligence, or mistake, of a workman.

Fig. 62. Dominique Francois Arago (1786–1853)
Secretary, French Academy of Sciences, from 1830 until his death; although he condemned use of powdered alum to hasten settlement of turbid water in 1837, by the following year he helped in advancing the water works art by promoting the Fonvielle filtering apparatus (see Chap. IV)
(From print lithographed at Paris after drawing from life by Maurier)

COAGULATION 307

... [These and other reasons not cited here led Arago to condemn] every means of clarification which would introduce into river water any new substance, that it does not originally contain; and therefore the most recent trials of engineers have all been directed to the employment of inert materials, or those which cannot add anything to the water. These materials are gravel of different sizes, ... sand of different degrees of fineness and pounded charcoal.

French aversion to coagulation was taken advantage of by the promoters of the Puech multiple filters (see Chap. IV) who claimed that by means of their system not only coagulation but also sedimentation was unnecessary. Despite this assertion both have been introduced in some of their installations.

Outstanding among the nineteenth century studies of coagulation by alum was one made by M. C. Jeunnet, at the Central Chemical Laboratory of Algeria, at Algiers, and reported by him in 1865 (15). These studies seem to have been overlooked by all writers on coagulation except Austen and Wilber, twenty years later (16). Had they been seen and studied without preconceived notions there might have been much less opposition to the use of alum or sulfate of alumina in England, France and the United States in the early years of rapid filtration. "The clarifying action of alum, applied at the rate of 200 to 500 ppm., upon muddy waters," wrote Jeunnet, "is an established fact and has been known for a long time; yet it would seem that this method has always been looked upon with misgiving, even when any other method is difficult to practice. ... I am now in position to show upon what slight grounds the apprehensions of the hygienists rest." Jeunnet experimented with waters ranging from distilled water to liquid mud. In no case was it necessary to add more than 400 ppm. of alum to produce clarification in seven to seventeen minutes. Subsidence unaided by coagulation required from one to eight days "and the supernatant was still cloudy." Humus bodies and organic matter seemed to be removed completely. Summarizing his findings, Jeunnet said that

——turbid water, no matter what be the nature or the quantity of the earthy material it holds in suspension, can be made potable within a period of from seven to seventeen minutes by adding to it finely powdered alum at the rate of 400 ppm., provided that the whole bulk of the liquid is, at the time when the reagent is added, in vigorous agitation. In this operation, the alum dissociates instantly into its constituent salts; the potassium sulfate is found quantitatively in the treated water; the aluminum sulfate decomposes and thereby brings about clarification; its base separates in insoluble form and carries down with it the suspended solids and the humus bodies, while the acid attacks the alkaline and alkaline earth carbonates and converts them into sulfates. Therefore the purified water shows enrichment in the sulfates of potassium and calcium, and in free carbon dioxide and sometimes also in bicarbonate; and it has lost its organic content.

Alum in quite a large excess will react in just the same way; it will be completely decomposed without any ill consequences to the water, other than trifling additional quantities of the sulfates of potassium and calcium. When the latter salt already is present in concentration approaching saturation, the reaction is not interfered with, but some calcium sulfate is precipitated.

Sodium alum reacts exactly like ordinary alum, there being no appreciable gain in time from its greater solubility. Aluminum and ferric acetates react only slowly and incompletely and are therefore unsuitable. Although slower in action than alum, aluminum biphosphate might be preferable if it were not that the liberated carbon dioxide redissolves notable quantities of earthy phosphates, which cannot be got rid of, even by boiling.

Aluminum sulfate is equally effective with alum; less of it is needed in ratio of 7 to 10; and it introduces no alkaline sulfate. (15)

Frank Hannan, translator of the above passage, and former chemist for the water works at Toronto, Ont., makes the following comments on some features of Jeunnet's paper:

The pH concept might help to explain certain things. Owing to temperature, pH of Algerian natural waters is probably high. By "alum," Jeunnet always means $KAl(SO_4)_2 \cdot 12H_2O$. We have drifted into calling aluminum sulfate "alum." Superiority of aluminum sulfate as a coagulant is clearly recognized.

The Netherlands and Belgium

A report on the efficiency of alum as a coagulant when applied to water carrying both finely divided clay and organic matter was made by a Netherlands Commission in 1869 or a little earlier. Some members of the commission proved, according to an abstract in a London journal, that the turbidity of Netherlands waters was "due to extremely minutely divided clay," which so held up organic matter that the combined material would "pass through filters and not deposit, even after many days of rest." Alum at the rate of 10 to 20 ppm. produced "a flocculant precipitate" which took up the turbidity of the water and left it perfectly clear (17).

The commission had been instructed to find means of improving the potability of waters where needed, particularly those drawn by villages and towns on the lower Maas, where "from time immemorial" persons not accustomed to using the water daily suffered from diarrhea. Repeated chemical and microscopical analyses had never "revealed the precise cause of this peculiar property, which is not possessed by the water of the same river, higher up" (see also *Iron and Its Salts* below).

Groningen, Holland, seems to have led the way in the use of alum for a municipal water supply. It built water works in 1879–80. While plans for a filter plant were under way, B. Salbach, engineer of Dresden, Germany (18), realized that pretreatment would be needed to remove all color from the peaty water of the River Aa. With the approval of Professors Huizinga and Van Calker, of the University of Groningen, he decided to use coagulation in fill-and-draw settling basins before the water went to slow sand filters. Tests showed that a minimum of 120 ppm. of "sulfated potassium aluminate" was necessary for complete decolorization. At each stroke of the pump which delivered water to the settling basins a solution of alum was also delivered by a small pump, ensuring the mixing of a proportionate amount of the coagulant to the raw water.

Allen Hazen, in the first edition (1895) of his book on filtration (19), stated that alum had been used repeatedly at Leeuwarden, Groningen and Schiedam, Holland, where the river waters are "colored by peaty matters which cannot be removed by simple filtration" but its use had been "generally abandoned, or at least restricted to time when the raw water is unusually bad." At Antwerp, Belgium, he added, resort was made to alum during the drought of 1893 to obtain "a better filter effluent, especially as there was some fear of cholera."

The Leeuwarden plant, apparently built in 1891, combined alum-coagulation with sedimentation, cascade-aeration and filtration (20).

America

Two early American improvised uses of a coagulant illustrate what has probably been done many times in other countries with means readily available. Abraham Booth, in his little book of 1830 (8), states: "Some toasted biscuits put into the water of the St. Lawrence were found serviceable in preventing its bad effects in the fleet of Sir Charles Saunders." Apparently Booth had in mind the large British fleet that arrived near Quebec in June 1759 before Wolfe fell in his gallant exploit. Use of Indian meal, sprinkled on the surface of a pail of water by boatmen on the Mississippi, was reported in 1819 by Schoolcraft, an American mineralogist, who was, however, better known for his writings on the American Indian (21).*

Use of coagulation on a public water supply in America was begun in 1885 by the Somerville & Raritan, N.J., Water Co. The process was firmly established as an adjunct to rapid filtration by that company's project (see *The Hyatt Mechanical Filter*, Chap. VII). Since then it has been, with few exceptions, part and parcel of rapid filtration in America.

Notable American studies of alum as a coagulant were made in January 1885, by Professors Peter T. Austen and Francis A. Wilber of Rutgers University, New Brunswick, N.J. (16). Theirs was the first scientifically conducted American investigation of the subject; it followed closely the granting of the Hyatt patent on simultaneous coagulation and filtration; it fixed a reasonable limit on the alum dosage; it suggested that coagulation as an aid to sedimentation without filtration was a possible treatment method but would require a long period of sedimentation; it also indicated, but with inadequate proof and without emphasis, that alum could be applied to filters without intermediate coagulation basins. The authors stated that of the many reagents suggested and tried for water purification none seemed to offer the advantages of alum. They also noted Jeunnet's paper of 1865, already summarized above, and mentioned the use of alum in the Hyatt filter. They added that

———by the addition of a minute amount of alum, water is rendered capable of a most perfect mechanical filtration. . . . *If it can be proven that alum not only clarifies water, but also removes from it disease germs and ptomaines, its use will prove of incalculable value to the human race, for facts begin to indicate that a vast number of diseases are communicated through drinking water* [author's italics] (16).

In an article by Austen it is stated that alum might be used where purification of the water is needed only at some seasons of the year

---

* At my suggestion, A. V. Graf, Chief Chemical Engineer, St. Louis Water Works, in 1936 tested the efficiency of corn meal in reducing the turbidity and bacteria in Mississippi River water. Two tablespoonfuls of corn meal were added to 2 l. of water, the water stirred, allowed to settle for 30 minutes, and then the top liter siphoned from the bucket. The turbidity of the river water was reduced from 1,900 to 150 ppm., the bacterial count from 5,600 to 2,500 per ml. and the *Esch. coli* index from 10,000 to 9,000. A parallel test with 2.5 g. of prepared mustard per liter gave less reduction of turbidity and total bacteria, but higher reduction of the *Esch. coli* index.—M. N. B.

COAGULATION

and where the expense of large pumps and filters would be a burden (22).

Soon after Austen and Wilber suggested the use of coagulation without subsequent filtration, this practice was followed at several water works, in some cases years before filters were built.

At Omaha, Neb., the water company introduced coagulation 34 years before rapid filtration. On August 1, 1889, when beginning to draw water from a new intake, upriver from the old one, the use of sulfate of alumina was begun. "It was used somewhat crudely for several years," wrote Theodore Leisen (23) long afterwards, "by injecting it into the discharge lines of the low-service pumps and dumping additional amounts into the weirs between the settling basins. Manufacture of aluminum sulfate at the plant was begun in August 1916 by the Metropolitan Utilities District. Rapid filters were put into full operation in October 1923."

The next known American use of a coagulant before filtration was in 1901, at Chester, Pa., when the New Chester Water Co. built a pumping station on the polluted Delaware River and two 7-mil.gal. settling reservoirs on Harrison Hill. Rapid filters were added the next year (24). In 1902, a coagulant was applied to settling reservoirs at Kansas City, Mo. (see Chap. XII).

The next recorded instance of the use of coagulation before filtration, and in some respects the most notable, was at St. Louis, Mo., early in 1904; but instead of sulfate of alumina, lime and sulfate of iron were used (see Chap. VII). Coagulation without filtration was put into use at Selinsgrove, Pa., in 1905 (25); Iola, Kan., 1906 (26); and Nashville, Tenn., in 1908 (27). Thirty-seven instances of coagulation without filtration, scattered through America, were listed in 1915 (28). By that time, rapid filtration was in full swing, generally combined with coagulation (see Chap. VII).

Antedating these plants, there appeared what seems to be the first American water works to employ coagulation as an aid to sedimentation. Although the water was passed through a filter, the filter was so small that it may be ignored, except as a curiosity.

The Vicksburg, Miss., Water Co. built works in 1887–88. Clarence A. Delafield, New York City, was engineer. Water was taken from the Mississippi River. Writing of the character of the water from the point of view of a Northerner, accustomed to clear or relatively clear water, Delafield stated, in a description of the water works:

"This turbid water is thoroughly enjoyed by those who are accustomed to it . . . ; but to those unfamiliar with its use the appearance of the liquid is disgusting, and a clear bright water is insisted on" (29).

The Vicksburg settling reservoir was of the flowing-through type, in three compartments, with combined capacity of 1 mil.gal. "Alum," at the rate of about 1 grain per gallon, was pumped to a long, narrow influent chamber and passed through numerous port holes, 4 ft. above the bottom of the chamber, into the first settling compartment. From the top of this compartment the water went down through passages in the division wall and was discharged into the

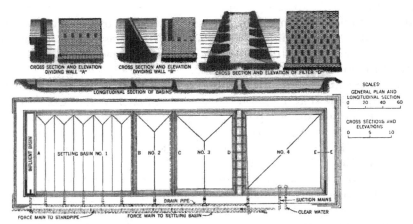

FIG. 63. COAGULATION AND SEDIMENTATION BASINS AND LATERAL-FLOW FILTER AT VICKSBURG, MISS.
Plant designed and built in 1887–88, by Clarence Delafield, Engineer
(From Delafield's "The Vicksburg Settling Basin," *Trans. A.S.C.E.*, August 1889)

second compartment 3 ft. above its bottom. Discharge from the second to the third compartment was in a thin sheet over the division wall. From the final settling compartment, water passed to a small lateral-flow filter located in a wall and thence to a clear-water chamber. Although Delafield stated that after the treatment plant had been in use several months the final effluent was giving "perfect satisfaction," four Jewell rapid filters were installed in 1890. Postchlorination was begun in 1918. The city acquired the works in 1915. At an unstated date the clear-water basin was covered to prevent algae growths (30, 31).

Lime for Clarification

Notwithstanding the extensive use of lime during the heyday of chemical precipitation of sewage, it has seldom been used, alone, for water clarification, although it has been used with sulfate of iron at a relatively few water treatment plants, large and small. Abraham Booth, in his book of 1830 (8), states: "At Senegal [West Africa] where the water is extremely unwholesome, unslaked lime has been used to purify it." At Sandhurst, Victoria, Australia, about 1873, use of lime was adopted to precipitate clay before slow sand filtration; it was used also at a second plant, without filtration (32). William Ripley Nichols (33) wrote in 1883 that "at several other works [than Sandhurst] lime is used . . . sometimes followed by filtration." No hint as to the location of these works is given. Studies of the excess-lime method of water treatment are detailed by Sir Alexander Houston in his London Metropolitan Water Board reports for 1912 (34) and later. In the following year (35), he notes various conditions in which the use of lime "appears to be especially attractive" and mentions the adoption of "liming" by Aberdeen, Scotland, in 1913. In his 1930 report (36), he tells that the excess-lime method was used by the Southend Water Co., England, in 1929—and the company appeared to be still using it in 1939 (37).

In 1943, Wattie and Chambers studied the relative resistance of coliform organisms and certain enteric pathogens to excess lime treatment in order to evaluate the frequent claim that this treatment will make the water bacteriologically safe. They reported that disinfection can be accomplished by such treatment within definite limitations of contact time and pH value (38).

Iron and Its Salts

Iron, either as an element, or as a part of a salt, was covered in many treatment patents and was employed as a coagulant at several water treatment plants in the last half of the nineteenth century. Outstanding among the patentees were: Medlock in 1857, with a proposal to produce iron oxide by merely suspending iron in water (see below); Spencer, also in 1857, with his magnetic carbide of iron as an oxidizing filtering material; Bischof in 1870, with spongy iron, also as a filter medium; Anderson, in the 1880's and 1890's, with his revolving cylinder in which metallic iron was constantly dropped into water as it passed through, the water then being aerated and filtered; and Sellers, who followed Anderson closely in time and method, using, perhaps unwittingly, the suspended iron rods of Medlock's patent, but making the rods revolve and rub on each other. All these schemes, except Medlock's, were put into use, but the total number of plants was small and all were soon abandoned. Better success attended perchloride and some other salts of iron.

Henry Medlock's British patent of January 21, 1857, on treating water with metallic iron, was notable because it aimed at oxidizing organic matter in solution in accordance with a growing conception of the dangers from organic matter or its products of decomposition. He proposed to put "scrap iron, iron turnings, iron wire, or sheet iron" in a vessel, using about 1 lb. of iron per gallon of water. After contact of 24 to 48 hours, the precipitate would be removed by filtration through sand. His specifications describe the chemical reaction of the iron on the water.

Six months after Medlock's patent was granted (June 19, 1857) one was issued to Thomas Spencer, of London, for the removal of "organic color and deleterious gaseous bodies" from water, using an oxide of iron either independently or in combination with subsequent filtration. A second British patent (June 23, 1858) covered the manufacture of "magnetic carbide of iron by heat." Ground to the size of ordinary sand, the product was to be used as a filtering material. If turbidity as well as color were to be removed, fine sand was added to the magnetic carbide. A dissertation on coloring and organic matters in water and their removal was embodied in the specifications of the first patent. A paper by Thomas Spencer, "On the Supply and Purification of Water," published in abstract in 1859 (39), told how he came to develop magnetic oxide or protocarbide of iron and explained its action on water. His correlation of the oxidation by iron oxide and by ozone is noteworthy in a paper published more than 85 years ago.

Unlike most of the multitudinous methods of water purification patented in England and elsewhere, Spencer's filter was adopted for several water works. A brief trade report in *Engineering* (London) (November 16, 1866) under the title, "Filtering Water," stated that filters of magnetic oxide of iron "overlain by sand, are now in use at Wakefield, Southport, and Wisbeach" (England). A week later, Spencer stated (39) that his process was to be applied to the Hooghly River water supply of Calcutta, India. Kirkwood's St. Louis report of 1869 (40) describes the Spencer filters at Wakefield, which he said produced satisfactory results with the water drawn from the Calder, characterized as "of a dark, inky hue, slightly offensive to the senses." A less favorable picture of the Wakefield water supply, with condemnation of the source rather than the filters, appeared in a report of the Rivers Pollution Commission (41). The filters were used for two decades.

Gustav Bischof, Professor of Technical Chemistry, Andersonian University, Glasgow, devised the spongy iron process of water treatment in 1870. Like Spencer, he placed a layer of his patented material below a layer of sand. Bischof, who had the advantage of the early knowledge of bacteria, claimed that his spongy iron destroyed disease germs. After establishing a large business in domestic filters he or his associates organized the Bischof Spongy Iron Water & Sewage Purifying Co. Seven British patents were granted to Bischof in the period, 1870–87. The first and basic patent (September 19, 1870) concerned the use of spongy iron for filtering and purifying water—preferably spongy iron made from purple or spent ore of iron pyrites. Notwithstanding high praise for the spongy-iron process, based on laboratory tests made in 1873–74 by the Rivers Pollution Commission (41), the material seems to have been used by only one municipal water supply in Great Britain—Stamford. A more important adoption was by a British-controlled water company of Antwerp, Belgium, in which William Anderson of London held an important position (42). After laboratory experiments, spongy iron was included in a permanent plant put into use on June 21, 1881. Water from settling reservoirs was passed through three prefilters and three final filters, the former containing spongy iron. Provision was made for aeration between the two sets of filters. In a lengthy discussion on Anderson's paper (42), Dr. Edward Frankland, whose opinion carried great weight in England for many years, emphasized the removal of bacteria by the filters but gave no supporting data. Anderson, in closing the discussion, said it had centered mainly on whether the "bacteria and their germs were or were not destroyed by the spongy iron; but what was wanted was a substance which would take out the colour of the water, which would remove the turbidity and which would reduce the organic contamination to an amount which would admit to be reasonable." Strange doctrine to relegate germs to the rear, but that was early in 1883!

Cementation of the gravel and spongy iron layers of the Antwerp filters, only two years after they were put into use, led Anderson to make experiments which resulted in the "Anderson Process." This was merely a device for producing by attrition finely divided particles of metallic iron, which were to be dropped into water flowing to a slow sand filter. It eliminated the spongy iron prefilter of the Bischof process, served the same general purpose as Spencer's carbide of iron, and was a more efficient method of producing oxide of iron than suspending iron wire idly in water as patented by Medlock in 1857. That the Anderson process did not originate with Anderson, although the revolving purifier did, is shown by the words of both Anderson and his latter-day partner, Easton Devonshire. In a paper read in 1883 (43) Anderson said that he, in association with Henry Ogston, studied the possibilities of agitating finely divided iron and water. It was decided to follow a method suggested by Sir Frederick Abel, "who in Medlock's time had already had considerable experience in the use of iron for purifying water," and had decided in "favor of simple agitation of a comparatively small quantity of iron with the water to be treated." No sooner did Anderson adopt Abel's suggestion than he began to take out patents, first in England, then in the United States. His first English patent, dated November 23, 1883, covered the purification of water by passing it through spongy iron placed in a horizontal cylinder, revolving on hollow trunnions which served as water inlets. To keep the iron constantly falling, longitudinal shelves were attached to the inside of the cylinder. In a second British patent, as also in most of the eight American patents granted to Anderson or his associate, Easton Devonshire, in 1885–91, plain rather than spongy iron was specified.

A permanent installation was put in use by the Antwerp Water Works Co. early in 1883. The entire supply of about 2.6 mgd. (U.S.) passed through three revolving cylinders before going to slow sand filters (42). By May 1904, stated Devonshire (44), the number of revolving cylinders had been increased to five. Air was forced through holes in a false bottom of the cylinders "thus aerating the water and changing the ferrous to ferric oxide." Owing to variations in the character of the water, the iron did not always "precipitate completely before passing through the filters, making it necessary to throw the revolving purifiers out of service." Rusting tanks were tried as a substitute for the cylinders (44). Apparently this arrangement was unsuccessful for in 1908 Puech-Chabal multiple filters were installed (45). On January 1, 1930, a second Puech-Chabal plant was also in use, much farther up the Nethe. At each plant precoagulation with sulfate of alumina was being used—unusual with Puech-Chabal filters (46). The Anderson process was used at a few other places on the continent of Europe and a plant was installed at Worcester, England, in 1892. Allen Hazen, in his book on filtration (1895), reviewed the Anderson process and expressed himself unfavorably regarding it (19). Judging from opinion expressed late in 1899 by several European water works men in discussing a paper by Weyl, the Anderson process had by then nearly run its course on the Continent (47).

A campaign to introduce the Anderson "revolving purifiers" in the United States, extending through several years, was a complete failure. In the early 1890's, demonstration cylinders were set up at Boston, Philadelphia, Allegheny (48), and St. Louis, and what was intended to be a permanent plant, which never functioned, was installed at Poughkeepsie, N.Y. A notable account of the trial at St. Louis, with reasons for its failure, appeared in the annual report of the water works for 1893–94.

By far the most ambitious attempt at coagulation with metallic iron was made at Wilmington, Del., by George H. Sellers, beginning in the 1890's. Following the construction of emergency plants built for the Tacoma, Wash., Water Co., of which Sellers was chief engineer, he obtained a contract for an elaborate plant at Wilmington, which was operated until 1903 and then abandoned as useless when Theodore Leisen became chief engineer of the Wilmington water works. Bundles of revolving iron rods were supposed to produce a coagulant applied to an upward-flow filter working at a high rate. The process was combined with aeration. The general idea both at Tacoma and at Wilmington was suggested by the Anderson process. Sellers, whether knowing it or not, harked back to the Medlock patent of 1857 in using iron rods, but went one better by having them revolve. He took out four United States apparatus patents and one process patent on July 24, 1896, and a British patent three days later, and organized the United States Filtering & Purifying Company of which he was manager.

Perchloride of Iron.—Perchloride of iron for the purification of water was studied in Europe and America between 1850 and 1900.

Its use was proposed and may have been adopted by Amsterdam in the late 1860's. L. H. Gardner induced St. Louis to try it in a large reservoir in the early 1880's. He experimented on a small scale and, at his suggestion, it was tried by Isaiah Smith Hyatt at New Orleans in 1883. This trial seems to have led to the Hyatt coagulation-filtration patent of 1884 (see *The Hyatt Mechanical Filter,* Chap. VII).

The earliest study of perchloride of iron for water treatment found was reported by Peligot early in 1864 (49), in a memoir on the composition of water. After two paragraphs describing the waters of the Seine and the Canal de l'Ourcq at or near Paris, Peligot included perchloride of iron in a discussion of several coagulants that would clarify swamp or putrescent waters. The passage reads:

> Copper sulfate, ferrous sulfate, ferrous chloride and especially ferric chloride produce in [water], when added in suitable quantity, flocculent precipitates which settle out more or less rapidly at the bottom of the vessel. In the case of ferric chloride, the deposit, in the form of ochreous floc, settles out in a few minutes. In the case of copper sulfate, settlement of the greenish precipitate which is formed is not complete until after standing for twelve to fifteen hours.

Seventy-three days after the publication of Peligot's memoir, Carl J. A. Scheerer, of Freiburg, Saxony, obtained a British patent (July 7, 1864) on "an improved process for cleansing and clarifying water by adding to it a very dilute solution of the neutral sulfate of peroxide of iron ($Fe_2O_3 + 3SO_3$)." Shortly after this agent is added to the impure water

> ———it becomes decomposed, and forms, with some of the impurities contained in the water, a basic salt, which is insoluble in water. The solid and insoluble particles of this new salt are precipitated, and, together with the impurities in the water, form a sedimentary deposit from which the purified water may be allowed to run off, leaving the sedimentary deposit in the tank or reservoir.

Perchloride of iron was given much attention in a report on water purification made by a Netherlands Commission in 1869 (17). Dr. J. W. Gunning of Amsterdam is cited as having shown that 0.032 g. of perchloride of iron, added to 1 l. of water from the River Maas, or to even fouler water, made it "perfectly wholesome and even agreeable for use." After setting forth more testimony of the same tenor the report says that water treated with perchloride of iron and carbonate of soda, then filtered through fine sand, would be supplied to

COAGULATION

Amsterdam through the water works then proposed for that city. The report also stated that the committee felt justified in issuing an order for the use of the process on the bad waters existing in many parts of the kingdom (17). (For a portion of the report dealing with alum, see above.)

Overlooked by previous writers on coagulation are the studies of the 1880's on perchloride of iron as an aid to sedimentation made by L. H. Gardner, Superintendent of the New Orleans Water Works Co. for many years before water works were built by the New Orleans Sewerage and Water Board. In May 1883, he briefly outlined his small-scale experiments with perchloride of iron on the muddy Mississippi at New Orleans (50). In October 1885, he published a summary of methods of water clarification used or proposed during the two decades past, including his office experiments at New Orleans and his tests on a 12-mil.gal. reservoir at St. Louis. He outlined Spencer's magnetic carbide and Bischof's spongy iron filters and the pre-coagulation methods of Anderson (see above). After trying all known precipitants, he "was convinced that a solution of iron is of easier mechanical application to water" than any of the others tried; that the "rapidity of action narrows the necessary area of settling basins materially"; that "mechanical devices of the simplest character" would admit to the influent water $\frac{1}{5}$ lb. of iron to 1,000 gal. of water; and that "settling reservoirs can be made self-cleaning." In this paper (51), Gardner related his earlier suggestion to Isaiah Smith Hyatt that led to the coagulation-filtration patent of 1884 (see Chap. VII).

Gardner's advocacy of iron as a coagulant went unheeded. Alum or sulfate of alumina was adopted in the Hyatt and other rapid filters and for two decades held undisputed sway. Then sulfate of iron, supplemented by lime, entered the field and gained a strong foothold. In limited use today are other iron compounds such as ferric chloride, chlorinated copperas (ferrous sulfate reacted with chlorine) and ferric sulfate, but sulfate of alumina still dominates the field.*

The earliest explorer that the author has found in the sulfate of iron field was somewhat casually mentioned in 1872 by Dragendorff †

---

* The nature, reactions, advantages and disadvantages of the iron compounds and of sulfate of alumina, as understood in 1940 by specialists in water treatment, are concisely reviewed in *The Manual of Water Quality and Treatment*, American Water Works Association, New York (1940).

† G. Dragendorff was a professor in the university town of Dorpat, 165 miles southwest of Leningrad.

thus (52): "Years ago, as is well known, Scheerer proposed for purification of water, especially when it contains a large quantity of organic substances, the use of a solution of ferric salt, especially of the sulfate."

"About two years ago," said Dow R. Gwinn, Superintendent of the Quincy, Ill., Water Co., in 1900 (53), "William Jewell of Chicago, with the assistance of the writer, began some experiments on the use of iron as a substitute for alumina at Quincy." The coagulated and settled water was pumped to mechanical filters. Among eight patents granted to William M. Jewell on July 17, 1900, three were for producing a coagulant by the reaction of sulfurous acid, iron and water. Looking back in 1930 upon the early events at Quincy, Wolman, Donaldson and Enslow (54) wrote that "as early as 1903 William B. Bull used ferrous sulfate and lime coagulants at Quincy, Ill." Bull was one of the owners and the leading figure at the Quincy water works for many years.

The first large city to adopt ferrous sulfate and lime was St. Louis, Mo., where an installation was put into use on March 22, 1904. This was done in great haste in order that visitors to the St. Louis Exposition might not be repelled by the turbid water which St. Louisans had put up with since Kirkwood's recommendation for sedimentation and slow sand filtration was ruthlessly thrust aside in 1866. A permanent plant was soon built. Next among the large cities to use sulfate of iron and lime were Cincinnati, beginning on November 1, 1907, and New Orleans, commencing in February 1909.

CHAPTER XIV

Disinfection

Disinfection, or germ-killing, first more or less incidentally and with vague or no ideas as to the how and why, then with intent and under scientific control, has been practiced for milleniums. Broadly, the agents employed have been heat, copper, silver, chlorine, ozone and ultraviolet rays. The greatest of these is chlorine.

Boiling to improve the quality of water was probably employed from the beginning of civilization. The earliest written mention of boiling that has been found is in Herodotus (484(?)–425 B.C.) (1) but the passage refers to Cyrus the Great, who lived a century before him:

> The Great King, when he goes to the wars, is always supplied with provisions carefully prepared at home, and with cattle of his own. Water too from the river Choaspes, which flows by Susa, is taken with him for his drink, as that is the only water which the Kings of Persia taste.* Wherever he travels, he is attended by a number of four-wheeled cars drawn by mules, in which Choaspes water, ready boiled for use, and stored in flagons of silver, is moved with him from place to place.

Hippocrates, the father of medicine (460–359(?) B.C.) (2) declared that boiling and straining rain waters was necessary to prevent them from having a "bad smell" and causing "hoarseness and thickness of voice to those who drink them." Aristotle (384–322 B.C.) is said to have advised Alexander the Great (356–323 B.C.):

> Do not let your men drink out of stagnant pools. Athenians, city born, know no better. And when you carry water on the desert marches it should first be boiled to prevent its getting sour.†

Much currency has been given to alleged passages from the Sanskrit advising that water be treated by boiling and plunging hot metal into it. Place, writing from India in 1905 (3), cites two medical maxims

---

* This statement by Herodotus is echoed by various writers (Plutarch, *de Exil.*, Vol. II, p. 601 D; Athenaeus, *Deipnosoph.*, II: 23, p. 171; Solinus, *Polyhist.*, XLI, p. 83; Eustath, *ad Dionys. Perig.*, 1073, etc.). The water under consideration is said at the present day to be excellent, and the natives vaunt the superiority of these two rivers over all other streams or springs in the world (Jour. Geog. Soc., Vol. IX, Part I, p. 89).—*From footnote in Rawlinson's translation of Herodotus* (1).

† Search of Aristotle's works, of the lives of Alexander by Arrian and by Quintus Curtius Rufus, disclosed no such statement. Inquiries of specialists in Aristotle have brought assurances that his works contain no such passage.—*M.N.B.*

from the Sanskrit, of about 2,000 B.C., one advising that water be exposed to sunlight and filtered through charcoal; the other directing that "foul water" be treated by boiling and "by dipping seven times into it a piece of hot copper," and then filtering it. *Sus'ruta Samhita*, attributed to the fourth century A.D. (4), reiterates these directions and specifies that filtration be through sand and coarse gravel (see Chap. I).

Like Cyrus the Great of Persia, centuries earlier, the Roman Emperor Nero (reigned 54–68 A.D.) had a predilection for boiled water but as befitted a more luxurious age he had it cooled. Pliny (5) states, "It was the Emperor Nero's invention to boil water, and then enclose it in glass vessels and cool it in the snow. . . . Indeed, it is generally admitted all water is more wholesome boiled."

Pliny (5), expressing his own opinion about 77 A.D., wrote, "The best correction of unwholesome water is to boil it down one-half." Plutarch (6), a few years after Pliny, mentions boiling and cooling of water somewhat incidentally, with an interesting reference to heat transference. He says:

All water, when it hath been once hot, is afterwards more cold; as that which is prepared for Kings, when it hath boiled a good while on the fire, is afterwards put into a vessel set round with snow, and so made cooler; just as we find our bodies more cool, after we have bathed, because the body, after a short relaxation from the heat, is rarefied and made porous. . . .

Corroborating the foregoing quotation is the following statement by Ellen C. Semple, in an article on ancient water works in Mediterranean lands: "The Roman plutocrats had their water first boiled, then chilled by mountain snow; if the snow was not clear, it was strained through fine linen cloths" (7).

All these references are to water boiled for kings or plutocrats. If a cat could look at a king so might the common people, even though poor, boil their drinking water—provided they thought it worth while to gather a few sticks for that purpose or to put a vessel of water over their charcoal braziers. No specific evidence that they did so has been found, but from the time that the Chinese began to drink no water except that boiled to infuse tea they, knowingly or not, had provided themselves with a large measure of protection from the water-borne diseases.

Paulus Aegineta, in his compendium of medical lore written late in the seventh century A.D. (8), says that "waters which contain organic impurities, [and] have a fetid smell or any bad quality may be so improved by boiling [or by mixing with wine] as to be fit to be drunk." In Adams' commentary on Aegineta, it is noted that Rhazes [or Rasis, an important Mohammedan physician of the ninth century] directed that water drawn from a deep well be boiled before use. Avicenna, an early eleventh century Persian physician (9), discusses boiling and distillation with apparently conflicting opinion as to which was preferable. His ideas on why either should be used are somewhat vague.

Passing over a long blank period, we come to Boerhaave (10), the noted Dutch chemist (1668–1738). Writing of "putrid water," he says, "But when this Water has thus spontaneously grown putrid, it may easily be rendered wholesome again, and may be drank without being offensive; for if you give it only one boil on the Fire, the Animals that are in it will be destroyed, which, with the rest of the impurities, will subside to the bottom"—but, as a finishing process, he advises adding a small amount of acid. Experience under the equator, he states, "where the Waters putrify horribly, and breed such quantity of insects, and yet must be drank," had proved this treatment to be effective.

In what appears to be the first American book on public health (1835), Dunglinson (11) says, "Whenever water is unusually contaminated, it may be boiled, filtered and agitated. . . . There are many valetudinarians, and some whole nations—as the Chinese—who never drink water that has not been boiled."

With the knowledge and acceptance of the fact that drinking water is a vehicle for spreading typhoid and cholera, together with the widespread use of filtration, private means of treating water for domestic consumption became less necessary. There was still a considerable demand for the use of heat to disinfect water before the advent of chlorination. Boiling was too costly and complicated for public use but became more feasible for special cases when the principles and practice of heat transference were developed. According to Samuel Rideal (12), sterilization by heat was first practically applied about 1888 by Charles Herscher, while other means, such as the Vaillard-Desmoraux apparatus and, in the United States, the Waterhouse-Forbes apparatus were introduced later. At Brest in 1892, say Samuel and Erik Rideal (13), there was a test of heat sterilization apparatus, supplied by Rouart-Herscher & Co., in which water was not exposed to contamination and heat interchange was employed. They also mention the Forbes apparatus in which water boils only a few seconds and so "retains most of the original gas and taste (U.S. Patent, December 13, 1898)."

M. Bechmann (14), in 1904, while chief of the water service of Paris, stated that "a single city, Parthenay [France, with] 7,500 inhabitants, has sterilized its water by heat, treating 3 cu.m. (793 gal.) per hour with Rouart sterilizers." Imbeaux's *Annuaire* (15) states that this installation was abandoned on the introduction of water from springs in 1905, shortly after the date of Bechmann's paper. Imbeaux's data show that the capital and operating cost for the Rouart installation were very high. He does not say whether the sterilizing equipment was put in when the river supply was introduced in 1895. After 1905, he states, Parthenay had a double system of water supply, including pumps and distribution system. No other example of heat sterilization for a municipal water supply has yet been found but in the discussion of distillation (see Chap. XV) two small plants in Texas will be noted.

Copper and Its Compounds

Abundance of copper in some parts of the world and the ease with which the metal can be worked have led to its wide use for water containers and cooking utensils. This use has extended through thousands of years in India and other Eastern countries. The passages from Place already quoted here and at more length in Chapter I have been used by several later writers to indicate that copper was regarded as a disinfecting agent centuries ago in India. These have included Professor J. J. Hinman Jr. of the University of Iowa (16), Samuel and Erik Rideal of London (13), Lt.-Col. C. H. H. Harold of the Metropolitan Water Board, London (17), Rideal and Baines (18), and Henry Kraemer (19). Hinman wrote in 1918: "It seems that colloidal copper is given off into water that is kept in a bright copper vessel." Kraemer in 1905 gave 28 references to papers on copper as a germicide. From these papers and from his own experiments he concluded that copper in small quantities is harmless to human beings. Apparently he did not know of the fanatical campaign against the use of copper vessels as filter containers and for cooking utensils conducted by Amy at Paris in the 1750's (20) (see Chap. IV).

After weighing all the citations I have examined, I have concluded that, although under some conditions the use of copper vessels may have had germicidal value, they were not used for that reason, but rather because copper was locally available, easily worked into desired shapes and could be easily cleaned and polished.

Concern over using copper in any form as a bactericide subsided soon after the epochal Moore-Kellerman studies on copper sulfate both as a germicide and as an algaecide (21, 22). Papers by D. D. Jackson (23) and a symposium (24), both given in 1905, dealing with copper sulfate and water supplies, describe experiments and opinions on copper sulfate as a bactericide. The experiments on "copper-iron sulfate" as a disinfectant which were made at Anderson, Ind., by C. Arthur Brown, were described by him in a paper presented in 1905 (25) (see also Chap. XVII).

Silver

Silver in minute quantities will destroy some macro- and microorganisms in water. Attempts to devise a practicable method of utilizing this agent to disinfect water have resulted in a few installations for special purposes where proprietors were concerned with avoiding the real or imagined objections to chlorination, or were enticed by the claims of a new magic process, in either case with little regard to cost. Many examples of the application of silver to swimming pools but few to city water supplies have been found.

Forerunner of many investigators of silver as a destroyer of waterborne organisms was Carl von Nageli, a Swiss botanist. In 1880 he observed the disappearance of algae, particularly *Spirogyrae*, from water containing minute quantities of copper or silver or their salts. The only paper on the subject recorded as published by him appeared in 1893 (26). Nageli named the action of silver "oligodynamic" or "forces of trifles." The field seems to have lain fallow until 1899 (27) and again until 1917 (28). After that many papers appeared up to 1935.

The first paper in English appeared in 1932, at London (29). All but one of the papers up to that time were printed in German; the exception was in Russian. In 1934–36 four or five papers on silver treatment were published in the United States. Two of these, although written in English, originated abroad: one was by S. V. Moiseev, of the Leningrad Branch of the Union of Scientific Research of the Institute of Water Supply and Sanitary Engineering (30), and one by Just and Szniolis, Assistant and Chief Engineers of the Department of Sanitary Engineering of the State School of Hygiene, Warsaw (31). Each of these papers reviewed the subject up to its time and listed authorities—about 25, excluding duplicates. Moiseev gave an account of laboratory experiments with silver-coated sand, conducted by him on water from the River Neva in 1930–31. Just and Szniolis described their own laboratory experiments, started at Warsaw in 1929, "with filters containing metallic silver, with solutions of silver salts and with electrocatadynization."

Between the dates of publication of the two papers just noted one appeared in New York. It dealt chiefly with silver treatment as developed by Krause in Germany. It named several small installations in that country, including one for drinking water, and also one in this country at a swimming pool in Washington, D.C. (32). In 1936, Shapiro and Hale published results obtained in both laboratory and swimming pool tests of the Katadyn Process (electrolytic method of treating water with silver) made in New York City with the cooperation of American representatives of Katadyn, Inc., in 1934–35. These authors reported (33) a number of reasons for their conclusions "that in the present state of development the Katadyn Process cannot be approved for use in swimming pools."

Several successful instances of the use of silver disinfection of swimming pools were cited and some test results reported in a paper by J. H. Dorroh, of the Department of Civil Engineering at the University of New Mexico (34), in 1936. Fifteen swimming pools in England had been equipped with silver treatment apparatus in 1934. One, for a swimming pool at the Congressional Country Club, Washington, was put into use on July 4, 1935.

Conclusions regarding disinfection by silver compounds, in the American Water Works Association's *Manual of Water Quality and Treatment* (35), are that "the silver process has not proved to be equal to other methods" and that "its use is not justified for public water supplies."

Chlorine

Nothing in the field of water purification came into use as rapidly and as widely, once it got a good start, as chlorination. Its impetus sprang from its adoption on a large scale in 1908 at the Boonton Reservoir of the Jersey City Water Works. It had been used before on small water supplies in the United States and abroad, in most cases experimentally. Still earlier, it had been used on sewage, also in America and Europe, but tentatively.*

The earliest proposals to chlorinate water were made before there was knowledge of water-carried disease germs. First of these found on record is a statement by Dr. Robley Dunglinson in his *Human Health* (11), published in 1835 at Philadelphia. To make "the water of marshes potable," he says, "it has been proposed to add a small quantity of chlorine, or one of the chlorides; but a quantity sufficient to destroy the foulness of the fluid can hardly fail to communicate a taste and smell, disagreeable to most individuals." When and by whom this proposal was made the Scotch-American physician did not say. His recognition of taste-and-odor difficulties arising from chlorination is significant in view of later experiences with the use of minute quantities of chlorine compared with the large amount he had in mind.

Patented Processes.—During the last 60 years of the nineteenth century there were granted fifteen British patents on water treatment by voltaic action, magnetic action, electric currents or the addition of a chemical oxidizing agent. Summaries of British patents on a wide range of methods of water treatment are given in a series of pamphlets (43). The earliest of these on chlorination and related processes are here outlined:

Voltaic action between unnamed filter media was claimed at least as early as 1839 for the Royal Patent Filters of George Robins. [Ure's *Dictionary of Chemistry*, 1839; *Encyclopedia Britannica*, 1842. The Britannica questioned the voltaic action but gave space to exterior views of household filters.]

Pocock took out a patent November 27, 1852, which called for precipitation by a "salt," followed by the addition of a "hypochloride," then filtration through charcoal.

Harrison patented November 16, 1863, a magnet adapted to filters of asbestos, talc, sponge or carbon, to polarize water.

Kühne, August 16, 1866, patented a method of disinfection and purification by such oxydizing substances as "chlorine-permanganates."

* Chronologically arranged references and other data on the discovery and early manufacture of chlorine, its adoption for bleaching in the late eighteenth and early nineteenth centuries, trials of the application of chloride of lime to London sewage in 1854 and 1884 and the promotion of electrolytic methods of sewage treatment by Webster in England, Hermite in England and France, and Woolf in the United States are found in Race (36), Hooker (37), and an A.P.H.A. Committee Report (38). A comprehensive exposition of the Webster process, by Webster himself, is given in Crimp (39) and abstracted at considerable length by Fuller (40). See also Metcalf and Eddy (41) and my summary of 1912 (42).—*M.N.B.*

Davis's patent, August 29, 1866, called for destruction of "infusoria and fungi together with animalculi and other insects" by the use of "caustic alkali"; then filtration through sand, gravel or vegetable charcoal to render the water "bright and limpid" (see also Chap. XVIII).

Pope and Sawyer, on May 7, 1873, patented the idea of passing electric current through metallic wire arranged in horizontal coils in water, thus precipitating any substance in solution.

Webster, January 27, 1887, patented "methods of purifying sewage and other impure liquids by electrolysis, applicable also in connection with filters, for purifying potable waters."

Destruction of dangerous germs by passing an electric current through an electrode immersed in water, it may be interjected here, was briefly described in 1874 by a French journal, *La Nature,* which credited its information to a statement by Dr. Dobell in the *Times* [London] and said that the same idea had been conceived by Dr. Stephen Emmons. The article characterized Dr. Dobell's process as a "new application of electricity" whereby cholera and typhoid germs would be destroyed by nascent oxygen. Whether the process had been covered by patents and, if so, when and where the patents had been granted the article does not say.

In the United States, from 1887 to 1898, a half dozen patents on water treatment by electrolysis were granted. Webster's British patent of January 27, 1887, may be considered the forerunner if not the model of later British and American patents on the use of electrode-generated electric current for purifying sewage and water. Although water treatment was claimed in the Webster patent, no record of such use of the process has been found. Nor does it appear that any process for the direct application of electricity to water has ever been permanently adopted. As will be shown, none of the electrolytic processes, whether proposed for sewage or water, does more than produce a chlorinating or oxidizing agent that can be applied to water or sewage.

The first American patent on chlorination of water was granted to Albert R. Leeds, Professor of Chemistry at Stevens Institute of Technology, Hoboken, N.J., on May 22, 1888. His application, however, was filed in mid-September 1887 or nine months after Webster's patent.* Leeds applied for a process patent on September 15, 1887, and for an apparatus patent four days later. His process application

---

* Ahead of Leeds, but for treating sewage, was the American patent of J. J. Powers, granted May 10, 1887. His claims were for apparatus generating gaseous chlorine from manganese dioxide, sodium chloride and sulfuric acid. He built six sewage works, all in New York State, of which four were within the present limits of New York City (44).

seems to have been denied—perhaps because of anticipation by Webster. He was granted a British patent, apparently on apparatus only, on May 22, 1888, the date of his American patent. Three paragraphs from Leeds' American patent specifications, which demonstrate his conception of how water purification may be induced by an electrical current, are the following:

> In the art of purifying water it has been found that many waters contain certain organic impurities, which it is highly important should be removed in order to render the water fit for drinking and many other uses, but which cannot be removed by ordinary mechanical filtering. In the treatment of waters containing impurities of this class many attempts have been made to purify the water by the use of chemicals, which acted either to precipitate the organic impurities or to reduce them to such condition that they could be readily removed by filtration.
>
> I have discovered that the organic impurities which are contained in large quantities in many waters when in their natural condition, as well as in factory slop and sewage, can be readily and economically removed, so as to render the water pure and wholesome, by treating the water with the gases obtained by the decomposition of water containing an acid or salt in solution, the decomposition being effected by means of an electric current. The acid employed may be hydrochloric, nitric, phosphoric, chromic, or sulfuric; or the salts of these acids may be employed, or a mixture of these acids or salts or acids and salts may be employed. The best results are, however, obtained by the use of hydrochloric acid.
>
> The present invention relates particularly to an apparatus for effecting the purification of water by means of gases generated as above stated. . . .

Leeds' apparatus was, in substance, a tank containing an acid solution; electrodes in the tank, through which an electric current passed; pipes leading from the solution tank and the raw-water tank or filter to a contact chamber; and a filter or a final tank to receive the electrically treated water. That is, the water could be filtered before or after it had been subjected to electric action, or it could be doubly filtered. The gases of electric decomposition "will have the effect," the specifications stated, of destroying "the organic impurities contained in the water, so that they will be precipitated or reduced . . . [so] that they can be readily removed by passing the water through an ordinary filter."

Electrodes combined with a mechanical filter were patented in Great Britain and the United States by Omar H. Jewell of Chicago in 1888 (British patent, February 7, 1888; American patent, July 10, 1888). Application for an American process patent was filed December 7,

1887, in the names of Omar H. and William M. Jewell * (father and son) but was not granted. Application for an apparatus patent was made ten days later by the father only. The applications were filed three months after the Leeds' applications and nearly a year after the date of the British patent to Webster.

The electrolytic portion of the Jewell apparatus consisted of electrodes set vertically in the dome of a mechanical filter. Current was supplied from a battery or other generator. It was assumed that carbonic acid gas and common salt would be used to form "an insoluble bicarbonate of soda, which is precipitated or caught by the filter material below, through which the water subsequently passes. Pure fresh water will be discharged" from the filter. For process claims the specification refers to the application for a process patent, which was not granted.

Laboratory experiments on passing low-voltage current through water were reported early in 1891 by R. Mead Bache. Although the results were not convincing, he suggested further study on the Philadelphia water supply (45).

Early Application of Electrolysis.—The first use of electrolysis in the field was in the Croton gathering ground of the New York City supply. There, on July 1, 1893, electrozone produced from salt brine by apparatus devised and installed by Albert E. Woolf was added to the sewage of the village of Brewster before it was discharged through perforated pipe into the East Branch of the Croton River. A pipe from the main electrozone conduit led to a second pipe discharging into Tonetta Brook about 500 ft. above its junction with the Croton River. The Brewster electrozone plant continued in operation until it was destroyed by fire in 1911. Subsequently bleaching powder was used.

To determine the value of electrozone in purifying Croton water and sewage, a laboratory was installed by Edward A. Martin, Chemist in the Department of Health, New York City. His investigations were begun in August 1893 and continued several months. In a re-

* The younger Jewell, who graduated from the University of Illinois School of Pharmacy in 1887, was chemical engineer of the Jewell Pure Water Co., to which the patent in question was assigned, and it may be assumed that the germ of the patent was his conception. Ten years later a different form of electrical apparatus, devised by William M. Jewell, was given a brief test at the Louisville Filtration Experiment Station, and a little afterward he applied chlorine to the effluent of a demonstration filter at Adrian, Mich.

port made public early in 1894, Martin stated that the addition of electrozone in suitable quantities rendered water sterile, induced sedimentation, removed taste and odor due to decomposition of organic matter in stored water and tended to decolorize water coming from peaty bogs. The report gave many and various data on the tests (46).*

Bleaching Powder.—Chlorination with bleaching powder (chlorine de chaux) was invented by Traube in 1894, states Dr. Edouard Imbeaux (48). In 1896, a typhoid epidemic at Pola on the Adriatic Sea (one-time chief naval station of Austria-Hungary) was stopped by the use of bleaching powder. The excess of chlorine was neutralized by sodium sulfite.

Professor Drown's Dicta.—"The so-called electrical purification of water by treating it with an electrolyzed solution of salt is thus seen to be simply a process of disinfection by sodium hypochlorite; electricity, as such, has nothing to do with it. There is nothing peculiar in the sodium hypochlorite produced by electrolysis. . . ." Thus declared Professor Thomas M. Drown in 1894, after having reviewed the subject (49).

So far, so good! Then came a question and answer strange enough in the light of subsequent events:

Finally, is it desirable in any case to treat a city's water supply with a powerful disinfectant like the hypochlorites? When the question is put in this bald way I cannot think it will receive the approval of engineers and sanitarians . . . in cases where a water supply has got into such a hopelessly bad condition that nothing will render it safe but disinfection by chloride of soda or chloride of lime, it is high time, I think, to abandon the supply, and in this opinion I feel sure most water works engineers will coincide.

In his advocacy of naturally pure water and his zeal against humbug exploitation of the magic powers of electricity, Drown did not foresee the vast legitimate field of water chlorination. He did point the way to the early mass production of sodium hypochlorite by the

---

* Immediately after the Brewster plant was put into use, the inventor escorted a delegation of New York City officials, including Hugh J. Grant, then Mayor, on a tour of inspection. I went with the party. We saw tanks of salt brine in which electrodes were immersed and the treated brine which was applied to sewage and to a stream, as stated in the accompanying text. The Mexican government sent an engineer to New York to investigate electrozone. He made a favorable report (47) but said that the "Woolf system" was nothing more than the one used by M. H. Hermite at Rouen in 1889.—*M.N.B.*

use of the new electric process then on trial, which was to make "bleach" available to water works at a low price. He could not have been expected to visualize that the powder would soon be superseded by the less costly and more convenient liquid chlorine and the ingenious apparatus for its accurate application to water at any desired rate. Like Dunglinson 60 years earlier, Drown did not foresee that minute quantities of chlorine would be sufficient.

Clemens Herschel, in discussing Drown's paper, deprecated Drown's slighting remarks on chlorination as discouraging to new processes, but he did not advocate chlorination. John C. Chase agreed with Herschel's plea against putting a brake on the development of new processes but said that he understood from Drown's paper "that, as a practical matter, purification of water by electricity is a humbug." And so it was, but in fairness to the long line of inventors and promoters of electrolytic treatment of both water and sewage it should be pointed out that most of them did not claim the process involved direct electrocution of bacteria.

French and English Skeptics.—Drown was not the only eminent scientist to be skeptical of what have later become revolutionary changes in water treatment. In France, Arago condemned the addition of alum to water a half century before the official Massachusetts taboo was placed on coagulation. In England, Edward Frankland testified in the late 1860's that although Clark's water softening process was "beautiful" it was impracticable—although soon afterwards he endorsed it (50). J. A. Wanklyn, sticking to his theory that dead organic matter rather than living organisms caused the spread of cholera and typhoid, "waxed sarcastic" over the botanical gardens of the bacteriologists (51, 52).

Chlorination and Electrolysis at Louisville, Ky.—William M. Jewell applied chlorine gas in January 1896 to the effluent from the Jewell rapid filter at the Louisville, Ky., testing station. "I set up this chlorine equipment on my own initiative," wrote Jewell to me in 1933 (53), "as we were not getting 97 per cent bacterial reduction in spite of all the coagulant we could use, even overrunning the alkalinity at that time." After the equipment "had been tested for about a week or two," Jewell states, he received a letter from Charles A. Hermany, President of the Louisville Water Co., "requesting the discontinuance of chlorine, on the ground that it was detrimental to the whole proceedings, would never come into general use and was unfair to our competitors in the test. Of course I acceded but protested their views."

The accompanying sketch shows one of the twelve chlorine gas generators used at Louisville. It was sent to me by Jewell in 1933 and is here published for the first time. The 12 U-tubes, wrote Jewell,

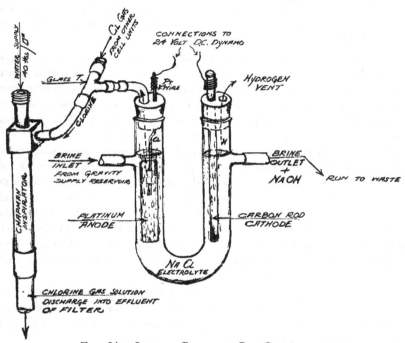

FIG. 64. JEWELL CHLORINE GAS GENERATOR

One of twelve generators installed by William M. Jewell for trial at Louisville Filter Testing Station in 1896; it was operated about ten days, applying chlorine at the rate of 0.25 ppm. to effluent from the Jewell filter
(Reproduction of sketch made by William M. Jewell)

———were connected to a gas header leading to the water inspirator which drew the gas from the cells and delivered it into the open gravity-type controller on the discharge side of the filter.

Mr. Fuller thought I used the cathode liquor along with the gas solution but I did not, as there was too much salt in it. It took about 10 lb. of salt to produce only 1 lb. of chlorine. This much salt would increase the chlorides and be objectionable, so I did not use [the cathode liquor].

Jewell's adventure in chlorination was given only these few lines in Fuller's Louisville report (54):

Jewell device for the application of chlorine . . . consisted of a set of small U-tubes, in which a common salt solution was decomposed by an electric current. A constant flow of water was maintained through the tubes. The water dissolved the hypochlorites and carried them with it to the water in the top of the filter. The apparatus was never used regularly, but was tried on January 21 and 22, and for very short periods at later dates. On January 22 available chlorine was applied in this way during the morning at the rate of 0.1 ppm. by weight of applied water.

The Jewell and Fuller statements differ as to whether chlorine gas or hypochlorite solution was applied to the water and as to the point of application but Jewell's sketch and statement are in agreement. The point of application is of minor importance. It seems safe to say that this was the first use of chlorine gas to reduce the bacterial content of the effluent from a working-scale water filter.

The Harris Magneto-Electric Water Purification System was tested at length in 1896–97 under the direction of George W. Fuller, at the Louisville testing station. Previously, small demonstration installations had been operated by Harris at a point on the badly polluted Passaic River in New Jersey and also in Brooklyn, N.Y. Fuller described at length Harris apparatus as first submitted and subsequently modified, and concluded that "the direct application of electricity and electro-magnets, as used in these devices, produced no substantial purification of the Ohio River."

The original Harris apparatus consisted of magnets charged with high voltage electric current discharged into water that subsequently passed through tanks, each of which contained electrodes. The fundamental principles of the system, Fuller says (54),

———were never explained to me. Electro-chemical action was considered to be an important factor in the destruction of the bacteria and organic matter in the water. It was intended that all suspended matter would be repelled by the action of the magnets situated at the top of the three tanks; and the magnets were to force the suspended matters, including the bacteria, to the bottoms of the tanks, where pipes leading to the sewer were provided.*

* Years later, Harris installed a sewage treatment plant at Santa Monica, Calif., which, after changes by associates or successors, remained in use for many years. A half dozen sewage treatment plants based on his system were built in Oklahoma. One of these, at Oklahoma City, was described at length by Hinckley, the description accompanying my review of attempts to treat sewage by electricity (42). A few years later all but one of the Oklahoma plants had been abandoned.—*M.N.B.*

Professors Palmer and Brownell of the Louisville Manual Training High School, stimulated by the trials of the Harris apparatus, set up experimental electrical devices in their laboratory. The Louisville Water Co. was induced to try their system to determine the relative merits of iron and aluminum electrodes to provide a coagulant for treating Ohio River water before filtration. Seventeen pages of Fuller's Louisville report (54) are devoted to these tests, made in the early part of 1897. It was found after work had been started that the devices made under plans provided by Brownell represented the Palmer and Brownell Water Purifier—patent applied for. The application was dated February 2, 1897; a patent granted February 15.

Besides data pertaining to the Harris and to the Palmer and Brownell (or Mark and Brownell) electrical devices, Fuller's report contains an analysis of electrolytic action and comparative data on coagulants produced by the electrolysis of iron and aluminum electrodes and commercial sulfate of alumina, with conclusions favorable to the latter for use with subsidence and mechanical filtration.

Chlorination at Adrian, Mich.—Chlorination was used in 1899 at Adrian, Mich., under the direction of William M. Jewell. In the spring, when unable to obtain bleaching powder, he improvised apparatus to produce chlorine gas in conjunction with a demonstration filter, using a horse sweep for power. In the autumn he used bleaching powder on the effluent from a permanent installation of filters, in order to bring the effluent from Jewell filters up to the Michigan standard of chemical purity. How long the bleaching powder was used is unknown, but in a report on analyses of the raw and filtered water made in April 1899 by Professor Victor C. Vaughan, of the Laboratory of Hygiene, University of Michigan (55), mention is made of a taste in the effluent due to the use of "sodium chloride." Chemically, Vaughan stated, the effluent did not come up to the standard shown on the blank form used in making the report, but the standard was an ideal one and the water was considered safe for use. Bacterial counts were 144 per ml. before filtration. The unfiltered water had been dosed with sodium hypochlorite. Recollections of these activities at Adrian by F. B. Smart, a company employee of that period (56), are in general agreement with the foregoing statements. Sodium chloride, he says, was never permanently adopted by the company and was soon discontinued. Since about 1915 the city has used chlorination continuously.

Permanent Chlorinating Plants.—The first permanent water chlorination plant anywhere in the world was put into use at Middelkerke, Belgium, in 1902. It was brought to the attention of water works men in America by George C. Whipple in 1906 (57) and again in 1931 by William Boby (58). From these sources and by later correspondence with Boby (59), the following notes have been taken: The installation was "invented" by Dr. Maurice Duyk, then chemist to the Belgian government. Chloride of lime and perchloride of iron (0.2 and 8 ppm.) were fed through drip cocks into badly polluted water before it passed through four gravity filters, each 5 ft. in diameter. This was continued until about 1921 when "pure spring water" was introduced. Mr. Boby states that in 1904, acting in collaboration with a consulting engineer, he delivered chlorinating apparatus to water works at Christchurch, England, but was informed by the local medical officer of the district that he would not allow any form of chemical treatment. "Thus," says Boby, "although a certain amount of contamination was known to be present in the water, I was debarred by the limited foresight and education of the medical officer in question from being the pioneer of chlorination in England."

The second permanent use of chlorination on a municipal water supply was begun early in 1905 at Lincoln, England, and continued until 1911. An alkaline solution of sodium hypochlorite, having the trade name "Chloros," was added to the water on its way to slow sand filters. It contained about 10 per cent available chlorine. The work was done under the direction of Dr. Alexander Houston, "with Dr. McGowan on the chemical side" (60). Considerably earlier (1897) Dr. Sims Woodhead used bleaching powder for water-main disinfection after the serious typhoid epidemic at Maidstone, England, attributed to pollution of the water by hop pickers.

Boonton and Bubbly Creek.—Impetus to chlorination came in 1908 from its introduction on a large scale at the Boonton Reservoir of the water works of Jersey City, N.J., and on a small scale at the Bubbly Creek filters of the Union Stockyards at Chicago. Chlorination at the Boonton Reservoir was not an outgrowth of chlorination at Bubbly Creek, as has been said repeatedly and as would be naturally inferred from George A. Johnson's statement (61):

> The first demonstration in this country in a practical way of the usefulness of hypochlorites in connection with water purification was made in the fall of 1908 at the filter plant of the Chicago stockyards, on the recommendation

and under the direction of the writer. Following directly on the heels of the spectacular results obtained at Chicago, came the adoption of the process for the sterilization at Boonton, N.J., of the impounded and unfiltered water supply of Jersey City, with which work the writer was also connected.

Careful study of available data shows that although the Bubbly Creek plant was put in operation a few days before the one at the Boonton Reservoir, the decision to use chlorination at Boonton was made first, and preliminary tests were made there first as well. The Boonton plant was the conception of Dr. John L. Leal. It was the outcome of litigation brought by Jersey City for alleged non-fulfillment of a contract made in 1899 for a new water supply. Under this contract water of the Rockaway River was to be impounded by a dam and conveyed to Jersey City by a conduit. Owing to litigation instigated by rival water supply interests, delays and financial embarrassment postponed the beginning of construction until 1902. Meanwhile the contract was taken over by the rival interests and executed by the East Jersey Water Co. The supply works were completed in 1904.

The contract provided that the water should be "pure and wholesome and free from pollution deleterious for drinking and domestic purposes" (62). Rightly or wrongly, it may be interjected, the people of Jersey City had come to believe that the sewage of a number of small towns above the dam would be diverted from the river or at least treated to avoid infection of the water supply. However this may have been, the city sued for non-fulfillment of the quality clause of the contract and sought to prove that the contractor should filter the water.

On the basis of evidence submitted to it the court ruled, on May 1, 1908, that although "perhaps two or three times a year" the bacterial count was too high and coliform organisms were found in too small samples of water, and although filtration would prevent this, yet evidently the contract did not contemplate the construction of filters. The court suggested, on the basis of evidence submitted by the city, that diversion sewers and sewage treatment works be constructed by the contractor.

Dr. Leal, sanitary adviser to the contractor, a few days after the court ruling, "strongly advised" the water company to suggest its own method for the complete fulfillment of the contract. "On June 4, 1908," says Leal, "the court authorized the company to submit to it within 90 days plans for meeting the terms of the contract." Accordingly, Leal recommended the use of sodium hypochlorite as a disinfectant, and the company agreed. "On June 16, 1908," says Leal, "I engaged the firm of Hering & Fuller to design the necessary works and Mr. George A. Johnson of said firm to operate the same" (62). Design of the plant was begun by Hering & Fuller three days later, testified George W. Fuller before the court (63).

As early as 1897 or 1898, wrote Leal in his report of the case (62), "I had made rather extensive experiments with electrolytical solutions of salt and also with solutions of bleach in connection with the proposed purification of another water supply. The results were most favorable from a bacterial standpoint, although the method was not used because it did not fulfill all the requirements of the water under consideration." Leal's first idea was to use an electrolytic solution at the Boonton Reservoir, but, being unable to find a suitable cell, chloride of lime was adopted. He knew by 1909 that chlorination had been tried as a disinfectant, generally on sewage, at a dozen places in England, France, Belgium, Germany, India, Mexico and the United States, including the Woolf process in the Croton watershed and at Jerome Park Reservoir, New York City, and the Duyk process at Middelkerke and Ostend, Belgium. "No special discovery," he said, "is claimed in connection with the process in operation at Boonton, N.J." He added: "I do claim, however, that this is the first time it has been used on any such scale, or as a continuous or permanent system of water purification. I also claim that, as a result of the investigations made by us, certain facts in connection . . . with the process have been obtained, which had not been heretofore recognized."

The claims for size and for useful information on chlorination were justified but apparently Leal had overlooked the fact that the installations he mentioned at Middelkerke, Belgium, and Lincoln, England, were also permanent.

The Boonton chlorination plant went into use September 26, 1908, said Leal in his court testimony. Johnson testified that he was at Boonton "getting things ready" for a week prior to September 26, and was in charge of the operation of the plant until the end of the year. Johnson also testified that for about a month before March 20, 1908, he operated an electrolytic cell obtained from the National Laundry Machine Co., Dayton, Ohio, and that from March 20 to 23, 1908, he applied electrolytically prepared hypochlorite of sodium to the water at Boonton in order to compare the relative efficiency of that agency and of bleaching powder. Comparative tests extending through three weeks showed no difference in the efficiency of the two (63).

The Boonton plant was approved May 9, 1910, by a special master in chancery. He reported the first cost of the plant as $20,546 and the operation and maintenance expense as $2,100 a year. Sewage works and the necessary trunk sewer to divert sewage from the water supply, he said, would have cost several hundred thousand dollars and required considerable capital and operating charges. Late in 1910, the New Jersey Supreme Court accepted the master's finding as to satisfactory character of the chlorination plant. Almost a year afterward the New Jersey Court of Errors and Appeals sustained the finding of the lower court (64).

Thus ended the litigation but the litigation is not all the story. Some years later the city completed at its own cost an activated-sludge plant to treat the sewage of some of the towns above the dam and a trunk sewer leading to the plant.

Chlorination at Bubbly Creek was adopted to bring the effluent from mechanical filters up to the contract guarantee. Data to establish the nature and course of events that led to chlorination for a short time have been obtained for use here from Arthur E. Gorman, Engineer of Water Purification, Chicago (65), and from Charles A. Jennings, an associate of Johnson in the tests on disinfection at Bubbly Creek (66).

In 1907, the Norwood Engineering Co. completed a 5-mgd. mechanical filter plant for the Union Stock Yards Transit Co., at Chicago. The filters were located on the south bank of Bubbly Creek, near Halstead St. Sulfate of iron and lime were used as coagulants. In 1908, George A. Johnson, of Hering & Fuller, New York City, made a series of four tests to determine whether the filter effluent complied with the contract guarantees. The test periods, says Gorman, were April 7–20, June 1–14, July 27–August 2 and September 3–17. Jennings, who assisted Johnson, states that sulfate of iron and lime worked nicely as coagulants and that the effluent looked well and was low in bacteria. But the organic matter was high and after the water had stood in the clear well the bacterial count "would jump very rapidly into the thousands and higher." First, copper sulfate was tried as a germicide. In the test runs of July 27–August 2 and September 3–17, hypochlorite of lime was used. The last-named run, says Jennings, may be considered as an acceptance test of the filter plant by the Stock Yards Co.

In February 1909 the city council passed an ordinance prohibiting the taking of water from the Chicago River or any of its branches for watering livestock or for use in preparing meats, poultry or provisions for human consumption or for any other use which endangers the public health. Use of such waters for motive power was not prohibited. In June 1909, sulfate of alumina displaced lime and iron as a coagulant at the Bubbly Creek filters. Later that year the city of Chicago brought suit against the Stock Yards Co. for violating the ordinance. A thousand pages of testimony were taken in the municipal court. In April 1910, states Gorman, partly as a result of the city's opposition and partly to avoid prejudice that might affect its business, the company discontinued the use of filtered water for its livestock. About November 1, 1915, a softening process was adopted at the filtration plant.

Race states that the hypochlorite of lime was applied $7\frac{1}{2}$ hours before filtration, at the rate of 45 lb. per mil.gal. (36, 64).*

Other Hypochlorite Applications.—Among the first American cities to adopt chlorination on a permanent basis was Poughkeepsie, N.Y., which began application of chloride of lime on February 1, 1909, and installed permanent apparatus on March 17 of the same year. It was the failure of combination treatment by sedimentation, coagulation and slow sand filtration to render the Hudson River raw water supply potable following the drought summer of 1908 that prompted the city authorities to consult George C. Whipple, and, on his recommendation, to substitute chloride of lime for alum in the treatment process. Introduced first into the low lift pump suction line by means of the regular coagulant apparatus, the chloride of lime was regularly applied at the inlet to the sedimentation basin as soon as a satisfactory dosing appliance could be devised (67, 68) (see also Chap. VI).

Largest city to adopt disinfection at the time was Philadelphia. Sodium hypochlorite, produced from electric cells, was applied in September 1909 to water in 200-mgd. prefilters at Torresdale on the Delaware River. Hypochlorite was again used in December 1910

---

* Articles on the Bubbly Creek plant were published in *Engineering Record*, 58: 659–68 and 58: 703–05 (1908); and in *Engineering News*, 64: 245 and 64: 342 (1910); the last of these was a 5,000-word letter from George A. Johnson, on litigation with the city.

but as the bacterial efficiency declined in cold weather it was decided to apply chloride of lime to the water in the clear-water basin. The chloride of lime treatment was discontinued in April 1911 but then resumed in December of the same year and maintained continuously until February 1913 (69).

The Application of Liquid Chlorine.—Liquid chlorine was first produced as an article of commerce in the United States in 1909 (70) and was used experimentally in 1910 for water disinfection by Major C. R. Darnall, U.S.A. Medical Corps, at Fort Myer, Va. Seven years earlier Lieutenant Nesfield, of the Indian Army Medical Service, reported the result of numerous experiments on the destruction of pathogenic organisms by chlorine and proposed its application to military use. "This," says Race (36), "was the first suggestion of the possibilities of compressed chlorine gas in steel cylinders."

In June 1912, Dr. Georg Ornstein of the Electro Bleaching Gas Co. experimented with liquid chlorine and developed the first equipment to employ the solution-feed process in which chlorine gas is dissolved in a minor stream of water and the solution is introduced into the major flow of water. At Philadelphia, liquid chlorine was used experimentally in September 1912, under the direction of Seth M. Van Loan, then Assistant Chief, Bureau of Water, assisted by George E. Thomas. They applied the gas directly to water in the clear-water basin of the Belmont filters.

The first full-scale application of liquid chlorine for water disinfection was made with the Ornstein equipment in November 1912, at the Niagara Falls filter plant of the Western New York Water Co. by Dr. Ornstein, who acknowledged the valuable technical assistance of H. F. Huy, Principal Assistant Engineer of the plant (36).

In December 1912, John A. Kienle, engineer of the water works of Wilmington, Del., began experiments with liquid chlorine. In a paper on the subject he states that the first results were rather discouraging, and then mentions profiting by experience at Philadelphia and acting on some of Dr. Ornstein's ideas (71).

Subsequently Dr. Ornstein further developed the solution-feed process and the Electro Bleaching Gas Co., which acquired the rights to Ornstein's U.S. Patent 1,142,361, marketed the equipment; Kienle became associated with the company to manage its sales. In 1917, Wallace & Tiernan Co. became sole licensee of the patent.

The first permanent liquid chlorine plant was installed at Philadelphia's Belmont filters in September 1913. During October and November of that year, additional installations were made at all the other Philadelphia plants. The largest of these, capable of treating 200 mgd. was put in use November 25, 1913, at the Torresdale plant on the Delaware (69).

Early in 1941 chlorination in the United States was being used by 4,590 of the 5,372 water works using any kind of water treatment (72). Hypochlorite of lime had largely given way to chlorine gas.

Chlorination Abroad

In Great Britain, the use of chlorination spread slowly. Aside from the instances already cited "perhaps the earliest example of its use was as supplementary routine treatment in connection with Paterson's rapid filtration plant installed in 1911 for the Cheltenham" water works on the River Severn (73). In 1916, the London Metropolitan Water Board began to apply bleaching powder to water from the Thames before it entered the Staines Aqueduct on its way to slow sand filters. This was a war-economy measure which, with subsequent adoptions of chlorination, saved many thousands of pounds by obviating pumping water to storage to get the benefit of bacterial reduction before filtration. In the summer of 1917, the board began to apply liquid chlorine to one of its supplies, using a British proprietary dosing apparatus then coming into use. Early in 1919 the board installed this apparatus to treat New River water. The plant had a capacity of 48 mgd. (U.S.) and "was for many years the largest gaseous chlorine installation in the United Kingdom" (73). It was not until 1921, Lieutenant-Colonel Harold tells in his report for 1936 (74), that chlorination of filtered water was begun, "with its attendant bogey of taste." Subsequently the use of ammonia to control taste was adopted. "The close of 1936 marks the advent of a new epoch, when all filtered water receives chloramine treatment before being passed into the supply." To that end chlorine was being applied at 28 points and ammonia at 45 points. Pre- and postchlorination combined, 300 mgd. (360 mgd. U.S.) were being treated.

Summaries of a questionnaire sent out by the British Waterworks Association in 1939 or 1940 show that in a group of over 600 local authorities about 40 per cent of the water supplies were not chlorinated, most of them being small. Of 134 supplies, each serving a population of 50,000 or more, only 26 did not chlorinate and, of those, twelve were in Scotland and three in Wales (75).

In France, chlorination goes under two names: javellization, named after *eau de Javelle,* first made in 1792 at the Javelle works near Paris, and verdunization, so named by Philippe Bunau-Varilla after he chlorinated the water supplied to French troops in 1916 at the Battle of Verdun. In a book by Bunau-Varilla (76) he states that at Verdun he used hypochlorite of soda in doses considerably smaller than had formerly been used in France. "Our experiences proved in the autumn of 1916," says Bunau-Varilla, "that when a water is clear it is not necessary to render it nauseating to make it safe for use." In January 1917, Bunau-Varilla treated clear water with chlorine at the rate of 0.1 ppm. Commenting on these statements, Francis D. West wrote in 1931 (77): "It is quite evident that they were playing safe with the water supplies for the army, by soaking them heavily with chlorine until they became so strong of chlorine that they could not be used. B-V., to my mind, went a little too far the other way."

In 1934, wrote Dr. Edouard Imbeaux (48), many French cities chlorinated their water supplies. Gaseous chlorine was used by Montpellier, St. Nazaire and Fremay. Many other cities used hypochlorite of soda or *eau de Javelle.* "Bunau-Varilla uses the term 'verdunization,' saying he invented the process at Verdun, during the war; it is not true, and we cannot accept the word 'verdunization' as scientific."

In 1935, S. McConnel (78) stated that French engineers preferred hypochlorite of soda or javellization to any other form of chlorine. Since Bunau-Varilla introduced verdunization he has pushed its use elsewhere, claiming no royalties. The process was being used in Paris, Lyons, Avignon, Amiens, Rheims, Bordeaux, Monte Carlo, Brussels, Geneva, Lisbon and Seville. Bunau-Varilla claims, "and is supported [in his claims] by French scientists," says McConnel, "that instantaneous sterilization is due to the emission of ultraviolet rays."

In Germany, wrote Dr. Karl Imhoff early in 1941 (79), chlorination apparatus was in place at that time on at least 30 per cent of the city water works, but in most cases was used only temporarily.

Ozonation

Ozonation began about the same time as chlorination. For some years it made more rapid progress, then chlorination shot ahead and left ozonation far behind. In America and Great Britain, ozonation

FIG. 65. GEORGE A. SOPER (1868–1948)
First American engineer to investigate ozonation of water; undertook research as subject for doctorate at Columbia University shortly after completing work as operator of Warren Filter at Louisville Filter Testing Station
(From photograph taken about 1896)

of public water supplies had hardly gained a foothold up to 1941. On the continent of Europe it fared better from the start, but only in France, with its strange complex of ideas on water purification, have many ozonation plants been established.

Handicapped by cost of both installation and operation, ozonation cannot compete with the equally efficient, simpler and far less expensive chlorination. One advantage of ozonation over chlorination is that it does not create new and increase existing taste-and-odor troubles. In America, taste-and-odor control has of late given promise to ozonation which it lacked or lost during the earlier years of its trial and abandonment. Even so, it must be regarded as an uncertain competitor of ammonia compounds and activated carbon. This phase of ozonation is reviewed by John R. Baylis in his monograph on the whole subject of taste-and-odor control (80). His section on ozone contains a succinct review of the history of ozonation with descriptions of the various types and makes of apparatus promoted up to 1935. This section therefore will be confined chiefly to brief mention of the earlier papers introducing ozonation to American water works men; to a chronological summary of American trials of the process; and to a brief summary of the scanty available figures on the number of ozonation plants in various other countries.

A few British and American patents on the production of ozone preceded by an early mention of ozone as a bactericide follow:

C. J. Fox, in 1873, reported experiments showing that ozone destroyed bacteria in fluids containing organic matter [Citation from Baylis (80)].

E. H. C. Monckton, in a British patent granted January 21, 1874, states "Water is purified by ozonizing it by passing electric currents through it in tubes or channels of special construction, which is at the same time being acted on by electric current."

In the American aeration process patent granted to Albert R. Leeds of Hoboken, N.J., May 6, 1884 (No. 298,101; application filed November 3, 1883), the claims are for purifying water by "saturating it with oxygen or ozone by causing the water to come in contact, while under artificial pressure and in motion, with compressed air." The specifications mention "destroying any deleterious substances."

On December 29, 1896, a British patent was granted to J. Y. Johnson (Electric Rectifying & Purifying Co.) on the production of ozone by passing oxygen from a reservoir through a cooling vessel into space between two connecting cylindrical vessels, across which there was a brush discharge of electric current.

Experiments and Comments.—The first American investigator of ozone as a water disinfectant was George A. Soper. As a part of work for his doctor's degree in the late 1890's he made laboratory tests at Columbia University, reviewed the literature of ozonation and visited the Tindal plants applying ozone to water at Blankenberg, on the coast of Belgium, and at Joinville le Pont, in the environs of Paris. His conclusion was that "drinking water can be sterilized, and that unpleasant colors and odors arising from organic impurities can be removed by ozone" (81).

Very soon after the appearance of Soper's thesis, Robert Spurr Weston published an abstract of a German paper on ozonation, read in 1899 by Theo. Weyl before the German Society of Gas and Water Works Engineers (82). The abstract summarized results at a small test plant at Charlottenburg, near Berlin. In appended critical comments, directed pointedly at Weyl's assertion that "sand filtration is uncertain" while ozonation renders water "germ free," Weston said that both Weyl and Dr. Soper "assume that the enginemen and electricians in charge of an ozone plant would be more careful than men with the same degree of training in charge of a filter plant."

Sharp dissent to Weyl's paper was voiced in the discussion which followed and which was abstracted by Allen Hazen for the benefit of American readers (83). William Lindley, English-born and -trained engineer located for many years at Hamburg and at Frankfort, Germany, speaking for those in charge of large water works, differed strongly with Weyl's contention that "sand filters are among the most dangerous appliances to be found in the control of cities." He added: "In scientific circles it will not do to set up the results of a little first experiment against a well-proved system which, on an enormous scale and through decades, has been carried out with success."

Another early contribution to the literature of ozonation, written for the particular benefit of American water works men, was made early in 1901 by the Russian engineer, Nicholas Simin, who was chief engineer of the water works at Moscow (84). After visiting experimental plants in Europe, the Russian engineer suggested the use of rapid filters and a small amount of coagulant to prepare water for ozonation. In discussing the paper, Allen Hazen aptly summed up the role of ozonation by saying that it must be considered as an auxiliary to sedimentation and filtration, "simply removing the relatively small number of bacteria remaining after these processes had been employed, together with the odor and, if enough ozone were used, the color."

In 1906, a masterly review of ozonation to that date was submitted to the American Water Works Association by George C. Whipple (57). In 1910, the subject was reviewed at length by *Engineering News* (85). This article described foreign apparatus, including the Vosmaer system, the American rights of which had been acquired by a Philadelphia company. A few foreign municipal installations, all small, were described, as were also three plants, also small, on this side of the Atlantic.

Shortly after the appearance of this article, two New York engineering journals described the ozonation plant completed in 1910 by the city of St. Petersburg [Leningrad] (86, 87). This seems to have been the largest installation up to that time in any country. Its capacity was 11 mgd. It consisted of 126 Siemens and Halske ozonators, five Otto emulsifiers and five sterilizing towers. Current stepped up to 7,000 volts was employed.

In America the earliest significant commercial attempts to apply ozone to municipal water supplies were made by the United Water Improvement Co. of Philadelphia. It acquired the American rights to the Vosmaer system, which had been previously installed at Schiedam and Nieuwersluis, Holland, and had been tested at the latter place, but soon took up the American patented system of J. Howard Bridge, who became associated with the company.

Ozonation tests were made in 1906 at an experimental filter plant operated by the Department of Water Supply at Jerome Park Reservoir, New York City. Apparatus was installed at the expense of the United Water Improvement Co. of Philadelphia. It included a Hungerford mechanical filter for the removal of bacteria, a refrigerator to remove moisture from the air used in the production of ozone, and an ozonizer of the "H. Blanken System," in which 60-cycle, 10,000-volt a-c. was used. It was operated May 7–31, 1906, by engineers acting for the city. This appears to have been the first disinterested American experimental test of water ozonation. A summary of the conclusions, by I. M. DeVarona, Chief Engineer of the Department of Water Supply, New York City, approved by Rudolf Hering and George W. Fuller, consulting engineers (88), stated that apparently the color in the water might be reduced from about 15 to 5 and the bacteria from about 100 to 7, at a cost of about $20 per mil.gal., but the plant "did not run one single day without stopping, a fact of itself sufficient to demonstrate the impracticability of using the proposed method of ozone purification as an adjunct of our filtration plant, regardless of its cost."

Five ozonation plants on municipal water supplies were installed in the United States and Canada within the 1908–12 period. All were eventually abandoned. As nearly as can now be determined, dates of service of the first three plants were: Lindsay, Ont., 1908–1910; Ann Arbor, Mich., 1909– or 1910–1914; Baltimore County Electric Water Co., near Baltimore, Md., 1910–1918. Not all were used continuously. There were some reconstructions. In 1927, an ozonation plant was installed by the borough of Ogdensburg, N.J., and in 1928, by the village of Delhi, N.Y. The first of these was given up in 1930, on the completion of rapid filters. The second was used until destroyed by a flood in 1935, and was followed by a chlorinating apparatus (89–96).

A year's experimental work at Milwaukee, Wis., in 1919, led to the conclusion that apparatus for water sterilization by ozone had not yet been developed to the point where it could compete with chlorine. These studies (97) were made at the Milwaukee filter testing station by Ernest F. Badger, under the direction of J. W. Ellms, Consulting Engineer, who reported to H. P. Bohman, Superintendent of Water Works. The ozone apparatus was made by the Ozone Co. of America, apparently a Milwaukee concern. Good bactericidal results were obtained on both raw and filtered water but the cost was considered prohibitive.

Recent American Installations and Experiments.—A new phase of ozonation in the United States was begun with the completion of plants in 1930 at Hobart, Ind., and in the spring of 1932 at Long Beach, Ind., for small privately owned water works controlled by a holding company interested in both water and lighting companies. Control of tastes and odors as well as bacteria were objectives of these ozonation plants. An official statement, written late in 1939, regarding these ozone plants, their designer, the company that built them, and its successor is here summarized by the director of trade relations of the new company (98):

The Hobart and Long Branch ozonation apparatus was designed by J. M. Daily, who "spent several years in France with the leading ozone company there." In his apparatus Daily incorporated "several

DISINFECTION

major improvements over its European prototypes." The equipment at the two Indiana towns was built by the American Ozone Co. The patents of that company were taken over in 1938 by Ozone Processes, Inc., organized "as a member of a group of operating companies owned by Welsbach Street Illuminating Co. Ozone Processes, Inc., is adequately financed and equipped to do necessary research and engineering work . . . never done previously in this country."

Besides an experimental laboratory at Moorestown, N.J., Ozone Processes has operated small "pilot plants" at Whiting, Ind., in 1938–39, and at the New Brighton station of the Beaver Valley Water Co. of Beaver Falls, Pa. A 1-mgd. testing plant was put into use early in 1941 at the Lower Roxborough station of the Philadelphia water works. Filtered water from the Schuylkill River was being ozonated there in cooperation with the Philadelphia Bureau of Water, of which Seth M. Van Loan was Chief. Manganese was quantitatively removed at a pH of 6.8 to 7.0 (99). The tests at New Brighton were given up at the request of the water company because it concluded that ozonation, like other processes tried, was unable to cope with the peculiarly troublesome tastes and odors of Beaver River water. At Whiting, the tests were followed by a contract with the city under which the ozone company installed a 3.5-mgd. ozonation plant that was put into use in July 1940. A 0.3-mgd. ozonation plant was completed at Denver, Pa., in March 1940.

Of the three Indiana towns, Long Beach is a small summer resort on Lake Michigan. It is supplied with water from the lake, collected by open-joint tile buried a few feet in lake sand. Ozonation apparatus was installed in 1930. After a series of tests by a private laboratory the water company began applying chloride of lime, in batches, to the pump well. The chlorinated and ozonated water had always met Treasury Standards up to the close of 1939. Occasional unsatisfactory samples had been attributed to the makeshift method of chlorination.

Hobart had a population of 5,800 in 1930, when the water works were constructed. The supply is taken from Deep River, a sluggish stream into which the sewage of Crown Point (4,000 population) is discharged 15 mi. above a small impounding reservoir. The water is coagulated, settled, filtered, ozonated and chlorinated. Ozonation was begun in the spring of 1932 because of odors in the water due to algae. At first, chlorine was applied to the filtrate before the water went to the ozonator but soon the chlorine was added to the raw water. At both Long Beach and Hobart ozonation was adopted some years ago, when the Daily apparatus was in its early stage. The water works at both towns are controlled by the Northern Indiana Public Service Company.

Whiting is located at the southern end of Lake Michigan. The population of about 11,000 is supplied with water from the lake which is badly polluted with domestic sewage and the wastes from oil refineries, steel mills, chemical plants, soap works and corn products factories. Coagulation, filtration and chlorination are said to have made the water hygienically safe but chlorination increased taste-and-odor troubles despite ammoniation. Encouraged by the ozonation trials of 1938–39, the city contracted October 23, 1939, with Ozone Processes, Inc., for the installation of an ozonation plant. This was put into use in July 1940. Contrary to what might be expected, ozone is applied to the raw water, which then passes through the old treatment plant.

The treatment plant, which had proved itself unable to control taste and odor before pre-ozonation was introduced, had a designed capacity of 8 mgd., but the abnormal and increasing pollution had limited it to a maximum of 4 mgd. The normal consumption is 2 mgd., with hot-weather peaks up to 4 mgd. This plant included: ammoniation just ahead of prechlorination; coagulation with alum; baffled flocculation; baffled sedimentation in two basins; rapid filtration; and postchlorination as needed. Activated carbon was not used in normal plant operation because no effective dosage was possible. The output was "generally very unpalatable, due to kerosene and phenolic tastes accentuated by heavy chlorine dosages necessary to cope with the normal heavy primary pollution" (100).

The pre-ozonating plant includes: electrostatic air filter, air dryer, air compressors, ozonators, meters and two ozonizers. Raw water is pumped to the top of the ozonizers. Ozonated air is delivered to the bottom of the ozonizers through porous tubes and passes up through the raw water. Once it has been ozonized, the water is dosed with chlorine, ammonia and alum. It then passes through a bubble-type flocculator and through two settling basins between which the activated carbon is or may be applied, and finally it is put through gravity rapid filters. The filter effluent is then chlorinated on its way to the clear-water well (98, 100).

Arthur W. Consoer, Consulting Engineer of Chicago, and James C. Nellis, Whiting City Engineer (101), state that six months' operation (October 1940–March 1941) showed a cost of $5.49 per mil.gal. as contrasted to $3.50 for chemicals alone for six previous months—but capital charges at 6 per cent would add about $2.00 to the cost of ozonation. The chemicals used were alum, chlorine and ammonia.

During the six-month period, activated carbon as well as ozone was used quite extensively, but with little reduction in threshold odor numbers over ozone alone, although the pilot tests indicated that the combination would be of value. Results of the six months of operation as well as comments of water consumers "indicate quite clearly that ozonation has done more than any other treatment in reducing tastes and odors in the water supply."

The contract between the city of Whiting and Ozone Processes, Inc., contained a guarantee centering on reduction of odor during a three-year period, backed by a surety bond. It provided that the contractor should not be released from the bond by "entry of any lawful order of the Board of Health of Indiana" during the three-year period "requiring a discontinuance of ozonation" (101). This, it may be interjected, reflected the fact that the board had withheld a permit for ozonation other than as a trial of the process.*

At Denver, Pa., a 0.30-mgd. ozonation plant was completed for the borough in March 1940, by Ozone Processes, Inc. The water supply is from Cocalico Creek, a small stream draining an agricultural area. Before ozonation the water is coagulated, settled and passed through a rapid filter. The borough adopted ozonation for sterilization, removal of tastes and odors and reduction of color. The Pennsylvania

*The industrial pollution of water in the lower end of Lake Michigan, and especially at Whiting, is extremely heavy. Changes in oil refinery practices, derived from the increased production of high octane gasoline during and after World War II, introduced new wastes into the lake and thereby lessened the degree of satisfactory removal of tastes and odors effected by ozonation. In April 1948 B. A. Poole, Director, Bureau of Environmental Sanitation, Indiana State Board of Health, reported that up to that time the board "had not been shown by plant results that ozone alone could be depended upon for disinfection of water to be used in a municipal distribution system." At the time of this report, Whiting was using ozonation to improve tastes in its raw water supply only when threshold odors reached a high point. Meanwhile, at Hobart, the ozonators, which had been rebuilt after the war, were not in routine use, primarily because escape of ozone into the plant limited operation to periods when the windows could be opened for ventilation; and at Long Beach, where two ozonators were available, ozonation had been discontinued, at least temporarily.—*Ed.*

Department of Health authorized the installation of the ozone plant for demonstration purposes, provided the final effluent was chlorinated. Tests were run until July 1, 1941, jointly by the borough and Ozone Processes, with checks by the State Department of Health. Late in September 1941, the Department approved the ozone plant as an agency for removing tastes and odors and the reduction of color, but insisted on a continuation of postchlorination (99–100).

Summary.—Despite promotion work and a few experiments, no ozonation plant was built in America until 1908. In the period 1908–1942, plants were built at eight water works in the U.S. and one in Canada. The first five of these have been abandoned. In the U.S. the general attitude of health departments that have had occasion to pass upon ozonation projects for municipal supplies has been to regard them as experimental and to grant tentative permits only.

Recent Ozonation Plants in Europe

In England, ozonation of municipal supplies did not begin until 1936. There were only four such plants in use early in 1940. At least two of the four British plants treat only the water of some of the sources drawn upon. The first and largest of the plants was put into use at Brighton in 1936 and has a capacity of 5 mgd. [All the capacities here given are in U.S. gallons.] In January 1937, the South Staffordshire Waterworks Co. put in use a 1.2-mgd. ozone plant of the Van der Made type at its Huntington pumping station, one of seven stations lifting water from the red sandstone (102). In April 1937, a 1.7-mgd. plant was put into operation by the Ashton-under-Lyne, Stalybridge and Dunkinfield Water District to treat water from the Knott Hill impounding reservoir, which was subject to seasonal infestations of algae causing tastes and odors (103). The fourth ozonation plant in England was completed in 1939 by the Colne Valley Water Co. It has a capacity of about 4 mgd. Early in 1941, Lt.-Col. G. Ewart Morgans, of the Water Purification Department of British "Otto" Ozone (104), stated that he had just made a contract with the city of Manchester, England, for a 3.6-mgd. filtration and ozonation plant from which water will be supplied in bulk to Salford. This plant was to be used for "exhaustive tests." It was expected that it would be extended to a capacity of 90 mgd. "This," says the representative of the "Otto" process, "is my fourth big ozone plant to be installed in Great Britain."

DISINFECTION

Prefatory to his paper on the Ashton-under-Lyne, Lancashire, ozonation plant, M. T. B. Whitson, Engineer and Manager there (103), reviewed the development of European ozonation processes and illustrated the main features of the various methods used to mix ozone and water, particularly the Otto process of 1900, 1914 and 1940, the Siemens-DeFrise process at Paris, of 1900, the Marmier-Abraham process at Chartres, of 1908, and the Van der Made process of 1940, installed at Ashton-under-Lyne. So far as known to him, Whitson said, "of the many types of ozone producers developed during the last 50 years, only two [were in use in 1940] on a commercial scale." Both had been developed from the Werner von Siemens experimental ozonizer of 1857 and depended for the production of ozone on the discharge of high voltage electricity between stationary electrodes. The companies formed by Siemens and DeFrise, Marmier and Abraham and Dr. M. P. Otto were amalgamated in 1910 and 24 plants were erected in France in the next four years. In 1940, said Whitson, there were in use on public water supplies (apparently meaning all types still in service), 90 installations in France, fourteen in Italy, five in Belgium, four in England, three in Roumania, two in the United States and one in Russia (103). "Ozonation, in France," Whitson said, "has come to stay." He believed it deserved more consideration in England than it had received.

The four largest ozonation plants in France, early in 1940, were: Toulon, 26.4 mgd. (U.S.), installed 1910–35, treating impounded surface water after filtration; Nice, 21.17 mgd., installed 1906–30, treating unfiltered water from Vesubie Canal; Nancy, 21.17 mgd., installed 1933, treating water from a filter gallery and from the Moselle River, filtered; Villefranche-sur-Mer, installed in 1910, treating filtered water from Vesubie Canal (105). Under construction for the city of Paris were plants with a contract capacity of 79.26 mgd., designed to ozonize filtered water from the River Marne. A. Gury, formerly Chief of the Paris Municipal Water Service (106), states that ozonation of the Paris supply at St. Maur was stopped during the first World War. Lt.-Col. C. H. H. Harold, Director of Water Examinations on the Metropolitan Water Board, London (17), said that Paris, in 1910, decided that 24 mgd. (U.S.) of Marne water should be ozonated, half by the DeFrise and half by the Otto process. He also mentioned an inspection, made in 1934, of ozonation plants in France, from Nancy in the East to

Finistère in the West and along the Riviera from Mentone to Toulon. The majority of the old installations in France, he stated, were located where hydroelectric power could be used. His general conclusion was favorable to ozone treatment, "provided that at all times an absence of breaches in the filter barrier can be guaranteed. On the other hand, chloramine offers an all-round security under every possible condition and is the Safety First water works valve."

In Germany, wrote Dr. Karl Imhoff early in 1941 (107), ozonation was used at some water works many years ago but not at all now.

Ultraviolet Rays

Shortly after chlorination got its running start and while ozonation was entering the lists as a rival, experimental trials of ultraviolet rays as a means of killing bacteria in public water supplies were inaugurated in France. No permanent adoptions for public water supplies in France or elsewhere in Europe have been found on record. In the United States, there were four adoptions on municipal supplies in the period 1916-28, and as many more by industrial concerns for their plants and villages more or less under their ownership. All four municipal installations and two of the industrial-communal plants have been abandoned. In Central America, ultraviolet ray apparatus was reported as installed in 1926, but its capacity was not given. Hundreds of swimming pools and other private or semi-private establishments in the United States have been equipped with ultraviolet ray lamps.

"The earliest positive mention" of the successful use of the bacterial action of ultraviolet rays, stated Kenneth C. Grant, civil engineer in Pittsburgh (108), in 1910, "[was] in 1877, by two English scientists, Downs and Blunt." He adds: "The first attempt to put this method of purification into use is at Marseilles, France, where an experimental plant is being operated in competition with several types of filtration and purification works." Grant's article was chiefly concerned with ultraviolet ray experiments he had seen at the physiological laboratory of the Sorbonne, in Paris. In those tests there was used a quartz-tube mercury arc lamp of the Westinghouse type.

An article on the Marseilles tests (109) makes it apparent that in 1909 or 1910 the Marseilles authorities invited proprietors of water purification apparatus to install and operate, at their own cost, plants to treat 200 cu.m. (52,840 U.S. gal.) of water per day. The Westinghouse-

Cooper-Hewitt Co. of Paris installed ultraviolet ray apparatus which treated water that had been passed through Puech-Chabal filters. During a month's run at the rate of 600 cu.m. or 158,500 U.S. gal. per 24 hours, the article states, the treated water showed no coliform organisms. The effluent from the Puech-Chabal filters, however, is said to have contained only 22 to 24 bacteria per ml. Details of all the competitive tests at Marseilles were given early in 1911 in an article by Walter Clemence, London agent of the Puech-Chabal filter interests (110).

Prompted by an article in *Eau et Hygiene, Engineering News* (111) said, "If the merciless hunt goes on, bacteria will become as scarce in France as snakes in Ireland." It noted that the water company at Rouen, although supplying water of good repute from springs back of the city, was carried away by "the current French craze for absolute sterility in water-supplies" and had experimented with ultraviolet rays from Westinghouse-Cooper-Hewitt mercury lamps. These were applied to the effluent from Puech-Chabal filters. All the tests showed that the filters alone gave a high reduction in total bacteria, a complete removal of coliform organisms, and a complete absence of bacteria in the water subjected to ultraviolet rays. Commenting editorially on the tests, *Engineering News* said: "Rouen turns ultraviolet rays on a corporal's guard of bacteria that, now and then a day, manage to pass the tortuous zoogloea-lined channels of a multiple battery of Puech-Chabal filters." The editor then asks: "Is this worth while? . . . have Rouen and other French cities which are so hard on the trail of the last solitary germ in their water-supplies ever approached a like state of perfection in all other or any other matter affecting public health? Are their milk supplies absolutely pure? . . ."

In the United States, the first known installation of ultraviolet ray apparatus on a municipal supply was put in use at Henderson, Ky., toward the end of 1916 (112). It was given up in 1923 or 1924. The population of Henderson in 1920 was about 12,000. This was the largest city in America to use ultraviolet rays. Berea, Ohio, installed ultraviolet ray apparatus in August 1923, and abandoned it in July 1936 (113, 114). Similar treatment was also begun in 1923 at Horton, Kan.; it has since been abandoned (115). The fourth and last of the municipal water works known to use ultraviolet rays was Perrysburg, Ohio, where apparatus was put into use in 1928 and abandoned in 1939 (114).

All four of these plants were equipped by the R.U.V. Co., Inc., of New York. In 1931 that company sent the author a list that included a considerable number of other water supply installations, of which a few were for industrial plants and related communities and others for hotels and institutions. Of four industrial-communal plants, information from company officials and state health departments shows that one installed in 1926 at Parr, S.C., by the Broad River Power Co., was in use in 1940; one completed in 1919 at Mascott, Tenn., by the American Zinc Co., was apparently in use in October 1939. Of the two not in use, one was installed in 1918 and given up in 1932 by The Washington Mills Co., at Fries, Va.; the other at the Gill Hill, Kan., Camp of the Cities Service Oil Co., was operated from 1919 to 1929. Reasons for abandoning six of the eight municipal or industrial-communal plants were: cost of operation compared with other means of water disinfection available; the belief of state officials that chlorination would have greater sanitary efficiency (116), and unsuitableness of the electric current in the case of Henderson, Ky. Late in 1941, an R.U.V. representative stated that no attempt to obtain municipal water supply installations had been made for some years, due to the small size of quartz lamps available.

A few vessels on the Great Lakes have used ultraviolet rays for water sterilization since 1916 or 1917. In 1928, when Frank H. Shaw took charge of U.S. Public Health Service control of water supplies in the Great Lakes district, bacterial samples were collected from 230 of 643 interstate vessels in use. Seventeen had R.U.V. installations, eleven used chlorination, two used ozone, 167 used stills and 33 took water from certified shore sources. During 1933, ozone on the two vessels was replaced by chlorination. Writing early in 1940, Shaw (116) stated that ultraviolet rays were still being used to sterilize lake water that had previously been filtered but probably some of the R.U.V. installations had been replaced by other methods of disinfection since 1928. Meanwhile, the number of vessels using chlorination had increased to 31.

CHAPTER XV

Distillation

Knowledge that vapors rising from bodies of water were purer than their origin was recorded by many ancient writers. Rain water, regarded as condensed vapors, was by some given first place among natural sources of supply. Aristotle (384–322 B.C.) notes that pure water can be obtained by evaporating sea water (1). St. Basil (c. 330–370 A.D.) in his fourth Homily on Genesis (2) says, "Sailors, too, boil even sea-water, collecting the vapor in sponges, to quench their thirst in pressing need."

The first known treatise on distillation is attributed to Geber or Jabir, an Arabian chemist who was born in 721 or 722 A.D. (3). Geber defined distillation as "an elevation of aqueous *Vapours* in their *Vessel*," some by and some without fire. The special object of distillation "which is made by ascent into the Alembeck, is the desire of acquiring Water Pure without *Earth* . . . for the *Imbition* of *Spirits,* and clean Medicines. When We need Imbition, We must have pure Water, which leaves no *Feces* after its *Resolution;* by which *Feculency,* Our Medicines and cleansed *Spirits* might be infected and corrupted."

"Johannes Gadeeschen, sive Johannes Anglicus, anno 1516," according to Stephen Hales (4), "says that Sea-Water may be sweetened four ways, *viz* by filtrating through Sand; By Clean Linen laid over a Boiler, and squeezing the Moisture out, as from Sponges; By Distillation: As also by their Bowls made of white Virgin Wax, which 'tis said will free the Water from its Saltness, and from some part of its nauseous Bitter."

In 1595, *A newe booke of Destyllation of Waters, called the treasure of Evonymous,* by Conrad Gesner, appeared (5). Among other woodcuts it contains one showing a still inserted in a vessel beneath which sticks of wood are burning. Sir Richard Hawkins, in a sea voyage begun in 1593, distilled sea water to obtain fresh water (6). In his *Sylva Sylvarum,* published posthumously in 1627, Sir Francis Bacon (7) stated that "the Taste of Water, in Distillations by Fire, riseth not." His explanation was untenable. He also noted that "Distilled Waters of Wormwood, and the like, are not Bitter" (Expt.

881) (7). Two books on distillation were published in 1651: one by Glauber (8); the other by French (9).

Interest in distillation had reached such a point by the close of the seventeenth century that it gave rise to commercial exploitation of distilling apparatus under the protection of patents. Forerunner of numerous patents on distillation and notable as the first-known patent on water purification of any type was one granted October 28,

FIG. 66. STILL AND WICK SIPHON
Ancient distillation equipment, *circa* 8th Century A.D.
(From *The Works of Geber*, London, 1928)

1675, to William Walcott on the "art of making corrupted water fit for use, and sea water fresh, clean and wholesome in very large quantities, by such wayes and means as are very cheap and easy, and which may be done and practiced with great speed." This was soon followed by a patent, also British, issued to Robert Fitzgerald and four associates for an "Engine for rendering salt water sweet and fit for cooking, washing and other purposes" (January 9, 1683). These grants were made the personal concern of King Charles II. Whether he was fully informed as to the nature of the methods proposed, and particularly the "ingredients" to be used is not made clear. To others, they were veiled in mysterious secrecy which, a half century later, appeared to Dr. Stephen Hales to be humbuggery. This, together with much information on the patents, the ensuing rivalry between the patentees and the attempt of the Fitzgerald group to exploit their apparatus, was told by Dr. Hales later in the preface to his *Account of Some Attempts to Make Distilled Sea-Water Wholesome* (4). Hales describes two allegedly successful installations of Walcott's stills on shipboard, one of which was in 1683, and urges that the space required to store coal for the stills was much less than that required to store casks of fresh water. He suggests the use of tinned instead of plain copper for stills to obviate "an ill Quality of Vomiting" given to the water by plain copper. Hales noted that "after several Tryals at Law between the Patentees, Mr. Walcott's Patent was superseded and laid aside; against which Mr. Walcott brought a Bill in Parliament in the year 1694, which passed the Commons, but not the House of Lords." Thus did litigation over water treatment patents begin as early as two and a half centuries ago.

The rise of Fitzgerald to fame, his government orders and the sad end of his partnership and apparatus are thus narrated by Hales:

Mr. Fitz-gerald's method met with such great applause that a Poem was published to celebrate his Praise, and silver Medals were made, representing and illustrating the Art of this new Inventor. A Still of his was set up at Hull and Sheerness: and by Order of the Council in the Year 1692, two of them were to be set up in the Islands of Jersey and Guernsey; but with no good Effect: The distilled Water was fiery, harsh and corroding. And in a little Time the Persons concerned with Mr. Fitz-gerald, finding themselves extreamly disappointed in their expectation, withdrew from any Partnership with him: Insomuch that his Instruments, which were dear enough before their Effect was known, were soon sold for old Goods, for want of a vent for them at Sea. (4)

Hale's own method of "procuring wholesome Water from the Sea," based on experiments and observations of the ideas expressed by others, was "first to let it putrify well, and then become sweet before it is distilled."

Interesting both as a human document and as a scientific paper is Robert Boyle's account of his investigation of the Fitzgerald process for distilling sea water (10). Charles II had doubts about the efficiency of the apparatus and the quality of the distillate. He called on Boyle for advice. The scientist passed on to the inventors all questions regarding the "engine" or distilling apparatus. He convinced the king and a group of nobles that the distillate was free from salt. Later, with the permission of the king, he disclosed that his tests for saltness were made with silver dissolved in *aqua fortis* and noted that it did not produce a cloud or precipitate. A spirit of nitre test was also used.

Pamphleteering on behalf of the Fitzgerald stills was continued in 1684 by Dr. Nehemiah Grew with a statement of the approbation of the College of Physicians (11).

Not until nearly a century after the Walcott and Fitzgerald patents were additional patents for water purification granted in England. There were two issued, both relative to distillation. One was taken out by Bartholomew Dominiceti, December 6, 1770, and covered "many particulars," including apparatus for disinfection. The other was granted to Alexander Mabyn Bailey on July 19, 1777, for "A Machine for making fresh water from sea water or brine springs without boiling"—heat was used, so presumably this was distilling apparatus. Up to at least 1805, no further patents for distillation were granted in England (12).

During the nineteenth century, particularly during the last half, distilling apparatus was developed to a high degree of perfection, in the character of the distillate, compactness of apparatus and unit cost of installation and operation. Its use continued to be limited primarily to ships and naval bases. Coincident with the opening of the Suez Canal, the British government installed sea-water distillation plants on or near the canal (13). Of these, one at Aden seems to have afforded the main supply of potable water at least until 1927, when a deep well was completed (14). So far as found, distillation has never been applied to municipal supplies of fresh water except at three lumbering camps in Texas—Diball, Manning and Wiergate. At these camps the supply is drawn from ponds in which logs are retained for some time. Water for cooking and drinking was distilled at least as early as 1932 and the practice was in use in 1939, according to V. M. Ehlers, Director of the Bureau of Sanitary Engineering of the Texas Board of Health (15).

CHAPTER XVI

Aeration in Theory and Practice

Water aerated naturally by flowing over sandy or pebbly beds or rocky falls has been extolled by writers of all ages and countries. Only a few of these enthusiasts realized that the waters they so highly praised were clear, bright, sparkling, tasteless and odorless when they reached the streams. In the eighteenth century, artificial aeration was directed at making up the oxygen deficiencies of distilled water and of rain water that had been stored in household cisterns. Toward the end of the eighteenth century and early in the next, aeration was applied to a few public water supplies carrying decomposed vegetable or animal matter. Not until the last half of the nineteenth century did aeration become a marked feature of municipal supplies. Even then, the number of applications was small and pertained chiefly to stored surface waters subject to taste and odors from algae growths. In this period, aeration was applied here and there, generally to ground waters, for the removal of iron, and then of manganese, and also to eliminate malodorous gases from sulfur-bearing ground waters.

There are two methods of aeration—the water may be discharged into free air, or air may be forced into a body of water. Apparatus used includes: low cascades; multiple-jet fountains throwing water to considerable heights; multitudinous spray nozzles discharging not far above the surface of a reservoir; superimposed trays or shelves; submerged perforated pipes; and porous tubes and plates. Motivation has been by gravity head for water, pumping head for water and pumping head for air. Chronologically, working installations consisted, first, of cascades and gravity-operated multiple-jet fountains; then forced aeration for a few years of commercial exploitation; then low-throwing spray nozzles; and, latterly, diffusion of air through porous tubes and plates in water.

Obsessed by the notion that removal of organic matter was the chief end and aim of aeration, many inventors and promoters centered their energies there. Chief among these was Professor Albert R. Leeds who patented and joined in exploiting a forced-aeration process in the 1880's and 1890's. In contrast, Professor Thomas M. Drown, after two years of laboratory experiment, announced that removal of organic matter from drinking water by aeration was not feasible and that elimination of gases causing bad odors and tastes was the desideratum. It is Leeds' due to tell that his method of forced aeration was used successfully by a few water works to prevent or counteract algae growths before the advent of copper sulfate treatment.

Concepts Before the Eighteenth Century

Theophrastus, an early Greek philosopher (c. 371–287 B.C.) (1), explains that running waters are generally better than standing water, adding, "and when aerated are still softer"—apparently smoother, less harsh. Columella (first century A.D.) (2) ranked as best, next to spring and rain water, running water from mountains, "provided it tumble down headlong over the rocks as at Guercenum." The *Charika-Samhita,* of about the same date in written form but probably much earlier in origin (3), said that the waters of rivers obstructed by rocks became clear and transparent by "constant beating." Pliny, late in the first century A.D. (4), noted that medical men justly condemn stagnant waters and approve running water as "being rendered lighter and more salubrious by its current and continuous agitation." Avicenna, a highly-esteemed Persian physician (10th century) (5), broadened the conception of nature's beneficent influence on water by putting spring waters first, "provided they flow from and over open ground exposed to the sun and wind." Contrariwise, he objected to water from wells or aqueducts because they are confined—"lacking oxygen" would be the modern term. Strangely, he deprecates the shaking up water gets in being drawn from wells or flowing down the slope of aqueducts. Later, he says well water "is cleaned . . . by the gases which bubble out of it in constant (molecular) motion." "Molecular" was apparently interjected by the translator.

Seventeenth and Eighteenth Century Ideas

Sir Francis Bacon (early 17th century) uttered the famous dictum "Running Waters putrefy not," in his *Sylva Sylvarum* (6), where he gave as the fourth way of preventing putrefaction, *"Motion or Stirring;* for *Putrefaction* asketh rest."

Clifton Wintringham, in 1718 (7), declared that nothing preserves "Water from corrupting and acquiring the most mischievous Qualities, so well as a brisk and rapid Motion." In the following year,

Dr. Edward Baynard, in his jovial poem, *Health* (8), has this to say of water:

> Give it Motion, Room and Air
> Its purity will ne'er impair.

Dr. John Armstrong in his didactic poem, *Art of Preserving Health*, published in 1744 (9), wrote

> ——the lucid stream,
> O'er rocks resounding, or for many a mile
> Hurl'd down the pebbly channel, wholesome yields
> And mellow draughts . . . .

Between the two poets in point of time, and broader in conception of nature's own methods of making impure waters fit for use than any of the citations thus far given, is the assertion made in Plüche's *Spectacle de la Nature* (1732) (10) that the self-purification of rivers is effected by sedimentation, agitation and aeration.

At the very close of the eighteenth century the *Encyclopædia Britannica* (11), after noting that "Rivers which run through great towns are loaded with animal and vegetable substances," makes the questionable statement that those more remote "are purer than most springs," giving as the reason that "they run with more rapidity . . . and a great part of their impurities are thus vitalized." Confidence in the self-purification of rivers continued widespread until near the close of the nineteenth century.

Stinking Water Sweetened.—After two thousand years of recognition of the good effects on water of natural aeration, experiments on artificial aeration were reported. The first of these that has been found was in a paper on blowing showers of air through water being distilled, read by Dr. Stephen Hales on December 18, 1755 (12). Mr. Littlewood, a shipwright, came from Chatham, wrote Dr. Hales,

> ——purposely to communicate to me an ingenious Contrivance of his, soon to sweeten stinking Water, by blowing a Shower of fresh Air through a Tin Pipe full of small Holes, layed at the Bottom of the Water. By this means, he told me, he had sweetened the stinking Water in the Well of some Ships; and also a But of stinking Water in an Hour, in the same manner as I blow Air up thro' Corn [wheat, etc.] and Gunpowder, as mentioned, in my Book on Ventilation (12).

For his experiments, Hales used "a Tin or Copper Air-box" 6 in. in diameter and $1\frac{1}{2}$ in. deep, with its top perforated "full of holes $\frac{1}{20}$-in. diameter," about $\frac{1}{4}$-in. apart. Rising from this was an air-

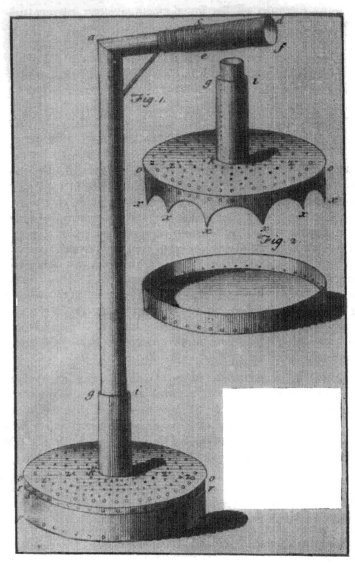

FIG. 67A. DIAGRAM OF LITTLEWOOD-HALES AERATOR FOR STINKING WATER
Bellows force air into perforated "box" immersed in water
(From reprint of Hales' papers read before Royal Society, London, 1756)

59

Explanation of the FIGURES.

Fig. 1. (o o p r) a Tin or Copper Air-box, fix Inches Diameter, and an Inch and half deep from (o to p.)

The Lid of the Box full of Holes, one twentieth Inch Diameter, and about a quarter of an Inch diftant from each other.

(g i k l) a Nozel foldered to the Lid of the Airbox, into which the Tin-pipe (a g i k l) is fixed fo as to take in and out; this Pipe to be two Feet long, and fix-tenths Inch Diameter.

(a b) a Bend in the Pipe five Inches long, to which is faftened the leathern Pipe (c c d f) fix Inches long; to which the Nofe of the Bellows is fixed at (d f.)

Fig. 2. (g i k l o o, x x) the Lid of the Box, whofe Rim (o x o x,) is a quarter of an Inch deeper than the Box (o p Fig. 1.) that the Airholes (o) may be pierced in its Upper-part; and the Lower-part is fcoloped with wide Scolops for the Air to pafs through the Holes (p p Fig. 1.)

FIG. 67B. EXPLANATION OF DIAGRAM OF LITTLEWOOD-HALES AERATOR
(Reproduced from page 59 of Hales' papers)

supply pipe to which was attached a leather hose connected with "the nose of the Bellows" used to force in air. Successful experiments with this device on milk and on water were reported by Dr. Hales in another paper read before the Royal Society, December 18, 1755. "Putrid water in marshy aguish countries," wrote Dr. Hales in the same paper, "may be the Cause of Ague, as well as the putrid Air carried into the Blood through the Lungs. . . . Blowing showers of Air up through the stinking Water of some aguish places may be beneficial." He also said: "Live fish may well be carried through several Miles, by blowing now and then fresh Air up through the Water, without the trouble of changing the Water. . . . But stinking Water will kill fish" (12).

In contrast with the Littlewood-Hales method of blowing air through water was the method of dropping water through the air, put into practical use apparently a little earlier. "Exposure to air in divided currents," says Dr. Edmund A. Parkes, Professor of Military Engineering in the Army Medical School (13), was

———proposed by Lind,* for the water of the African west coast more than 100 years ago. The water is simply passed through a sieve or a tin or wooden plate, pierced with many small holes so as to cause it to fall in finely divided streams. A similar plan, devised by Mr. Osbridge, has been used in the British Navy. A hand pump is inserted into a cask of water, and the water is pumped up, made to fall through perforated sheets of tin. It soon removes hydrosulphuric acid, offensive organic vapors, and, it is said, dissolved organic matters.

Postaeration in "a reservoir exposed to a current of air" was included in Montbruel and Ferrand's project of 1763–64 to supply Parisians with filtered water taken from the Seine above sources of pollution (see Chap. IV).

Aeration in Early Nineteenth Century Europe

At the Quai des Celestins water treatment plant, Paris, put in use by Happey in 1806, it is stated by Dunglinson (1835) that after the water had been settled, then filtered twice, it was aerated by being dropped like rain from the bottom of the second filters into clearwater tanks. This use of aeration is not mentioned by the earlier writers (see Chap. IV).

* James Lind, M.D., a British naval surgeon and hygienist. He wrote a book on scurvy (1753) and one on the health of seamen (1757) (see Chap. III, Refs. 11 and 12).

AERATION

For restoring to water that has been boiled or distilled "the beneficial qualities of the atmosphere," Sir John Sinclair in 1807 (14) mentioned the well-known method of pouring water from one vessel to another. He proposed agitation in a "common barrel churn," and "machines on the principle of shower-baths." In large towns, aerated water "may be prepared in considerable quantities and sold so cheap as a half penny a bottle."

Of the many British patents on aeration taken out in the nineteenth century, the first was granted February 8, 1812, to Robert Dickinson and Henry Maudslay on "a process for sweetening water and other liquids." The process consisted "simply in forcing a stream or streams of air through the foul or tainted water." A bellows or preferably a pump could be used, the air being forced to the bottom of a water cask through a tube or hose ending in a tube of iron or copper perforated with small holes "to divide the air into numerous small streams, that the surface of water brought into contact with the air may be greater." The effect of the air is "that the offensive gas held in solution . . . will . . . be in a short time expelled from the water; after which the water should be left at rest for a little time, to allow its insoluble impurities to subside." Both the process and the apparatus are substantially the same in principle as those described in 1755 by Dr. Stephen Hales and patented again and again in England and America during the nineteenth century.

Of a dozen other British patents granted during the nineteenth century eight were for use on distilled water (generally sea water). Fraser took out a patent January 15, 1818; Peyre, February 23, 1836; Clark in 1843; Clegg and Fell, March 18, 1859; Normandy, March 27, 1860; Chaplin and Russell, October 23, 1862; Starnes on June 29, 1871.

Four other British patents on aerating fresh water, two of which were in general principle anticipations of American patents or practice, included: one by Theodore Cotelle December 1, 1838, which covered admitting air to a filter through tubes in the sides of the container; one by Richard Johnson, September 5, 1857, which covered dropping water for some distance in "jets, sheets or streams" upon a filter of broken slate, stone or other material so contact with atmospheric oxygen would cause "mineral particles held in solution by carbonic gas" to precipitate on the surface of the bed [similar to some American practices]; one by J. Storer on June 9, 1880, which covered rapidly rotating screw blades attached to a shaft; one by W. F. B.

Massey-Mainwaring and J. Edmunds on March 4, 1855, which covered "oxygenated" water or impressed air nearly down one side of a deep well of relatively small diameter, divided by a vertical partition, spreading it nearly at the bottom and letting the water rise through the other half of the well. In substance, this last process anticipated John W. Hyatt's United States patent of July 14, 1885.

The earliest known of the cascade type of aerator, working in series, was put into use in 1848 by the Gorbals Gravitation Water Co. to supply a district afterwards annexed to Glasgow, Scotland. Water from a large settling reservoir cascaded into a basin and from it into the first of three filters, arranged in steps. Similarly, there was a cascade between the first and second filters, the second and third filters and the last filter and a clear-water reservoir. The first three cascades were 9 in. and the fourth 12 in. high (see Chap. V).

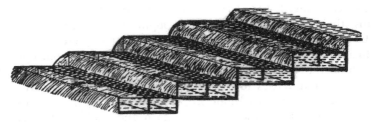

FIG. 68. RUSSIAN AERATOR, CIRCA 1860
Built for government paper mills on River Neva near St. Petersburg; water cascades upon and through wire network into trough with longitudinal partition; after several repetitions it falls upon sand filter
(From Humber's *Water Supply*, London, 1876)

In Russia, a dozen years later, an aerator was included in a water treatment plant built to supply a government mill on the River Neva at St. Petersburg (Leningrad). A detailed but vaguely worded description of the aerator was given in 1867 by C. E. Austen in a discussion of a paper by Edward Byrne describing experiments on the removal of organic and inorganic material from water (15). A condensation of the communication appears in Humber's *Water Supply*, accompanied by a sketch apparently made up from Austen's description (16). From these sources and from a memoir of Bryan Donkin (17), it appears that an engineering firm, which had been established in 1803 by the first Bryan Donkin, contracted with the Russian government in 1858 to build "a mill to supersede the existing hand-mill for the manufacture of State papers and Government bank notes." The contract included pumps, settling reservoirs, an aerator and sand filters. Young Donkin was sent to St. Petersburg in 1859 to superintend construction of the works. On "their successful completion in 1862 he received the personal thanks of the Czar through the Minister of Finance." Austen, who calls the aerator a strainer, describes the water of the Neva as soft but holding "in solution a large quantity of decomposing vegetable substances." The "strainer," he says, was made up of four "troughs" arranged in the form of steps. Each of these "was divided longitudinally by a partition which did not reach the bottom, into two compartments, the inner one of which was covered by wire gauze, and received the water as it fell from the step above, and the outer of which contained the horizontal-tipped orifices through which the water escaped as it flowed to the step below." Each step was 2 ft. high. The lowest step was superposed to one of four sand-and-gravel filter tanks, "each 558 ft. in area." After passing downward through filter tanks, the filtered water flowed to "deep wells in which it was stored for use." The rated capacity of the treatment plant was 100,000 cu.ft. in 10 hours, or about 750,000 gal. (U.S.). Strange as it now seems, Austen made this statement, which was apparently based on personal observation:

——the water entered the first step in a perfectly pellucid state, but before it had passed through two sheets of gauze it became turbid, and deposited a black scum on the wires which required constant cleaning, so great was the quantity of the deposit. In the first [filter] tank the water was partially covered with a black scum or froth, sometimes more than an inch in thickness, and a thin scum, having a metallic lustre, appeared on the surface of the second reservoir [filter].

Aeration in Twentieth Century Europe

Puech-Chabal Cascade Aeration.—The system of aeration most widely used on the Continent was that of cascades placed between Puech-Chabal multiple filters. The method was introduced early in the twentieth century, shortly after Puech-Chabal filters in series came into use. A notable example was afforded at Magdeburg, Germany, about 1910. Low cascades between the roughing filters were supposed to maintain the supply of dissolved oxygen (see Chap. IX). Over a half century earlier, cascades for aeration were placed between multiple filters in the Gorbals district of Glasgow (see Chap. V) and in 1892, as already mentioned, at Tacoma, Wash.

Candy's Compressed Air and Oxidizing Water Works Filter.—Candy's apparatus, described in a paper read by Don early in 1909 (18), is entitled to notice as a curiosity among both aerators and mechanical filters. Water admitted beneath the filtering material, after the filter has been drained, is passed up through the material. This, it is claimed, forces the interstitial air into the space between the top of the material and the dome of the filter tank and compresses it. For a filter 6 ft. in diameter, with a 5-ft. depth of filtering material and a 2-ft. space between the filter media and the top of the dome, 250 cu.ft. of free air is assumed to be compressed. Raw water is sprinkled into the compressed air through four radial arms just beneath the dome of the filter tank, drawing the compressed air with it into the filter. There are three layers of filtering media: at the top, crushed silica, graded from coarse downward to fine, designed to remove most of the suspended matter; "oxidium, for oxidizing and purifying," wherein the air contributes to oxidation; fine silica sand; and grit for final filtration. The layers are supported on grids. The filter is washed by reverse flow. When the air gage shows that the "impressed air is becoming exhausted through solution in the water" the air supply is renewed by recharging, as at the outset, generally once in 24 hours. Don suggested that doing this once a day, and wasting 1,000 gal. of water, might be obviated by using an air pump. This type of aerated filter, Don stated, was new and not widely used.

Aeration in America

The first known aerator * on an American water supply was a part of works built in 1860–61 by the Elmira, N.Y., Water Works Co. As described by a report of the *Tenth Census of the United States* (20), water from an impounding reservoir was admitted to a distributing reservoir "through a fountain discharging-trough—a cluster of holes for aerating and purifying." The *Manual of American Water Works* for 1897 (21) stated that on entering the distributing reservoir the conduit "turns up," making a fountain jet.

* Aeration was proposed but not adopted for Toronto, Canada, in 1854. Water pumped from Lake Ontario was to be delivered into a reservoir, above the water surface, through a perforated pipe "extending along one side of the reservoir." The plan was submitted by Henry Y. Hand, Professor of Chemistry, Trinity College, Toronto, and Sandford Fleming, a railway engineer, and won a second premium of £50 (19). None of the competitive plans was adopted.

James Caird, Troy, N.Y., consultant to the company and city since the late 1890's, states that "on the top of the riser was the figure of a crane with its neck stretched out and water spouting through its mouth." The original works for Elmira were designed by Alphonse Fteley and Francis Collingwood in 1860.

The next American aerator of record was of the single cascade type. It was part of the water works of Lawrence, Mass., completed in 1875. Water from the Merrimac River or an adjacent filter gallery was discharged from the force main through a bell-shaped mouth onto a stone platform from which it fell over six granite steps, each 10 ft. wide, into the reservoir (21).

Early Multiple-Jet Aerators.—Earliest of a number of American multiple-jet fountain aerators was a notable one for the water works of Rochester, N.Y., completed in 1876. Ten years later, J. Nelson Tubbs, Engineer and Superintendent of Water Works, wrote that the value of a thorough aeration of water had been an accepted fact for many years. Accordingly, when the works were constructed a device for aerating the supply during the summer had been provided (22).

In the center of the Mount Hope or Highland Park distributing reservoir, the 24-in. supply main bringing water from the Rush storage reservoir was carried up through a masonry tower and capped with an enlarged cast-iron dome. The dome was pierced with 20 circular 2-in. holes, arranged in two concentric circles around a central 6-in. hole. Into these holes were screwed threaded brass caps for the addition of reducers so the openings could be "adjusted and graded to any sizes desired to give adequate supply, and also to cause the ascending jets of water to assume varied and symmetrical forms." The head on the jets was the difference in elevation between the storage and distributing reservoirs, or about 118 ft. The height of the jets ranged from 60 to 100 ft.

The water returns in finely-divided particles, almost in the form of spray, and in its passage through the air is thoroughly aerated. The fountain when in operation is a most beautiful and conspicuous object, being visible in some directions at a distance of at least twelve miles, and attracts a vast number of visitors during the season in which it is in operation (22).

A similar aerator was put into use at the Cobb's Hill Reservoir in 1908. During the winter season, the incoming water at each reservoir is discharged near the bottom of the reservoir. In 1941, both aerators were still operated from April to November (23).

372   THE QUEST FOR PURE WATER

Most elaborate of the early American fountain aerators was one put into use October 26, 1890, by the Utica, N.Y., Water Works Co. Seventy-six vertical pipes, with perforated caps, 1 ft. above high-water level, discharged into a distributing reservoir. This reservoir was fed from another, under a 44-ft. head when the upper reservoir was full. The risers were fed from 12-in. pipe laid in a quadrangle formed on each side by three 12-ft. lengths of cast-iron pipe joined by

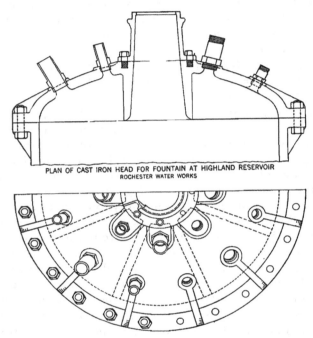

FIG. 69. MULTIPLE JET AERATOR AT HIGHLAND RESERVOIR, ROCHESTER, N.Y.
Cast-iron fountain head on top of riser from gravity supply main; large jet in center, twenty smaller jets set at angles in two concentric rings
(From drawing of March 29, 1875, supplied by Rochester Water Dept.)

quarter bends. This square was laid on the bottom of the reservoir in water 10 ft. deep. Attached directly to this quadrangular manifold were 71 2-in. vertical pipes about 2 ft. apart. Five branch pipes led to risers. Of these, four had diameters of 2 in. and led one from each corner of the square. A 6-in. branch led to a 4-in. riser at the center of the square. The caps of the 76 risers were perforated to give orifices of the following diameters: 16 1-in.; 16 ¾-in.; 16 ⅝-in.; 56 ½-in.; 80 ¼-in.; and 52 ⅛-in. The relation of these sizes to each other and their purpose is not stated. The jets converged toward the center of the fountain. Their total discharge capacity was 4,000 gal. per min. By means of screw joints the upper part of each riser could be removed to avoid ice damage.

The Utica aerator was designed by H. J. Wood, one-time secretary of the water company. Its object was to eliminate tastes and odors. When installed, the water supply was impounded in three reservoirs from which it passed by gravity to the distribution system. With the growth of the city the capacity of the aerator became inadequate, reservoirs were built at a higher level and the aerator was dismantled —probably in 1902 (24).

A fountain aerator with 70 ring nozzles to give aeration before slow sand filtration was put into operation as a part of water works for Ilion, N.Y., completed in September 1892. Water from a creek having a drainage area of cultivated farm land was carried through an 8-in. gravity pipeline 6,000 ft. to a storage reservoir having a flow line 180 ft. below the lip of the diverting dam. This main came into the bottom of the reservoir and was carried up through a masonry tower to a point a little above the surface of the water in the reservoir. The uptake pipe terminated in a cap having 70 ring nozzles, arranged concentrically, thus: at the center, a single ½-in. nozzle, then 6 ⅜-in., 14 ¼-in. and 49 $\tfrac{3}{16}$-in. nozzles. The center nozzle discharged vertically, the others at angles of 5, 10, 15 and 20 degrees, proceeding outwardly. The maximum compound discharge was nearly 0.9 mgd., and the maximum observed height of the central jet was 90 ft. "The water discharged is broken up into a multitude of small drops and falls into the reservoir in the condition of fine rain, so that aeration is very complete." Earle J. Trimble, Supervising Engineer, Ilion Water Works, wrote on July 8, 1942, that the fountain was still used, but in summer only. A similar jet fountain was installed for the water works of Frankfort, N.Y., about the same time (25).

An aerating canal or a series of cascades was included in the water works of Little Falls, N.Y., completed in 1888 with Stephen E. Babcock as engineer. The supply was an "exceptionally pure mountain stream," stored in a reservoir, but was conveyed 8¾ miles through a closed conduit. Just before reaching the distributing reservoir, the water was conveyed 1,600 ft. in an open trapezoidal-shaped canal, in which were sixteen weirs, each 2 ft. high and 10 ft. across. From the lower end of the aerating canal a cast-iron pipe was carried to the bottom of the distributing reservoir, turned up vertically and discharged over the top and sides of a low pyramid of masonry into the reservoir —"giving a final oxidation" (26).

At Tacoma, Wash., a few years later, aerating cascades and "windrows" of scrap iron were placed in a wooden flume ahead of upward-flow filters, simulating the Anderson process of producing metallic iron as a coagulant (see Chap. IX).

An air-induction aerating device was included in water works built in 1888 by a private company to supply Jefferson City, Mo. Professor J. B. Johnson, then of St. Louis, Mo., was engineer for the works. Water was pumped from the Missouri River to a 0.5-mgd. aerating and settling basin, 20 ft. deep. The force main discharged water downward into the settling basin through a flaring-mouthed pipe "containing bent tubes through which air was drawn" (27). L. W. Helmreich, Vice-President of the Capital City Water Co., stated on July 30, 1940, that notwithstanding extensive inquiries he could find no one who remembered an aerating device.

Aeration to remove sulfur taste from artesian well water at Jacksonville, Fla., has been practiced since about 1889. Water from a 10-in. artesian well, 1,020 ft. deep, sunk in 1888–89 to supplement smaller wells of lesser depth was discharged through a pipe, turned downward, 10 ft. above the surface of an octagonal settling basin 50 ft. across and 30 ft. deep. In 1896, two aerating basins, 50 ft. in diameter by 10 ft. deep, were put into use. These were open but have since been covered to exclude dust from nearby streets. Water from the artesian wells was discharged vertically at the center of the basins making a fountain aerator. In 1915, two new basins were built to aerate water from the old and from new wells. Each basin was about 50 ft. in diameter and 10 ft. deep. Water was discharged from their centers upon a table sloping gently in all directions and dotted with baffle blocks so spaced as to break up the water as it dropped from the table into the basin. In 1927, two additional aerating basins, similar to those constructed in 1915, were made. These receive water from wells at another location. All the aerating basins are screened and ventilated.

Forcing air up through a slow sand filter bed and also through the filter effluent at Nantucket, Mass., was provided for in a treatment plant designed in 1891 and completed in 1892, with J. B. Rider as engineer. The object was to remove tastes and odors caused by excessive growths of micro-organisms in Wannacomet Pond. Since no trouble occurred in 1892, the filter and aerator were not used until 1893. Air was delivered upwards into the filter through vitrified underdrains. W. F. Codd, Superintendent, Wannacomet Water Co., in a letter dated February 27, 1894 (28), stated that the air delivered through the underdrains caused air passages in the filter through which the water passed down without much filtration; and that it also caused craters at the surface of the filter thus making the sand of uneven thickness. In consequence, air was shut off from the underdrains. *Anabaena* appeared August 8, 1893, says Codd, whereupon the filter and aerator were put into use with good results. Contrariwise, the records of the Massachusetts State Board of Health, says Arthur D. Weston, Chief Engineer there (29), showed that the plant was not effective in removing tastes and odors caused in 1893 by large growths of *Anabaena*. The plant was used from time to time as late as the summer of 1919, says Weston, but "as it had never removed the tastes and odors to any considerable extent it was considered as abandoned." Since 1929, water has been taken from tubular wells at Wyers Volley. In 1939, J. H. Robinson, Manager at Wannacomet (30), wrote that "the filter and aerator were not entirely successful and after a few years were abandoned; cost of operation and lack of capacity were contributing reasons."

Professor Drown on Aeration.—The inability of man-made aeration to accelerate nature's method of oxidizing organic matter in potable water and the uses to which artificial aeration can be effectually applied were announced by Thomas M. Drown in 1892. His paper (31) was based on two years of experiments in the laboratory of the Massachusetts State Board of Health. The study was designed to test the "theory of accelerated oxidation." Drown said in part:

―――it is not uncommon in water works practice to aerate water by causing it to flow over a series of steps, . . . by forcing it into the air as a fountain, or by pumping air under pressure into the distribution system. Whatever the method the idea behind it is that the water will thereby be purified by oxidation to a degree beyond that which would take place if the water were exposed to the air on the surface only. . . . The question, let it be clearly understood, is not one of supplying oxygen to an impure water, like sewage, which contains no oxygen, but this: Will impure water, which contains at all times more or less free oxygen in solution be purified more rapidly by oxidation if the amount of oxygen is increased by spraying the water or by pumping air into it; can the natural process of oxidation be hastened by these means? It is to this question that the experts give a negative answer. . . . [Conclusions] (A) The oxidation of organic matter in water is not hastened by vigorous agitation with air or by air under pressure. (B) The aeration of water may serve a useful purpose by preventing stagnation, by preventing the excessive growth of algae, by removing from water disagreeable gases, and by the oxidation of iron in solution.

The significance of Drown's experiments and conclusions will become more apparent after considering the section on Leed's and Hyatt's patented processes of aeration, below.

Other American Aerators.—At Wilmington, Del., in 1894, George H. Sellers attempted aeration as a part of his elaborate water treatment plant modeled on the Anderson process (see Chaps. X and XIII). At Butte, Mont., in 1895, an air compressor, mounted on a flat boat propelled by a stern paddle wheel, was tried in combatting the tastes and odors caused by micro-organisms in a large reservoir. Air from the compressor was discharged through perforated pipe hung over the sides of the flatboat. "A few weeks experiments with this homemade contrivance convinced me that it was doing no good and the project was abandoned," Eugene Carroll, General Manager, reports (33).

Pre- and postaerators were put into use in March 1895, in connection with new gravity rapid filters on the water works of Lexington, Ky. This aeration-filtration system was adopted, wrote S. A. Charles, Secretary and Superintendent of the Lexington Hydraulic and Manufacturing Co. (34), "to remove algae and vegetable growths, principally, and also some mud from impounded surface water. . . . We think that we have provided for aeration to a greater extent than has previously been done." Pre-aeration was effected by letting the raw water fall 7 ft., in about 1,600 fine jets from perforated pipe, to the surface of the mechanical filters. This aerator was beneath "a sort of dome with adjustable openings designed to create a draft and assist in carrying off any gases which escape during the spraying." Effluent from the filters passed through a "vat of charcoal" 10 ft. square having a zinc bottom perforated with $\frac{1}{8}$-in. holes, from which the water fell in showers into the clear-water basin.

An aerator for iron removal was completed at Reading, Mass., in July 1896 (35), to treat water from a filter gallery near the Ipswich River. The aerator basin was 33×19 ft. in plan $\times 4\frac{1}{2}$ ft. deep, and was placed above the coagulation basin. Water was delivered to the aerator through the bottom of a hopper at one end of the basin. In rising, the water met a descending supply of milk of lime. The limed water then passed into one of four longitudinal compartments in the settling basin and around end-to-end baffles, giving a total travel of about 130 ft. During this journey the water was violently agitated by air blown up through 23 $1\frac{1}{2}$-in. nozzles placed at intervals in the bottom of the compartments. Air was supplied by a blower with a guaranteed capacity of 150 cu.ft. per min. at a maximum pressure of 3 psi. or about $2\frac{1}{2}$ psi. available pressure. Each particle of water, a contemporary description stated, received 23 successive aerations, thoroughly oxidizing the iron in the supply. On emerging from the basin the aerated water was dosed with alum and passed to the coagulation basin and then to Warren gravity mechanical filters. The entire plant was designed by F. L. Fuller, after experiments by Desmond FitzGerald and Thomas M. Drown (36, 37).

Forced Aeration Designs by Leeds

Nearly all of the aerators thus far described worked under gravity heads and discharged water into the air. In the last two decades of the nineteenth century, Professor Albert R. Leeds and John W. Hyatt patented forced aeration by means of which air or oxygen was discharged into water. Their systems were generally operated in conjunction with rapid filtration (38–43).

Leeds' system was initiated at Philadelphia, where water drawn from the Schuylkill River to supply Philadelphia in January and February 1883, was noxious to smell and taste. This was attributed to the ice cover from the Fairmount Dam to the headwaters of the stream which prevented restoration of oxygen absorbed by organic matter in the water. Leeds, who was Professor of Chemistry in the Stevens Institute of Technology and also chemist to the Philadelphia Water Department, showed by experiments in the institute laboratory at Hoboken that aeration of samples of water from the Schuylkill restored dissolved oxygen and decreased the organic content (38).

These results were submitted to Messrs. Chesborough, Merrick and Graff, members of a special water commission, and to Col. William M. Ludlow, chief engineer of the water works, with a proposal that all the water from the Schuylkill be purified by introducing air under pressure into the force mains. The idea was so novel to these men that they were unwilling to incur the responsibility of advising its adoption. But in November 1884, Colonel Ludlow converted one of the turbines at the Fairmount pumping station into an air pump. Delivery of 20 per cent by volume of free air into the 48-in. force main leading 3,000 ft. to the Corinthian Basin increased the oxygen in the water 17 per cent and decreased the free ammonia 80 per cent.

Subsequently, Ludlow procured air-compressors for all of the Schuylkill pumping stations but only at Belmont was the compressor used and there not after 1886. Conflicting reasons for the abandonment of aeration at Philadelphia were given. Leeds, in a lecture delivered before Franklin Institute, December 23, 1886 (39), stated that the other force mains were "too leaky to permit" their use for aeration. But the Philadelphia water works report for 1886 (40) declared: "This is not the reason, as the other pumping mains are all in good order." It added: "They are laid in such a manner [on such a profile?] that no engineer would care to assume the risk of damage to engines and mains that would probably result from the use of this process."

Whatever may have been the reason for ordering air compressors at all the Schuylkill pumping stations, putting them in use at only one and abandoning that one after some months of use, the dream of aerating a large part of the Philadelphia water supply soon faded.

Directly after his experiments on the Philadelphia water supply in 1883, Professor Leeds applied for apparatus and process patents on saturating water with oxygen or ozone by introducing air under pressure into water under pressure and in motion. These were granted in 1884.*

The object of his invention, Leeds stated in his specifications, was "to restore by a rapid and powerful method of aeration," oxygen lost by water when stored in reservoirs or when covered by ice, as in a stream, this loss causing the water to become foul and unwholesome. Furthermore, his object was to utilize the power of oxygen to destroy deleterious substances.

* Apparatus application filed September 17, 1883; granted April 8, 1884. Process patent, filed November 3, 1883; granted May 6, 1884. British patent, dated November 10, 1884. Leeds' American patents were soon assigned to a company (name unknown); they passed to the National Water Purifying Co. in 1886, the New York Filter Co. in 1891. Leeds or one of these companies acquired patents on aeration by introducing air directly into reservoirs, notably one or both of the D'Heureuse patents of 1871 and 1884.

AERATION

In all the earlier artificial processes of oxidizing water for domestic supply, Leeds said, use of air under atmospheric pressure was proposed and during treatment the water was confined within wells, cisterns or reservoirs, as in the American patents of D'Heureuse, February 28, 1871; Collins, May 3, 1881; and McCurdy, October 7, 1882.

Professor Leeds' theory was that if air could be forced into water under high pressure the oxygen in the compressed air would "go into solution in the water to the almost entire exclusion of the nitrogen." His process could be applied by any convenient method, he said. The one shown in the patent was to force air into the lower end of a pump line delivering water into a reservoir through a submerged bell mouth. All air in excess of that absorbed by the water was to be "detained" in the closed system by means of "air-chambers . . . on the upward bends of the pipe."

Leeds did not limit his patent to the use of "oxygen contained in the air" but included either "ordinary pure oxygen," or else "the allotropic form of oxygen commonly called 'ozone' as well as any mixture of oxygen or ozone with air." The claims allowed in the patent were:

1. In the art of purifying water, the process of saturating water with oxygen or ozone, consisting of introducing into water while in motion, under pressure, compressed air also in motion, substantially as described.

2. In the art of purifying water, the process of saturating it with oxygen or ozone by causing the water to come in contact, while under artificial pressure and in motion, with compressed air in a system of pipes and air chambers, permitting both air and water to enter under pressure, to move through said system while under pressure, and to be discharged into a suitable reservoir, substantially as described.

These process claims were put into use at several places. In addition, compressed air was applied directly to the bottom of several reservoirs. Although in papers that he wrote Leeds claimed that these reservoir installations were under his system, they seem to have been reversions to the so-called inventions of the earlier patents acquired by the companies that took over Leeds' patents.*

Hackensack, N.J.—Largest and apparently the longest-lived installation of the Leeds system of aeration was made by the Hackensack

---

* Robert H. Thurston, well-known mechanical engineer, who filled important chairs of engineering at Stevens Institute of Technology, 1871–85, and at Cornell University, 1885–1903, took out an aeration patent on July 14, 1885. This was soon after the date of the Leeds' patent and may have been suggested by it, as they were both at Stevens Institute at the time. Thurston's single claim was for automatic apparatus to force air into the suction pipes of a pumping main.

Water Co. on the water supply of Hoboken and adjacent towns in New Jersey. Aeration was begun in September 1884. Positive evidence on how long it was used is lacking but it was used in 1897 and may have been continued until 1905. At first compressed air was admitted into the force main leading to a 15-mgd. distributing reservoir in Weehawken. Later, this was supplemented by an air compressor at the reservoir from which air was delivered into the bottom of the reservoir. When a second reservoir was built, this was supplied by air piped over from the compressor of the first reservoir.

Leeds wrote in 1892 (41) that during one of the hottest periods of the unusually warm summer of 1884, the water supply of Hoboken acquired "a very unpleasant vegetable odor and taste," originating at Weehawken reservoir. In "less than 24 hours, the previously clear water became covered with a thick coat of bluish-green algae." He proposed "that an air-compressor should be attached to the 30-in." force main at the New Milford pumping station. The air would thus have an opportunity to act on the water during its long flow to the reservoir, 100 ft. above the river.

Having in mind the statement in the 1886 report of the Philadelphia water works (40) (see above), Leeds noted that the force main passed over "two summits and through two submergences." He added (41):

> Some distinguished hydraulic engineers predicted failure on the ground that the air would fill the summits, and would act as a cushion against which the water might be pumped without advancing further in the main. But with the least possible delay, Mr. Charles B. Brush, the Chief Engineer, began the aeration. The growth of algae ceased immediately and the water was restored to its usual palatable condition. During the eight years that have elapsed, the trouble of either air-cushions or air-leaks in the main or distribution service has never occurred.

Prolonged heat in 1886, "brought about the formation of patches of algae in the corners of the distributing reservoir." The incoming water "contained its normal percentage of dissolved oxygen and carbon dioxide" but on standing oxygen was lost, the carbonic acid was increased, and "stagnation appeared." Thereupon, said Brush,

> I applied another patented feature of my aeration system,* which is the driving in of air through a pipe carried around the four sides of the reservoir

---

* Leeds' patents covered only forcing air into water in motion under pressure. The earlier patents of D'Heureuse (1871), Collins (1881) and McCurdy (1882), Leeds noted, were for admission of compressed air into wells, cisterns or reservoirs. These patents were acquired by the company that promoted Leeds' patents.

and terminating in many rose-jets placed at the ends of terminals going down nearly to the bottom. The air boils up at any or all of these points through the water, and the breaking up and disappearance of patches of algae take place forthwith. [Monthly analyses] show the beneficial effect of this system and the entire avoidance of the trouble, which it was introduced to remedy.

Chief Engineer Brush stated on May 1, 1889 (42): "We forced fresh air into our pipes in 1884 and up to this time we have had no repetition of trouble. We are now aerating the water in the reservoirs as well as in the pipes. The public sees the air bubbling up in the reservoir. It looks like springs and creates a favorable impression." In 1891, Brush said that air had been supplied to the force mains for seven years; and that one main was fifteen and another sixteen miles long (43). In 1893, he wrote that compressed air under 90 to 125 psi. pressure had been forced into the rising mains at the pumping station (44). On January 7, 1897, Brush reported that aeration had been used from April 15 to November 15 each year since 1884 (21). Each of two of its reservoirs was "encircled by 3-in. wrought-iron pipes from which at intervals of about 100 ft. a 2-in. wrought-iron pipe is laid down the slope of the reservoir, to which is attached at right angles a perforated pipe about 6 ft. long lying on the bottom of the reservoir. Each discharge pipe has a gate near its top for regulating the air." Writing in 1935 and 1936, M. W. Cowles, Health Officer of the Hackensack Water Co., stated that the company's files contained little information about aeration. So far as he could learn, aeration at the first reservoir was continued until the completion of the rapid filtration plant at New Milford in the fall of 1905 (45).

Champaign-Urbana, Ill.—Notwithstanding indisputable evidence that aeration in connection with National filters was put into use in June 1887, and was being used the following January at the water works supplying Champaign and Urbana, Ill., the officials of the company now owning the works can find no record of either filtration or aeration having been used there at the time. But a letter from S. L. Nelson, Superintendent, Union Water Supply, dated Champaign, January 14, 1888, to the National Water Purifying Co. (46), begins: "The filter plant which you placed in our works, June last, has been working satisfactorily, and the combined influence of Aeration, Lime Precipitation, and Filtration renders our water clear and bright, free from odor and vegetable matter and sparkling in appearance, resembling the namesake of our town (Champagne). It also removes the hardness of the water." Further on, Nelson uses the phrase "six months tests." Publicity matter issued by the National Water Purifying Co. and its successor in 1889 and 1893 cite these works among those having aerators for mains and reservoirs. The water works in question were completed in 1885 by the Union Water Supply Co. Water was taken from an abandoned coal prospecting shaft, 8 by 12 ft., by 40 ft. deep. It was pumped direct to the mains and to a 0.25-mil.gal. brick-lined reservoir in excavation and embankment, 60 by 60 ft., by 16 ft. deep. The source of supply was soon changed to driven wells. Years later, aeration was employed as a part of an iron-removal plant.

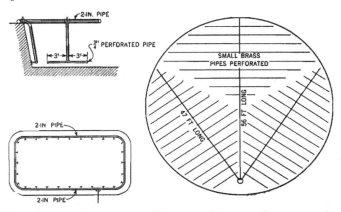

FIG. 70. FORCED AERATION AT NORFOLK, VA., AND BROCKTON, MASS.
Left: Norfolk (1888): compressed air delivered at bottom of reservoir through perforated iron pipes
Right: Brockton (1891): air delivered at bottom of 1.3-mil.gal. tank, 62 ft. in diameter by 59 ft. high, through grid of 39 small brass pipes with $\frac{1}{4}$-in. perforations 3 ft. apart, supplied by 2-in. iron feed pipes
(Redrawn from (Norfolk) diagram dated June 5, 1888 and (Brockton) Leeds' "Mechanical Aeration of Water," *Stevens Indicator*, 1892)

Norfolk, Va.—A plant for aeration by compressed air in accordance with the Leeds patent was put into operation on the water works of Norfolk, Va., July 1888. It was installed by the National Water Purifying Co. in place of filters that had been recommended. The water supply was from impounding reservoirs, the bottoms of which were not stripped before being flooded. When the reservoirs were drawn down, vegetable growths occurred, followed by tastes and odors.

AERATION 383

Aeration helped eliminate odors but the pumping engineer, in his report for 1889-90, stated that the compressor was too small for the average consumption of 3 mgd. (47). As first operated, compressed air was delivered through outlets from a pipe paralleling the inside of the reservoir. Replying to an inquiry, R. W. Fitzgerald, Chief Chemist, Norfolk Water Works, stated December 3, 1935 (48), that the installation consisted of "an air-pressure and perforated-pipe system which aerated the water in the pump suction and basin. It was operated intermittently until 1896 when a connection was made from the compressor directly to the delivery main. The records do not show how long air was pumped into this main but this system was not used after the installation of mechanical filtration [in August 1899]."

New Rochelle, N.Y.—Persistent attempts to aerate the water supply of New Rochelle, N.Y., were made during an uncertain number of years, probably beginning in 1890. The 1893 catalog of the New York Filter Co. cited New Rochelle as an example of water works employing aeration, under its (Leeds) patents, by injecting air under pressure into a continuous body of moving water, also under pressure. It did not tell when the system was installed. The New Rochelle Water Co. built works in 1885–87, with J. J. R. Croes as designing engineer and Charles W. Hunt as constructing engineer, the latter becoming operating engineer and superintendent. The supply was taken by gravity from an impounding reservoir on Hutchinson Creek. Late in 1896 there were two impounding reservoirs, both on the same creek. Water from the reservoirs was delivered in part by gravity, but some was pumped to standpipes (21).

E. T. Cranch, Manager of the New Rochelle Water Co., in 1941 (49) quoted John G. De Veau, a retired employee, who began working for the company at the pumping station in 1891. Cranch wrote:

Mr. De Veau advises us that at the time he started with the Company there was a Clayton Air Compressor driven by a Backus Water Wheel which discharged air through a 2-in. line into the gatehouse from Reservoir No. 1. This method of aeration was found unsuccessful and discontinued. The next attempt at aeration was by means of a riser pipe installed in a standpipe at the Union Corners Pumping Station located at the foot of Reservoir No. 2. The riser pipe extended 5 ft. above the standpipe and all the water entering the standpipe was pumped through the river pipe. This method also proved unsuccessful and was discontinued. Westinghouse air compressors were then installed at the Union Corners Pumping Station and air forced into the gatehouse at the dam. This air was distributed by means of a ring of perforated pipe at the inlet from the gatehouse to the pumping station. We understand that this equipment was kept in use for a number of years but was finally discontinued. All of the above work was done under the direction of Mr. Charles B. Brush. At no time have we had any filters.

Brush stated (43) that in 1890 he was called on to advise on the water supply of a small town, the water works of which were practically owned by a "wealthy, busy gentleman" [apparently Adrian Iselin, President, New Rochelle Water Co.]. Brush states that he set up an air compressor in a pipeline from a 100-mil.gal. reservoir, the compressor being driven by hydraulic pressure from the supply main—in general agreement with De Veau's statement concerning the first aerator. Air was forced back up the main, came out into the reservoir in large quantity and agitated the water with "considerable violence." Algae trouble was stopped.

Brockton, Mass.—In the spring of 1891, disagreeable tastes and odors in a large standpipe at Brockton, Mass., led consulting engineer Phineas Ball to confer with Leeds on aeration. As a result, states Leeds, Brockton purchased the right to use his system of aeration and Ball designed apparatus. The standpipe was 59×61 ft. and had a capacity of 1.25 mil.gal. Water from an impounding reservoir on Salisbury Brook was pumped to the standpipe for high service. An air compressor delivered about 172,000 cu.ft. of air per day into three 2-in. pipes radiating from one side of the bottom of the standpipe. From these, numerous $\frac{3}{8}$-in. and $\frac{1}{4}$-in. brass pipes, with $\frac{1}{2}$-in. perforations, 3 ft. c. to c., extended over the whole bottom of the standpipe. Superintendent W. F. Cleaveland stated that aeration had greatly improved the quality of the water (41).*

Charleston, S.C.—Algae growth in water from artesian wells led to installation of aeration apparatus at Charleston, S.C., late in 1891 or early in 1892. The wells were $2\frac{1}{2}$ to 5 in. in diameter and about 2,000 ft. deep. They discharged into a reservoir from which water was pumped to a standpipe. The reservoir was about 135×195 ft. in plan and 22 ft. deep. A 6×10-in. air compressor in the adjacent

---

* In 1941, George E. Bolling, Director of Health and Water Laboratories, Brockton, was unable to give any definite information regarding the installation. Tradition had it that the system was installed without cost to the city and after a considerable period of trial was given up as not affording results commensurate with the cost of operation (50). It was not mentioned in a report for the *Manual of American Water Works* submitted in 1896 by the late Horace Kingman. I found nothing about it in gathering data on the early Brockton filter (see Chap. VI).

pumping station delivered air to a receiving tank from which a 1½-in. pipe led to and around the top of the reservoir. Twelve branch pipes, with a valve at the top of each carried air down the inner slope of the reservoir to its bottom, where it was delivered through three $\frac{1}{16}$-in. holes in the cap of each branch pipe. Superintendent C. A. Chisholm stated in June 1892 that a pressure of 15 to 20 psi. in the receiver was sufficient and that volume rather than pressure was needed (41). In 1941, J. E. Gibson, Manager and Engineer of the Charleston Water Department (51), stated that no records of aeration at Charleston were available. He thought that aeration might have been stopped in the 1890's when all the wells were equipped with air lifts. In the winter of 1926–27, an aeration system was installed in the sedimentation basin of the Goose Creek impounded supply. It has a capacity of 10 mgd. and consists of 200 Yarway involute-type cooling spray nozzles. These are placed about 2 ft. above the surface of the water and spray the water up about 2 ft.

The Charleston aerator seems to have been the last one of the Leeds type installed. About that time the National Water Purifying Co. was merged into the New York Filter Co., which seems to have given up aeration attempts.

Hyatt's Induced-Air Aerator

Scarcely had Leeds been granted his first aeration patent when John W. Hyatt, rapid filter pioneer (see Chap. VII), applied for a patent on a device for sucking air into and mixing it with water in its downward passage through a group of tubes. Soon afterward he applied for a patent on another air-induction aerator. Both patents were granted in 1885.* Apparatus in line with these patents was soon put into use at three water works. In 1887, Hyatt took out two other aeration patents.

All of Hyatt's patents were on apparatus. In his specifications of his first patent he said that by passing the combined water and air through a filter the particles of filtering material would finely subdivide the air and enhance its action. When the air and water were thus intimately combined, the water would absorb the oxygen of the air and the impurities in the water would be consumed or rendered inert. Any materials remaining, while probably not injurious until

* First application filed February 28 and granted September 15, 1885; second application filed May 2, 1885 and granted July 14, 1885.

after further decomposition, would be offensive to the eye unless removed by the filter.

In one of his devices, water was to be passed down through an inverted cone-shaped vessel pierced with holes articulated above and with a group of Sprengel air pumps. Water falling through these induction tubes was to suck in air and mingle it with the water. To mix the air and water still more, the combined fluid was to be passed over one or more such devices as small stones, horizontal perforated plates or baffles attached to the inside of the left arm of the U-tube containing the Sprengel pumps. The water was then to be passed up through the right hand arm of the U-tube, which might also be baffled, and into the top of the filter.

Hyatt's other patent of 1885 consisted of a closed tube sunk mostly in the earth and a small inner tube extending nearly to the bottom of the outer tube. The inner tube had a bend at the ground level, with a horizontal extension to a tank or reservoir. Slightly above the ground level, a perforated plate concentric to both tubes, was placed. From the perforations, small vertical pipes extended downward a short distance. Water under pressure, discharged into the closed space above the outer tube, passed through the small tubes down through a succession of concentric perforated plates to the bottom of the outer tube, then up through the central tube and to the bottom of the water in the tank. By this means the air sucked in was intimately mixed with the water. Any surplus of air escaped from the surface of the water in the tank. The pressure of the air could be increased as the water and air descended, diminishing the size of the air bubbles. [Compare with British patent issued to Massey-Mainwaring and Edmunds, March 4, 1885, two months before Hyatt filed the application for his tubes-in-the-ground patent (see above).]

Two other patents on aeration devices were granted to Hyatt, both in 1887, one on April 5, the other on July 5. The first of these was on apparatus for automatically mingling air with the contents of a conduit leading from an elevated reservoir without material loss of head. It was based on the principle that contraction of a descending fluid vein tended to produce a partial vacuum therein, drawing air into the current of water if the mouth of the conduit were parallel with the surface of the reservoir at a suitable distance below it. The fourth Hyatt patent was on injectors for aerating water. Sheet metal funnels or thin conical discs, slightly separated from each other by projections, were set with pipe nozzles at the opposite ends.

In a trade catalog of 1886 Hyatt stated that his aerating system combined 25 per cent or more of atmospheric air with water under static pressure, "oxidizing the impurities, destroying the conditions favorable to germ propagation, and so regenerating the water that it will keep sweet much longer in pipes and reservoirs than water not so treated." Hyatt seems to have been the only aeration patentee who claimed that aeration affected water-borne germs and he claimed inhibition rather than destruction.

Three aerators of the type of one or the other Hyatt patent of 1885 were installed: one in 1886 for the City Water Co. of Belleville, Ill.; one in 1887 for the Greenwich Water Co., supplying Greenwich, Conn., Rye and Port Chester, N.Y.; the third in 1888 for the Long Branch Water Supply Co. in New Jersey. The first two of these were of the tower and the third the underground or well type. All these companies are now controlled by the American Water Works & Electric Co., of New York City. No descriptions of the aerators are available in the company's records, nor are dates of installation and abandonment given (52).

Belleville, Ill., and Greenwich, Conn.—Meager descriptions of the Belleville and Greenwich aerators, cited in *Sanitary Era,* a Hyatt house organ of the eighties, have been found. In addition, brief statements are made of the nature of the water supplies concerned (53–55). The Belleville *Daily News-Democrat* of October 13, 1886, reporting the testing of the new filters on the previous day, said that "the water is first carried to the top of the tower through the aerator, by means of which it is charged with air [and then] passes through two filters" (53). The supply was Richland Creek, which in time of freshets was turbid and impure with organic matter from cultivated land (54). This source was abandoned in 1895 or early 1896 for deep wells, just after the installation of Jewell filters.

The Greenwich Water Works Co. contracted for Hyatt filters in March and put them into use in July 1887. The Port Chester *Journal* stated February 9, 1888, that four filters had "recently" been put in at the Collequam Reservoir. Two days later the Greenwich *Graphic* said that water passed up to the top of a tower to the height of the reservoir (50 ft.) then down again, sucking in air, then to and through an alum container and four filters (55). The supply was from impounded brooks and springs, never wholly free from discoloration, due to organic matter, presumably caused by slow disintegration of vegetable growths in swamp areas (55). In 1928, a new filter plant was put in use by the Greenwich Water Co. Forty-eight Sacramento-type aeration nozzles were placed on top of the settling basin (52).

Long Branch, N.J.—Best known of the Hyatt aerators was the one put in use June 28, 1888, by the Long Branch Water Supply Co. Water from springs and a small somewhat discolored creek was gathered in a pond, passed down one half of a 16-in. pipe sunk 100 ft. into the ground, and up through the other side of a vertical partition. As described in an engineering journal soon after the aerator was put into use (56), water passed down the open half of the pipe with great velocity, sucking in a large volume of air. The mingled air and water rose through the closed half of the pipe and entered the pump well from which, after receiving a coagulant, it was forced through a battery of Hyatt filters composed of sand and coke. Aeration and filtration combined, the article stated, removed all impurities in suspension and also a large part of the objectionable matter in solution.

Writing in 1895, Professor Leeds stated that on his recommendation combined aeration and filtration was adopted in 1887 for the 2-mgd. water supply of Long Branch. The object of aeration, there as elsewhere, he said, "was to charge the water itself with oxygen to the maximum and then allow this oxygenated water to purify the filter bed" (38). The Hyatt filters of 1888 still gave summer service in 1940 at the West End Station of the Monmouth Consolidated Water Co., but for some years had been operated by gravity. The change was probably made in 1929 when a clear-water basin was installed. The date on which aeration was abandoned cannot be ascertained (52).

Aeration in Twentieth Century America

Before summarizing twentieth century aeration developments in the United States, a few of the more interesting plants will be mentioned.

Alternate sprays and cascades produced by discharging water over the edge of plain pans and through perforated pans, superimposed, were produced by an aerator put in use at Winchester, Ky., late in 1900 or early in 1901. It was equipped with a ball float and cone adjuster. William Wheeler was designing engineer (57).

Double aeration and double filtration were put into use at South Norwalk, Conn., June 22, 1908, to treat impounded surface water subject to organic growths and tastes and odors. Water was aerated before and after filtration. Both aerators and the first filters were still being used early in 1940 but the final filter had been converted into a clear-water basin. Each aerator was a steel box, $6\frac{1}{2} \times 9\frac{1}{4}$ ft., by 4 ft. deep, with bottom perforated by 6,836 3/16-in. holes, 1 in. c. to c. The plant was recommended in 1904 by Harry W. Clark and built four years later after designs by him and William S. Johnson, both of Boston. The aerators were based on experiments by Clark to find the best method of making up deficit of free oxygen due to much organic matter in a state of change (58) (see Chap. IX).

Unique aerators, designed by Malcolm Pirnie, at that time in the office of Hazen and Whipple, were put into use at Providence, R.I., and Poughkeepsie, N.Y., in 1926. A cluster of spray nozzles is controlled automatically by "utilizing that portion of the head between the water on the filters and in the clear well which remains after the water passes through the filters." These and similar aerators at Rahway, N.J., and West Palm Beach, Fla., were described by Pirnie in 1927 (59).

At Waukegan, Ill., air at atmospheric pressure is sucked in

————from an intake manifold by the velocity of water flowing past hundreds of air tubes—on the injector principle—into a specially designed orifice or mixing throat.... The mixture of air and water (and a chemical when introduced) then rises through a U-tube and discharges in violent turbulence over the circular, herringbone-baffle-studded discharge plate, a few inches above the water level in the basin. The entrained air is released, sweeping out with it the volatile gases—including CO_2—and supplying fresh oxygen.

This apparatus ("Aer-O-Mix"), says Herman Anderson, of the Vogt Brothers Manufacturing Co., was first used in January 1929, at Waukegan (60). The air-induction element of this apparatus has a family resemblance to the device patented by Hyatt in 1885, based on the use of Sprengel pumps (see above).

Porous tubes or plates for the diffusion of air through water, taken over from the activated-sludge method of sewage treatment, have been used at a few water works, beginning at Brownsville, Texas, in 1931. Under a patent granted to Henry E. Elrod, this type of aeration apparatus is being promoted as the "Aerator-Mixer" (61).

General summary.—In the four decades of the present century the proper objectives of aeration have been defined and various means of adapting apparatus to those objectives have been devised. Aeration for the reduction of organic matter, although widely used in sewage treatment, has been given up for water supplies. Its use continues for the reduction of odors and as an aid to the removal of iron and manganese.

Aeration by spraying into the air dominates. Other methods call for jets; weirs and pans; showers through small, closely-spaced perforations; coke trays; and compressed air admitted to the water at the bottom of basins. In no case reported is high-pressure air used; nor is there a single instance of compressed air admitted to a force main —two major principles laid down by Professor Leeds in his patents and papers (see above). St. Paul affords the most notable instance of the use of compressed air, where equipment was installed in 1926 after experiments made in the previous year (62, 63).

By far the largest aeration plants in the world are those below the Ashokan and Kensico reservoirs, in the Catskill supply of New York City. Each has 600 nozzles. At Ashokan the entire reservoir discharge of 600 mgd. passes through the aerator. These aerators were installed primarily for ornamental purposes (64).

The number of water works in the United States using some method of aeration early in 1942 is unknown. In 1931, the American Water Works Association's Committee on Water Purification and Treatment listed aeration at more than 100 works in the United States (65). The list seems to have been incomplete, even for that date, and since then there have been many installations. The United States Public Health Service Census as of early 1941 (66) lists aeration plants, but gives no totals except where aeration is an adjunct to "simple chlorination," the total of such being 83. In the tables by states, the aerator devices are classed as overflow trays, cascade or other splash types, contact beds, coke or other material, spray, patented, and other types, but no summaries under these heads are given.

The basic principles of aeration are discussed by Baylis in *Elimination of Taste and Odor in Water* (67). He gives tabular data on: odor control at 36 water works; reduction of carbon dioxide at fifteen works; removal of iron and manganese at eighteen works; and removal of hydrogen sulfide at four works. Duplications reduce the number of works to about 60. References are given to 68 original sources of information. Most of these are to articles published between 1920 and 1935; few go back of 1910; none cited appeared earlier than 1900. Broadly, all deal with current practice.

CHAPTER XVII

Algae Troubles and Their Conquest

Four chapters of unsurpassed interest in the history of water treatment have been unfolded between the discovery of "little animals" in 1675 by Leeuwenhoek and the present widespread utilization of copper sulfate as an efficient algicide. The chapters may be characterized as (1) discovery, observation, speculation and rationalization; (2) European troubles and remedies, the latter centering on exclusion of sunlight from small reservoirs containing ground waters; (3) American plagues and studies, for the most part baffling because pertaining to surface waters stored in reservoirs too large to be covered; and (4) discovery of copper sulfate as an algicide and the evolution of various means for its application to large reservoirs.

Discovery, Observation, Speculation and Rationalization
Leeuwenhoek to Dwight, 1675–1796

Leeuwenhoek.—"In the year 1675, I discover'd living creatures in Rain water, which had stood but a few days in a new earthen pot, glazed blew within." So wrote the self-educated Dutch naturalist, Antony van Leeuwenhoek, in 1676. Of several kinds of organisms described by Leeuwenhoek in his letter to the Royal Society of London (1), those he observed in rain water "put forth two little horns, continually moving themselves." This led him to call them "living Atoms," little "animals" and, individually, "Animalcula."

King.—In a series of observations on animalculae, reported in 1693, Dr. Edward King (2), noted that after steeping oats in rain water some nine or ten days he could easily see, by his best microscope, "seven or eight sorts of animalcula . . . all very nimble in their motions." In a decoction of herbs, he "saw little creatures like Eels . . . with a wriggling motion." In remarks directed to those "that disbelieve Microscopical Experiments: . . . who may as well deny the use of spectacles," King noted that a minute quantity of spirit of vitriol, put on a fine needle and introduced in a drop of water containing "some hundreds of these animalcula . . . very nimbly frisking about [causes them] to spread themselves, and tumble down seemingly dead."

Harris.—In observations made in 1694–96, John Harris, an English clergyman, saw what appear to have been green algae in one case and blue algae in another (3). On April 27, 1696, with a "much better microscope" than the one previously used, he "look't on a small Drop of green surface of some Puddle-water, which stood in my Yard. This I found to be altogether composed of Animals of several shapes and magnitudes; but the most remarkable were those which I found gave the Water that Green Colour, and were Oval Creatures, whose middle part was of a Grass Green but each end clear and transparent. They would contract and dilate, tumble over and over many times together, and then shoot away like Fishes." May 18, 1696, Harris "look't on some of the Surface of Puddle-water which was bluish, or rather of a changeable Colour, between Blue and Red." There follow these reflections on the origin of the organisms seen by the microscope:

> How such vast numbers of Animals can be thus (as it were at pleasure) produced, without having recourse to Equivocal Generation, seems a very great difficulty to account for. But tho' the resolving it that way makes short work of the mystery (for 'tis easie enough to say they are bred through putrefaction) yet the asserting Equivocal Generation, seems to me to imply more absurdities and difficulties, than perhaps may appear at first sight: I wish therefore, that this matter would a while employ the thought of some ingenious and inquisitive Man (3).

Rutty.—"Seeds of the alga fluviatilis" are mentioned in 1757 by Dr. John Rutty of England (4) as often occurring in rain water; also seeds of "mosses, and little mushrooms, which last appear to the naked eye in the form of slime or mouldiness." Rain water, he said, was greatly variable in its solid contents, depending upon the "different degrees of heat of the exhaling sun,, the winds that bring it, the different soils from which it was exhaled, and different seasons of the year." In spring and summer "it commonly contains the little eggs of animalcules." Its aptness to putrefy "is easily amended by boiling, which presently destroys the animalcules, which, with other sediment, drop to the bottom." Three drops of spirit of vitriol, or five drops of spirit of salt per quart of water, would prevent putrefaction. Standing water of pools or reservoirs was not considered by Rutty, but in *A Methodical Synopsis of Mineral Waters* (4), he discusses "common water," in the form of rain, snow and dew, as "nature's own distilling."

Priestley.—In one of the volumes recounting his extensive studies of air, Joseph Priestley, English chemist (5), asserts that in water, upon exposure to the sun, and particularly spring water, a green substance was formed that was a copious source of vital or dephlogisticated air.* He believed that this substance belonged neither to the vegetable nor to the animal kingdom, but was unorganized filmy matter deriving its color by exposure to the sun.

Ingenhousz.—A few years later the Dutch royal physician and naturalist, Jan Ingenhousz (6), wrote that after three years of study he decided that Dr. Priestley examined the green substance only when it was in an advanced stage, and that had he examined it from its origin, he would have seen it giving indications of animal life. Ingenhousz confirmed Priestley's observations that the green matter on the surface of water was produced much more copiously and rapidly when animal or vegetable substances were added to the water. After noting that other species of insects were sometimes intermixed with those of a green color and expressing the belief that only the latter could produce dephlogisticated air, he concluded that probably the green species were the result of putrefaction of organic matter in the water. That, he thought, explained why the green kind was not produced spontaneously in water that had been boiled; but he acknowledged, says a reviewer of Ingenhousz's *Experiments* (16), "that they are generated in fixed air, notwithstanding it be putrescent."

In concluding his notice of the *Experiments,* the reviewer expressed surprise "that this intelligent and respectable philosopher has so strong a propensity to revive the exploded doctrine of equivocal or spontaneous generation." The reviewer thought it was absurd to maintain "that *corruption*" or the dissolution of animal and vegetable bodies into their "respective elements, should become the immediate parent of organization. . . . Surely it were much easier to believe the existence of *ova* . . . or of germs inconceivably minute, making every part of nature their nidus, and waiting to be developed by putrefaction, and by various other circumstances."

Dwight.—The citations thus far made will suffice to give the earliest observations of algae and related organisms in water and speculations on their mode of generation. Late in the eighteenth century Timothy Dwight, President of Yale College from 1795 until his death in 1817, made observations on organisms flourishing at times on natural bodies

* "Phlogiston. Principle (1635–1743) assumed to form a necessary constituent of all combustible bodies and to be given up by them in burning."—*Std. Dictionary.*

of water in New England, their probable origin and theories correlating them to the prevalence of certain diseases in the neighborhood (7).

Late in September 1796, Dwight saw a narrow "lake" in Marlborough, Mass. He was told by a Mr. Williams, owner of one of the farms on the margin of the lake, that "no endemic prevailed there" (Vol. 6; 1: 346). "It has been commonly supposed," Dwight continued, "that standing waters are insalubrious in countries subjected to such intense heat as that of a New England summer. The supposition is almost, if not quite, absolutely erroneous, so far as New England is concerned." Dwight remarked on the abundance of ponds and lakes in New England, and added:

> I suppose vegetable putrefaction to be especially considered the cause of autumnal diseases. That [it] may be an auxiliary cause of these evils may, I think, be rationally admitted. But that it is the sole cause, or even the principal cause, may be fairly questioned. This putrefaction exists regularly every year; the diseases, in any given place, rarely. The putrefaction exists throughout the whole country; the diseases, whenever they exist, are confined to a few particular spots. [They cannot be due to stagnant waters, because they] are found on plains, in vallies, on hills and even on the highest inhabited mountains.

Dwight then relates experiments "a number of years since" with ground pepper put into a tumbler of water. After a few days, the microscope showed "an immense number of living animalcules"; two or three days later, the microscope showed none "in some scum." After two or three days more, they reappeared. "This astounding process continued until the water became so foetid as to forbid further examination." From these observations Dwight concluded that there was a succession of eggs laid by the organisms. Returning to the spread of diseases he concluded:

> Whatever instrumentality vegetable putrefaction may have, I am inclined to suspect, for several reasons, that animalculine putrefaction is the immediate cause of those disorders, whatever they are, which are usually attributed to standing waters. It will, I believe, be found universally, that no such decease [sic] is ever derived from any standing waters which are not, to a considerable extent, covered with a scum; and perhaps most, if not all of those that have this covering, will be found unhealthy. The New England lakes, so far as I have observed, are universally free from the thinnest pellicle of this nature, are pure potable water, are supplied from adjacent springs, and are, therefore, too cool, as well as too much agitated by winds, to permit, ordinarily, the existence of animalcules (7).

European Troubles and Remedies: 1825–55

Toulouse, France.—Aquatic plants attributed to the strong heat of the sun's rays appeared in the water of an infiltration basin built in the sand at Toulouse, France, soon after it was put into use in the early 1820's (see Chap. XI). Year after year, the growths—described as reptiles, plants, animals—died and putrefied in the lukewarm water until it became intolerable. Various efforts to stop the growths (unfortunately not described) being futile, the open basin, at the suggestion of D'Aubuisson (8), was converted into a "little aqueduct" or filter gallery, after which the water was "an agreeable drink." This is the first record of an algae-infested public water supply, and the first specific example of an effective preventive.

Nottingham, England.—Owing perhaps to a cooler climate, Thomas Hawksley, noted English engineer (9), had less trouble with conferva in a filter basin completed in 1831 at Nottingham, England, than did D'Aubuisson and his associates at Toulouse. Hawksley removed the growths about once in three weeks in summer and in six weeks during the winter, by pumping out the water and sweeping the bottom of the basin with a broom. This basin was, however, supplemented by a closed filter gallery.

Warrington, England.—A troublesome growth of "animalcules" at Warrington was stopped in the 1840's by covering a reservoir. The supply, says J. F. Bateman, a civil engineer (10), was gathered from slopes of cultivated land. After the water was "rendered perfectly pellucid" by filtration through sand it was delivered to a reservoir not over 6 ft. deep. This "very soon became filled with animalcules." He then installed a copper wire strainer of 60 strands to an inch to intercept the organisms but clearing it was so difficult that he ultimately covered the reservoir with flagstones supported by beams and pillars. This was entirely successful.

London.—Partly to prevent organic growths, all reservoirs storing filtered water within five miles in a straight line from St. Paul's Cathedral, London, were required by Parliament in 1852 to be covered from and after August 31, 1855. Filtration was also required of all water for domestic consumption supplied within the metropolis from and after December 31, 1855, except water pumped from wells into a covered reservoir or aqueduct, without exposure to the atmosphere. In the statistical tables for the eight metropolitan companies contained in his book of 1884, Colonel Sir Francis Bolton, water examiner for the London metropolitan area (11), lists 53 covered reservoirs for filtered water. None held more than about 9 mil.gal. (U.S.). No greater improvement in water works construction, wrote Bolton, was ever effected than covering the London reservoirs, thus "protecting the water from all atmospheric impurities, as well as from light and heat. Reservoirs which when open required cleaning out twice a year, owing to vegetable growth, aerial impurities, and animal life constantly accumulating therein, were found to be perfectly free from any objectionable deposit for five years after being covered over."

Berlin.—In Berlin, according to William Lindley, Engineer of the Berlin Waterworks Co., filtered water from the River Spree, on standing in tanks became "covered with confervae and vegetation of various kinds; the water lost its transparency and became so turbid as to resemble the slush of the London streets." The managers thought it necessary to stop water distribution once a month and clean the tanks. On Lindley's recommendation, the tanks were covered and the growths stopped immediately. This was brought out in a discussion of a paper read in 1867 (12). The date of covering was not mentioned but the supply from the Spree was introduced in 1853.

American Plagues and Studies: 1845–91

Boston, Mass.—American annals of algae control begin at Boston November 18, 1845, with a report by John B. Jervis, Chief Engineer of the Croton Water Works, New York City, and Prof. Walter R. Johnson, of Philadelphia (13). These "two important engineers from abroad," were engaged to report on various possible sources of water supply for Boston. Long Pond, Spot Pond and the Charles River were considered, temperatures at various depths taken, samples of water collected and subjected to both microscopical and chemical analyses. In water from all three sources, "infusorial insects" were found. After saying that animalcules, in themselves, were not harmful, the engineers named conditions under which they might be troublesome, anticipating to a considerable degree the conclusions drawn decades later by FitzGerald of Boston, Forbes of Brookline and Rafter of Rochester. In summer, said Jervis and Johnson, few water sources are without animalcules. They did not intend "to assert that a source *may not,* from its shallowness, stagnancy, high temperature, and other causes, become offensive on account of its excessive productiveness of animalcules." None of the sources considered was objectionable in this respect. Diking off shallow areas of Spot Pond, in case it was utilized, was discussed. Long Pond, afterwards known as Lake Cochituate, was recommended as a source of supply. Appended to the Jervis and Johnson report was one by Professor J. W. Bailey, dated September 30, 1845, containing a tabulation of "the various species of infusoria" found by him, keyed with "orders, plates and figures of Ehrenberg's *Infusionsthierchen.*" This monumental work had been published only a few years earlier—in 1838.

It seems safe to say that the Jervis and Johnson and the Bailey reports presented the earliest notable studies of algae and related troublesome organisms in public water supplies made in America. This assumption is all the safer because at the close of 1845 there were only 70 water works in the United States.

Apprehensions of tastes and odors, reviewed by Jervis and Johnson, proved to have been too lightly dismissed, when, six years later, the Lake Cochituate supply was put into use. In October 1854, the water had a marked peculiar taste—fishy to some, cucumbery to a great majority of consumers. A disagreeable smell sometimes accompanied the bad taste. E. N. Horsford and Charles T. Johnson were engaged to ascertain the cause (14). Johnson stated that the trouble did not originate in the pipes, but in Lake Cochituate; that it was not due to animal putrefaction but to vegetable fermentation and that the water contained nothing deleterious. A remote cause for this fermentation was thought to be the long drought, summer heat and unusually low water in the lake.

Due to the vagaries of nature, or because the attention of the water authorities was centered in other matters, twenty years passed before the annual reports of the Cochituate Water Board recorded further serious troubles from organic growths. Early in October 1875, wrote Joseph P. Davis (15), engineer for an additional supply, complaints of bad tastes were received—the water tasted like fish oil to some, dead leaves to others but it was a "cucumber taste" to most. On October 23, samples of water were taken from Lake Cochituate, at all depths, but no peculiar taste could be detected. The next day Brookline and Chestnut Hill reservoirs were visited. At the latter only a slight taste was found, but overnight the taste spread through 500 mil.gal. of water in one division of the reservoir. On turning off the reservoir, taste in water from the mains stopped. On resuming use of the reservoir April 1, 1876, there was no trouble from taste.

Davis engaged William Ripley Nichols, Professor of Chemistry at the Massachusetts Institute of Technology, who was then rising to deserved fame as an authority on the quality of water, to study the problem. Nichols called to his aid Dr. W. G. Farlow, Assistant Professor of Botany at Harvard University, and Edward Burgess, Secretary of the Boston Society of Natural History. All three were baffled. Farlow in a report dated December 15, 1875 (15), named various organisms found in the Chestnut Hill reservoir. He concluded that the cucumber taste was not caused by "any living plant undergoing decomposition that could be detected by the microscope." Burgess reported (15) that no "microscopical animals of any kind" had been found in large numbers. Nichols, in his covering report of April 1876 (15), confessed that he was "quite in the dark as to the cause." There was no proof, he said, that the water would injure a healthy person. There the matter stood—beyond the ken of a chemist, a botanist, and a zoologist!

Thinking it advisable to make one more effort to find the cause of contamination of the Boston water supply, a committee of the city council engaged Ira Remsen, Professor of Chemistry at Johns Hopkins University, to investigate the subject. This he did November 4–17, 1881 (16). In his report he stated that the cucumber taste in the water, which had occurred several times from 1854 on, had affected the Croton water supply, New York City, in the winter of 1881, in apparently the same way. It had given trouble at New Haven, Conn., in 1864, 1865 and 1872; at Hartford in 1871; at Norwich, Conn., for several years in succession. Other places similarly affected at one time or another were Keene, N.H.; Holyoke and Lynn, Mass.; Albany, N.Y.; York, Pa.; and Jacksonville, Ill. In the winter of 1881, Remsen said, he had investigated similar trouble at Baltimore; but his results like those of all others who had carefully studied the subject had been unsatisfactory.

At Boston, careful chemical examinations made by Remsen showed that of five sources containing the largest amounts of albuminoid ammonia, only the Bradlee basin of the Chestnut Hill Reservoir and Farm Pond were being used. Peculiar substances retained by screens at the Farm Pond gate house were submitted to Professor Farlow who identified a freshwater sponge. Professor Alpheus Hyatt, of the Boston Society of Natural History, pronounced the organism to be *Spongilla fluviatilis* Auct. Remsen said the evidence was almost conclusive "that this sponge is the whole cause of the present difficulty" at Boston, and probably of many other cases of trouble. He suggested drawing off the water of Farm Pond, searching diligently for the sponge, removing it, then refilling the pond from Basin No. 2 through a restricted channel. The committee to whom Remsen reported stated

FIG. 71. FIRST BIOLOGICAL LABORATORY ON AN AMERICAN WATER WORKS SYSTEM, CHESTNUT HILL RESERVOIR, BOSTON WATER WORKS
Opened September 1889 under direction of Desmond FitzGerald; later supervised successively by George C. Whipple, F. S. Hollis and Horatio N. Parker (From an 1892 photograph supplied by Parker, City Bacteriologist, Jacksonville, Fla.)

that the city engineer found it impracticable to empty Farm Pond completely. The committee advised that a permanent conduit should be built across or around the pond so it could be cut out of service in case the trouble was repeated. Instead of acting on the recommendation, the city constructed a flume and a ditch.

The Sudbury Aqueduct, fed by huge impounding reservoirs, was put into use in 1878. Desmond FitzGerald, who had been construction engineer on this project and was for many years operating engineer of the western division of the Boston water works, conducted filtration experiments and studies of micro-organisms in the waters under his charge from 1888 to 1894 (17).

FIG. 72. INSIDE VIEW OF CHESTNUT HILL RESERVOIR LABORATORY
Originally only examinations for micro-organisms made, but, beginning in latter part of 1891, bacterial determinations also undertaken; note date on calendar is November 1892
(From photograph supplied by Horatio N. Parker; made at the time that he was assistant to George C. Whipple)

At the Chestnut Hill Reservoir he established what is believed to be the first biological laboratory connected with an American water works system. In his report for 1889, FitzGerald said that experimental filters of sponge and other materials had been operated for the removal of micro-organisms and that construction of additional filters had been ordered. In his report for 1890, he stated that weekly biological examinations had been made of the water in all storage and distribution reservoirs, at surface, mid-depth and the bottom, including number and kind of organisms. In addition, during that year 90 special investigations had been made of the water in brooks feeding the reservoirs and the effect of swamps. These data, he said, would be useful in planning improvements of the water whenever undertaken. No major improvements of existing supplies were made nor was his vast accumulation of data on micro-organisms published. This was largely due to the construction of works for an additional supply. Some of FitzGerald's data, presumably, led to the expenditure of millions for stripping the site of the Wachusett Reservoir on the Nashua River to remove organic matter which might cause disagreeable tastes and odors.

Baldly put, the immediate practical lessons derived from a half century of studies of tastes and odors in the surface water supplies of Boston by FitzGerald and his predecessors, were to avoid shallow flowage and to strip reservoir sites of organic matter. More far-reaching results were the correlation of various micro-organisms with tastes and odors, which became useful after the advent of copper sulfate treatment for algae control in the first decade of the twentieth century.

Albany, N.Y.—Bad tastes and odors in small impounding reservoirs of the water supply of Albany, N.Y., began in 1852 and recurred at unpredictable intervals thereafter (18). They were studied with untiring zeal by George W. Carpenter in his long superintendency of the water works. In August 1852, said the water commissioners in their annual report, the water became unfit for use. The cause assigned by Carpenter was animalcules which overspread the bottom of one of the reservoirs and decomposed there. Tastes and odors recurring in the next three years led a committee of the city council to state October 29, 1855, that "no more alarming event, short of the actual visitation of a pestilence, can befall a large city than the sudden poisoning of its water supply at the commencement of the hot season."

After a comparative respite of ten years the plague recurred in 1865. In his report for that year, Carpenter said that it was "impossible to convince some that water so impregnated can possibly be innoxious." The local health board invited Professor Ten Eyck and several physicians to visit and examine the reservoir. On August 11, 1865, they reported that although they appreciated the inconvenience caused by the bad water they could state that thus far it had not been injurious to health. There was no more sickness than usual in the city and no more in those parts where bad water was distributed than in the rest of the city.

Frequent microscopical examinations of the trouble-giving water, Carpenter stated in his report, were made by Professor Ten Eyck, who found the impurities to be minute vegetable organisms, similar to those described by James R. Chilton and John Torrey in their reports to the Croton Aqueduct Board, New York City. Examined with a microscope, wrote Carpenter (18), "the water appeared to be filled with minute organisms, resembling fine threads of glass or lines of light," extending in all directions, with beautifully developed structures. They were motionless. A plate in Carpenter's report for 1865 reproduced a drawing by Professor Ten Eyck, showing filaments magnified 1,000 diameters. Carpenter's final words were: "What the origin of this particular form of algae may be, or whether its development can be checked by counteracting influences, are questions which cannot, as yet, be satisfactorily answered. We know, however, its form, size and laws of growth; and, what is far more important, that it is not deleterious to health."

By 1866, Carpenter was convinced that the processes of nature finally leading to the destruction of algae and infusoria "cannot be hastened by any artificial means yet discovered." All the means of destruction, "some feasible and some chymerical," yet suggested, he declared in his 1866 Report (18), were based on the assumption that the water in which the organisms grow lacks oxygen. But "if it were possible to charge the water with an additional quantity of oxygen, by a force-pump or by exposing it, through the action of a large waterwheel to the atmosphere (both of which have been suggested) the foreign matter, the real cause of the offensive taste and odor, would remain."

In his report for 1872 (18), Carpenter said that the impurities occurring in the Rensselaer lake supply were not confined to any particular season, having appeared in spring, summer and autumn, while in November 1872, for the first time in years, the cucumber taste appeared for a short time in Lower Tivoli Lake. In this report, Carpenter summarized "well-established facts" as to taste and odors gathered from many cities. His tentative conclusion was "that the impurities were climatic," and that the atmosphere was the great reservoir of spores, which develop under favorable conditions of air and water, thus accounting for their erratic appearance.

Commenting in his report for 1875 (18) on recommendations for filtration, Carpenter noted that "some filtered water, when exposed in uncovered basins, is more liable to become offensive than when it is turbid." Carpenter's summing up of his experiences with taste and odors during 20 years was a "counsel of despair" as to remedies, tempered with the statement that "however offensive the impurities, they are not deleterious to health." Introduction of a supply from the Hudson River in 1875, to supplement the gravity sources, lessened when it did not eliminate algae troubles. But unfiltered Hudson water substituted for an occasional nuisance a scourge of typhoid, not checked until filters were put into use in 1899 (see Chap. VI).

Trenton, N.J.—"Unpleasant smell and taste" of the water supply of Trenton, N.J., in the summer of 1855 caused uneasiness among the citizens. State Chemist Wurtz was called on to make an investigation. Chemical analyses of samples of the Delaware River water then being pumped to a 1.4-mil.gal. reservoir, of water from the reservoir itself and of water from springs which had recently been delivered to consumers, assured Wurtz that the river water was chemically satisfactory. He then had a pupil examine the contents of the reservoir. The water showed several varieties of animalcules, lichens and minute plants. Sediment at the bottom of the reservoir was

———almost wholly composed of forests of minute plants through which roamed herds of such animals as *volvox globator,* 'globe jelly,' *vibric anser,* or 'goose animalcule' and several specimens of *Bacillaria* and *navicala.* On the surface of the water he found a slight green scum which when magnified resolved itself in collections of *cercalia mutabilis,* an animal production characteristic of stagnant water. Numerous large green water weeds may also be found floating in the reservoir. . . .

Open reservoirs, in which water is kept standing for several days to stagnate in the heat of the sun are perfect hot-beds for the growth of animal and vegetable life. . . . The breeding of those microscopic creatures, under favorable conditions, is so rapid that in a very few hours the water will become alive with them. It was to one of these animals, a species of cyclops, that the so-called 'fishy' taste and smell of the reservoir water, which has at two or three periods been found so annoying, was due (19).

Wurtz then suggested that the reservoir be covered, thus depriving the "organic germs of the light and heat of the sun, which constitute their means of life, and they will cease to germinate" (19). No earlier

American specific suggestion for covering a reservoir to prevent algae growths has been found.

New York City.—Professor John Torrey, leading American botanist of his time, made a classic report in the summer of 1859 on tastes and odors in the Croton water supply of New York City, after a visit to Croton Lake (20). He attributed them to a "microscopic conferva-like plant, which abounds in a volatile odorous principle, soluble to some extent in water." This, he said, "is the first time it has been offensively brought to our notice. Even when it was most abundant in the Croton, I do not believe it communicated any unwholesome quality to the water."

Schenectady, N.Y.—William J. McAlpine, a leading American water works engineer for many years, proposed a simple plan in 1867 to avoid the organic growths that were more and more frequently causing worry and disgust to American water works officials and citizens. He wrote:

> Stored water is sometimes defiled for a few days by the rapid generation and decay of vegetable matter and animalculae. This requires the conjunction of a high temperature and quiet atmosphere, *and perhaps a certain electric condition of the latter* [author's italics]. These conditions occur . . . after long intervals of time. The plan proposed would enable the supply [from Sand Creek] to be taken directly from the stream without storage (21).

Springfield, Mass.—Notorious among cities periodically afflicted with intolerable tastes and odors from algae growths in surface water supplies is Springfield, Mass. When, in July 1873, Phineas Ball, of Worcester, advised taking a gravity supply from Higher and Broad Brooks, rather than pumping it from the Chicopee River, he gave as reasons the "present purity" of the water from the brooks. The water commissioners, in their first report (22), said that the greater part of the Ludlow Reservoir basin was swampy and covered with sprouts and small wood. The site was not cleared except for the removal of old stumps from a part of the area. Sand was spread over a few acres near the outlet of the reservoir. Removal of all stumps and grubbing, ploughing, cleaning and sanding the entire bottom—445 acres—the water commissioners stated, "was beyond the means at command." Moreover, they knew of no city, supplied in like manner from artificial ponds of this size, where such a work has been attempted, and they have not been able to learn that any serious trouble has arisen where similar sources of supply had been drawn upon (22).

In support of the belief or hope that the organic matter in the bottom of the reservoir would cause no trouble, the commissioners cited a letter from F. T. Stanley, President of the New Britain, Conn., Water Co., written late in 1874. He stated that the bottom of his company's reservoir was not cleared when constructed because it would have been financially impracticable. After having been filled for seventeen years the bottom was still covered with stumps. For the past twelve or fourteen years it caused but little complaint. On a Thursday in September 1874, the shores were lined with green vegetable matter, which disappeared the following Saturday. Whether "discoloration" was caused by the decomposition of aquatic plants along the shores of the reservoir or of stumps in the bottom was unknown. After quoting this report from New Britain, the Springfield commissioners stated that they anticipated no annoyances from the impurities mentioned. Three brooks and many springs discharged into the reservoir. Its large area gave opportunity for winds to sweep over its surface, agitating and aerating the water more than would be possible in a smaller reservoir.

Ball's construction report described a "filter" built across the lower end of the reservoir, consisting of excelsior in frames extending 100 ft. between masonry walls, abutted by parallel walls of dry rubble, filled between by sand. Water was turned into the mains December 31, 1874. The filter, which was not completed until August 7, 1875, was soon abandoned as useless. At least that was the official explanation. Rumor had it that the excelsior portion of the filter collapsed during the first night it was in use. However that may have been, the self-assurances of the water commissioners that there would be no algae trouble were swept away with the coming of warm weather in 1875. The Ludlow Reservoir was cut out from the system. As companions in greater misery, the water commissioners cited Boston, Worcester, Holyoke, New Britain and other towns. They appealed to Professor William Ripley Nichols of the Massachusetts Institute of Technology for advice. His studies, reported January 1, 1876, did not extend beyond a visit to the Ludlow Reservoir, and chemical analyses of the water. He stated that Engineer Ball had observed, the previous summer, "a peculiar alga belonging to the *Nostoc* family." Nichols did not think that the organism need cause the slightest anxiety. He did not hesitate to commend the water for general use. He believed that if the reservoir were kept full the condition of the water would improve in time.

Although Nichols' report was of little practical value, the water commissioners, as if to check up his consolatory pious hope, engaged him to make a continuous study of their troublesome water problems, which he carried on for two years.

Because of heat and drought in 1876, the feeders to the Ludlow Reservoir delivered but little water. In the city the water was at times disagreeable in taste, odor and color. In the hope of remedying this, an 8-in. pipe was laid from the Ludlow supply main and discharged into a brook feeding the Van Horn distributing reservoir. By this means, the report for 1876 said, the water was agitated, and largely divested of its disagreeable qualities, so that little or no complaint was made.

Nichols reported January 1, 1877, that during the past year he had made weekly chemical analyses of Ludlow Reservoir water and also microscopic examinations of the minute vegetable organisms in the water. Dr. W. G. Farlow, Professor of Botany at Harvard University, identified the few organisms found as *Clathrocystis aeruginosa*. Nichols had seen abundant growths of that plant in other ponds and reservoirs. The New Britain Reservoir had had a similar growth annually. He was still hopeful that if the Ludlow Reservoir was kept reasonably full while the water was unpleasant, the inconvenience would become shorter. Filtration, as a small experimental filter had shown, would remove much of the troublesome matter and could be satisfactorily done by individual consumers. To make it efficient on a large scale the effluent should be piped to consumers without exposure to air.

After two or three years of comparatively little trouble, the character of the water became town talk in 1881. The subject had been investigated from geological, botanical and zoological viewpoints by the most accomplished savants in the country, the annual report for 1881 said. Trials of filtration showed that the multitude of minute particles in the water soon choked up the sand. In its report for 1882, the water board summarized answers to a questionnaire sent out by the New York health board. Of 143 reporting water works in the United States and Canada, 60 had experienced more or less trouble from algae.

Flirtations with the possibility of filtration and other palliatives went on for some years. Superintendent Hancock, in his report for

1886, expressed the opinion that the troublesome vegetable life was brought into Ludlow Reservoir from the Belchertown Reservoir, which flooded a swampy area and sent down a vast amount of decaying vegetation. After noting that the Ludlow Reservoir had been ponded for twelve years, which he thought was long enough to eliminate a large part if not all of the original organic matter in the reservoir, he said if nothing but good water were put into the reservoir all would be well. The water commissioners, in their report for 1886, stated that an inspection of the Belchertown Reservoir showed that water from it was unfit for use, since in its passage through two miles of mucky swamp it gathered much vegetable matter and germs of algae. They proposed bypassing a brook that emptied into the Belchertown Reservoir. This was done in 1887. Other diversions were made as was also provision for draining water from the Ludlow Reservoir at different levels.

The taste-and-odor problem was so far from being solved in 1889 that the water board turned its attention to mechanical filtration. A report by the State Board of Health, September 13, 1889, discouraged the adoption of mechanical filtration and suggested turning the streams entering the Ludlow Reservoir directly into the mains or else substituting water from a new supply. In 1891, the waters of several brooks were turned into a main leading into the city. Ludlow Reservoir was emptied of water and fish in 1893 and a new gate installed for drawing water from the bottom of the reservoir.

No algae appeared during 1894. The "Ludlow odor" was absent. Analyses by the State Board of Health placed the water high among the water supplies of Massachusetts. But alas! On September 15, 1895, the thermometer at Ludlow Reservoir fell to 33°F. and the fall turnover came earlier than usual. On September 20–22 the temperature rose to 90°, the highest of the season. A rapid growth occurred and for several weeks the water was in bad condition. By drawing on water stored in recently acquired reservoirs, it was possible to empty the Ludlow Reservoir and refill it with fresh water, after which the water was again good.

Algae were rampant for two months in 1896, rendering the water unusable. Again the Ludlow Reservoir was shut off. During the hot seasons of 1897 and 1898 this experience was repeated. In 1899, the reservoir was emptied, staked off in squares of 100 ft. and the bottom sounded. Mud and other objectionable material was found to depths of 20 to 46.5 ft. To remove all this muck would be impracticable. Conditions undoubtedly would have been improved by covering the bottom with sand and dividing the reservoir into two basins so that one could be emptied each year, but that would have reduced the storage capacity.

Water brought into the Ludlow Reservoir from additional sources, including mill ponds never properly prepared for storing water, had made the supply worse rather than better. This the State Board of Health reported on July 24, 1899, in response to a request for advice. Neither removal of mud from the bottom of the reservoir nor from parts of it, with the remainder covered with sand, the board said, was feasible. It had recommended filtration experiments in 1897 but the water board had done nothing. The state board knew of no experiments on water like that in question. Therefore it gave general approval to a plan submitted by Percy M. Blake (23) for a supply from a branch of the Westfield River, provided organic matter were removed from the bottom of the proposed reservoir.

Unwilling to give up its old supply, the water board decided to experiment on rectifying it by filtration. For that purpose the city council, in July 1901, authorized the expenditure of $12,000. Under Blake's direction, gravity sand filters were built and operated. With water in its usual summer condition, a single filtration failed to remove the taste and odor, but aeration of this effluent, followed by a second filtration, was successful. After more than a year of experimentation, in which the state board collaborated, Blake reported in 1902 (23) that it had been shown that all the taste, color and odor could not be removed from Ludlow water at all times. He advised the city to go to the Westfield River for a new supply and the State Board of Health concurred. In October 1902, the city council created a special water commission. It reported to the council on March 23, 1904, transmitting a joint report by Samuel M. Gray and George W. Fuller and a separate report by Elbert E. Lochridge dealing with filtration and aeration experiments (24).

Anabaena, said the engineers, had always been the main cause of tastes and odors in the existing supply. These organisms, which with others had been mentioned by Blake, seemed to overgrow and dominate others in the summer, but in winter animal organisms had "produced somewhat disagreeable odors." Remedial measures mentioned and dismissed by Gray and Fuller were: the removal of 2,000,000 cu.yd.

of the worst part of the reservoir bottom; the piping of air into the bottom of the Ludlow reservoir to prevent stagnation; ozone treatment; and seeding of the reservoir with organisms antagonistic to *Anabaena*. Copper sulfate, studied at that time by the U.S. Department of Agriculture, was considered but data were not available when requested in June 1903. Gray and Fuller concurred in Blake's opinion that when *Anabaena* were not present in "epidemic form," single filtration would produce a satisfactory supply from the Ludlow Reservoir, but when such epidemics continued for one to five months it would not be possible to eliminate them by filtration and aeration without excessive cost. Filtered water from other sources would be financially practicable. A supply from the Westfield River was advised.

Grasping at a straw to save abandonment of the Ludlow supply, the water board wrote to the State Board of Health on May 31, 1904, for approval of experiments with copper sulfate. The state board arranged with Dr. Moore, of the U.S. Department of Agriculture, to experiment with that agent at a reservoir, not in use, where *Anabaena* had appeared. The organisms disappeared after the reservoir was thus treated, but the state board, always ultra conservative, found almost as much copper in the reservoir 24 hours later as had been applied. It regarded it "essential to determine what became of the copper." It could not advise its application to the Ludlow Reservoir "until its probable effect" was more definitely known. Meanwhile it was trying copper sulfate at abandoned reservoirs.

Clinging to the Ludlow supply, the city council and mayor, in August 1904, requested the water commissioners to construct immediately a single filtration plant at Ludlow, even though Gray and Fuller had advised that it would be inadequate when *Anabaena* were at their worst. The cost was limited to $300,000. Allen Hazen and George C. Whipple were engaged to prepare plans and specifications. Hazen reported on November 30 that single filtration would not supply good water during the *Anabaena* season; that prefilters would raise the cost to $350,000; and that at times pumping would be necessary to supply the proposed filters. He advised postponement of the filtration project until an expected report from the Massachusetts Board of Health was received and studied.

Doggedly, the commissioners instructed Hazen to proceed but Hazen stood his ground. On December 2, the water board requested the city council to authorize an investigation of the possibilities of Little River, as advised by Hazen. Obstinately, the council, on December 19, ordered "further development of the Ludlow sources." The mayor, now convinced of his previous error, vetoed the order on December 27. The council being unable to muster votes enough to override the veto, left the whole matter in the air at the close of 1904.

At last, the water commissioners, in their report for 1904, threw up the sponge, putting the blame for 30 years of trouble on "the assurance of the expert authorities of the earlier years," that the "water of Ludlow Reservoir would undergo self-purification." *Anabaena* had appeared from July 22 to October 7, 1904, "calling forth again the oft-repeated demand of our citizens" that "water that can be used in the house, store and factory, be provided, at whatever cost."

Allen Hazen was engaged to investigate Little River as a source of supply. He soon reported that without filtration it would be better than Ludlow water with a single filtration. The city council, after waiting until October 9, 1905, hung like a dog to a bone to the Ludlow system. Plans for intermittent filtration of Ludlow water in summer were made by Allen Hazen. The filters were put in operation July 7, 1906. They were used in summer for some years, first for Springfield, then for an adjacent town. Finally, water from a new gravity source, developed with Hazen as engineer of design and E. E. Lochridge as chief construction engineer, was delivered to the city December 21, 1909. The water was settled, aerated, passed through covered slow sand filters, then, without exposure to light, flowed eight miles to a reservoir. Thus by abandoning an old supply for a new one, rather than the promised self-purification of the reservoir or any of various palliatives tried, did Springfield escape from the algae nuisance which for decades had made Ludlow Reservoir water notorious.

Poughkeepsie, N.Y.—At Poughkeepsie, N.Y., tastes and odors in the filtered water were noted in the annual report of the water works for 1875. The filter was bypassed for a time. To forestall a repetition of the trouble the temperature of the raw water was watched so direct service could be used. In 1877, microscopic studies made by Cornelius Van Brunt were of material aid to Superintendent Davis in operating the works. Algae growths in the clear-water basin led in 1891 to roofing it with timber and covering the roof with earth. Clogging of the filter by algae was mentioned by Superintendent Fowler in a paper read in 1898 (25) (see Chap. VI).

Hudson, N.Y.—Observations at Hudson, N.Y., on summer tastes and odors, caused by micro-organisms, were recorded by Engineer J. B. G. Rand in a report made in 1875, directly after the construction of slow sand filters by the city. He correlated these data with air and water temperatures at the river pumping station and at the filters, 300 ft. higher. He recommended that both the filters and the clear-water reservoir be covered to prevent trouble from ice and from aquatic growths (26) (see Chap. VI).

Brookline, Mass.—All the American troubles with algae, thus far reviewed, were in surface water supplies. At Brookline, Mass., however, water from a filter gallery was the source of trouble. The works were put into use on May 27, 1875. In 1878, there were complaints of the quality of the water. The trouble was attributed to the entry of swamp water into a vitrified clay conduit. The report of the water board for 1885–86 noted a peculiar taste and odor in the water, occurring after November 1. On that date, pumping sixteen hours a day was stopped, as had been done in the two years preceding. Algae were found in the open receiving reservoir six days after it had been cleaned to the bottom stone. Growth of algae, the report said, was always more rapid after a cleaning, because in the process a multitude of small fish that fed on the algae were lost, while the germs of a new crop of algae were present when the reservoir was refilled, but the fish to destroy them were still lacking. From May 1 to November 1 of both 1884 and 1885, when the pumps were worked sixteen hours a day, and water was supplied direct from the galleries, there were no complaints of the quality of the water. From this, it was inferred that exposing the water to the sunlight caused the troublesome growth. Timidly, the board suggested that the pumps be run sixteen hours a day the year around, keeping the reservoir constantly full for emergencies, but it did not feel justified in adding $75 to $80 a month to operating expenses. Two years later, the board noted that a special committee had stated that primarily the water was satisfactory but could not be kept free from taste and smell as the works were then being operated. Remedial alternatives were: running the pumps 24 hours a day, with all connections between the reservoir and consumers shut off, which was not recommended; or, covering the reservoir, which was advised. To support its advice, the committee said that roofing the high-service tank had ended all complaints from consumers in that district.

Fayette F. Forbes, Superintendent, who previously has kept modestly in the background, asserted that the unpleasant taste was due wholly to algae growth. Frequent microscopical examinations detected growths and correlated them with changes in pumping or drawing water from the reservoir. "With our present knowledge, nothing but a total exclusion of light" from the reservoir "can wholly stop this trouble." Forbes, who had begun microscopical examinations in 1887, was probably the first water works superintendent to do so regularly and use his observations in operating the works in his charge. He published three papers on the subject and one on covering the Brookline reservoir (27–31). After some delay, a covered reservoir was put into use January 1, 1893. Eleven years later, a second one was completed. In 1911, a third reservoir, for high service, went into operation. All three were designed by Forbes (28).

Denver, Colo.—Charles P. Allen, Chief Engineer of the Denver Union Water Co., stated in 1896 (33) that algae growths in reservoirs storing water from a filter gallery had been stopped by roofing the reservoirs with 12-in. boards spaced about 1½ in. apart, laid north and south. With boards laid east and west, algae grew as fast as if the reservoir was not covered. About a third of the roof area was composed of doors which could be kept open or closed. Three reservoirs, with capacities of 6, 12 and 15 mil.gal., had been covered. Before adopting this plan, experiments were made with small reservoirs, some lined with concrete, some with asphalt, some with earth; some covered and some not. The kind of lining used made little difference in algae growths.

Copper Sulfate: 1904–42

After 80 years of failure by water works engineers to prevent tastes and odors from algae, except by covering small reservoirs, George T. Moore and Karl F. Kellerman proved that copper sulfate was an efficient algicide. Their reports published in 1904–05 (34, 35) on studies begun in 1901, were given publicity in the technical press and before chemical and water works associations (36). Within a few years, the first appearance of algae organisms in water supplies was speedily followed by the application of copper sulfate by means of boats, either by spraying it on the surface or, more commonly, from bags immersed in the water. Row boats soon gave way to motor boats, particularly for large bodies of water. Various methods of applying copper sulfate were described by Hale in 1930 (37). Among these was wholesale treatment of water flowing through enormous aqueducts, first used late in 1919. The paper was a comprehensive review of the whole subject of controlling micro-organisms. It included a classified tabulation of organisms, their characteristic tastes and odors, the dosage of copper sulfate required for their control, and also, for some organisms, the chlorine dosage required.

Artificially created turbidity to prevent or lessen filter clogging by algae growths was introduced at Huntington, W.Va., in 1924; Evanston, Ill., in 1925; and Louisville, Ky., in 1927. At Louisville, wrote W. H. Lovejoy, Superintendent of Filtration in 1928, the sediment in one compartment of the settling basin was dredged out and delivered to the other, from which the water went to the filters (38). But "since 1936 and up to the present time," wrote Lovejoy, August 28, 1942, "pre- and superchlorination have supplanted turbidity for combating algae and lengthening of filter runs. Prechlorination at 20–30 lb. per mil.gal. has been highly successful in solving this problem."

Prechlorination of water applied to slow sand filters was introduced by Ilion, N.Y., in 1929. The object was to prevent algae growths on the filters, increase the length of the filter runs between cleanings and permit bypassing the filters during fires. Postchlorination had been practiced for many years. Copper sulfate has been applied to the storage reservoir by means of a hydraulically operated hypochlorinator since November 1941. In previous years the height of the algae season in the reservoir had been January, but there was no trouble in January 1942. In the old days, states Supervising Engineer Earle J. Trimble (39), the filters were scraped about once in three weeks. Prechlorination and continuous application of copper sulfate has increased the runs to at least ten weeks.

The Butte, Mont., Water Co. had algae troubles in its main reservoir from its construction in 1892 until the announcement of the copper sulfate treatment in 1904 by Moore and Kellerman. Numerous experiments, including aeration, during that period were ineffectual in removing the odor and taste of vegetable algae during the summer months. The reservoir was treated with copper sulfate in the summer of 1904 and the treatment has been continued with great success ever since. In 1900, an additional supply was procured from the Big Hole River, from which it is pumped across the Continental Divide. passing through two reservoirs of 13 mil.gal. each before it reaches the distribution system. This supply is treated with chlorine and ammonia at the pumping station, but during the low water season of the river, algae trouble developed in the 31 miles of influent pipeline, causing a slight taste and odor in the city water. To overcome this, Superintendent M. W. Plummer designed an automatic machine to inject a regulated amount of copper sulfate and lime at the intake of each reservoir. Manager Eugene Carroll writes (40) that the water supply has been entirely satisfactory, without taste or odor, ever since the use of copper sulfate treatment for vegetable algae was first applied in 1904.

Broadcasting copper sulfate in the form of crystals was introduced by the Los Angeles Bureau of Water Supply early in 1935. The chemical was distributed from a specially designed apparatus mounted on a boat. By using crystals of smaller and smaller size the copper sulfate could be supplied to the top layer and succeeding lower depths of water. When reported in 1935 the method was in the development stage but had been used extensively. Reduced amounts of copper sulfate and increased efficiency were indicated (41). On January 12, 1942, R. F. Goudey, Sanitary Engineer for the Water Works Bureau, wrote, for use here, that this method of "copper sulfating reservoirs" had been adopted not only by Los Angeles but also by the city of San Diego and the San Diego County Water Department; and by three eastern cities, which obtained equipment from the Utility Fan Corporation. That concern states that it shipped one of its copper sulfate distributors to Newburgh, N.Y., in 1941.

A change in the method of applying copper sulfate to Skaneateles Lake was made in 1938 by the water division of Syracuse, N.Y. Instead of dragging the agent in bags fastened to a motor boat, a concentrated solution of it was sprayed from nozzles mounted on outriggers attached to a motor boat, thus covering 50-ft. lanes (42).

CHAPTER XVIII

Softening

Until the middle of the nineteenth century the chief objective of water purification was clarification. When softening was introduced it made little headway until well into the twentieth century. Once accepted, it forged ahead but in the early 1940's only one water works used softening out of every dozen that could employ it advantageously. The belated introduction of softening was not due to lack of knowledge of its possible benefits nor to ignorance of its underlying principles.* In effect if not in name it had been practiced on a small scale by rule of thumb for ages—ever since soap came into use. Potency of certain wood ashes and of earthy alkaline salts to make hard waters soft was noted by several eminent scientists during the century preceding the announcement in 1841 of what became the well-known Clark process. Most of the writers who have sketched the history of softening prior to Clark's patent have singled out Cavendish of the 1760's and Henry of the 1780's, neither of whom was primarily concerned with softening, and have overlooked Home and Rutty, of the 1750's, whose researches in this field entitle them to high standing as pioneers. Clark's lime process and its subsequent modifications, with important mechanical improvements, have been used for a century, but since 1925 have had a rival of ever-increasing importance—zeolite or base exchange.

More than two hundred years ago, Dr. Peter Shaw (1) stated, in one of his London chemical lectures of the early 1730's, that hard water becomes softer on adding to it alkaline salts.

In a classic treatise on bleaching, published in Edinburgh in 1756, Dr. Francis Home (2) described 129 experiments made by him, of which the objective of 45 was how to soften water. These tests ran from June 15 to the end of an unstated year, presumably 1755.

---

* This was well expressed by Baylis in notes sent to the author in 1936: "There is no reason why chemists in the latter part of the eighteenth century could not have softened public water supplies by chemical precipitation, for it appears that they understood all of the reactions involved." To which I add that although soft water was considered desirable and was sought after, particularly for industrial use, there was no insistent public demand for softening such hard waters as were being supplied to municipalities.—*M.N.B.*

FIG. 73. DR. FRANCIS HOME OF EDINBURGH
Made earliest detailed experiments on water softening in the 1750's
(From portrait in John Kay's *A Series of Original Portraits and Caricature Etchings*, Edinburgh, 1838)

Home's method was, first, to learn what made waters hard, then to search for what could best soften them. His main standard was the soap curdling point. Of eleven hardening agents tried, he found that "the soluble part of lime" stood first, at a "soap curdling point" of 45. Then came a sharp drop to 18, for oil of vitriol, and a steady decline as follows: spirit of sea salt, 15; salt of amber, 10; spirit of nitre, 9; blue vitriol, 7; sugar of lead, 5; salt of steel, $4\frac{1}{2}$; alum, 4; epsom salt, 3; cream of tartar, $1\frac{1}{2}$. He ranked the softening powers of the following salts thus: fixed alkaline salts, though not of the strongest kind, 2; volatile salt of hartshorn, 1. He added: "Filtration through sand softens in proportion to the length of its course. Putrefaction softens in proportion to its degree." Summarizing, Dr. Home (2) stated:

> This method which we have discovered of softening hard waters, is easy, expeditious, and cheap; qualities absolutely necessary to render it useful to the public. It is easy, as the most ignorant can do it; expeditious, as it becomes fit for all family-uses immediately, and for drinking in half an hour; and cheap, as the material costs but a mere trifle; nay, may be prepared by any person. By this change, the hard water not only becomes fit for all the common uses of life, but as beneficial as it was before hurtful to the health of man. . . . I may venture to affirm, that no other material can ever be found capable of softening hard water: and tho' one was discovered endued with the same property, it could not be of the same use to mankind, as there is none, alkaline salts excepted, to be had every where. A particular substance or plant was only to be found in particular places, but this material is to be got where-ever plants grow. So kind is the general parent of nature, that he has provided a remedy, every where to be found, for so common an evil; but, at the same time, has left the discovery to our own industry.
>
> How much we stood in need of such a discovery, most great towns, especially those on the sea-coast, nay the greatest part of some counties, can testify. *Newcastle* is a remarkable instance of this distress. In all the pants [sic] or pipes there, two excepted, the water is hard; and to such a degree, that it is three times more so than the hard water which I have examined.*

Two years after the appearance of Home's book, Dr. John Rutty (3) published his large work, *A Methodical Synopsis of Mineral Waters,* which opened with a section on "common water." His first two chapters dealt with "distinguishing characters, effects and uses" of soft and of hard spring waters. They were prefaced by 38 pages of tabular data on various hard and soft waters, including "Experiments

---

*Dr. Home here states that since "these papers were in press" he had found "that Dr. Shaw, in his chymical discourses, has given an imperfect hint of this quality of alkaline salts, but does not inform us of the manner of doing [utilizing] it, or reasons on which it depends, or qualities of the water after it is softened" (1).

in Concert" on 38 springs of hard water in Dublin. These tables gave for each water: hydrometric readings; taste; reactions to soap, solutions of silver, lead and alum, lime water, acids, milk, flesh, syrup of violets, galls, sumach, logwood, rhubarb, ash-bark, and cale; also "contents in grains per gallon"; quality of contents (marine and other salts); and effect on bowels of human beings; besides various reaction-tests of residues on evaporation. This was certainly a remarkable exhibition of the qualities of hard and soft waters.

After noting that, without due regard to the natures of different waters, "we could not but at random be supplied with gruel, puddings and even a smooth mixture of milk and water," Dr. Rutty (3) added:

> It is of great importance in building a town to chuse a proper situation with regard to the quality of springs: Our common spring-waters, if not immoderately hard, will become soft by standing a few days; such are those at Henley, and divers others, whose waters by being exposed two days, become soft and fit to wash with; but the situation of Thame (Plot's *Nat. Hist. of Oxfordshire*) in the same county, is much worse, for there waters will not grow soft by standing two days, as the others.
>
> The sacred records (2 Kings ii. 19) mention a city, the situation of which was pleasant, but the springs naught, and the land barren; which waters were not amended but at the expence of a miracle; tho' it is observable, that this was not wrought without means, viz. by salt put into a new cruse, and cast into the spring of the waters by the prophet, whereby they were healed.
>
> And indeed, in the natural way, one method of softening hard water is by means of an alcaline salt, e.g. by putting into it, in a bag, the ashes of green ash or beech burnt to a whiteness; an experiment not only very useful, but illustrative of the nature of the mineral matter impregnating hard waters, viz. as being an acid united to a terrestrial matter.

Although sometimes mentioned as one of the earliest "discoveries" in water softening, that was not the object of the experiments on Rathbone Place (London) pump water made by Cavendish in September 1765 and published in 1767 (4). His experiments were made chiefly to learn why calcareous earth remained suspended in water. His main conclusion was that "the unneutralized earth, in all waters, is suspended by being united to more than its natural proportion of fixed air." The "unneutralized earth," he found, was "entirely precipitated" from Rathbone Place and other London pump water "by the addition of a proper quantity of lime water," and the exposure of the water so treated long enough "for all the lime to be precipitated." This, in effect, had been noted previously by Doctors Shaw, Home and Rutty.

The last of the eighteenth century "discoverers" of the potency of lime to soften water was Thomas Henry, in or about 1781. He opened a paper (5) on preserving sea water from putrefaction with this significant statement: "It has been frequently remarked by chemical and philosophical writers, that a new experiment is seldom made in vain. Though the operator may even fail of attaining immediate object of his pursuit, he may yet, fortuitously, acquire the knowledge of some new fact, which may be productive of improvement and advantage to science." The object of the experiment thus introduced was an appeal from "a Gentleman who had obtained a quantity of sea water, for the purpose of bathing a child, asking me to think of some expedient" to keep the water from becoming putrid. This was about the time that Henry had published a description of his method of preserving fresh water from putrefaction at sea (6).

Summing up his thirteen experiments, Henry said "It appears that quicklime, dissolved in water, precipitates the magnesian earth from the marine acid, with which it is united in the sea water, and uniting with that acid, is retained in the water, under the form of a marine selenite. What the water loses, therefore, of one salt, it gains of another. At the same time, the magnesia, being precipitated by a caustic calcareous earth, falls in a state similar to that to which it is reduced by calcination, viz. void of fixed air. In this state, I have formerly proved, by a train of experiments, that it is strongly antiseptic." (Henry's *Experiments and Observations*, p. 58.)

At Black Rock, near Cork, Ireland, water was softened by means of potash of soda, in or before 1818. This was done on the suggestion of Edmund Davy, Professor of Chemistry at Cork Institution. Later, he made experiments showing that the water of limestone districts could be softened with potash of soda or by boiling for twenty minutes (7). No earlier instance of softening water for public use has been found.

Softening, and the characteristics, advantages, disadvantages and particular uses for both hard and soft water, are given considerable space in a unique little treatise of 1830 on the properties of water by Abraham Booth, styled "Operative Chymist, Lecturer on Chymistry, Pharmacy, etc." (8). "Simple boiling," he notes, will soften waters "whose hardness consists of the carbonates of lime and magnesia . . . for as the carbonic acid is expelled . . . the earth subsides," but this "will not remove sulphate of lime, and, as this is almost constantly present in water, boiling is but a partial mode of purification. . . .

All the earthy salts which oppose the solution of soap may be decomposed by the addition of an alkali . . . the ashes of fern or wormwood, which contain a good deal of carbonate of potash, are often used for softening hard water for the purpose of washing."

Clark's Softening and Purifying Process

In 1841, Thomas Clark, Professor of Chemistry in Aberdeen University, Scotland, announced to an unready and, for a long time, an indifferent world his method of not only softening but also otherwise purifying public water supplies. In time, it came into use. He did not claim to remove permanent hardness. This was achieved later by the use of soda or by soda and lime. Mechanical improvements were provided by inventors and manufacturers. All the early plants using the Clark lime process were designed by Samuel Collett Homersham. It has been generally overlooked that Clark originated what much later became known as the excess-lime method, both for softening and purification, claiming destruction of "insects" long before the germ theory of the spread of certain diseases was accepted. Homersham's plants employed an excess of lime.

On March 8, 1841, Clark was granted a British patent on "A New Mode of Rendering Certain Waters (including the Thames) Less Impure and Less Hard, for the Supply and Use of Manufactories, Villages, Towns and Cities." His claim that the process would render waters less impure as well as less hard was elaborated by Clark in a pamphlet of 1841 (9) and the importance of purification was emphasized in a paper read in 1856 (10). In the pamphlet Clark said that, besides softening, his process would separate vegetable and coloring matter, "destroy water insects" and convert water from the Thames into a better supply for London than the new supply projected by Thomas Telford. In the paper of 1856 he declared that: "Freedom from organic matter is of still more importance than freedom from hardness. It seems a fact well established by observation, that some of the poisons producing epidemic disease find a congenial habitat in waters contaminated with organic matter."

Clark's patent covered the use of lime as a precipitant, followed by subsidence or by subsidence and filtration. "It is a triumph of chemical over mechanical art," Clark wrote in 1841, "that, by adding chalk, water would be freed from chalk, itself the largest impurity" in the London supply. Concerning other processes for softening water Clark said that boiling and distillation were impracticable while "carbonate of soda," which "on a small scale is used to prepare water for washing," would cost the London water companies £1,000 a day against £10 for lime (9).

In testimony given on June 3, 1843, before the Commissioners on the State of Large Towns, Clark described his process and deplored the indifference to it shown by the London water companies (12). In 1851, he repeated the testimony before a Commission on Water Supply to the Metropolis. This commission was composed of three men, filling chairs of chemistry in as many London colleges, all Fellows of the Royal Society. In their report (11), the commissioners stated that the Clark process was limited to the precipitation of carbonate of lime, "with a portion of the organic and colouring matter," by means of caustic lime. The commissioners had first seen the operation of the process at the Mayfield Print Works near Manchester. There 0.3 mil.gal. (Imp.) of water treated "daily at a trifling expense and with little trouble, but more for discolouration [decoloration] than softening." The water thus treated was passed through sand filters.*

The chemical commission made experiments on water then being supplied London from the New River and the Thames. These were followed by observations of a large-scale trial of the Clark process at the works of the Chelsea Co., London, directed by James Simpson Jr., Resident Engineer. The Chelsea tests and the operations at the Mayfield Print Works led the commission of chemists to conclude that the Clark process was practicable. It also was of the opinion "that no sufficient grounds exist for believing that the mineral contents of the water supplied to London are injurious to health." The General Board of Health, to whom the chemists reported, showed more concern over the hardness of the London water supply than over its pollution. It advised the use of Clark's process until a supply of naturally soft water could be obtained.

The Clark Scale of Hardness.—In a paper read May 14, 1856, Clark stated that: *"Each degree of hardness is as much as a grain of chalk, or the lime or the calcium in a grain of chalk, would produce in a gallon*

---

* Apparently these observations were made late in 1850. The Mayfield works seem to have been the same as "Hoyle's works at Manchester," regarding which S. C. Homersham testified in 1868. At these works, he said, he first saw the Clark process in use. These cloth printers used "fine spring water." After it had become "discoloured in some of their processes" they added lime to the water and filtered it to make it "clear enough for some of their rougher processes."

[Imp.] *of water, by whatever means dissolved.*" [Italics in original] (10).

In his paper of 1856 Clark stated that the first installation of his process for municipal supply was at Plumstead where a plant had been in "successful operation for the last year-and-half" and that "Mr. Homersham is the engineer who planned these works with success." Clark might have added that Homersham consulted him. This plant, completed at the close of 1854, was the first softening plant for municipal supply in the world. Others followed slowly elsewhere in England: 1861, 1868 and 1870. All these and also plants for a castle and a hospital, states Humber, were designed by Homersham (13). All treated "spring or well water. None used filters. None attempted to remove permanent hardness."

The plant of 1854 was built by the Plumstead, Woolwich & Charlton Consumers' Water Co. to compete with the Kent Waterworks Co. operating in parts of the London metropolitan area. Both companies supplied water from wells in the chalk. Despite its softened water, the Plumstead Co. became bankrupt and was absorbed by the Kent Co. which immediately abandoned the softening plant. The Kent Co. claimed that the plant was a failure. Homersham, who designed it, testified before the Royal Commission on Water Supply in March 1868 that the Kent Co. did not operate the softening plant an hour and abandoned it for fear of a demand to provide softened water throughout its whole water supply area (14).

In his testimony, Homersham stated that in the Plumstead plant cream of lime was used as the reagent, instead of milk of lime, as used at later plants (14). Lime and water were passed through three agitators in succession, each consisting of "a pipe enlarged for a short distance, and four plates with small holes" in them. The mixture then went to a catch basin or grit chamber, then to another agitator and finally to a "depositing reservoir." The ratio of cream of lime water to the untreated water was 1 to 8 or 1 to 9. There were open depositing reservoirs, each holding about 0.22 mgd. (Imp.), each filled in about $3\frac{1}{2}$ hr. and emptied in succession, the cycle taking about 10 hr. From the settling reservoirs it was lifted 160 ft. to a service reservoir on Plumstead Common. The nominal capacity of the softening plant was 0.6 mgd. (Imp.) but before its abandonment in 1861 it was worked at 1 mgd. (Imp.) in summer. The "whiting" or lime sludge was sold in its wet state to bristle manufacturers at 7 shillings a long ton (14).

The next Clark plants for municipal supply were built in 1861 by the Caterham Spring Water Co. to supply Caterham and vicinity, in 1868 by the Chiltern Hill Water Co., and in 1870 by the Canterbury Gas & Water Co. The last, says Humber (13), treated water from "boreholes" 490 ft. deep. Lime water was run into one of two reservoirs until it was 20 in. deep, then raw water was delivered through eight nozzles until the 15-ft. level was reached. From five to ten hours was allowed for sedimentation. The rated capacity of the plant was 1.44 mgd. (Imp.) but the consumption was far less.

The Colne Valley Water Co. included a Clark-process plant in works put into use in 1873 to supply suburbs of London from wells in the chalk. With enlargements this plant was used until 1925 when lack of room and difficulty of chalk disposal led to building a zeolite base-exchange plant with a capacity of 3 mgd. (U.S.). To this was added 5.4 mgd. capacity in 1932 (15).

The cases for and against water softening were reviewed by the Royal Commission on Water Supply in its report of 1869 (14). After summarizing the data given in the report of the Chemical Commission of 1851 (11) and the testimony given by a galaxy of witnesses at its own hearings, the Royal Commission concluded that "there is no doubt . . . as to the advantage of soft over hard water for washing and, with some few important exceptions, for general manufacturing purposes." Softening would be advisable for towns in the manufacturing districts, but in the metropolis there were no large demands for soft water to be used in manufacturing. The commissioners did not think that the advantages of soft water for the London district would "justify going to a great distance to obtain it, in place of the ample supply nearer at hand."

As to applying the Clark process to the metropolitan supply, the Royal Commission said that, apart from the great expense entailed, "it does not appear to be applicable to the Thames waters on a large scale. It appears more suitable for small districts supplied from chalk wells, or for use in manufactories where soft water is specially required." In reaching this conclusion, the commission seems to have been largely influenced by the testimony of Homersham, engineer for the only municipal softening plants built up to the date of the report, all of which were treating water from wells and none of which used filters. The commission may also have been much impressed by a doubting chemist who testified before it on February 28, 1868. He was Professor Edward Frankland, then examiner of metropolitan water supplies and for years one of the highest authorities on the chemistry of water. Asked for his opinion of the Clark process, Frankland said: "It is a beautiful process and a comparatively simple process, but I believe that it never could be carried out for softening such vast volumes of water as are required for the supply of London." Little did he dream of achievements to come when chemists and engineers joined hands in softening immense volumes of water—but that was decades later and in the United States. In justice to Frankland, it should be noted that six years later, as member of the Rivers Pollution Commission, he signed a report containing an estimate that £600,000 would cover the cost of softening 100 mgd. (Imp.) for the London metropolitan area by the Clark process and expressing the belief that the process was practicable for use wherever softening was needed.

"The Alleged Influence of the Hardness of Water on Health" was given eighteen pages in the Sixth Report of the Rivers' Pollution Commission, dated 1874 (16). The commission went so far as to gather general death rates for over 200 British cities and towns. These were classified by relative hardness of the corresponding water supplies and by various environmental conditions. The commission concluded that neither hardness nor softness of the water consumed affected the general death rate. This is not to be wondered at in view of the fact that the average death rate of the towns investigated, outside London, ranged from 24 (Liverpool) to 17 (Isle of Wight) with only twenty towns below 20 per 1,000. For the London metropolitan district the showing was worse: a general death rate of 24.6.

After tests of the Clark process on samples of water from the Thames, the Lee and deep wells supplied in 1870–71 by three of the London water companies, the Rivers Pollution Commission (16) concluded that "Clark's method is equally efficacious in softening all three kinds of water." It also removed "a considerable proportion of organic impurities carried by the Thames and Lee, as indicated by organic carbon and organic nitrogen" (16). Although the commission did not think hard waters detrimental to health, it strongly favored the use of "this simple and inexpensive process" of softening because of the vast saving of soap it would effect. It listed 87 British towns, including London, where softening could be used advantageously, giving the hardness of their supplies and the reductions that might be effected by softening (16).

Despite Clark's hope of 1841 that his process would be adopted by the London water companies and the favorable opinion of the process expressed by inquiring commissions, none of the companies adopted that or any other water softening process; nor has the Metropolitan Water Board done so since taking over the works in 1905. Thanks to various methods of water conservation and treatment, the vast population of the Metropolitan Water District is still supplied from the Thames and other old sources of supply.

Other British Patents

For over 40 years after 1841, the date of Clark's patent, his process held the field. This was less a tribute to Clark than proof of indifference to water softening.* There was no lack of other patents meanwhile.

In an article on softening published early in 1885, Baldwin Latham, British sanitary engineer (18), said that "as any of the alkaline earths may be used instead of, or in addition to, lime it is not surprising that, since the date of Clark's patent, numerous patents have been taken out for softening and purifying water in which lime, in combination with other alkaline earths, have been proposed." Latham mentioned about 35 British patents. First on his list of those issued up to the close of 1883 was one granted to John Horsley on April 26, 1849.†

Horsley named as reagents, "calcined or caustic barytes, phosphate of soda, oxalic acid, or one of the various preparations of those substances." In a promotion pamphlet of 1849, Horsley said that Clark's process was incomplete because it left sulfates and muriates in solution whereas by using "a solution of baryta," sulfates as well as carbonates would be extracted. Horsley stated that "there is more than enough evidence to prove that earthy or calcareous matter held in

---

* A popular exposition of the disadvantage of hard water and the consequent cost to Londoners appeared in the *Ladies Companion* (London) in 1850 and was reprinted in the first number of *Harper's Magazine* (New York). This was perhaps the first attempt to inform the general reading public of England and America on the evils and costs of hard water. It made little impression on either side the Atlantic (17).

† Latham's paper gave a comprehensive review of water softening up to the close of 1884, including a summary of ancient practices. It dealt, as no one else seems to have done, with the use of soap as a softening agent through the centuries, and emphasized its wastefulness compared with the far less costly modern processes of softening.

solution is the true matrix of the animalculae, ova, or germs," and that by his process "the germs are at once liberated and instantaneously destroyed, as has been experimentally determined." This went beyond Clark's claim of destruction of "water insects." No evidence of a plant following Horsley's patent has been found.

John Henderson Porter, a London civil engineer, made mechanical improvements in means for utilizing the Clark process. British patents on softening were taken out by him in 1876, 1879 and 1881 and jointly with Herbert Porter in 1884. "My invention," said Porter in his patent of April 21, 1876, "consists in the utilization of the precipitate of carbonate of lime resulting from the [Clark] process as the medium of filtration." This he proposed to do by retaining the lime on filter cloths. In his patent of May 20, 1879, he claimed mixing the reagent in a receiver and then passing the mixture into a closed vessel where the mixing was continued by causing the water and the reagent to follow a circuitous course through the vessel. The chemical reaction thus produced could be promoted, if desired, by mechanical means or by a current of air or of water. In February 1884, a London chemical journal (19) published an article on the Porter-Clark process. Installations for sugar refineries, paper mills, dyeing plants and railways were mentioned. Plants for the Northwestern Railway at Camden, Willesden and Liverpool were mentioned and one of them described. In Latham's paper on water softening, already mentioned, he said that recent inventions for "carrying out the Clark process may be described as the application of machinery to the saving of time, space and labor." Of these, the Porter-Clark process comes first in time (19).

Removal of permanent hardness from water already subjected to Clark's process of 1841 or preferably to the Porter-Clark process patented in 1876, was claimed in a British patent granted to A. Ashby on November 29, 1878. The method called for was the addition of enough soda or potash to precipitate the soluble salts of lime, magnesia, and iron, other than those which cause temporary hardness. The water might be heated when necessary.

After a half century of invention and promotion only a few municipal water softening plants had been built in England, and, as far as is known, none were built elsewhere. In 1888 one was put into use by the borough of Southampton, England. Although its capacity was only 2 mgd. (Imp.) [2.4 mgd. (U.S.)] it seems to have remained for some years the largest softening plant in the world. Chemically, it followed the Clark process, using lime as a precipitant. Mechanically, it was based on the Atkins patent of December 31, 1881. A 10 per cent solution of lime was prepared in cylinders, equipped with stirrers for occasional use. The solution then passed to a baffled mixing tank and to a softening basin. From this it went to Atkins filters consisting of "a fine layer of carbonate" gathered on woven wire cloth stretched over a disk of perforated zinc supported by a cast-iron plate covered by radial and circumferential grooves. The filter cloth was cleaned by sprays of water. The plant was designed by William G. Atkins, one of the patentees of 1881, and built by the Atkins Filter & Manufacturing Co., under the supervision of William Matthews, Superintendent of the Southampton Waterworks (20). In the early 1890's, states George W. Fuller, the Atkins wire cloth filters were replaced by cloth filters designed and patented by C. J. Harris, resident engineer of the water works (21).

Among the widely used softening processes adopted in England in the 1890's was the Archbutt-Deeley. According to a paper read in 1898 by Leonard Archbutt, who was a chemist for the Midland Railway Co. (22), this process had been in use since January 1892, by the Midland Railway at Derby, England, "clarifying and softening the sewage-polluted water of the River Derwent, reducing the hardness from 15 to 5 degrees and giving considerable purification." It had been adopted at almost 50 works in England and abroad. Lime was slaked in a tank in which the water was boiled by means of a steam coil. Anhydrous carbonate of soda was then added, the mixture boiled and stirred until the soda was dissolved. The reagent thus formed was injected through perforated horizontal pipes into a softening and settling tank.

The only municipal plants named were one at the new water works of Swadlincote and Ashby and an adoption in 1897 at St. Helens. The former plant treated a very hard well water containing "in solution a considerable amount of iron, which precipitates on exposure to light and air." By using lime only, all the iron was removed and the hardness reduced from 22 to $8\frac{1}{2}$ degrees. The iron in the water aided precipitation. The softened water was "bicarbonated by means of coke." The "engineering firm" of Mather & Platt, Manchester, England, installed Archbutt-Deeley plants and controlled the American patents.

Charles P. Hoover (23) says, "The recarbonation process devised by Archbutt and Deeley . . . was employed at several plants in England to overcome difficulties due to excess causticity, such as incrustation of filter sand, clogging of service pipes and meters and unpalatable water due to excess of alkalinity." He added that the first use of recarbonation in the United States, "on successful plant scale, was in 1921, at Defiance, Ohio [Nicholas S. Hill Jr., Engineer]. Since that time practically all water softening plants have been equipped with recarbonation devices." Elsewhere, Hoover notes (24) that "recarbonation for lime-softened water was provided in the very first municipal plant built in North America [Winnipeg, Canada]."

Clark and Similar Processes in America

Except for a short-lived plant at Champaign, Ill.,* hitherto overlooked, no softening plant for a municipal supply was built in America until early in the twentieth century. Then Winnipeg, Manitoba, built a plant which was soon followed by one at Oberlin, Ohio. These were preceded by many plants for industrial supply.†

The Illinois Central Railroad equipped some of its locomotives with water softeners in or about 1879. A filter was placed in the forward dome of the locomotives. Oyster shells were found to be the best filtering medium but it was difficult to get enough of these (so far inland, in 1879). "A good substitute was found in rough scrap

* The earliest known evidence of an attempt to promote water softening in the United States is an advertisement by the American Soft Water Co., of Chicago, published early in 1887 (25). The company claimed to have "the only reliable method for softening hard [lime] water for preventing scale in steam boilers." The system was also "a perfect one for purifying water holding earthy, vegetable and other impurities in suspension." Neither process nor apparatus was described, beyond saying that filters, adapted to any pressure, were used. Designs and estimates were offered "for purifying water in large quantities for cities and villages and manufactories." No evidence has been found that this company ever installed a plant.

† For at least six months hardness was removed from the water supply of Champaign and Urbana, Ill., seat of the University of Illinois. In a letter dated January 14, 1888, S. L. Nelson, Superintendent, Union Water Supply (26), said that the National filter plant, placed at the works in June 1887, had worked satisfactorily ever since. The "combined influence of Aeration, Lime Precipitation, and Filtration," he said, "renders our water clear and bright, free from odor and vegetable matter, and sparkling in appearance. . . . It also removes the hardness of the water." At that time, the supply of Champaign and Urbana was taken from an abandoned coal mining shaft. No record of the use of a National filter can be found by the water company in question nor in the public libraries of either city. Conceivably, the filter was installed for demonstration purposes only.

iron, which is now used exclusively." Raw water was delivered on top of the filter through a rose spray (27). No record of these softeners could be found in 1938.

In September 1897, Rudolph Hering advised the city of Winnipeg, Manitoba, to get water from artesian wells and soften it instead of going 50 miles to the Winnipeg River. By the solicitation of Col. H. N. Ruttan four bids were received. The contract was awarded to the Pittsburgh (Pa.) Testing Laboratory, of which James O. Handy was chief chemist. The plant had a guaranteed normal capacity of 2.4 mgd. (Imp.) [2.88 mgd. (U.S.)] and was put in use in May 1901. The plant was cited as using the Clark or more strictly the Porter-Clark process, but, besides slaked lime, caustic soda was used, thus reducing permanent as well as temporary hardness. After receiving these reagents, the water went to filter presses. Owing to "slight but annoying incrustations on the valves and condenser of the pumping engine . . . and in a few instances deposits on meters," after the plant had been accepted (October 1902), and after a series of experiments by Handy, carbonating apparatus was installed. In this, coke was burned in a brick furnace. The resulting gases were drawn by a blower and exhauster through a water-jacketed condenser or cooler and then passed to a washing tank where sulfurous acid was removed by spraying water upon coke placed on a shelf. These and other details, including changes in the plant made up to 1904, were described by Handy in a paper prefaced by a historical review of water softening processes abroad and in the United States (28). Handy stated that the Winnipeg plant was "the first municipal softening plant in America and one of the two largest in the world." There were in the United States and Canada (early in 1904), Handy said, about 275 water softening plants, of which over 100 were in railway service. As the aggregate capacity was given as about 65 mgd., most of the plants must have been small.

The college town of Oberlin, Ohio, was the first municipality in the United States to build a water softening plant (see footnote regarding Champaign, Ill., above). It was put into use December 23, 1903. Credit for the venture at Oberlin, a town of 4,000 population, is due to W. B. Gerrish, city engineer and superintendent of water works and to the water board that backed him. Of the three members of the board, one was a professor of science and another a professor of chemistry in Oberlin College. The plant was designed by C. Arthur

Brown. Oberlin built water works in 1887, taking its supply from the Vermillion River. The water was stored in a large reservoir, from which it flowed into a small "settling reservoir," then to a pump well from which it was lifted to a steel tank supported by a masonry tower. Gerrish wrote in 1905 (29) that the water supply was of surface origin, from an agricultural district, so hard "that with possibly a half dozen exceptions each family was provided with a rain water supply in addition to the city supply."

Brown's plan, thanks partly to local conditions, was simple. The small settling basin was divided into two compartments of 0.33-mil. gal. capacity each, the daily consumption then being only 0.165 mgd. Raw water from the adjacent storage reservoir was passed through a chemical mixing box, admitted to the bottom of the first softening basin, drawn from its top through a float arm, entered the bottom of the second basin and was finally taken from the top of the latter to the pump well—all by gravity. The softened water was then forced to and through rapid pressure filters to the water tank, the filters being installed within the stone tower that supported the tank.

Average counts made November 28 to December 15, 1904, showed an extensive reduction of bacteria effected by the long period of storage (theoretically some 90 days). Softening (subsequent to storage) reduced the bacteria from 371 to 13 per ml. or about 95.5 per cent. After passing through the rapid filters, the elevated tank and the mains, there was an average of less than 5 bacteria per ml. in the water delivered to consumers. A State Department of Health count on August 18, 1904, showed 490 bacteria per ml. before and 33 after softening, a reduction of about 90 per cent. Tests for coliform organisms were positive for the unsoftened and negative for the softened water in both local and state counts.

Excess-chemical treatment for control of caustic alkalinity was introduced at Oberlin two or three years after the plant was put into operation. To this sulfate of iron was soon added. Softening was still practiced at Oberlin early in 1940, apparently with the lime and soda-ash process. Years before, however, the rapid sand filters gave way to upward-flow excelsior filters on account of cementing of the sand.

Incidental to sedimentation, from 1904 to 1915, and since then as a part of a rapid filtration plant, St. Louis, Mo., has had the benefit of a considerable degree of water softening. Within these limits, St.

Louis was the first large American city to soften its water supply (30) (see also Chaps. VI, XII and XIII).

An oft-overlooked American installation of apparatus designed for softening was put into use on April 7, 1906, at Lancaster, Pa. It was part of a purification plant built by the Pennsylvania Maignen Filtration Co., of Philadelphia, under a contract to deliver doubly filtered water from Conestoga Creek to the pumping station of the city of Lancaster. In discussing a paper on softening in 1906, P. A. Maignen (31) said that after reagents had been mixed with the water, the latter passed through two 50-ft. cylindrical tanks with concentric baffles; then upward through scrubbers and finally downward through slow sand filters covered by a filtering membrane. J. E. Goodell, Chemist at Lancaster (32), in 1940 added to Maignen's description that the amount of soda ash used was too small for softening the water being treated. When the city bought the Maignen plant in May 1924, it gave up the use of the apparatus designed for softening. The double filtration plant was abandoned when the city completed rapid filters in November 1933.

Cursed for over forty years with a hard water supply, the city of Columbus, Ohio, put a 30-mgd. water softening plant in use in September 1908. It was then by far the largest softening plant in the United States, if not in the world. It was enlarged to 54-mgd. capacity in 1923. In 1938, said Charles P. Hoover, Chemist (33), it still had the "distinction of being the largest softening plant in the world using both lime and soda ash." So objectionable had been the water supply of Columbus for years past that when the softening plant was put into use in 1908 there were over 17,000 cisterns in the city (181,511 population in 1910). Many private wells were used for drinking water.

The Columbus water works was put into use by the city on May 1, 1871. Water was taken from a filter gallery. In his first annual report, J. L. Pillsbury, Engineer, proposed that the water be softened. He suggested the use of "a solution of caustic lime," which had been approved by "Professor Wormsley." Ten years later, after an unsuccessful experience with a filter basin, Superintendent Doherty reported that office tests had shown that 1 oz. of lime added to 36 gal. of water made the city supply "superior for washing to the rain water obtained from cisterns." The water works trustees, he said, would give a sample package of lime to citizens. Plans for a supply from the Scioto

River were shelved in 1886 in favor of an extension of the filter gallery.

Subsequently, use of the river was proposed at intervals. In 1901, Samuel M. Gray again advised its use, providing it was subjected to softening and rapid filtration. Funds for this purpose were not authorized until 1904. When John H. Gregory, with Hering & Fuller as consulting engineers, took up the design of the Columbus plant, the basic principles of rapid filtration had been established at the Louisville and Cincinnati experimental plants in the late 1890's but no comparable experiments on water softening had been made. Incidentally, the Cincinnati experiments had shown that lime would not only soften water but would also reduce bacteria. At the Columbus sewage testing station, bottle and barrel studies of water softening, under the direction of George A. Johnson, then associated with Hering & Fuller, were utilized as a guide in designing the chemical phase of the softening plant. The entire purification plant, as completed in 1908, was described at length by Gregory in 1910 (34).

The plant, with facilities for recarbonation added, and with enlargement in 1923 to a capacity of 54 mgd., was described by Hoover in 1927 (35). A marked feature of the enlarged plant is the division of the entire flow of raw water into two parts, one of which flows without treatment to the baffled mixing tanks while the other constantly receives soda ash and alum needed to soften and clarify the entire volume of water. From the mixing tanks, the water goes to settling basins. Just before passing from these to the rapid filters, carbon dioxide gas is added to neutralize any excess lime. This, says Hoover, converts normal carbonates to bicarbonates, and prevents deposits in the filters and distribution system.

New Orleans, La., included softening as a part of its purification plant of 1909. Cincinnati, Ohio, provided for softening when it enlarged its filtration plant in 1936–38, but up to September 1939 softening had never been used there. It would not be needed continuously. Minneapolis, Minn., completed a softening plant in 1939, using lime and recarbonation. It is designed for an ultimate capacity of 120 mgd., has radial baffles in the precipitation tanks, extending buttress-like from the outside of the inner inverted cone (36). At St. Paul, Minn., an enlargement of the existing rapid filtration plant, completed in 1940, included a provision for softening, put into use on January 6, 1941. Lime is used (37).

SOFTENING

Upward-flow reaction and sedimentation is one of the most notable of the recent features of water softening practice. The idea is not new but like many other conceptions it was long dormant.*

In its modern application the water is treated in a specially designed unit. Raw water and softening chemicals are mixed by mechanical stirring in the presence of previously formed sludge; the softened water is clarified in rising to the top of the clarifier where it is discharged and the sludge or precipitate is drawn off at the bottom, both continuously. Broadly, the basic principle is similar to that of the activated-sludge process of sewage treatment, but the activation in this method of water softening is chemical instead of bacterial—the formation of floc by the newly applied reagent being quickened by that remaining in the old sludge. Two types of apparatus are in vogue: The Green-Behrman Accelator (U.S. patent, December 20, 1927) and the Spaulding Precipitator (U.S. patent, November 19, 1935). In the "Accelator" the raw water is introduced on one side and the chemicals on the other side, both near the bottom of the tank below the suspended sludge level; the water and reagents rise up through a central mixing compartment, containing an agitator, then pass down through a concentric compartment, then up through an outer clarifying compartment. In the "Spaulding Precipitator," the raw water, with chemicals added, passes down through a truncated cone below which there is an agitator, then upward through a concentric inverted truncated cone at the base of which there is suspended sludge. Schematic cross-sections are given in a paper by Spafford and Klassen (38). The paper contains tabular data for accelerators and precipitators at several softening plants in Illinois and is followed by general discussion. An exposition of the "Accelator," with a cross section showing details and half-tone views of installations at Anna, Ill., and Williams Bay, Wis., are included in an article by the patentees (39).

The evolution of Spaulding's upward-flow precipitator, and its installation as part of the second water softening plant at Springfield, Ill., and the large time saving it effected compared with the first Springfield plant were told by Spaulding and Timanus in 1935 (40). Two years later, Spaulding went into the subject in more detail regarding both theory and practice (41).

* Antecedents of upward-flow precipitation were included in water softening patents taken out in England by William Lawrence (December 24, 1891) and in the United States by Herschel Koyl (July 3, 1900 and July 2, 1901).

Excess Lime

Credit has been widely given to the late Sir Alexander Houston of London for introducing excess-lime treatment of water for softening. It seems to have been entirely overlooked by Houston and by writers crediting him with having originated excess-lime treatment that all the Clark-process plants from 1854 until 1870 used what Clark himself called "excess lime" and that the same term was used by Baldwin Latham in referring to these plants in his paper of 1885. Also generally overlooked has been the adoption of excess-chemical treatment at Oberlin, Ohio, after the softening plant completed late in 1903 had been in use some three years.

In Dr. Houston's Eighth Research Report (42), he says that his studies of 1911, which led to his "discovery" of excess-lime treatment, were prompted by the failure of Parliament to authorize the full water storage program of the Metropolitan Water Board. He therefore studied means to make up deficiencies in storage at times of bad water. This led to a "new way of adding lime ('excess lime')." Quoting further:

> With hard waters it is a case of adding an excess of lime to the major proportion of the total volume, rendering the minor proportion "safe" by adequate storage, ozonization, chlorination or other method, and mixing the two together so as to neutralize the excess of lime and render the whole perfectly innocuous. With soft waters, the procedure is to treat the whole bulk of water with an excess of lime and neutralize with carbonic acid or sulphate of alumina or acid. The former operation incidentally involves "softening" and the latter may involve "hardening" the treated water.

So far as has been found, the excess-lime method has never been applied to soften London metropolitan supply, but precoagulation has been used for some years to lessen the filter burden.

In his Twenty-fifth Annual Report (1930), Dr. Houston quoted at length a description of an excess-lime softening plant treating a new water supply for Southend, England (43). The plant was put into use in September 1929. It had a capacity of 7 mgd. (Imp.) [8.4 mgd. (U.S.)]. It treated stored river water which had a temporary hardness of 94 ppm. and was liable to sewage pollution. By use of the excess-lime method nearly all the temporary hardness was removed, the water clarified and a "pronounced" disinfection effected. Whether the excess-lime process has been adopted elsewhere cannot be stated.

In the United States, Houston's experiments were immediately confirmed by Hoover and Scott in studies made at Columbus (44) and subsequently, says Hoover, at a number of other softening plants, special reference was made to bacterial efficiency. Heavy reductions in coliform averages for the year 1929 at the softening plants of Columbus, Youngstown and four other Ohio cities are listed, with associated data, by Hoover's Lime Association Bulletin 211 (23). In this booklet, he defines "excess treatment" as "overtreating with lime . . . then neutralizing the excess lime with soda ash." He defines "split treatment," for plants not equipped for excess treatment as "overtreating as large a portion of the hard water as possible to get maximum reduction of hardness and then neutralizing the excess with raw water." This was substantially what was done in the Clark-process plants designed by Homersham.

In a letter written for use here Hoover said in part (33): "I think Sir Alexander Houston deserves credit for observing the effect of lime treatment in reducing bacteria, . . . [but not] for excess lime treatment in water softening." Hoover raised technical questions concerning Dr. Houston's methods of determining excess lime in water, and in conclusion expressed the belief that "the late C. H. Koyl, Superintendent of Water Supply of the Chicago, Milwaukee & St. Paul R.R., was the man who made excess-lime water softening practicable."

That Dr. Houston was primarily concerned with bacterial reduction rather than water softening, the above quotations from him make evident. That Dr. Clark was also largely concerned in other improvements to water than softening is also evident. In his patent of 1841 he declared that his process would render water "less impure and less hard." In a promotion pamphlet of the same year he claimed destruction of "numerous water insects." Bacteria and their removal were beyond his ken, as they were beyond the dreams of Thom and Simpson fifteen years earlier when they perfected their filters. All three built far better than they knew.

Zeolite in Europe and the United States

Sixty-seven years after Clark took out his British patent on the use of lime to soften water, the first of several German patents was granted to Robert Gans for what was to become known as the zeolite or base-exchange process of water softening. After being used for a few years to treat water for industrial purposes, the process was applied to municipal supplies. By 1925, it began to rival the lime and lime and soda-ash processes. By 1940, the newer process was widely used. The antecedents and basic principles of the zeolite process have been summarized by H. M. Olson (45, 46), by a court judge (47) and by Boris N. Simin (48). Early investigators of zeolites mentioned by Olson were: Cronstedt, a Swedish geologist, in 1756, Thompson, in 1845, Way in 1850 and 1852, and Eichorn in 1858.

On April 6, 1908, says a United States District Court Decision (47), Dr. Robert Gans, of Pankow, near Berlin, was granted German Patent 197,111 for his invention of "a form of artificial zeolites" created by fusing "clays and soda ash, and hydrating them, to which he gave the arbitrary name 'Permutit.'" He "found that hard water could be continuously softened by filtration through them and that the artificial zeolites could be regenerated by washing them with a salt solution after the exhaustion of their softening bases. . . . The device first used by Gans," says the court decision, "was defective." United States patents were taken out by Gans December 14, 1909 (reissued February 17, 1914), June 7, 1910, and August 22, 1916. By a disclaimer of February 26, 1920, the third of these patents was limited, says the decision, to "a filter composed of a layer of zeolites resting on a layer of sand and quartz, downward filtration, means of cutting off the supply of water on exhaustion of the zeolites, and means of passing through the zeolites a solution of salt capable of regenerating the zeolites."

In 1911, Boris N. Simin (son of Nicholas Simin, one-time Chief Engineer of the water works of Moscow, Russia) reviewed the development of the Gans process abroad (48). He gave references to seven articles by Gans, all in German. Two early industrial installations of softening plants are mentioned by Simin. A small one in a flour mill at Kirsanaff, Russia, supplying a 110-hp. boiler, had been working since the middle of 1910. The largest zeolite plant in Germany was at a textile mill in Bremen and treated 300 cu.m. (79,000 gal.) per hour. Apparently, the earliest zeolite plants in Germany were installed in 1908. Of about 250 installations in Western Europe, most were in Germany. There were fifteen in Russia. These data by Simin do not even hint at the failure of the early form of the Gans process asserted by the court decision.

The largest zeolite plants known to D. D. Jackson, when he read a paper in May 1915 (49), were at Dresden, Germany, and Hooten, Eng-

SOFTENING

land, the first for the removal of manganese and the second for iron removal. There was also at Hooten a 1.25 mgd. (Imp.) zeolite plant for softening the municipal water supply. The latter reduced the hardness from 20 degrees to 10 degrees by bringing half of the volume to 0 degrees, then mixing it with the other half.

England led America by a decade in putting into use a zeolite plant for municipal supply. "The first public water supplies in Great Britain to be softened by this [base-exchange] process were those of the West Cheshire Water Board (1912) and the Colne Valley Water Co. (1924) to whose engineers credit is due as pioneers" (50).

The West Cheshire Water Board supplies parts of the county boroughs of Birkenhead and Wallasey and a half dozen urban districts nearby. The supply is taken from "boreholes in the New Red Sandstone," at two localities. At each of these there is a Permutit softening plant and at one a "Candy iron extracting plant."

The Colne Valley Water Co., which supplies the whole or parts of twenty London suburbs in Middlesex and Hertford counties, installed a 2.5-mgd. (Imp.) base-exchange softening plant in 1924 and added a 4.5-mgd. plant in 1932. The first of these replaced a Clark process plant installed in 1873 and subsequently enlarged. "The whole installation," it was stated in 1935, "is believed to be the largest single base-exchange plant . . . in Europe." The water treated is from wells in the chalk (15).

In 1939 there were at least twenty municipal base-exchange softening plants in Great Britain and others under construction.

In the United States, the first community to be supplied with zeolite-treated water, says Olson (45), was Wyomissing, Pa., in 1922, where soft water was delivered to city mains from a plant in a textile mill. Laurens, Iowa, is credited as having been the first American town to be provided with softened water from a plant built for city supply. This was in 1924. The first large zeolite plant in this country was put into use August 25, 1925, by the Ohio Valley Water Co. (later called the Pittsburgh Suburban Water Co.) to supply Avalon, McKees Rocks and other suburbs of Pittsburgh. Its capacity in 1938 was 7 mgd., according to H. E. Moses, Chief Engineer of the Pennsylvania Department of Health (51). The McKees Rocks plant, as this installation is called, was for a time the largest U.S. zeolite plant.

The Metropolitan Water District of Southern California, late in 1939, awarded contracts for a 100-mgd. unit of a 400-mgd. plant to soften water from the Colorado River for Los Angeles and vicinity. Lime-zeolite treatment was to be used at the start but if found advisable excess-lime and soda-ash treatment can be substituted (52). Delivery of softened water to a part of the district was begun on June 18, 1941.

Recent Summaries

The Streeter Water Purification Census of 1930–31, listed 144 softening plants in the United States but it is now known that some were omitted (53). The United States Public Health Service Census of Water Treatment Plants, issued in early 1941, showed 680 softening plants of which 510 used the lime or lime-soda and 170 the zeolite process (54). An Olson census made as of July 1, 1941, showed fewer plants: 576, of which 377 were of the older and 197 of the newer type. The North Central States (Ohio, Indiana, Illinois, Michigan and Wisconsin) led the various groups with 250 plants of all types (55). A supplemental summary and census made by Olson as of January 1, 1945, showed a gain of 89 plants to make a total of 665, of which 427 were reported as chemical precipitation plants and 238 as zeolite (56).

Canada had four softening plants in 1941. The largest of these was at Edmonton, Alberta, where lime and soda-ash were used (57).

England and Wales reported softening plants, early in 1940, in 46 towns, most of them small. Of the 33 places for which the type of process was given, about half used lime (including a few lime and soda-ash) while the other half used base exchange or zeolite. No softening plants were reported for Scotland, Northern Ireland or Wales (50). Thus, a century after Clark's softening process patent was granted in 1841, and 45 years after the Rivers Pollution Commission (16) listed 87 towns in Great Britain where softening might be used advantageously, there were less than half that number in England and Wales. For this, war and rumors of war are partly responsible. Meanwhile a number of hardening plants have been installed. These treat very soft water from moorland or other catchment areas to prevent attacks on water pipes. The Sheffield supply, which is taken from elevated moorlands and the Derment Valley, and has a hardness per 100,000 of 1.1 degrees temporary and 2.7 permanent is limed and passed through rapid and slow sand filters.

A complete bibliography of the literature of water softening would include scores of reports and papers in addition to those named in

SOFTENING

the appended list of references. *The Index of the Proceedings and Journal of the American Water Works Association, 1881 to 1939,* has about 150 entries under "Softening," some of which are duplicates of those given here.

Collins, Lamar and Lohr, of the U.S. Geological Survey, give nearly 700 chemical analyses of various public water supplies of the United States and a map showing "Weighted Average Hardness, by States, of Water Furnished in 1932 by Public Supply Systems in Over 600 Cities in the United States" (58).

Only one book devoted exclusively to water softening has been found and this is small and old (59). It contains no historical data. The only apparatus described is that of one British company. A large part of Hoover's booklet (23) is devoted to softening. A concise review of softening, containing high-spot historical data and useful references, is given in *The Manual of Water Quality and Treatment* (60). Baldwin Latham's article, "Softening of Water" (18), contains an historical summary from ancient times to 1855, including an outline of British and of a very few French and German patents, issued between 1838 and 1883.

CHAPTER XIX

Cause and Removal of Color

Color removal was rarely if ever attempted until the latter part of the nineteenth century. Even since then it has seldom been accomplished except when incidental to some other objective of treatment or as an important adjunct to storage. In the early days of filtration of public water supplies, the term "color" was often used when turbidity would have been more accurate.

A clear concept of the difference between turbidity and color was shown by Dr. John Bostock * in a paper published in 1830 (1). He stated that New River water at London after heavy rains in December, 1827, was "very turbid and dark coloured," but after some hours of sedimentation, although the water was nearly transparent, "the dark colours still continued." Neither boiling nor filtration through sand and charcoal removed the color, but "alum and certain metallic salts, especially when heated, threw down a precipitate, and left the water without colour." The most efficient of the metallic salts "appeared to be the sulphate of iron," but the water treated by it had been boiled.

William West, also in 1830 (2), wrote that color was "quickly and completely separated by aluminous earth in a state of minute division." It was also separated by muriate of tin. It "obstinately resists mere filtering . . . yet sand, containing, as I apprehend, some alumine [sic] is effectual in separating it." Exposure to air in a reservoir also removed color.

A notable essay on color in water was published in 1862 by Professor Wilhelm von Beetz, a German physicist (3). He reviewed theories advanced by his immediate predecessors and also the much earlier discussion by Sir Isaac Newton (paper on light and color contributed to the Royal Society in 1675, included in Newton's *Optics*, 1704). In his review of the conclusions of his immediate predecessors, which he endorsed in general terms, Beetz wrote:

[Robert Wilhelm] Bunsen (Liebig's *Annalen*, 62: 44) was the first to state, and establish experimentally, the simple proposition that "chemically pure

---
* See also statement by Robert Spurr Weston near end of this chapter.

water is not, as commonly assumed, colourless, but naturally possesses a blue colour." He observed this coloration on looking at a piece of white porcelain through a column of water two yards long. He explained the brown to black colouration of many waters, especially of North German inland lakes, as arising from an admixture of humus; the green colour of the Swiss lakes, and, still more so, the siliceous springs of Iceland, as arising from the colour of the yellowish base, and of the siliceous sinter surrounding the springs, and which is caused by traces of hydrated oxide of iron. Wittstein (*Sitzungsber. der K. bayer,* Akad. der Wissensch, in München, 1860, p. 603), by careful chemical investigations, has quite recently shown that the green colour also derives its origin from organic admixtures. According to him, the less organic substance a water contains, the less does its colour differ from blue. With the increase of organic substances, the blue gradually passes into green, and from this, as the blue is more and more displaced, into brown. Water is softer the nearer it is to brown, and harder the nearer it is to blue; this does not arise from a greater or less quantity of organic substance, but of alkali, on which, again, the proportion of dissolved organic substance depends. This alkali dissolves the organic substance in the form of humic acid. If a water does not contain much humic acid, this is not caused by a want of humic acid in the ground, but by the fact that this ground did not give to the water an adequate quantity of alkaline solvent material.

From these results we may consider the question settled as to why, on chemical principles, some waters are blue, others green, and others brown.

The remainder of Beetz's essay describes his and other studies of color with reflected and with transmitted light, rather than with the organic or mineral contents of the water employed. His apparatus is illustrated by several drawings.

A long series of papers by Walthère Spring, a Belgian attached to Liege University, was published in 1883–1910. He reported in 1897 that ferric compounds, acting on humic bodies in upland bog waters would reduce color and that light aided the reaction and was possibly necessary.*

The first American recorded study of color removal was made in 1874 by a Medical Commission created to report on the relative merits of the Sudbury River and other possible sources of water supply for Boston (5). In a supplementary report the commission summarized laboratory studies on the effect of storage and exposure to light. The studies showed that the color diminished as the intensity

---

* These studies were called to my attention in 1940 by Frank Hannan of Toronto. He states that Spring's studies on the source of color in water make up a part of his collected works (4), the larger part of which is written in French while the remainder is in German.

of the light increased. This reinforced the commission's arguments on the benefit of storage of surface supplies in large reservoirs. Telescopic glass tubes and reflecting prisms were used in these studies.

Desmond FitzGerald directed more elaborate studies of color removal at the Chestnut Hill laboratory of the Boston water works in 1886–94 (6). The studies were in direct charge of F. S. Hollis. It was concluded that not iron, as was first assumed, but carbonaceous matter was the source of color in the surface waters studied. Exposure of water to the sun in gallon-sized glass bottles resulted in complete removal of color. These studies provided data in favor of long storage in reservoirs to improve the quality of water.

Color Removal at Greenock, Scotland

The earliest known instance of removing color from a public water supply occurred in 1827 in Greenock, Scotland. Water for a gravity domestic and power supply was impounded from an upland area much of which yielded "moss water" carrying color and "other dissolved matter." So far as feasible, the colored water was excluded from the reservoir storing the domestic supply. To remove color from any moss water reaching the filters, amygdaloid crushed to pea size or smaller was mixed with the sand used in Robert Thom's "self-cleansing" or reverse-flow wash filter. He was led to do this by observing that in the moss lands the water of springs emerging from trap rock or amygdaloid was clear as crystal. Such rock was abundant in the hills about Greenock but it was somewhat costly and in time became "saturated" and had to be replaced. Eventually, animal charcoal seems to have been substituted for amygdaloid (see Chap. V). Thom did not explain why amygdaloid removed color. It is an igneous rock containing almond-shaped cavities (hence the name) filled with such foreign matter as quartz, calcite or a zeolite.

John R. Baylis, Engineer of Water Purification, City of Chicago, Frank Hannan, formerly Chemist, Filtration Plant, Toronto, Canada, and Robert Spurr Weston, Consulting Chemist and Engineer, Boston, kindly read the first draft of this chapter and sent comments and supplementary data as follows:

Baylis: In regard to the use of amygdaloid for color removal at Greenock, I cannot see how the material we now know as amygdaloid would effect color removal. I suspect the material which Thom classified as this substance was some other material [see below].

There is the possibility that the color was in suspension and not in solution. I know of a few instances in which filtering water through a Berkefeld or a porcelain filter has removed color. Such removal I feel confident was by straining and not by absorption of the color on the surfaces of the filter media.

Hannan: [Amygdaloid.] I think that it is now generally admitted that the color due to humus bodies is colloidal and negatively charged. As such, it resists filtration; but when a suitable proportion of positively charged colloid (such as are the metallic hydroxides, as a class) is introduced, coagulation follows with resultant precipitation, removable by filtration. My *guess* would therefore be that the amygdaloid probably supplies in colloidal form either ferric or aluminum hydroxide in adequate quantity. (For decomposition of rocks, see *Data of Geochemistry*, U.S. Geological Survey Bulletin 770, 12: 479–542, by Dr. Frank Wigglesworth Clark.)

Weston: [Questions statement on use of amygdaloid at Greenock.] Regarding the distinction between turbidity and color, the A.P.H.A.-A.W.W.A. *Standard Methods of Water Analysis* use two terms, namely "true color" and "apparent color," the first being the color of water after filtration through paper. This color may be due in part to matter in colloidal suspension, that is, to particles too fine to be removed by short periods of subsidence.

Color is due primarily to chlorophyll which is a compound of iron and organic matter, sometimes in colloidal suspension, sometimes in true solution. The brown color of surface waters is due to the oxidation of the iron, the same that takes place when the sap is withdrawn from the leaves in the fall and the autumn colors brighten the landscape. This same brown iron-organic color is dissolved from the leaves on the surface of the ground, from the loam beneath, and from the mosses and peat in the swamps.

I am glad that you have emphasized the color removal by storage, which is of course due to bacterial decomposition, coagulation, and resulting subsidence, and also to the bleaching action of sunlight. A few years ago I looked up the matter of removal by storage and found that on the average half of the color was removed in 350 days and about 75% in 760 days.

Other Experience in Color Removal

Thomas Spencer, in a paper published at London in 1859, claimed removal from water of matters in solution: color, "other impurities and taste." After experiments with other materials he found protocarbide of iron the best for the purpose. By roasting red hematite of iron he produced carbide of iron, or magnetic oxide. This, mixed with fine sand, was used as a layer in a filter beneath a top layer of fine sand. Subsequently, the Spencer Process was adopted at Wakefield, Stockport and Wisbeach, England. At Wakefield the water treated received dye and other wastes from factories built on the River Calder after that stream was chosen as a source of supply. At Wisbeach the water was discolored by peat and at Stockport well water was colored by "iron rust" (see Chap. XIII).

With the advent of rapid filtration in the eighties and nineties color removal became increasingly common in the United States. Notable among the highly colored waters so treated was the supply of Norfolk, Va., where a rapid filtration plant was completed in 1899 (7). An exceptional case of decolorization is afforded by Springfield, Mass., where excess coagulation with sulfate of alumina was begun in 1911, a year after the sand filtration plant with precoagulation was put into use (8).

Ozonation for reduction of color, as well as tastes, odors and bacteria, was adopted at Long Beach, Ind., in 1930, and was being studied diligently in experimental and demonstration plants in 1941–42 (see "Ozonation" in Chap. XIV).

CHAPTER XX

Iron and Manganese Removal

Considering the widespread use of ground water for centuries and the possibilities of iron and manganese troubles from it and from some surface supplies, it is remarkable that so little attention was given to remedial measures until late in the nineteenth century.

In 1868, Salbach (1) announced that "certain ground waters could be freed from iron by aeration followed by filtration through gravel and sand." Charlottenburg, Germany, in 1874, seems to have been the first city to build an iron-removal plant. Among other German cities that followed suit were Breslau (see below), Dresden, Hamburg, Koenigsburg and Leipzig (2).* Hazen says (3), without giving dates: "Among the earliest plants for the removal of iron were the filters constructed at Amsterdam and The Hague, Holland (3) (see below)."

In the United States, the first iron-removal plants were installed at Atlantic Highlands in 1893 and Asbury Park, N.J., in 1895. Next came Reading, Mass., and Far Rockaway, N.Y., in 1896. Ground water was treated at all those plants. The iron at Asbury Park was in the form of sulfate of the protoxide. There and at Atlantic Highlands the iron was removed by double filtration (sand, then animal charcoal) in rapid Continental filters, without aeration or chemical treatment; but at Asbury Park the compressed air used for air-lift pumps supplied air (4). At Reading, water from a filter gallery contained sulfate of iron; at Keyport, water from artesian wells carried bicarbonate of iron. Each plant used milk of lime, aeration and rapid filtration, in the order named, the filters being of the Warren

* A summary by Goetz of the early history of deferrization in Germany follows: Salbach, at Halle on the Saale, while aerating and filtering water for removal of algae, found that iron was also removed. Anklam, of the Berlin water works, in 1880, showed that aeration and filtration would eliminate iron from the wells being studied, but Lake Tegel was adopted instead. Oesten, in 1886, was refused a patent for aerating water with atmospheric oxygen, followed by filtration; he continued experiments at Berlin in 1888–89 under Koch and Proskauer. In 1890, Piefke erected his first plant for aeration by trickling water down over coke. In 1893, Oesten obtained a patent for aerating water in the form of rain. After 1893 iron-removal developed rapidly, but the methods used were based more or less on those of either Oesten or Piefke. (Notes supplied by Frank Hannan, Toronto, Ont.)

type (5). At Far Rockaway slow sand filtration was employed (6). Seven iron-removal plants in Massachusetts were described in 1931 in a paper by Sterling and Belknap (7). In the illuminating general discussion that followed, Robert Spurr Weston stated that the tendency was toward the upward-flow type of aerator, which saves head and is easily cleaned.

In Canada, only five iron-removal plants were known to be in use in 1942, all recent: Etobicoke Township, London, Newmarket and Simco, Ont.; and Yorkton, Sask. The Etobicoke plant employs zeolite for both softening and iron removal. At London, Newmarket and Yorkton a modified form of the Reisert rapid filter is used.

Great Britain had 30 or more iron-removal plants on public water supplies in 1939, most of which used polarite in Candy rapid filters according to an advertisement in *Water and Water Engineering*, London, October 1939, and reports in *Engineers Handbook and Waterworks Directory*, 1939.

Poland, most of whose city water supplies are drawn from the ground, seems to lead the world in percentage of water works employing iron removal. Statistics furnished for use here in January 1937, by A. Szniolis of Warsaw, showed that of about 170 water works in that country (not including Upper Silesia) about 45 used deferrization, employing "exclusively aeration, open or closed, and filtration."

In French Indo-China at the twin cities of Saigon and Chalon (8), an American driven-well company adapted the German Reisert iron-removal system to local conditions in 1933–35 and subsequently applied it to ground-water supplies at three cities in Canada (9), already named, and on the works of the Jamaica Water Supply Co., Long Island, N.Y., and at Bridgeton, N.J. (10). Ten plants, with a combined capacity of 30 mgd. (Imp.) were built to serve groups of wells at Saigon-Chalon. They combine in a single closed tank: (a) contact aeration by passing water downward and compressed air upward through a layer of broken lava at the top of the tank and (b) filtration in the lower part of the tank. At Jamaica, the filter media are sand and calcite [98 per cent calcium carbonate], 1 to 1, mixed.

Demanganization

Frequently, but not always, both manganese and iron give trouble in the same source of supply. The earliest instance found on record is at Zutphen, Holland, where a pre-aeration and double filtration plant was completed in 1889. The water was aerated by cascading over a weir. The prefilters were of coarse river sand and the final filters of fine river sand. In 1923, this plant, which had become too small and exposed to possible contamination, was replaced by a larger one, inclosed, which was still in use in February 1940. In the new plant, a spray aerator was used, with fine gravel for the prefilters and sharp sand for the afterfilters. Aeration is designed to remove carbon dioxide and by oxidation change the iron and manganese compounds from a soluble to an insoluble state so they will be removed by filtration (11).

A sudden outburst of manganese and iron trouble at Breslau, Germany, in 1906, gave rise to the term "Breslau calamity." Before that time, wrote Weston (2) in 1914, "when the manganese in the well supply rose suddenly to 220 ppm., little attention was paid."

In the United States, Middleborough, Mass., put into use on September 26, 1913, a plant for removing both iron and manganese. Sprinkling nozzles discharged water upon a coke trickling aerator after which it was passed to a settling basin. Large iron-and-manganese-removal plants were put in use at Lowell and Brookline, Mass., October 16, 1915, and August 20, 1916. The Lowell plant had a capacity of 10 mgd. It included coke prefilters 10 ft. deep, upon which water was sprayed, settling reservoirs, and slow sand filters (12). Providence, R.I., began removing iron and manganese from surface water in August 1931.

In the Central West, plants for removing both iron and manganese were built at Brainerd, Minn., in 1931, and at Lincoln, Neb., in 1935. At the Brainerd plant, water from wells is forced up through finely ground pyrolusite, containing about 70 per cent of manganese dioxide, then sprayed upon a bed of coke, then passed through a filter composed of a layer of fine silica sand resting on crushed sandstone. A description of this plant by Carl Zapffe, Manager of the Iron Ore Properties of the Northern Pacific Railway and President of the Brainerd Water Board (13), contains a review of manganese in water supplies, with 74 source references, divided into five chronological groups: (1) Bacteriological, 1836–1906; (2) Inorganic Chemistry, 1906–14; (3) Physico-chemistry, 1914–22; (4) Catalysis, 1922–30; (5) Present Practice, 1930. The plant at Lincoln, put into use in December 1935, employs aeration, chlorination, upward flow through gravel contact beds, sedimentation and rapid sand filtration (14).

Unique among recent plants which utilize zeolite for softening and manganese removal is one put into use in 1937 by the Edgeworth Water Co. in Pennsylvania. The two objects are effected separately and by different procedures, after which the effluents are mixed and sent to consumers. Ordinary greensand zeolite is used in each case. Manganese is removed by upward filtration, at a rate of 10 to 12 gal. per sq.ft. per min. under a total head of a little over 5 ft. Regeneration is effected by drawing in permanganate of potassium and permitting it to remain in contact several hours while the plant is out of use at night. The regenerated unit can then demanganate for a long time. Softening is effected in another set of zeolite units regenerated with common salt. This removes all the manganese so long as the softener effluent is in the zero zone. When the demanganized water and the softened water are mixed the combination has about 85 ppm. of hardness, is free from manganese and is entirely potable (15).

Latest Summary for the United States

Early in 1941, plants in the United States for the removal of either iron or manganese, or both, totaled 598, located in 36 states. Fifty were combined with softening plants. The majority of the total number were small, as the combined output of the 598 plants was only 220 mgd., giving an average of only 0.37 mgd. By far the largest plants are those which serve Providence, R.I., and Memphis, Tenn., the populations of which in 1940 were 253,000 and 293,000 respectively. The output of the Providence plant averaged 26 mgd., and of the Memphis plant, 20 mgd., compared with rated capacities of 67 mgd. and 40 mgd. Surface water is treated for Providence and artesian well water for Memphis (16).

CHAPTER XXI

Taste and Odor Control

Aside from the elimination of disease germs from drinking water, no achievement in the history of water purification has been greater than the control of tastes and odors. Strange to relate, the crowning triumph in the conquest of water-borne disease germs, chlorination, accelerated to a high degree the tastes and odors borne by some waters, such as those laden with algae and, worse yet, those into which certain industrial wastes, particularly phenol, were poured. Efforts to cope with tastes and odors, chronologically arranged, have been: boiling, aeration, dosing with vegetable, chemical or mineral substances and ozonation. Some of these efforts have been direct attacks upon tastes and odors; some have been aimed at the prevention of growths of nuisance-producing substances, particularly algae and other organisms.

Attempts to prevent tastes and odors have been recorded from time to time during the past 2,400 years. They became more numerous toward the close of the nineteenth century, began to achieve notable success in the first decade of this century, and pointed the way toward victory at the end of the third decade.

In earlier chapters there are reviewed many aspects of taste and odor troubles and of control methods employed up to the early twentieth century.

Sanskrit lore, probably dating back to 2000 B.C., recommended filtration through charcoal and exposure to sunlight, as well as boiling, followed by exposure to sunlight and the dipping of hot iron into it seven times. The earliest authenticated written recognition of bad odor in water is accompanied by a prescription for removal by boiling and straining and is found in the books of Hippocrates, the Greek physician (460–357 B.C.). Additional correctives are found in later literature of Greek and Roman origin (see Chap. I).

Lowitz, the Dutch chemist, announced at St. Petersburg, Russia, in 1789–90, the results of experiments which demonstrated that powdered charcoal, either alone or supplemented by a few drops of sulfuric acid, would cure putrid water. Partly on the strength of these experiments, charcoal in various forms was often used in water treatment, usually as a filter medium, alone or with other materials, from the 1790's onward, until the general adoption of sand alone for both slow and rapid or mechanical filtration. The activated carbon of recent years is an improvement on the findings of Lowitz more than a century earlier. In 1807, Cavallo, Italian by birth, English by adoption, prescribed adding freshly-made powdered charcoal to water, followed by agitation and subsequent removal by filtration (see Chap. III).

Paulus Aegineta, seventh century A.D., recommended boiling to cure bad water. Boerhaave (1668–1738) wrote that when water has "spontaneously grown putrid, give it only one boil in the fire, [whereupon] the animals that are in it will be destroyed." Ozonation, originally and still used almost exclusively for disinfection in Europe, was applied to taste and odor control at Hobart and Long Beach, Ind., in 1930 and 1932, and at Whiting, Ind., in 1940 (see Chap. XIV).

Distillation of bad water for its "preservation sweet on shipboard," was the subject of British patents late in the sixteenth century. It has been used under special conditions on two Texas water supplies (see Chap. XV).

Two Thousand Years of Aeration

As with boiling, aeration has appeared in the literature on taste and odor removal or prevention for well over 2,000 years, and is attributed to much earlier Sanskrit traditions. For centuries, the benefits of natural aeration were the theme of many writers, culminating in Sir Francis Bacon's dicta: one negative—"Running Waters putrefy not"; the other positive—"Motion or stirring" prevents putrefaction, which "asketh rest." The first-known apparatus contrived for artificial aeration, a device to force air through water, was described by Hales in 1755. In contrast, Lind, at about the same time, described a method of dropping water into air. A number of patents on apparatus and processes for odor removal were granted in Great Britain in the first half of the nineteenth century, and in the United States during the second half. Of the latter the most notable were the Leeds and the Hyatt patents of the 1880's: the Leeds method was to force air through or discharge it into water in reservoirs; the Hyatt process covered the induction of air into tubes of water either below or above ground. Vigorous exploitation of the Leeds and the Hyatt systems resulted in their adoption by a few water works as described in Chap. XVI, but sooner or later their use was abandoned. Various types of cascade, jet and other open-air aerators are also described in Chap. XVI. Cascade aerators were used at the Quai des Celestins filters, Paris, in the 1820's or earlier; at the Gorbals water works, Glasgow, Scotland, in the 1850's; and at Little Falls, N.Y., in 1888. A fountain-discharging trough was installed at the Elmira Water Works Co. in 1860–61. Multiple-jet aerators were put into use at Rochester, N.Y., in 1876, and at the Utica, N.Y., Water Works Co., in 1890. There were many others. Pan aerators, with or without coke filling, were used at Winchester, Ky., in 1901 or 1902. A cluster of spray nozzles, controlled automatically by utilizing that portion of the head between the water on the filters and in the clear well was developed by Pirnie and used in 1927 at Providence, R.I., and at Poughkeepsie, N.Y. An induced-air system, known commercially as "Aeromix," was installed at Waukegan, Ill., in January 1929. It bears a family resemblance to the device patented by Hyatt in 1885. Porous tubes or plates for air diffusion, similar to those used in the activated-sludge process of sewage treatment, have been used at a few water works, beginning at Brownsville, Texas, in 1931 (see Chap. XVI).

Algae Troubles and Their Control

Since early in the nineteenth century, city water supplies have been growing in number and size, and there has resulted an increased use of storage for both surface and ground water supplies. As a rule, storage was in open reservoirs, thus exposing the water to sunlight. Underground waters, however satisfactory until brought aboveground, usually contain enough mineral matter to afford abundant food for algae and other odor- and taste-producing organisms when exposed to the light. With surface waters the danger depends largely upon the nature and amount of organic matter on the reservoir bottoms or growing in coves or other shallows along shore. Whatever the origin of the algae, their life- and death-processes may give rise to tastes and odors intolerable to water consumers, and not only intolerable, but producing fear of disease and death.

The earliest known experience with bad tastes and odors in a ground water supply, occurred at Toulouse, France, in the early 1820's and was recorded in detail by D'Aubuisson. Many decades later, at Brookline, Mass., F. F. Forbes, Engineer and Superintendent, and his microscope, had a long struggle with algae in ground water. In both cases exclusion of light from the water stopped the trouble. At Toulouse, an open filter basin infested with algae was converted into a covered filter gallery from which water went to pumps; at Brookline, the storage reservoir was covered to exclude light.

Notorious among instances of algae nuisances arising from surface waters stored in large open reservoirs is the experience of Springfield, Mass., with its Ludlow reservoir. After almost annually recurring plagues, a new supply, also of surface water, but filtered, was introduced. Many other American cities suffered from the algae plague before and after Springfield (see Chap. XVII).

Modern Developments in Taste and Odor Control

Two factors have intensified the need for improvements in water treatment practices related to taste and odor control. The first is the rapid increase in the amount and complexity of industrial wastes coupled with the general indisposition of industry to stabilize the waste material it produces. The rapid growth of the heavy chemical industry, especially in the field of organic chemistry, has resulted in the discharge of a great variety of waste products into the streams of Europe and America, the removal of which, from a water later handled by a purification plant, calls for a high degree of technical skill on the part of the water treatment plant operator.

The second factor responsible for the increase in taste and odor problems is chlorination. No material has contributed so greatly to the production of safe water supplies. Adequate removal of bacterial contamination from water has been achieved by chlorination to a remarkable degree. But with its ability to destroy bacteria, chlorine also tends to form objectionable compounds with organic materials in water and to produce odors which in some instances would not be present if the water had not been chlorinated.

Houston in 1912, was the first to find, during an emergency, that the addition of a large or *super* dose of chlorine could be used to destroy odors as well as to disinfect. The excess chlorine was then removed and a satisfactory water resulted. Not until 14 years later was the process applied on a large and permanent scale by Howard in treating the Toronto supply. This will be discussed later.

Chloramination.—Race (1), at Ottawa in 1917, applied ammonia with chlorine with resultant reduction of end tastes. During the early twenties, Harold (2) and Adams (3) in England found that the use of ammonia with chlorine not only corrected the unpalatability but suppressed the production of chlorophenol tastes which by that time had begun to affect many water supplies. Chlorophenols are unstable and highly odorous products derived from the addition of chlorine to water which contains phenols, etc., derived from coke oven wastes.

McAmis (4), at Greenville, Tenn., began the regular application of ammonia with chlorine in 1926 and was soon followed by Spaulding (5) at Springfield, Ill., Ruth (6) at Lancaster, Pa., Harrison (7) at Bay City, Mich., and Lawrence (8) with Braidech (9) at Cleveland. The growth of chloramination was rapid; in 1933, 35 per cent of all treatment to correct odor and taste in the Middle-West involved the addition of ammonia to water.

Later experience has shown that taste in water polluted by industrial wastes derived from coal-tar plants can be controlled by chloramination. Its value is limited when other taste-producing substances are present.

Superchlorination.—At Toronto in 1926, Howard (10, 11) began regular superchlorination of the supply. In its earlier form, the process involved the addition of approximately one part per million of chlorine to the water. A contact period of one to one-and-one-half hours was followed by dechlorination with sulfur dioxide. Variations in the dosage used have followed, but at this writing the procedure is essentially the same as it was first developed: high rate chlorination, contact, and then dechlorination. For the treatment at Toronto of water derived from Lake Ontario, the process has been a successful one. Only a few operators adopted superchlorination in the fashion practiced at Toronto. The process requires very careful control or the results will be unsatisfactory. Obviously also the conditions in many waters do not respond to this type of treatment.

In the late thirties, studies of Howard (10), Scott (12) and Gerstein (13) were reviewed by Faber (14) and Griffin (15) and later by Hassler (16). It was ascertained that the primary difficulty of high rate chlorination as previously practiced derived from the failure of the operator to satisfy the chlorine demand of the water completely. It has since been found that if chlorine is added to water to such an extent that practically all of the residual chlorine is "free" instead of "combined" many of the taste-producing substances will be destroyed and the remaining free chlorine does not, in many instances, manifest itself as odor.

This method of adding high doses of chlorine has been called "break-point" chlorination (15) because it was found that, in many supplies, after a certain amount of chlorine had been added, the residual curve "broke" and a lower residual of chlorine ensued. Successive additions of chlorine thereafter resulted in increased residuals. Scott (12) first reported this phenomenon in 1928, but not until more than ten years had passed was its real importance apprehended. Now it is becoming well understood that the significant factor is the complete satisfaction of the chlorine demand of the water and the presence in the finished product of free, rapidly reactive chlorine.

This form of chlorination has been widely applied since 1940 and the process, when better understood and properly applied, gives promise of correcting offensive tastes and odors in many water supplies which have previously not responded satisfactorily to earlier processes.

Activated Carbon.—The beginnings of the use of charcoal to improve the taste of water are lost in pre-history. The adaptation of charcoal in the modern form—activated carbon—has been both rapid and spectacular since the late twenties. While Baylis (17, 18) was conducting tests of its use in this country, Sierp (19, 20) in Germany was using activated carbon to remove chlorophenolic tastes in water. The material was used in granular form at the Hamm Water Works in 1929 and independently at Bay City, Mich., by Harrison (7) in 1930. Meanwhile at the Hackensack Water Company plant in 1929, Spalding (21) used powdered activated carbon to remove odors in water caused by wastes from an alcohol denaturing plant. By 1932, 400 plants in the United States were using activated carbon in odor control; and by 1943, nearly 1,200 plants were using it. Much of its popularity as an odor corrective lies in the fact that while the material is dust producing, it is easily applied. If it is added to the water in such a manner as to disperse it thoroughly, it is generally effective. Within limits of cost, it cannot be added to an overdose point. The material is nontoxic and is not a health hazard.

At this writing, the removal of taste and odor from water is the most severe problem facing the treatment plant operator. In extreme cases, reliance is not placed upon a single method or material. The water may be superchlorinated, dechlorinated, its pH adjusted, treated with carbon and even given final treatment after filtration by adding more chlorine. Ammonia may be used either as a phase of pretreatment or at the end of the process in order to form chloramine in the distributed water. No phase of water treatment is being given more intensive study and in no aspect of water supply practice does the consumer maintain a more active interest. Progress is being made and greater progress appears imminent.

This may be said in closing. Too much has been left to the magic touch of the water purification technician. Too little has been done to restrain industries and municipalities in their discharge of untreated wastes and sewage into streams which later must be used as sources of water supply. Water works engineers and chemists have achieved great results in their production of bacteriologically safe water. But there is a limit beyond which they cannot go in removing from waters the odorous materials derived from industrial wastes and municipal sewage. Nothing has so adverse an effect upon the attitude of the water user as a supply which has an offensive odor or taste. He may be told that it is safe but his sense of smell disagrees. Too long have modern cities demanded that water purification correct the evils of unrestrained stream pollution. The time has come for the pollution to be corrected.

CHAPTER XXII

Medication by Means of the Water Supply

In striking contrast to the removal of objectionable materials from water, described in the preceding chapters, is the addition of substances to make up deficiencies detrimental to health. The first such addition attempted was the periodic dosing of water with iodine to combat goiter; the second, the continuous application of fluorine to prevent decay of teeth or dental caries. Iodization was tried in the years 1923–33; fluorination was inaugurated by plant scale studies in 1945.

Although these practices are medication rather than purification, their possible foreshadowing of great events to come in the water treatment field, particularly if fluorination becomes established, warrants at least brief consideration in this book.

Iodization Against Goiter

From remote times certain glandular swellings on the neck have been attributed to the waters of widely scattered regions of the earth. In more recent times the swellings have been called goiter. Ascribed by many writers to snow waters and by others to ill-defined mineral contents of drinking water, it was finally agreed that goiter was due to a deficiency of iodine. Acting on the belief that a small amount of iodine added to the public water supply at considerable intervals would prevent goiter, three water works in the United States and one in England adopted this practice.

The lead was taken by Rochester, N.Y., in April 1923. The practice was continued for ten years (1, 2, 3).

Eight years after iodization was introduced at Rochester, G. W. Goler, Health Officer (3), wrote that it was still in use, giving a "reduction in simple goiter of more than one-half, as shown by annual medical school inspections. The iodide cost us less than one cent per person per year." He added: "I do not believe that many cities would adopt the procedure even if we could show that goiter had been abolished." No reason for this opinion was given. Presumably Dr. Goler was impressed by the fact that iodization had made no headway elsewhere and had been continued in Rochester out of respect for his long-continued able service as health officer and his winsome personality. Dr. Goler's successor in office wrote, in September 1933 (4), that in 1932, "because of economy, we used only one-half the amount of iodide previously used—limiting the number of days that additions were made—but this year we had to omit it entirely."

At Sault Ste. Marie, Mich., a plan was adopted in 1923 for dosing the water supply for a period of two weeks twice a year. This was begun in August 1923. In the fall of that year, "the Michigan Medical Association," City Manager Sherman wrote ten years later (5), "persuaded the salt companies of Michigan to put sodium iodide in table salt, not to remove it in the process of manufacture, after which we felt it was unnecessary to treat the water supply."

In England, the Ilkeston and Heanor Water Board, supplying several places in the county of Derbyshire, where "Derbyshire Neck" was prevalent, applied sodium iodide to its water supply during the year beginning October 1924. Continuous dosing at the rate of 3 oz. per mil.gal. (Imp.) [2.5 oz. per mil.gal. (U.S.)] was practiced. A day's supply of the agent was added to 1 gal. (Imp.) of water. This solution was introduced by drop feed from a suitable glass vessel into the main suction pipe of the main pump. The water came from Meerbrook Sough (a collecting tunnel) and two bored wells. Dr. Barwise, Medical Officer of Health for the county of Derbyshire, who advised iodization, died before the completion of the twelve-month test and the studies he was making (6).

The Anaconda Mining Co., reported H. M. Johnson, Superintendent (7), began iodizing the water supply of Anaconda, Mont., in April 1925. It continued to do so until the fall of 1933, then stopped for "economic reasons." Sodium iodide was added to the water for fourteen days in April and October. Dosing was at the rate of 0.644 lb. per mil.gal. or 0.0003 gram per gal. The salt was dissolved in the reservoirs and at a penstock where additional water was drawn for pressure purposes. At the penstock, solution feed was used. The local school board supplied the school children with chocolate-coated iodine tablets in the spring and fall of 1926, then discontinued the practice. Figures supplied by the school board showed that about six per cent of the children had simple goiter. After discontinuance of the services of a full-time school nurse, writes Johnson, the company was unable to obtain figures showing whether goiter among school children was on the decrease, but it continued to use iodine until the fall of 1933 (7, 8).

Valiant efforts to iodize the water supply of Minneapolis were made in the middle 1920's by Arthur F. Mellen, Filtration Engineer. He won the approval of the Hennepin County Medical Society and the Minneapolis Board of Public Welfare. The Committee on Water Supply of the City Council held hearings on the proposal. Although four of its five members favored the plan, it was not reported out of the committee. All members of a later committee journeyed to the Buffalo convention of the American Water Works Association of 1926 and listened to Mellen's comprehensive paper on the problem (9). Notwithstanding all this, the "strong opposition of the Christian Scientists, the American Medical Liberty League and others," wrote Mellen in 1933, "made it impossible to carry the plan through" (10).

Despite its warm-hearted support over a period of ten years, water iodization against goiter was applied to only four water supplies, soon given up at two and finally abandoned by all. Public apathy, opposition by secular and professional groups, the supply of chocolate-covered iodine tablets to school children, and the supply of salt high in iodine by salt manufacturers checked at the very outset and then brought to an end a promising revolution in water treatment.

A damaging setback to the iodization of public water supplies took place in the very year of its adoption at Rochester. A "most thoroughgoing discussion" of the practice took place at the Annual Conference on Water Purification held at Columbus, Ohio, in the fall of 1923. "The consensus of opinion of the health officers and water purification men attending the joint session," wrote Bolt and Wolman in their official review of the whole subject of iodization for the American Water Works Association (11) seemed to be that "iodine taken internally in almost any form in small amounts regularly constitutes a preventive of simple goiter." The best plan of administering iodine, the conclusions said, was to give it "individually through specially prepared chocolate-covered tablets known under the trade name of 'iodo-stanin.'" The Conference concluded that:

> Medical authorities who have made a study of goiter prevention are inclined to recommend that, if whole populations are to be reached, then manufacturers of common table salt should be required to leave in the salt prepared for usual domestic consumption, "normal" amount of iodine. This amount . . . is being investigated.

Application of iodine in any form to public water supplies for goiter prevention purposes is not recommended. It is a wasteful procedure; and from the standpoint of sound public policy the practice of using the public water supply as a medium of medication is undesirable (11).

In a vigorous analysis of the objections to iodization of water supplies made by the Ohio conference, Bolt and Wolman declared that its practicability had been shown by two cities; that it was "no more medication than the addition of chlorine, lime, soda, alum or iron salts to our water supplies and the day has passed since we objected to these." Twelve conclusions in favor of water iodization were submitted (11). Two were:

The iodization of public water supplies in those places where goiter was shown to be prevalent would be a logical measure for the protection of the whole community. As a *preventive* method against an important disease, it is comparable to chlorination . . . to prevent typhoid fever and other intestinal diseases. . . . Iodization . . . is relatively cheap, simple, can be easily applied and controlled. It has not yet [1925] been tried long enough . . . to furnish conclusive evidence of its efficacy under various conditions. . . . The whole matter of the prevention of goiter should be under the supervision of the health authorities of the community with the advice and assistance of the sanitary engineering department and the local medical profession.

Overlooked by most if not all the writers on iodization except Bolt and Wolman (11) are the studies of the iodine content of various natural waters begun by Chatin in 1850 and his advocation, "for goitrous regions, [of] the placing of small quantities of potash (containing iodine) in the rain water," and of "the use of drinking waters known to contain iodine and foods grown in goiter-free areas." Reports made by several commissions appointed by the Paris Academy of Science to examine Chatin's data and proposals "indicate that, while they accepted his chemical work as sound, they could not come to the point of believing that such minute quantities of iodine as he found could produce such profound physiological changes as are found in goiter. . . ."

Early in 1945 about a third of the table salt produced by the Ohio Salt Co., Wadsworth, Ohio, was being iodized by the addition of 0.01 per cent of finely powdered potassium iodide. It is assumed that the same percentage is applicable to the output of other salt producers and the practice has been increasing. Iodized salt is being added to animal feed supplied to some sections of the United States.

The Michigan State Medical Society appointed an Iodized Salt Committee in 1922. The committee advised iodization. The Mulkey Salt Co. was the first manufacturer to adopt the recommendation and other companies followed suit. In Michigan, a survey showed that in some areas the incidence of endemic goiter among school children of adolescent age dropped from 39 to 4 per cent. The work of the Michigan State Medical Society has been carried on by the American Public Health Association through a Study Committee on Endemic Goiter. On its advice the percentage of potassium iodide added to table salt was reduced to the 0.01 per cent already mentioned. This has been approved by the National Research Council and the Surgeon General of the United States Army. In 1945, the U.S. Army ordered only iodized salt for cooking and table use (12).

Fluorination to Combat Decay of Teeth

Unlike goiter, tooth decay or dental caries is a universal curse. Concern over its ravages and growing hopes of its control by medicating public water supplies led to the inclusion of four papers on the subject in the program of the 1943 Convention of the American Water Works Association (13, 14, 15, 16). Leader of this section of the program was Dr. H. Trendley Dean. His comprehensive review was supplemented by 86 references to the literature of the subject ranging in date from 1874 to 1943. Brief digests of many of these papers, with critical comments, were given by Dr. Dean (13). In opening his discussion, Dr. Dean said:

> Control of dental caries is the fundamental problem of modern dentists. . . . Among civilized people, few individuals escape its attacks. . . . Of the first 2,000,000 [draftees] examined for the Army, the chief cause for rejection was dental defects. During 1941, about $500,000,000 was spent for dental services in the United States . . . despite which these services were supplied to a portion of the population only, the majority receiving either no dental service or merely extraction.

Alarm over mottled teeth enamel in some parts of the United States led to studies in 1916 which later showed that the water supplies of those sections were high in fluorine. They also showed that teeth with mottled enamel were not affected by caries. After reviewing a number of studies of caries incidence in cities using water high and low in fluorides, Dr. Dean said (13):

Theoretically, the idea of fluorination of the domestic water supply for the reduction of dental caries prevalence appears sound. Because of its unusually high prevalence, dental caries seems particularly suited for control measures through a communal medium such as the water supply. It would not involve adding anything not already in water supplies used daily by more than a million people in this country. Furthermore, the amount [of fluorine] suggested, namely, 1 ppm., is considerably lower than many hundreds of thousands of these people are now using daily. Much investigative work, however, is necessary before serious thought can be given to a recommendation for its general application.

As a practical test of the efficiency and safety of fluorination, Dr. Dean (13) suggested that two cities of 40,000 to 50,000 population, using fluoride-free water from the same source, be chosen for study. To the water supply of one of these cities, the "fluoride concentration" would be brought to 1 ppm. A carefully controlled study of the population of each city, born after fluorination in one of them was begun, and continued for "a sufficient number of years, would be necessary to demonstrate that the addition of [1 ppm.] of fluoride to a fluorine-free water will actually reduce the amount of dental caries in the community."

If such an experiment were successful, as much presumptive evidence indicates, public water supplies would fall into three groups:

(1) Those carrying naturally the *optimal* concentration of F, i.e., about 1.0 ppm., and would therefore require no treatment;

(2) Those carrying an *excessive* concentration of F requiring the removal of the excess in order to protect the community against endemic dental fluorosis (mottled enamel); or

(3) Those *deficient* in fluoride, to which fluoride might be added to bring its concentration up to the optimal in order to inhibit dental caries attack.

Viewing the domestic water problem in this light, one might with justification expect that the community's public water supply is destined to play an important role in communal public health (13).

Following Dr. Dean's introduction of the subject, Dr. D. B. Ast of the New York State Department of Health outlined "A Program of Treatment of Public Water Supply to Correct Fluoride Deficiency" (15), which he had developed in his work on oral hygiene.

In the third paper presented on this subject at the 1943 Convention of the American Water Works Association, Harold J. Knapp, Health Commissioner of Cleveland (14), urged that "in any public health program, it is necessary that the several scientific professions concerned . . . each play its role in a cooperative endeavor." He added:

It is fitting that a testimonial be expressed to the imagination, the vision and the courage of the engineering profession. Through the years it has blazoned many a trail in public health. Certainly without its ingenuity and its continuing research, we should not now enjoy the benefits of the modern methods and appliances which protect us from disease. Would it not be tragic to be compelled to lapse back to the primitive conditions of yesteryear?

It is our official duty as health officers (and certainly it is our personal desire) to be alert to the research and to the progress in applied engineering practice; to the end that we shall neither be the first to reject nor the last to accept.

Abel Wolman (16) summed up the case for fluorination and echoed with emphasis the warning that there is urgent need for further study before recommending its general adoption. He agreed with Dr. Dean's conclusions that "a small amount of fluorine contributes to a high level of dental health" and that "there are sound reasons and sound hopes" for attempting "to solve an almost insoluble" problem "by the mass handling of dental caries, which so far has not succumbed successfully to individual patient treatment." Continuing, Wolman recalled the "severe medical, engineering and legal controversy," accompanied by acrimonious debate, over the addition of alum to water as a coagulant; and later over the addition of chlorine as a disinfectant. In both cases, he said:

We had an amazing set of controversies which ran through the courts, through most of the medical associations, and all of the engineering associations, in order to prove . . . that the balance of values was in favor of, and not against, the public. . . . That whole history, however, covered the efforts to eliminate products from water rather than [as now] to introduce new complexes of a chemical or biological nature.

In closing, Wolman again reiterated Dr. Dean's warning that much investigation would be necessary before fluorination of water supplies could be recommended for general adoption (16).

Prior to his participation in the forum, Dr. Ast had prepared a comprehensive thesis on the caries-fluorine hypothesis "to present the story of dental caries as it exists today, to present its public health significance and to suggest a study for its control," in which he "proposed to introduce nontoxic doses of sodium fluoride into public drinking waters to test the fluorine hypothesis, which points to an inverse ratio of caries to fluorides present in drinking water" (17).

In general accordance with Dr. Ast's plan, the New York State Department of Health made arrangements to apply sodium fluoride to the water supply of Newburgh, N.Y., beginning in or about May 1945. As outlined by C.R. Cox, Chief of the Bureau of Water Supply of the State Department (18), a small dry-feed dosing apparatus will be used to apply commercial sodium chloride to the filtrate of four filters on its way to the filtered water reservoir. The water supply of Newburgh has a natural fluorine content of from less than 0.05 to 0.17 ppm., varying with seasonal influences.

For control purposes, the city of Kingston, N.Y., will be used. Its population is about the same as that of Newburgh. The Kingston supply "contains no detectable fluorine" (18).

The result of the studies of teeth-decay control by adding fluorine to public water supplies will be awaited with interest. Indications are that many years will be required to determine its theoretical efficiency, its physical and financial practicability, and the willingness of city authorities and private water companies to accept the wholesale medication of public water supplies.

What the future holds in store for the water works man in the field of mass medication remains to be seen. But the fact that the public water supply can now be seriously considered as an appropriate and safe vehicle for the distribution of the physician's prescription is fitting tribute to the success of man's quest for pure water.

Epilogue

In the earliest days of the human race, water was taken as found. It might be pure and abundant, plentiful but muddy, scarce but good, or both scarce and bad. To get more or better water, man moved to other sources rather than transport better water to his own location or try to improve the quality of water at hand.

Man's earliest standards of quality were few: freedom from mud, taste and odor. When after many centuries watertight household vessels became available, water could be more or less fully clarified by storage in containers or by exudation through the pores of the receptacle, leaving sediment behind. After the lapse of many more centuries it was discovered that sedimentation could be aided by adding a precipitant or coagulant. Before this, probably, it was found that "bitter" water could be "sweetened" by placing in it some vegetable or mineral agent, the latter often given the generic name "salt." With the development of the culinary art, it was observed that certain vegetables and flesh cooked better and were more palatable in some waters than in others, and that this was attended with less or more incrustation on the vessels—that is, that soft waters were preferable to hard. The terms "hard" and "soft," however, were not applied nor was much use made of softening processes until comparatively recent times.

Filtration for the removal of turbidity, either through fibrous or finely granulated material, was known in ancient times, but little used. A process akin to filtration was one of the earliest and most effective means of obtaining water free from suspended matter, but it was very time-consuming. This was the utilization of capillary action, employing a thread or piece of cloth, to remove water from one vessel to another, leaving suspended matter behind. Siphoning by atmospheric pressure, workable on a large scale, was used at an early date to remove supernatant liquid from a settling vessel and leave the sediment undisturbed. Much later came distillation by artificial heat, an adaptation of Nature's immense-scale, constantly working method of evaporating and condensing the water of streams, lakes and oceans, depositing them again in highly purified form. Artificial distillation, like wick-siphonage, was limited in scope. For centuries it was employed chiefly in laboratories. Then its use was extended to freshening salt water, chiefly for use on ships, and, very rarely, to highly objectionable marsh water where no other could be obtained on shore.

These were the chief methods of improving the quality of water until recent times. Their use was confined to the household and the factory until the last two centuries. For municipal supply they were employed hesitantly and on a small scale until the first half of the nineteenth century in Europe and well toward its close in America.

In the last sixty years, with advances in the arts and sciences, including the acceptance of the germ theory of disease and of water as one of the chief means of spreading cholera and typhoid, standards for the quality of water have been raised. Most notably, they now include bacterial limitations. With the raising of standards of quality there have been devised many and various mechanical and chemical agencies for their attainment. Decades ago, the late Professor William T. Sedgwick noted the advent of the water refining age. Great as was his vision it is unlikely that he foresaw the advances in water purification that have since been achieved.

Although the purification of city water supplies was not undertaken until the latter part of the eighteenth century, and by then only haltingly, the earlier and isolated methods of treatment used during the previous three thousand years have also been reviewed. They illustrate man's never-ending quest for pure water. They foreshadow things to come, and they show what might have come much earlier if the demand for pure water had kept pace with increasing knowledge of means for its acquisition.

BIBLIOGRAPHY
and
INDEX

BIBLIOGRAPHY

The reference list which follows includes approximately 900 entries numbered consecutively by chapters and keyed to the text by number. The items relate to books, pamphlets, papers read before professional societies and personal letters from authorities in America and abroad. No pains have been spared to trace back to original sources casual citations of the various authors mentioned.

Most of the references given can be found in either the Engineering Societies Library of the United Engineering Trustees, Inc., or the New York Public Library, both in New York City, and, also, presumably, in most of the large, and many of the small, public and college libraries.

A number of rare books and pamphlets were gathered by the author for use in the preparation of this book. These, together with photostats of parts of books or pamphlets from sources not readily accessible, have been deposited in the Engineering Societies Library as the Moses Nelson Baker Collection on Water Purification.

Some of the technical societies referred to have published consolidated indexes of their publications. Of prime importance among these are:

Index to the Proceedings, Journal and Other Publications of the American Water Works Association, 1881–1939, Inclusive. American Water Works Assn., New York (1940).

Index to the Journal of the American Water Works Association, 1940–1944. American Water Works Assn., New York (1945).

[Since 1922 the *Journal* has carried abstracts of the world's water works literature. *Although their subjects are not classified in the above indexes, both subjects and authors are indexed annually in the *Journal* itself.]

Index to the Transactions and Journal of the New England Water Works Association, 1882–1939. New England Water Works Assn., Boston (1940).

Index to the Transactions of the American Society of Civil Engineers, 1867–1920. American Society of Civil Engrs., New York (1921).

Index to the Transactions of the American Society of Civil Engineers, 1921–1934. American Society of Civil Engrs., New York (1935).

The entire field of engineering and closely allied arts is covered in *Engineering Index*—issued annually, beginning with 1910, and previously for 1884–89, 1890–99, 1900–04 and 1905–09. The first volume gave references not only to the technical periodical literature of 1884–89, but also to the entire proceedings of the American Society of Civil Engineers, the American Society of Mechanical Engineers and the Association of Engineering Societies, "so far as they seemed deserving." Throughout, the *Engineering Index* entries are classified by subjects and briefly describe or summarize the articles or papers indexed. Gradually these indexes have been extended to cover engineering and allied publications throughout the world.

Engineering Index has been published successively by the Association of Engineering Societies (St. Louis, Mo.); Engineering Magazine; the American Society of Mechanical Engineers; and Engineering Index, Inc., the last three all of New York City.

Another general guide to technical periodical literature is the *Industrial Arts Index,* published by the H. W. Wilson Co. of New York. Cumulative annual volumes of this index, issued beginning in 1913, provide a ready guide to the articles published in the major American and foreign publications in the engineering field.

Chemical Abstracts and *British Chemical Abstracts,* covering articles published in the chemical professional journals and proceedings, have been published for some years past by the American Chemical Society, Washington, D.C., and by the Society of Chemical Industry, London. So numerous are the individual abstracts that the cumulative classified indexes published every few years fill huge volumes. Useful in referring to *Chemical Abstracts* is:

List of Periodicals Abstracted by Chemical Abstracts With Key to Library Files and Other Information. American Chemical Society, Ohio State University, Columbus, Ohio (1946).

Public Health Engineering Abstracts, a monthly publication of the United States Public Health Service, Washington, D.C., also includes water works literature.

Engineering News and its successor, *Engineering News-Record,* have published a series of cumulative indexes of their contents: 1874–90, 1890–99, 1900–04, 1905–09, 1910–17 and (*Engineering News-Record*) 1917–22, 1923–27 and 1928–43. *Engineering Record* issued a *Digest* and *Index,* Vol. 5–18 (Dec. 1881–Nov. 1888).

Key to Publication Abbreviations

Standard abbreviations for some of the engineering journals and society proceedings and transactions most frequently used are:

Eng.	Engineering (London)
Eng. News	Engineering News (New York)
Eng. News-Rec.	Engineering News-Record (New York)
Eng. Rec.	Engineering Record (New York)
Jour. A.W.W.A.	Journal American Water Works Association (New York)
Jour. Chem. Soc.	Journal of the Chemical Society (London)
Jour. Franklin Inst.	Journal of the Franklin Institute (Philadelphia)
Jour. N.E.W.W.A.	Journal of the New England Water Works Association (Boston)
Philosophical Trans., Royal Soc.	Philosophical Transactions, Royal Society (London)
Proc. A.S.C.E.	Proceedings of the American Society of Civil Engineers (New York)
Proc. A.W.W.A.	Proceedings of the American Water Works Association (New York)
Proc. Inst. C.E.	Proceedings of the Institution of Civil Engineers (London)
Proc. Inst. M.E.	Proceedings of the Institution of Mechanical Engineers (London)
Trans. A.S.C.E.	Transactions of the American Society of Civil Engineers (New York)

Key to Library Symbols

The letter symbols which appear at the end of most of the publication references included in the list refer to the location where the publications cited by the author may be found. Following is a guide to the use of these symbols:

- [A] New York Academy of Medicine, New York City
- [B] Baker Collection Deposited in the Engineering Societies Library, New York City
- [C] Chemists' Club Library, New York City
- [D] New York Public Library, New York City
- [E] Engineering Societies Library, New York City
- [F] Columbia University Library, New York City
- [G] Library of Congress Catalog, Washington, D.C.*
- [H] Bibliothèque Nationale, Paris, France
- [I] Cornell University Library, Ithaca, N.Y.
- [J] British Museum Catalog, London, England
- [K] College of Pharmacy, Columbia University, New York City
- [X] References which have not been located or verified

* The location of these books may be found by writing to the Union Catalog, Library of Congress, Washington, D.C.

CHAPTER I

From the Earliest Records Through the Sixteenth Century

1. PLACE, FRANCIS EVELYN. Water Cleansing by Copper. *Jour. Preventive Medicine* (London), 13:379 (1905). [A]
2. SUS'RUTA. The Hindu System of Medicine According to Sus'ruta. Tr. from the Sanskrit by UDOY CHAND DUTT. Vol. 95, Chap. 45, pp. 215–222, *Bibliotheca Indica, Royal Asiatic Society of Bengal.* J. W. Thomas, Calcutta (1883). [D]
3. WILKINSON, JOHN GARDNER. *The Manners and Customs of the Ancient Egyptians.* Rev. by S. BIRCH. J. Murray, London (1879). [D]
4. EWBANK, THOMAS. *A Descriptive and Historical Account of Hydraulic and Other Machines for Raising Water, Ancient and Modern.* D. Appleton & Co., New York (1842). [D]
5. HERO OF ALEXANDRIA. *The Pneumatics of Hero of Alexandria, From the Ancient Greek.* Taylor, Walton & Maberly, London (1851). [D]
6. *The Old Testament.* King James Version. Exodus, 15:22–27.
7. *Ibid.* Exodus, 17:1–7.
8. *Ibid.* II Kings, 2:19–22.
9. *Geoponica: Agricultural Pursuits.* Tr. from the Greek by T. OWEN. W. Spilsbury, London (1805). [B]
10. VITRUVIUS, POLLIO. *The Ten Books on Architecture.* Tr. by M. H. MORGAN. Harvard Univ. Press, Cambridge (1914).
11. PLINY THE ELDER. *The Natural History of Pliny.* Tr. by JOHN BOSTOCK & H. T. RIPLEY. (Bohn Classical Library) H. G. Bohn, London (1855–57). [D]
12. PALLADIUS, RUTILIUS TAURUS AEMILIANUS. *Paladius on Husbondrie.* Ed. by BARTON LODGE. N. Trubner & Co., London (1873 & 1879). [D]
13. PLUTARCH. *Vitae Parallelae.* Tr. called Dryden's, corrected from the Greek and rev. by A. H. CLOUGH. Little, Brown & Co., Boston (1859). [D]
14. HERODOTUS. *History.* English version, ed. by GEORGE RAWLINSON. D. Appleton & Co., New York (1859). Vol. I, p. 252. [D]
15. ATHENAEUS OF NAUCRATIS. *The Deipnosophists.* Tr. by CHARLES BURTON GULICK. (Loeb Classical Library) W. Heinemann, London (1927). [D]
16. PLATO. *Plato's Symposium; or the Drinking Party.* Tr. by MICHAEL JOYCE. J. M. Dent & Sons, Ltd., London (1935). [D]
17. ARISTOTLE. *De Generatione Animalium.* Tr. by ARTHUR M. PLATT. Clarendon Press, Oxford (1910). [G]
18. HIPPOCRATES. *The Genuine Work of Hippocrates.* Tr. from the Greek by FRANCIS ADAMS. Williams & Wilkins Co., Baltimore (1939). [D]
19. CAESAR, C. JULIUS. *Caesar's Commentaries on the Gallic War, With the Supplementary Books Attributed to Hirtius, Including the Alexandrian War.* Tr. by W. A. MCDEVITTE & W. S. BOHN. American Book Co., New York (190?). [G]
20. ORIBASIUS. *Oeuvres.* French tr. from the Greek by BUSSEMAKER & DAREMBERG. Daremberg ed., Paris (1851–76). [D]
21. FRONTINUS, SEXTUS JULIUS. *The Two Books on the Water Supply of the City of Rome.* Tr. by CLEMENS HERSCHEL. Dana Estes & Co., Boston (1899). [D]

22. NEUBURGER, ALBERT. *The Technical Arts and Sciences of the Ancients.* Tr. by HENRY L. BROSE. Macmillan, New York (1930). [E]

23. GALEN. *De Simplicium Medicamentorum Facultatibus.* Cited in *Medicina Aegyptorum,* by PROSPERO ALPINO, Venice (1591). [A]

24. AEGINETA, PAULUS. *The Seven Books of Paulus Aegineta.* Tr. from the Greek with a commentary by FRANCIS ADAMS. In *Greeks, Romans and Arabians on All Subjects Connected With Medicine and Surgery.* Sydenham Society, London (1844–47). [D]

25. JABIR IBN HAYYAN. *The Works of Geber Englished by Richard Russell, 1678.* New ed. with introduction by E. J. HOLMYARD. J. M. Dent & Sons, London (1928). [B]

26. IBN SINA, HUSSAIN IBN ABD ALLAH. *A Treatise on the Canon of Medicine of Avicenna.* Incorporating a tr. of the "First Book" by G. CAMERON GRUNER. Luzab & Co., London (1930). [D]

27. STOW, JOHN. *A Survey of the Cities of London and Westminster Brought Down From the Year 1633 to the Present Time by J. Strype.* London (1720). [J]

CHAPTER II

Seventeenth Century

1. BACON, FRANCIS. *Sylva Sylvarum, or a Natural History in Ten Centuries.* W. Lee, London (1670). [D]

2. DIGBY, KENELM. *Two Treatises: In the One of Which, the Nature of Bodies; in the Other, the Nature of Man's Soule.* John Williams, London (1645). [E]

3. ———. *A Late Discourse, Touching the Cure of Wounds by the Powder of Sympathy, Whereby Many Other Secrets of Nature Are Unfolded.* Tr. from the French by R. WHITE. R. Lowndes, London (1660). [G]

4. GLAUBER, JOHANN RUDOLF. *The Works of the Highly Experienced and Famous Chymist.* Tr. by C. PARKE. The Author, London (1689). [D]

5. RAPIN, RENÉ. *Of Gardens.* Englished by GARDINER. B. Lintott, London (1706). [G]

6. PORTIUS, LUCAS ANTONIUS. *The Soldier's Vade Mecum; or the Method of Curing the Diseases and Preserving the Health of Soldiers.* R. Dodsly, London (1747). [B]

7. HARE, AUGUSTUS J. C. *Venice.* G. Allen, London (5th ed., 1900). [D]

8. CORYAT, THOMAS. *Coryat's Crudities: Hastily Gobled up in Five Months Travells in France, Savoy, Italy.* Macmillan, New York (1905). [B]

9. MATTHEWS, WILLIAM. *Hydraulia.* Simpkin, Marshall & Co., London (1835). [D]

10. BLACKIE, WALTER GRAHAM. *The Imperial Gazetteer: A General Dictionary of Geography.* Blackie & Sons, London (1855). Vol. II, p. 1175. [D]

11. ANON. The Water Cisterns in Venice. *Jour. Franklin Inst.,* Third Series, 70:372–73 (1860). [E]

12. ANON. The Water Supply of Venice. *Eng. News,* 9:219 (1882); 14:308 (1885). [E]

13. COLLINS, W. D. Personal Letter. Washington, D.C. (October 24, 1940).

14. ANON. A View and Description of a Paper-Mill and Its Implements. *The Universal Magazine of Knowledge and Pleasure* (London), 30:260 (1762). [D]

15. HUNTER, DARD. *Papermaking Through Eighteen Centuries.* W. E. Rudge, New York (1930). [D]

Chapter III
Eighteenth and Early Nineteenth Centuries

1. LA HIRE, PHILIPPE DE. Remarks on Rain Water, Construction of Cisterns. Vol. II, pp. 50–64, *The Philosophical History and Memoirs of the Royal Academy of Sciences at Paris.* London (1742). [D]

2. MARSIGLI, LUIGI FERDINANDO. *Histoire Physique de la Mer.* Amsterdam (1725). [G]

3. ROCHON, M. Memoir on the Purification of Sea-Water and on Rendering It Drinkable Without Any Empyreumatic Taste by Distilling It in Vacuo. (From the *Jour. de Physique.*) *Repertory of Arts, Manufactures and Agriculture,* Second Series (J. J. Wyatt, London), 24: 297–302, 367–373 (1814). [D]

4. BOERHAAVE, HERMANN. *Elements of Chemistry.* Tr. from the original Latin by TIMOTHY DALLOW. J. & F. Pemberton, London (1735). [D]

5. HALES, STEPHEN. *An Account of Some Attempts to Make Distilled Water Wholesome, Containing Useful and Necessary Instructions for Such as Undertake Long Voyages at Sea. Showing How Sea-Water May Be Made Fresh and Wholesome, and How Fresh-Water May Be Preserved Sweet.* R. Mauly, London (1739). [D]

6. PLÜCHE, NOEL ANTOINE. *Spectacle de la Nature; or Nature Displayed.* Tr. from the original French by SAMUEL HUMPHREYS. J. & F. Pemberton, London (1736). [D]

7. VATER, ABRAHAM. Filtri Lapidis Mexicani, Examinatio et Comparatio Cum Aliss Lapidibus Facta. *Philosophical Trans., Royal Soc.,* 39:106–111 (1735). [D]

8. CHAMBERS, EPHRAIM. *Cyclopedia; or an Universal Dictionary of Arts and Sciences.* London (5th ed., 1741–43). [D]

9. DIDEROT, DENIS. *Encyclopédie; ou Dictionnaire Raisoné des Sciences.* Briasson, Paris (1761–65). [D]

10. CROKER, TEMPLE HENRY; ET AL. *The Complete Dictionary of Arts and Sciences.* The Authors, London (1766). [B]

11. LIND, JAMES. *A Treatise on Scurvy.* Edinburgh (1753). pp. 234–237. [D]

12. ———. *An Essay on the Most Effectual Means of Preserving the Health of Seamen in the Royal Navy.* D. Wilson & G. Nichol, London (1774). pp. 84–93. [B]

13. HOME, FRANCIS. *Experiments on Bleaching.* A. Kincaid & A. Donaldson, Edinburgh (1756). [B]

14. BUTLER, THOMAS. A Safe, Easy and Expeditious Method of Procuring Any Quantity of Water at Sea, and an Easy Method of Preserving Fresh Water Pure, Sweet and Wholesome During the Longest Voyage, and in the Warmest Climates. *Monthly Review* (London), 13:309 (1755). [D]

15. RUTTY, JOHN. *A Methodical Synopsis of Mineral Waters.* William Johnston, London (1757). [B]

16. HEBERDEN, WILLIAM. Remarks on the Pump-Water of London, and on the Methods of Procuring the Purest Water. *Medical Trans., College of Physicians* (London), 1:1–22 (1768). [B]

17. PERCIVAL, THOMAS. *Experiments and Observations on Water: Particularly on the Hard Pump Water of Manchester, England.* London (1769). [J]

18. BUCHAN, WILLIAM. *Domestic Medicine; or a Treatise on the Prevention and Cure of Diseases.* Thomas Dobson, Philadelphia (1795). [D]

19. BERTRAND, JEAN. *De l'Eau Relativement à l'Economie Rustique; ou Traite de l'Irrigation des Prés.* A. J. Marchant, Paris (1801). [H]

20. SCOTTE, J. P. Journal of the Weather at Senegambia, During the Prevalence of a Very Fatal Putrid Disorder, With Remarks on That Country. *Philosophical Trans., Royal Soc.*, Part II, 70:478–506 (1780). [D]

21. LOWITZ, JOHANN TOBIAS. Memoir on the Purification of Corrupted Water. *The Universal Magazine of Knowledge and Pleasure* (London), 96:22–25 (1795). [D]

22. FELDHAUS, FRANZ MARIA. *Die Technik der Vorzeit, der Geschichtlichen Zeit und der Naturvolker.* W. Englemann, Leipzig (1914). [D]

23. CAVALLO, TIBERIUS. *Elements of Natural or Experimental Philosophy.* London (1803). [D]

24. ANON. On the Methods Most Easily Practicable for Preserving and Purifying Water. *The Artist* (London), No. 15, pp. 1–2; No. 16, pp. 1–10 (1807). [I]

CHAPTER IV

Four Centuries of Filtration in France

1. TOMLINSON, CHARLES. *Cyclopedia of Useful Arts, Mechanical and Chemical Manufactures.* G. Virtue, London (185?). [D]

2. AMY, JOSEPH. *Nouvelles Fontaines Domestiques, Approuvées par l'Académie Royale des Sciences.* J. B. Coignard, Paris (1750). [B]

3. ———. *Nouvelles Fontaines Filtrantes, Approuvées par l'Académie Royale des Sciences. Pour la Santé des Armées du Roi sur Terre et sur Mer et du Public.* Antoine Boudet, Paris (1752). [B]

4. ———. *Suite du Livre Intitulé "Nouvelles Fontaines Filtrantes, Approuvées par l'Académie Royale des Sciences."* Antoine Boudet, Paris (1754). [B]

5. ———; ET AL. *Avis Utiles pour les Nouvelles Fontaines Filtrantes, Domestiques, Militaires et Marines.* Paris (1758?). [B]

6. BELGRAND, EUGENE. *Les Travaux Souterrains de Paris.* Dunod, Paris (1873–87). [D]

7. MONTBRUEL, JEAN BAPTISTE MOLIN DE & FERRAND, NICOLAS. Décret de la Faculté de Medicine, Rendu sur le Rapport qui Lui a été Fait du Projet de MM. de Montbruel & Ferrand, de Procurer, aux Habitants de Paris, l'Eau de la Seine, Filtrée & Prise au Pont-à-l'Anglais. *Rapport des Commissionaires*, Paris (1763). [B]

8. ———. *Eau de la Seine Filtrée et Epurée, Prise au Pont-à-l'Anglais, Pour la Consommation de Paris.* Paris (1763–64). [H]

9. GIRARD, PIERRE SIMON. *Recherches sur les Eaux Publiques de Paris.* Paris (1812). [D]

10. GURY, A. *L'Eau à Paris.* (Extraits d'un Manuscrit, écrit en 1913.) Paris (1939). [Mimeographed Abstract.] [B]

11. MIRABEAU, HONORÉ GABRIEL VICTOR RIQUETTI DE. *Réponse à l'Ecrivain des Administrateurs de la Compagnie des Eaux de Paris.* Bruxelles (2nd ed., 1786). [B]

12. WILSON, JOHN. A Method of Purifying, Clarifying Fluids. *British Specifications of Patents*, No. 2626 (1802). [D]

13. ROCHON, M. Memoir on the Purification of Sea-Water and on Rendering It Drinkable Without Any Empyreumatic Taste by Distilling It in Vacuo. (From the *Jour. de Physique.*) *Repertory of Arts, Manufactures and Agriculture,* Second Series (J. J. Wyatt, London), 24: 297–302, 367–373 (1814). [D]

14. ANON. Sur les Filtres Pour la Purification de l'Eau. *Annales des Arts et Manufactures* (Paris), 13: 288–305 (1804). [D]

15. PARMENTIER. Observations on Clarification. *Repertory of Arts and*

Manufactures (London), 16:130–138, 176–184 (1802). [D]

16. (a) ANON. On Purifying the Water of the Seine at Paris. *Technical Repository* (London), 1:316–317 (1822). [X]
    (b) ANON. Purification of the Water of the Seine at Paris. *Royal Inst. of Great Britain Quarterly Jour. Science, Literature and the Arts* (London), 13:423 (1882). [D]

17. ANON. *Recherches Statistiques sur la Ville de Paris et le Département de la Seine.* Imprimerie Royal, Paris (1826). [D]

18. ANON. Fontaine Dépuratoire. *Dictionnaire Technologique*. Thomine, Paris (1826). [D]

19. GENIEYS, RAYMOND. Clarification et Dépuration des Eaux. *Annales des Ponts et Chaussées* (Paris), Series I:56–768 (1835). [D]

20. MATTHEWS, WILLIAM. *Hydraulia*. Simpkin, Marshall & Co., London (1835). [D]

21. DUNGLINSON, ROBLEY. *Human Health; or the Influence of Atmosphere, Food . . . on the Healthy Man.* Philadelphia (1844). [D]

22. DARCY, HENRY. *Les Fontaines Publiques de la Ville de Dijon. Distribution d'Eau et Filtrage des Eaux.* Victor Dalmont, Paris (1856). [B]

23. HAUSSMANN, GEORGES EUGENE. *Documents Relatifs aux Eaux de Paris.* Dupont, Paris (1861). [E]

24. ARAGO, DOMINIQUE FRANCOIS. Report Made to the Academy of Science on the Filtering Apparatus of Henry de Fonvielle. Tr. by J. GRISCOM. *Jour. Franklin Inst.*, New Series, 22:206–212 (1838). [E]

25. ANON. Fonvielle's Filtering Apparatus. *Jour. Franklin Inst.*, New Series, 23:350–352 (1839). [E]

26. DELBRÜCK. Die Filtrazion des Wassers in Grossen. *Allgemeine Bauzeitung*, 18:103–129 (1853). [2 plates, Nos. 556 & 557 in separate atlas.] [D]

27. MALLET, CHARLES-FRANCOIS. *Notice Historique sur le Projet d'une Distribution Générale d'Eau à Domiciles dans Paris.* Carileau-Goeury, Paris (1830). [H]

28. KIRKWOOD, JAMES P. *Report on the Filtration of River Water, for the Supply of Cities, as Practiced in Europe.* D. Van Nostrand, New York (1869). [B]

29. IMBEAUX, EDOUARD. *Annuaire Statistique et Descriptif des Distributions d'Eau de France, Algerie, Tunisie.* C. Dunod, Paris (3rd ed., 1931). [D]

30. ———. Personal Letter. Nancy, France (August 20, 1934).

31. CLEMENCE, WALTER. The Water Supply of Marseilles and Trials of Filtering and Sterilizing Apparatus. *Eng.*, 91:139–142 (1911). [E]

32. HUMBER, WILLIAM. *A Comprehensive Treatise on the Water Supply of Cities and Towns, With Numerous Specifications of Existing Waterworks.* London (1876). [D]

33. DUPUIT, JULES. *Traité Théorique et Pratique de la Conduite et de la Distribution des Eaux.* A. Dunod, Paris (2nd ed., 1865). [B]

34. HAZEN, ALLEN. *Filtration of Public Water Supplies.* J. Wiley & Sons, New York (1st ed., 1895). [B]

35. METCALF, LEONARD & EDDY, HARRISON P. Disposal of Sewage. Vol. III, *American Sewerage Practice*. McGraw-Hill Book Co., New York (1914–15). [D]

36. IMBEAUX, EDOUARD. *Qualités de l'Eau.* Paris (1934). [X]

37. MARBOUTIN, M. Filtrées à Sable Non-submergés. *Le Génie Civil* (Paris), 54:247 (1908–09). [E]

38. ANON. Epuration Bactérienne des Eaux au Moyen des Sables Fins Non-submergés. *Le Génie Civil* (Paris), 45:220 (1904). [E]

38. ANON. Les Distributions d'Eau en France. L'Usine de Mézières-Sur-Couesnon. *L'Eau* (Paris), pp. 31–34 (March 1909). [E]

Chapter V

British Contributions to Filtration

1. PEACOCK, JAMES. Specifications of the Patent for His Invention of a New Method for the Filtration of Water and Other Fluids, December 23, 1791. *Repertory of Arts & Manufactures* (London), 11:221–225 (1799). [D]

2. ———. *A Short Account of a New Method of Filtration by Ascent: With Explanatory Sketches Upon Six Plates.* The Author, London (1793). [D]

3. GRAHAM, THOMAS. *Elements of Chemistry, Including the Applications of the Science in the Arts.* H. Bailliere, London (1850). [D]

4. CREBER, W. F. H. Personal Letter. Manchester, England (February 11, 1937).

5. WESTON, ROBERT SPURR. The Wheeler Filter Bottom. *Eng. News*, 72:22–24 (1914). [E]

6. DRYDEN, JOHN. *The Poetical Works of John Dryden.* (Aldine ed. of British Poets.) Bell & Daldy, London (1866). Vol. 2, p. 264. [D]

7. ANON. A Review of "A Short Account of a New Method of Filtration by Ascent," by JAMES PEACOCK. *Monthly Review* (London), Series II, 6:178–180 (1795). [D]

8. ANON. Sur les Filtres Pour la Purification de l'Eau. *Annales des Arts et Manufactures* (Paris), 13:288–305 (1804). [D]

9. SINCLAIR, JOHN. *The Code of Health and Longevity.* A. Constable & Co., Edinburgh (1807). [D]

10. CASHMORE, H. M. Personal Letter, supplying information re Boulton & Watt Collection. Birmingham Eng. Reference Library, Birmingham, England (1937). [X]

11. LEE, JAMES. Personal Letter. Paisley, Scotland (December 4, 1936).

12. COCHRANE, JOHN. Personal Letters. Glasgow, Scotland (1936, 1938).

13. TELFORD, BOULTON & WATT. Correspondence Between Telford, Boulton & Watt, 1800–1833. Birmingham Public Library, Birmingham, England (1937). [X]

14. MACKAIN, D. On the Supply of Water to the City of Glasgow. *Proc. Inst. C.E.*, 2:134–136 (1842–43). [E]

15. TELFORD, THOMAS. *Life of Thomas Telford.* Ed. by JOHN RICKMAN. Hansard, London (1838). [E]

16. GIBB, ALEXANDER. *The Story of Telford: The Rise of Civil Engineering.* A. Maclehose, London (1935). [D]

17. SIMPSON, JAMES. Discussion of "On the Supply of Water to the City of Glasgow," by D. MACKAIN. *Proc. Inst. C.E.*, 2:136–138 (1842–43). [E]

18. MALLET, CHARLES-FRANCOIS. *Notice Historique sur le Projet d'une Distribution Générale d'Eau à Domiciles dans Paris.* Carileau-Goeury, Paris (1830). [H]

19. DUPIN, CHARLES. *Two Excursions to the Ports of England, Scotland and Ireland in 1816 and 1818; Together With a Description of the Breakwater at Plymouth and of the Caledonian Canal.* W. Lewis, London (1819). [B]

20. ———. *The Commercial Power of Great Britain: Exhibiting a Complete View of the Public Works, Under the Several Heads: Streets, Roads, Canals, Aqueducts, Bridges, Coasts and Maritime Ports.* C. Knight, London (1825). [D]

21. MATTHEWS, WILLIAM. *Hydraulia.* Simpkin, Marshall & Co., London (1835). [D]

22. GALE, JAMES MORRIS. The Glasgow Water Works. *Trans. Seventh Session Inst. of Engineers, Scotland*, 2:21–72 (1863–64). [E]

23. STIRRAT, JAMES. Bleacher, Paisley Examined. General Board of

Health. Appendix II, pp. 77–87, *Report on the Supply of Water to the Metropolis* [Glasgow], Royal Commission on Metropolitan Water Supply (1850). [E]

24. GALE, WILLIAM. Letter and Plan. Appendix II, pp. 87–92, *Report on the Supply of Water to the Metropolis* [Glasgow], Royal Commission on Metropolitan Water Supply (1850). [E]

25. DARCY, HENRY. *Les Fontaines Publiques de la Ville de Dijon. Distribution d'Eau et Filtrage des Eaux.* Victor Dalmont, Paris (1856). [B]

26. GREAT BRITAIN RIVERS POLLUTION COMMISSION (1868). *Report of the Commissioners to Inquire Into the Best Means of Preventing the Pollution of Rivers.* George Edward Eyre & William Spottiswoode, London (1874). [E]

27. MARWICK, JAMES D. *Glasgow, the Water Supply of the City From the Earliest Period of Record.* Robert Anderson, Glasgow (1901). [B]

28. DODD, RALPH. *Observations on Water: With a Recommendation of a More Convenient and Extensive Supply of Thames Water to the Metropolis.* London (1805). [J]

29. BATEMAN, JOHN FREDERIC LA TROBE. *History and Description of the Manchester Water Works.* T. J. Day, Manchester (1884). [D]

30. HOPE, THOMAS C. & TELFORD, THOMAS. *Reports on the Means of Improving the Supply of Water for the City of Edinburgh, and on the Quality of the Different Springs in the Neighbourhood.* A. Constable, Edinburgh (1813). [E]

31. ANON. A New Mode of Forming Artificial Filters. *Glasgow Mechanics Magazine & Annals of Philosophy,* 3:7–9 (1825). [D]

32. THOM, ROBERT. *A Brief Account of the Shaws Water Scheme, and Present State of the Works.* Columbian Press, Greenock (1829). [B]

33. ———. On a Water Filter. *10th Report of the British Association for the Advancement of Science,* pp. 207–208 (August 1840). [E]

34. ANON. *Centenary of the Shaws Water Company's Works, 1827–1927.* James McKelvie & Sons, Greenock (1927). [E]

35. MCALISTER, JAMES. Personal Letter. Greenock, Scotland (December 14, 1936).

36. THOM, ROBERT. Principes de Filtration Applicables à l'Approvisionnement d'Eau des Grandes Villes. Lettre du 20 March 1829, de M. Thom à Sir Michel Shaw Stewart. Tr. by CHARLES-FRANCOIS MALLET. *Annales des Ponts et Chaussées,* Series I: 13:222–229 (1831). [D]

37. ANON. General Results of a Gardening Tour, in the Year 1831, From Dumfries by . . . Greenock to Paisley. *Gardener's Magazine* (London), 8:385–391 (1832). [D]

38. ANON. Shaws Water Works, Greenock. *Mechanics Magazine* (London), 17:305–313 (1832). [D]

39. BALDWIN, LOAMMI. *Report on the Subject of Introducing Pure Water Into the City of Boston.* John H. Eastburn, Boston (1834). [D]

40. STORROW, CHARLES S. *A Treatise on Water Works.* Hilliard, Gray & Co., Boston (1835). [D]

41. THOM, ROBERT. On the Means of Supplying Towns With Water at High Pressure. Vol. 1, pp. 17–22, *First Report of the Commission for Inquiring Into the State of Large Towns and Populous Districts, 1844.* Clowes & Sons, London (1844–45). [E]

42. SLOPER, B. G. Report on the Filtration of Water. Vol. I, pp. 162–166 (Appendix), *First Report of the Commission for Inquiring Into the State of Large Towns and Populous Districts, 1844.* Clowes & Sons, London (1844–45). [E]

43. HORSLEY, JOHN. *An Account of Several New Patent Processes for Purifying the Waters of Cities.* London (1849). [X]

44. TOMLINSON, CHARLES. *Cyclopedia of Useful Arts, Mechanical and Chemical Manufactures.* G. Virtue, London (1852). [D]
45. DELBRÜCK. Die Filtrazion des Wassers in Grossen. *Allgemeine Bauzeitung,* 18:103–129 (1853). [2 plates Nos. 556 & 557 in separate atlas.] [D]
46. ANON. Memoirs of Robert Thom. *Proc. Inst. C.E.,* 7:7–9 (1848). [E]
47. DUNBAR, WILLIAM. Personal Letter. Kilmarnock, Scotland (July 30, 1942).
48. HUMBER, WILLIAM. *A Comprehensive Treatise on the Water Supply of Cities and Towns With Numerous Specifications of Existing Waterworks.* London (1876). [D]
49. ANON. *The Dolphin; or Grand Junction Nuisance.* J. L. Cox, London (1827). [B]
50. ROYAL COMMISSION ON METROPOLITAN WATER SUPPLY. *Report of the Commission Appointed to Inquire Into the State of the Supply of Water in the Metropolis, 1828.* London (1828). pp. 1–155. [E]
51. SELECT COMMITTEE [Commons] ON THE SUPPLY OF WATER TO THE METROPOLIS. *Report of July 19, 1828.*
52. DAVIDSON, J. R. Personal Letters. London (1935 & 1939).
53. SIMPSON, CLEMENT P. Personal Letter. London (1937).
54. CHELSEA WATER WORKS. Extracts from *Court Minutes.* (November 1 & 15, 1827). [B]
55. ANON. Supply of Water to the Metropolis. *Royal Inst. of Great Britain Quarterly Jour. Science, Literature and the Arts* (London). pp. 354–355 (January–June 1829). [D]
56. WHITELY, DAVID. Personal Letter. Lincoln, England (1939).
57. TYSON, HENRY. Personal Letter. Boston, England (1939).
58. BRITTON, JOHN. *The Beauties of England & Wales.* Thomas Maiden, London (1807). Vol. 9, pp. 523–808. [D]
59. SIMPSON, JAMES. Evidence. Vol. 1, pp. 723–754, *Report of Committee [Commons] on the Metropolis Water Bill.* London (1851). [E]
60. ANON. An Early Chapter in the History of Water Filtration. *Eng. News-Rec.,* 105:609–610 (1930). [E]
61. HUMPHREYS, WILLIAM HENRY. *The York Water Works.* York [Eng.] Water Works (1910). [E]
62. HENDERSON, T. F. Personal Letter. Aberdeen, Scotland (November 10, 1936).
63. ANON. Memoir [to James Simpson]. *Proc. Inst. C.E.,* 98:440 (1888–89). [E]
64. ANON. Memoir [to Charles Liddell Simpson]. *Proc. Inst. M.E.,* 2:1080 (1925). [E]
65. SIMPSON, CHARLES LIDDELL. Discussion. *Proc. Inst. M.E.,* pp. 299–300 (April 1916). [E]
66. REES, ABRAHAM. *The Cyclopoedia: or Universal Dictionary of Arts, Sciences and Literature.* Longmans, London (1819–20). [D]
67. *The London Encyclopoedia or Universal Dictionary of Science, Art.* T. Tegg, London (1829). [D]
68. BOOTH, ABRAHAM. *A Treatise on the Natural and Chymical Properties of Water and on Various Mineral Waters.* George Wightman, London (1830). [B]
69. *Penny Cyclopedia of the Society for the Diffusion of Useful Knowledge.* C. Knight, London (1833–43). [B]
70. URE, ANDREW. *Dictionary of Arts, Manufactures and Mines.* Longmans, London (1839). [D]
71. *Encyclopedia Britannica.* A. & C. Black, Edinburgh (7th ed., 1842). [D]
72. CRESY, EDWARD. *An Encyclopedia of Civil Engineering.* London (1847). [D]
73. GRAHAM, MILLER & HOFMAN. Chemical Report on the Supply of Water to the Metropolis. *Jour. Chem. Soc.,* 4:375–413 (1852). [C]

74. HUGHES, SAMUEL. *A Treatise on Water Works for the Supply of Cities and Towns.* John Weale, London (1856). [B]
75. SMITH, ROBERT ANGUS. On the Air and Water of Towns. *18th Report of the British Assn. for the Advancement of Science,* pp. 16–31 (1848). [D]
76. ———. On the Air and Water of Towns. Action of Porous Strata, Water and Organic Matter. *21st Report of the British Assn. for the Advancement of Science,* pp. 66–77 (1851). [D]
77. WITT, HENRY M. On a Peculiar Power Possessed by Porous Media (Sand and Charcoal) of Removing Matter From Solution in Water. *The London, Edinburgh and Dublin Philosophical Magazine & Jour. of Science,* 4th Series, 12:23–34 (1846). [D]
78. BYRNE, EDWARD. Experiments on the Removal of Organic and Inorganic Substances in Water. *Proc. Inst. C.E.,* 26:544–555 (1866–67). [E]
79. FRANKLAND, EDWARD. On the Metropolitan Water Supply During the Year 1866. *Chem. News* (London), 15:151 (1867). [C]
80. ANON. Reports on the Organic Pollution of Drinking Waters. *British Medical Jour.,* 1:351 (1872). [A]
81. ANON. Report on the Action of Various Kinds of Filters on Drinking Water—Atkin's Moulded Carbon-Filter. *British Medical Jour.,* 2:338 (1872). [A]
82. CORFIELD, WILLIAM HENRY. *Water and Water Supply.* D. Van Nostrand, New York (1875). [D]
83. ANON. Filtration Not a Trustworthy Means for Purifying Drinking Water Contaminated With Choleraic Poison. *Chem. News* (London) 29:37 (1874). [C]
84. *Encyclopedia Britannica.* A. & C. Black, Edinburgh (9th ed., 1875–89). [D]
85. FRANKLAND, PERCY F. Water-Purification: Its Biological and Chemical Bases. *Proc. Inst. C.E.,* 85:197–219, Part 3 (1885–86). [E]
86. RIDEAL, SAMUEL. *Cantor Lectures on the Purification and Sterilisation of Water.* Trounce, London (1902). [E]

CHAPTER VI

Slow Sand Filtration in the United States and Canada

1. WHITFIELD, GEORGE H. Watering the City of Richmond. pp. 5–20, *Annual Report,* Richmond, Va. (1930). [D]
2. STEIN, ALBERT. Description of Water Works for Richmond, Va. pp. 21–32, *Annual Report,* Richmond, Va. (1930). [D]
3. DAVIS, JAMES L. A Brief History of the Origin and Erection of the Water Works of the City of Richmond. *Annual Report of the Superintendent of the City Water Works,* Richmond, Va. (1882–83). [E]
4. BALDWIN, LOAMMI. *Report on the Subject of Introducing Pure Water Into the City of Boston.* John H. Eastburn, Boston (1834). [D]
5. BOLLING, C. E. *Description of the Water Works of Richmond, Va., 1832–89.* (Pamphlet for the Richmond Water Committee.) Everett Waddey, Richmond (1889). [D]
6. ANON. Editorial tribute to Albert Stein. [From an unidentified newspaper published within a week of Stein's death, July 27, 1874.] [B]
7. SMITH, MARSDEN C. Plant Capacity Raised by Bettering Operation. *Eng. News-Rec.,* 112:688–689 (1934).
8. STORROW, CHARLES S. *A Treatise on Water Works.* Hilliard, Gray & Co., Boston (1835). [D]

9. DUNGLINSON, ROBLEY. *Human Health; or the Influence of Atmosphere, Food . . . on the Healthy Man.* Philadelphia (1844). [D]

10. ARAGO, DOMINIQUE FRANCOIS. Report Made to the Academy of Science on the Filtering Apparatus of Henry de Fonvielle. Tr. by J. GRISCOM. *Jour. Franklin Inst.*, New Series, 22:206–212 (1838). [E]

11. KIRKWOOD, JAMES P. *Report on the Filtration of River Waters, for the Supply of Cities, as Practiced in Europe.* D. Van Nostrand, New York (1869). [B]

12. NICHOLS, WILLIAM RIPLEY. Filtration of Potable Water. pp. 139–226, *Ninth Annual Report*, Massachusetts State Health Dept., Boston (1878). [D]

13. ———. *Water Supply Considered Mainly From a Chemical and Sanitary Standpoint.* J. Wiley & Sons, New York (1883). [D]

14. ———. Iron as a Material for Purifying Potable Water. *Sanitary Engr.*, 10:220 (1884). [E]

15. FANNING, JOHN THOMAS. *A Practical Treatise on Water Supply Engineering.* New York (1877). [D]

16. CROES, JOHN JAMES ROBERTSON. The Filtration of Public Water Supplies in America. *Eng. News*, 10:277–278 (1883). [E]

17. ———. *Statistical Tables From the History and Statistics of American Water Works.* Eng. News, New York (1885). [E]

18. ANON. Work at the Lawrence Experiment Station. pp. 34–66, *22nd Annual Report*, Massachusetts State Health Dept., Boston (1890). [D]

19. HAZEN, ALLEN. Some Physical Properties of Sand and Gravels, With Special Reference to Their Use in Filtration. pp. 539–556, *24th Annual Report*, Massachusetts State Health Dept., Boston (1892). [D]

20. HAZEN, ALLEN. *The Filtration of Public Water Supplies.* J. Wiley & Sons, New York (1st ed., 1895). [B]

21. WESTON, EDMUND B. Report on the Results Obtained With Experimental Filters. pp. 1–182, Appendix, *17th Annual Report*, Rhode Island State Board of Health. Freeman & Sons, Providence (1896). [D]

22. FULLER, GEORGE WARREN. *Report on the Investigations Into the Purification of the Ohio River at Louisville, Kentucky.* D. Van Nostrand, New York (1898). [B]

23. PITTSBURGH [Pa.] FILTRATION COMMISSION. *Report of January 1899.* [E]

24. FULLER, GEORGE WARREN. *Report on the Investigations Into the Purification of the Ohio River Water for the Improved Water Supply of the City of Cincinnati, Ohio.* Commercial Gazette Job Print, Cincinnati (1899). [B]

25. ANON. Water Power in the United States. *Tenth Census of the United States (1880).* Govt. Printing Office, Washington, D.C. (1887). [B]

26. *Manual of American Water Works.* Ed. by M. N. BAKER. Eng. News, New York (1888, 1889–90, 1891 & 1897).

27. ANON. The Jewell Mechanical Water Filter Plant at Wilkes-Barre, Pa. *Eng. News*, 35:330–332, 354–359 (1896). [E]

28. EDDY, ROBERT HENRY. *Report on the Introduction of Soft Water Into the City of Boston.* John H. Eastburn, Boston (1836). [D]

29. CARPENTER, GEORGE W. On Filtration to Remove Organic Growths From Water. pp. 39–45, *Annual Report*, Albany Water Commissioners (1875). [E]

30. MCALPINE, WILLIAM JARVIS. *Report Made to the Water Commissioners of the City of Albany, August 1, 1850, on the Proposed Projects for Supplying the City With Water.* Albany (1850). [E]

31. PHILADELPHIA WATERING COMMITTEE. *Report of May 23, 1854, With the Accompanying Reports of Frederic Graff on Filtration, and Professors Booth and Garrett on*

Schuylkill Water. Crissy & Markley, Philadelphia (1854). [D]

32. CHESBROUGH, E. S. New Water Works Intake vs. Filtration or Sedimentation. pp. 26–44, *19th Semi-Annual Report*, Chicago Board of Water Commissioners (1861). [E]

33. REYNOLDS, MYRON B. & GORMAN, ARTHUR E. Progress Toward Filtration in Chicago. *Jour. A.W.W.A.*, 24:965–983 (1932).

34. MEMBERS OF THE BUREAU OF ENGINEERING (Dept. of Public Works, Chicago). Conception, Design and Construction of the South District Filtration Project. *Eng. News-Rec.*, 128:995–1018 (1942). [E]

35. KIRKWOOD, JAMES P. Best Methods of Obtaining an Abundant Supply of Pure Water for the City of Cincinnati, 1865. *Report of the Water Supply Commission*, Cincinnati, Ohio (1865). [E]

36. MCALPINE, WILLIAM JARVIS. *Report on Supplying the City of Oswego With Water, Made to the Mayor and Common Council*. C. Morrison & Co., Oswego, N.Y. (1866). [D]

37. MCCAFFREY, W. A. Personal Letters. Oswego, N.Y. (April 26, 1940 & February 26, 1941).

38. MCALPINE, WILLIAM JARVIS. *Report of W. J. McAlpine, Civil Engineer, to the Water Commissioners of Schenectady*. Daily Union Printing House, Schenectady (1868). [E]

39. SAWYER, JOSEPH B. *Preliminary Report on the Water Supply for the City of Manchester (N.H.), Made to the Directors of the City Aqueduct Co., November 23, 1869.* J. B. Clarke, Manchester (1869). [D]

40. LOWELL (Mass.) WATER SUPPLY COMMITTEE. *Report of the Joint Special Committee on a Supply of Water for the City of Lowell, September 1869.* Marden & Rowell, Lowell (1869). [D]

41. ANON. Special Report on Mechanical Filtration. pp. 69–84. *16th Annual Report of the Lowell (Mass.) Water Board*, Lowell (1888). [D]

42. LESAGE, LOUIS. Filtering Beds. pp. 19–23, *Annual Report of Montreal Water Supply for the Year Ending December 31, 1875.* Louis Perrault & Co., Montreal (1876). [D]

43. U.S. PUBLIC HEALTH SERVICE. National Census of Water Treatment Plants of the United States. *Water Works Eng.*, 96:63–117 (1943).

44. POUGHKEEPSIE (N.Y.) WATER COMMITTEE. *Statement and Report on the Supply of the City of Poughkeepsie With Water, December, 1855.* Platt & Schram, Poughkeepsie (1855). [D]

45. ANON. The City of Poughkeepsie, Historical and Descriptive, October, 1889. (As published in the *Poughkeepsie Eagle.*) Platt & Platt, Poughkeepsie (1889). [D]

46. POUGHKEEPSIE (N.Y.) WATER COMMISSIONERS. *Annual Reports* (1870–95).
POUGHKEEPSIE (N.Y.) PUBLIC WORKS BOARD. *Annual Reports* (1896–1933). [D]

47. LAWLOR, THOMAS F. Personal Letter. Poughkeepsie, N.Y. (June 21, 1935).

48. RAND, J. B. G. *1st Report to the Board of Water Commissioners of the City of Poughkeepsie.* Isaac Platt & Sons, Poughkeepsie (1870).

49. ANON. Editorials. *Poughkeepsie Eagle* (July 30 & August 27, 1870). [X]

50. FOWLER, CHARLES E. The Filter Beds of Poughkeepsie, N.Y. *Eng. News*, 27:432 (1892). [E]

51. ———. The Operation of a Slow Sand Filter. *Jour. N.E.W.W.A.*, 12:209–244 (1897–98).

52. COLE, THOMAS A. Personal Letters. Poughkeepsie, N.Y. (1934–42).

53. OTIS, JOHN C. The Poughkeepsie Water Works. *Jour. N.E.W.W.A.*, 23:283–301 (1909).

54. PIRNIE, MALCOLM. Effluent Aerators Control Mechanical Filters. *Eng. News-Rec.*, 99:376–380 (1927). [E]

55. CONGER, W. H. Personal Letter. Poughkeepsie, N.Y. (June 10, 1935).

56. O'HARA, M. J. Hudson's Modern Water System. (75th Anniversary Edition) *Hudson Daily Star*, Hudson, N.Y. (December 1923). [X]

57. MCALPINE, WILLIAM JARVIS. *Report to the Water Commissioners of the City of Hudson*. Bryan & Webb, Hudson, N.Y. (1872). [E]

58. BRADBURY, ANNA ROSSMAN. *History of the City of Hudson, N.Y.* Record Print & Publishing Co., Hudson, N.Y. (1908).

59. HUDSON (N.Y.) WATER COMMISSIONERS. *Annual Reports* (1872–73 to 1894–95). [X]

60. ROSSMAN, CLARK G. *Municipal Water Supply of Hudson, N.Y.* Hudson, N.Y. (1900). [X]

61. WARDLE, J. MCCLURE. Personal Letters. Hudson, N.Y. (1934–35; 1941–42).

62. *The Caledonian*. St. Johnsbury, Vt. (1837). [X]

63. BOUSFIELD, E. H. Personal Letter. St. Johnsbury, Vt. (October 19, 1934).

64. OREBAUGH, R. W. Personal Letter. St. Johnsbury, Vt. (November 4, 1933).

65. PERHAM, JOHN M. Personal Letters. St. Johnsbury, Vt. (1934–42).

66. WHEELER, ROBERT C. Personal Letters. Albany, N.Y. (1934).

67. FAIRBANKS, CORNELIA TAYLOR. Personal Letters. St. Johnsbury, Vt. (1934–35).

68. (a) FAIRBANKS, EDWARD TAYLOR. *The Town of St. Johnsbury, Vt., A Review of One Hundred and Twenty-Five Years to the Anniversary Pageant 1912*. The Cowles Press, St. Johnsbury (1914). [D]
    (b) CHENEY, R. G. *Re* St. Johnsbury. Personal Letter. Springfield, Mass. (January 1, 1935).

69. MOAT, C. P. Personal Letters. Burlington, Vt. (1934–42).

70. ANON. Water Purification in America—Aeration & Continuous Sand Filtration at Ilion, N.Y. *Eng. News*, 31:466–468 (1894). [E]

71. ANON. Water Purification in America—Aeration & Continuous Sand Filtration at Nantucket, Mass. *Eng. News*, 31:336 (1894). [E]

72. ANON. Water Purification in America—Lawrence, Mass. *Eng. News*, 30:97 (1893). [E]

73. ANON. Water Purification in America—Aeration & Intermittent Sand Filtration at Mount Vernon, N.Y. *Eng. News*, 32:155 (1894). [E]

74. ANON. Water Purification in America—Covered Sand Filter Beds at Grand Forks, N.D. *Eng. News*, 33: 341–342 (1895). [E]

75. DANCKAERTS, JASPER. *Journal of a Voyage to New York and a Tour in Several of the American Colonies in 1679–80 by Jasper Dankers and Peter Sluyter*. Tr. from the original manuscript in Dutch for the Long Island Historical Society and ed. by HENRY C. MURPHY. Long Island Historical Society, Brooklyn (1867). [D]

76. MUNSELL, JOEL. *The Annals of Albany*. Joel Munsell, Albany (1869). Vol. 1, pp. 43–63; Vol. 10, p. 255. [D]

77. REYNOLDS, SCHUYLER. *Albany Chronicles: A History of the City Arranged Chronologically From the Earliest Settlement to the Present Time*. J. B. Lyon & Co., Albany (1906). [X]

78. WEISE, ARTHUR JAMES. *The History of the City of Albany, N.Y., From the Discovery of the Great River in 1524 to the Present Time*. E. H. Bender, Albany (1884). [D]

79. KALM, PEHR. *Travels Into North America: Containing Its Natural History*. Tr. by JOHN REINHOLD FORSTER. The Editor, London (1770–71). [D]

80. MORSE, JEDEDIAH. *The American Geography; or a View of the Present Situation of the United States of America*. S. Kollock, Elizabethtown (1789). [D]

81. ANON. Albany Water Works. A Hundred Years and More Ago. *Eng. News-Rec.*, 83:605 (1919). [E]

BIBLIOGRAPHY—CHAPTER VI 485

82. HORTON, THEODORE. *History of the Albany Water Works.* Unpublished MS. (n.d.) [X]

83. CUSHMAN, W. M'CLELLAND. *Report to the Albany Water Commissioners on the Aqueducts, etc., for Supplying the City of Albany With Water From the Mohawk River and From the Hudson River.* W. & A. White, Albany (1842). [E]

84. DOUGLASS, D. B. Report of March 19, 1846. In *A Plan and Estimate for Supplying the City With Water.* Report of Special Committee of Albany Common Council. Weed & Parsons, Albany (1846). [E]

85. CLAXTON, F. S. Report on Hudson River, Patroon's Creek and Mohawk River. Unidentified newspaper clipping (February 5, 1889). [E]

86. PERKINS, WILLIAM A. Report on Norman and Hunger Kills. Unidentified newspaper clipping (July 1849). [E]

87. ALBANY WATER COMMISSIONERS. *Annual Report.* Albany (1851). [E]

88. MCALPINE, WILLIAM JARVIS. *Report Made to the Water Commissioners of the City of Albany, August 1, 1850, on the Proposed Projects for Supplying the City With Water.* W. H. VanDyck, Albany (1850). [D]

89. ALBANY WATER COMMISSIONERS. *Report to the Common Council of the City of Albany Transmitting the Reports of James P. Kirkwood and C. F. Chandler for Procuring an Additional Supply of Water for the City of Albany.* Weed & Parsons, Albany (1872). [D]

90. CHANDLER, CHARLES FREDERICK. *Report on the Waters of the Hudson River, Made to the Water Commissioners of the City of Albany, N.Y.* Albany (1872). [E]

91. ALBANY INSTITUTE. Report on the Water Supply of the City of Albany, May 21, 1872. *Trans. Albany Inst.*, 8:218–227 (1872). [E]

92. HOGAN, PETER. *The Water Supply of the City of Albany, Together With the Report of the Special Committee and Resolutions to the Common Council.* Munsell, Albany (1873). [E]

93. NEW YORK STATE DEPARTMENT OF HEALTH. Report on the Albany Basin. Weed & Parsons, Albany (1885). [E]

94. FANNING, JOHN THOMAS. *Report on a Water Supply for New York and Other Cities of the Hudson Valley.* Two pamphlets (1881, 1884). [E]

95. (a) ANON. Water Supply From Lake George. *Eng. News*, 9:80; 104 (1882). [E]
    (b) ANON. The Adirondack Water Supply for New York and the Hudson Valley Cities. *Eng. News*, 13:58 (1885). [E]

96. CHANDLER, CHARLES FREDERICK. *Report on the Waters of the Hudson River, Together With an Analysis of the Same. Made to the Water Commissioners of the City of Albany.* Trowe's Printing & Book Binding Co., Albany (1885). [D]

97. MASON, WILLIAM P. *Report on the Albany Water Supply, Made to the Board of Health of the City of Albany.* Argus, Albany (1885). [D]

98. LEEDS, ALBERT RIPLEY. *Chemical, Biological and Experimental Inquiry Into the Present and Proposed Future Water Supply of the City of Albany.* Pohemus, New York (1886). [E]

99. ———. Discussion of "Filtration," by S. A. CHARLES. *Proc. A.W.W.A.*, pp. 124–129 (1896).

100. HAZEN, ALLEN. Report to the City of Albany Recommending That the Water Supply Be Filtered and the Construction of a Sand Filtration Plant. *46th Annual Report of the Water Commissioners of the City of Albany.* Albany (1896). [E]

101. ANON. The New Sedimentation Basin and Masonry-Covered Sand Filter Beds at Albany, N.Y. *Eng. News*, 39:91–92 (1898). [E]

102. ANON. Progress on the Water Purification Plant at Albany. *Eng. News*, 40:254 (1898). [E]

103. HAZEN, ALLEN. The Albany Water Filtration Plant. *Trans. A.S.C.E.*, 40:244–352 (1900). [E]
104. GREENALCH, WALLACE. Experiments on Double Filtration at Albany. *59th Annual Report of the Bureau of Water*. Albany (1908–09). [E]
105. HORTON, THEODORE. Investigations of the Water Supply of Albany. Vol. 2, pp. 153–168, *Annual Report*, New York State Dept. of Health (1920). [E]
106. ANON. Plan for New Water Supply for Albany Adopted. *Eng. News-Rec.*, 97:520 (1926). [E]
107. ANON. Court Holds Albany Water Caused Case of Typhoid Fever. *Eng. News-Rec.*, 101:221 (1928). [E]
108. HILL, NICHOLAS S. Report With Recommendations for an Improved Water Supply for the City of Albany. Unpublished report (October 14, 1926). [X]
109. HORTON, ROBERT ELMER. Report on Hannacrois-Catskill Gravity Water Supply for the City of Albany, to the Board of Water Supply. Unpublished report (October 11, 1926). [E]
110. ——— & SMITH, BENJAMIN L. The New Gravity Water Supply for Albany. *Jour. N.E.W.W.A.*, 45:89–135 (1931).
111. CHANDLER, H. C. Personal Letter. Albany, N.Y. (September 28, 1942).
112. HORTON, THEODORE. The Typhoid Outbreak at Albany, Due to Flooded Filters. *Eng. News*, 69: 1020–1023 (1913). [E]
113. GREENALCH, WALLACE. The Flooding of the Albany Filtration Plant and Previous High Floods at Albany. *Eng. News*, 69:754 (1913).
114. WILLCOMB, GEORGE E. Personal Letter. Albany, N.Y. (June 1935).
115. BERRY, A. E. Personal Letter. Toronto, Ont. (1942).

CHAPTER VII

Inception and Widespread Adoption of Rapid Filtration in America

1. GARDNER, L. H. The Clarification and Purification of Public Water Supply. *Scientific American Suppl.*, 20:8146–8147 (1885). [D]
2. GILL, HENRY. Ueber Filter-Anlagen zur Wasserversorgung mit Besonderem Bezug auf Berlin. *Deutsche Bauzeitung* (December 17, 1881). [F]
3. RICHOU, G. Appareils de Filtrage. *Le Génie Civil* (Paris), 2:545–546 (1881–82). [E]
4. ANON. Water Power in the United States. *Tenth Census of the United States (1880)*. Govt. Printing Office, Washington, D.C. (1887). [B]
5. CROES, JOHN JAMES ROBERTSON. Statistical Tables From the History and Statistics of American Water Works. Eng. News, New York (1885). [E]
6. ———. The History and Statistics of American Water Works. *Eng. News*, 8:413 (1881); 9:325 (1882). [E]
7. NEWARK FILTERING CO. Proposal to Albany (N.Y.) Common Council. pp. 616–617, *Proc. Common Council*, Albany (1885). [E]
8. ———. Proposal to Use Iron and Lime Instead of Alum as a Coagulant. p. 577, *Proc. Common Council*, Albany (1885).
9. ANON. The Multifold Water Filter. *Eng. News*, 9:1 (1882). [E]
10. ANON. Improved Filter. *Scientific American*, 46:4 (1882). [D]

BIBLIOGRAPHY—CHAPTER VII

11. NICHOLS, WILLIAM RIPLEY. Filtration at Berlin. *Sanitary Engr.*, 5: 299 (1882). [E]

12. NEWARK FILTERING CO. Letter re "Filtration at Berlin," by WILLIAM RIPLEY NICHOLS. *Sanitary Engr.*, 5: 384 (1882). [E]

13. HAZEN, ALLEN. *The Filtration of Public Water Supplies.* J. Wiley & Sons, New York (1st ed., 1895). [B]

14. WATSON, HAROLD E. Personal Letter. Newport (June 1936).

15. Extracts From Official Records of Rahway, N.J. (1938). [X]

16. *Manual of American Water Works.* Ed. by M. N. BAKER. Eng. News, New York (1889-91).

17. RAHWAY (N.J.) WATER COMMISSIONERS. *Minute Books.* Rahway (1876-80). [X]

18. MAHAFFY, J. LYNN. Personal Letter. Trenton, N.J. (December 12, 1933).

19. BATY, J. B.; CROFT, H. P.; GASTON, HUGH K.; & MILLER, LOUIS L. Extracts From the Minutes of the Somerville Water Co. [X]

20. ANON. The Hyatt Water Filtering Plant at Long Branch, N.J. *Eng. News*, 20:280 (1888). [E]

21. WICKERSHAM, L. E. Personal Letter. New York (January 3, 1941).

22. WEIR, PAUL. Personal Letter. Atlanta, Ga. (1942).

23. RICHARDS, WILLIAM. Practical Results of Mechanical Filtration. *Proc. A.W.W.A.*, pp. 148-153 (1888).

24. ———. The New Water Supply for Atlanta, Ga. *Proc. A.W.W.A.*, pp. 30-33 (1892).

25. WIEDMAN, HERMAN F. & WEIR, PAUL. The Modernization and Enlargement of Atlanta's Filter Plant. *Water Works & Sewerage*, 88:435-442 (1941). [D]

26. DECOSTA, J. D. Orinda Filtration Plant of the East Bay Municipal Utility District, Oakland, California. *Jour. A.W.W.A.*, 28:1551-1570 (1936).

27. WARREN, JOHN A. Excerpts from an unpublished diary (1884-89).

28. WARREN, JOSEPH E. Personal Letters. Cumberland Mills, Maine (1937-42).

29. ANON. The Warren Water Filter. *Eng. News*, 18:361-362 (1887). [E]

30. ANON. Water Purification in America—The Warren Mechanical Filter. *Eng. News*, 31:6-8 (1894). [E]

31. LEEDS, ALBERT RIPLEY. Discussion of "Filtration," by S. A. CHARLES. *Proc. A.W.W.A.*, pp. 124-129 (1896).

32. ANON. Incomplete Filter Systems—Fouling and Failure. *Sanitary Era*, 3:141-143 (1889). [E]

33. ANON. The National Filter. *Fire & Water*, 1:229 (1887). [X]

34. ANON. The Jewell Mechanical Water Filter Plant at Wilkes-Barre, Pa. *Eng. News*, 35:330-332, 354-359 (1896). [E]

35. ANON. The National Filter Plant at Terre Haute, Ind. *Eng. News*, 25:127 (1891). [E]

36. WILLIAMSON, L. L. Personal Letter. Terre Haute, Ind. (1931).

37. DURBIN, W. H. Personal Letter. Terre Haute, Ind. (October 10, 1941).

38. TAYLOR, W. E. Personal Letter. Terre Haute, Ind. (February 1941).

39. NEW ORLEANS SEWERAGE & WATER BOARD. *Semi-Annual Report.* New Orleans (1899-1900). [D]

40. ANON. The Mechanical Filter Suit Brought by the National Water Purifying Co. Against New Orleans Water Works Co. *Eng. News*, 34: 349, 385 (1895). [E]

41. ANON. National Water Purifying Co. vs. New Orleans Water Works Co. p. 773, *48th Louisiana Annual.* [X]

42. EARL, GEORGE G. Personal Letter. New Orleans, La. (June 26, 1896).

43. ANON. Mechanical Filters at New Orleans and Proposed Filter Plant for Minneapolis. *Eng. News*, 32: 174-175 (1894). [E]

44. FRIEDRICHS, CARL C., JR. Purification Methods at New Orleans. *Jour. A.W.W.A.*, 28:537–541 (1936).
45. BLAKE, PERCY M. Report to the Committee of Investigation of the Public Water Supply of Athol, Mass. (May 4, 1896). [X]
46. JEWELL, IRA H. Personal Letter. Chicago, Ill. (August 8, 1936).
47. ANON. Another Court Decision Regarding the Use of Coagulant in Mechanical Filters. *Eng. News*, 38: 221 (1897). [E]
48. WHEELER, ROBERT C. Personal Letter. Albany, N.Y. (September 1942).
49. CAIRD, JAMES M. Personal Letter. Troy, N.Y. (September 1942).
50. ANON. Settlement Between O. H. Jewell Filter Co. and New York Filter Mfg. Co. *Eng. News*, 39: 137 (1898). [E]
51. JEWELL, IRA H. Development of Rapid Sand Filtration to Increase Capacity. *Jour. A.W.W.A.*, 30:817–826 (1938).
52. ANON. The Removal of Iron From Artesian Well Water by Mechanical Filtration. *Eng. News*, 35:364–367 (1896). [E]
53. HODKINSON, GEORGE F. Personal Letter. Philadelphia, Pa. (April 16, 1941).
54. ANON. Bacterial Test of a Mechanical Filter at Louisiana, Mo. *Eng. News*, 42:318 (1899). [E]
55. ANON. Water Purification at Vincennes, Ind. *Eng. News*, 43:291–293 (1900). [E]
56. Defendant's Testimony—Ira H. Jewell *vs.* City of Minneapolis. U.S. Circuit Court (December 12, 1912). [X]
57. FULLER, GEORGE WARREN. The Filtration Works of the East Jersey Water Co. at Little Falls, N.J. *Proc. A.S.C.E.*, 29:153–202 (1903). [E]
58. LOUISVILLE WATER CO. *Annual Reports.* Louisville (1856–). [X]
59. LOUISVILLE WATER CO. Proposal of Extension of the Louisville Water Works Transmitted to the Mayor and Council of Louisville (March 29, 1876). [X]
60. FULLER, GEORGE WARREN. *Report on the Investigations Into the Purification of the Ohio River Water at Louisville, Ky.* D. Van Nostrand, New York (1898). [B]
61. SOPER, GEORGE A. *The Efficiency of the Warren Mechanical Filter as Developed at the Experiment Station of the Louisville Water Co.* Cumberland Mfg. Co., Boston (n.d.). [X]
62. ANON. A Proposed New System of Mechanical Filtration for Louisville, Ky. *Eng. News*, 45:52–53 (1901). [E]
63. RUSSELL, S. BENT. Personal Letter. St. Louis, Mo. (March 12, 1934).
64. ANON. Important Changes in the Water Filtration Plant of Louisville, Ky. *Eng. News*, 60:314 (1908). [E]
65. LOVEJOY, W. H. Reconstruction of Filters at Louisville, Ky. *Jour. A.W.W.A.*, 14:352–356 (1925).
66. ———. Filter Troubles Caused by Micro-organisms at Louisville, Ky. *Eng. News*, 64:664 (1910). [X]
67. STOVER, FREDERICK H. Micro-organism Troubles in the Operation of Mechanical Filters. *Proc. A.W.W.A.*, pp. 457–473 (1913).
68. LOVEJOY, W. H. Algae Control by Creating Turbidity at Louisville, Ky. *Eng. News-Rec.*, 101:505–507 (1928). [E]
69. CARPENTER, LEWIS V. Further Instances of Use of Artificial Turbidity for Algae Control. *Eng. News-Rec.*, 101:852 (1928). [E]
70. LEISEN, THEODORE A. *Souvenir of the Louisville Water Co.* J. P. Morton & Co., Louisville (1911). [D]
71. LOVEJOY, W. H. Personal Letter. Louisville, Ky. (August 26, 1942).
72. ANON. Typhoid in the Large Cities of the United States in 1940. *Jour. American Medical Assn.*, 118: 222 (1942); abstracted, *Jour. A.W. W.A.*, 34:449–457 (1942).

BIBLIOGRAPHY—CHAPTER IX 489

73. FULLER, GEORGE WARREN. Personal Letter. New York (March 12, 1934).
74. BENZENBERG, GEORGE H. Report to the Board of Trustees. A Brief History of the Old Water Works, 1897–1909. Ebbert & Richardson Co., Cincinnati (1909). [D]
75. FULLER, GEORGE WARREN. Report on the Investigations Into the Purification of the Ohio River Water for the Improved Water Supply of the City of Cincinnati, Ohio. Commercial Gazette Job Print, Cincinnati (1899). [B]
76. BAHLMAN, CLARENCE. Personal Letters. Cincinnati, Ohio (1934–42).
77. JOHNSON, GEORGE A. The Typhoid Toll. Jour. A.W.W.A., 3:249–326 (1916).
78. ST. LOUIS BOARD OF WATER COMMISSIONERS. Annual Reports. St. Louis (1865–). [D]
79. KIRKWOOD, JAMES P. Report on the Filtration of River Water, for the Supply of Cities, as Practiced in Europe. D. Van Nostrand, New York (1869). [B]
80. SEDDONS, JAMES A. Clearing Water by Settlement. Eng. News, 22:607–609 (1889). [E]
81. GRAF, AUGUST V. Notes sent to the author through the courtesy of Edward E. Wall, Director of Utilities. St. Louis, Mo. (n.d.). [X]

CHAPTER VIII

Upward Filtration in Europe and America

1. CROES, JOHN JAMES ROBERTSON. Methods of Filtration of Water in Use in the United States and Europe, With Details of Construction, Cost and Efficiency. Proc. A.W.W.A., pp. 65–72 (1883).
2. ANON. The Pawtucket (R.I.) Water Works: General Description. Eng. News, 21:493–504 (1889). [E]
3. HARDING, THOMAS E. Personal Letter. Pawtucket, R.I. (March 11, 1938).
4. ANON. Filter Bed at Storm Lake (Iowa). Eng. News, 2:519 (1891).
5. ANON. Coagulation, Sedimentation and Upward Filtration at Bartlesville, Okla. Eng. News, 9:212 (1894). [X]
6. PERKINS, C. E. Personal Letter. Bartlesville, Okla. (1933).
7. HUDDLESTON, THOMAS. Personal Letter. Grange-over-Sands, England (September 20, 1939).
8. BRACKETT, F. W. & Co. The Brackett Upward-Flow Filter. Colchester, England (Latest date of testimonials 1935). [X]

CHAPTER IX

Multiple Filtration: Seventeenth to Twentieth Centuries

1. BACON, FRANCIS. Sylva Sylvarum, or a Natural History in Ten Centuries. W. Lee, London (1670). [D]
2. ROCHON, M. Memoir on the Purification of Sea-Water, and on Rendering It Drinkable Without Any Empyreumatic Taste by Distilling It in Vacuo. (From the Jour. de Physique.) Repertory of Arts, Manufactures and Agriculture, Second Series (J. J. Wyatt, London), 24:297–302 (1814). [D]
3. PORTIUS, LUCAS ANTONIUS. The Soldier's Vade Mecum; or the Method

of *Curing the Diseases and Preserving the Health of Soldiers.* R. Dodsly, London (1747). [B]

4. ANON. Gerson's System of Filtration. *Eng.*, 44:534 (1887). [E]

5. DELBRÜCK. Die Filtrazion des Wassers in Grossen. *Allgemeine Bauzeitung*, 18:103–129 (1853). [2 plates Nos. 556 & 557 in separate atlas.] [D]

6. WILLIAMS, JOHN. *An Historical Account of Sub-ways in the British Metropolis for the Flow of Pure Water and Gas, Including the Projects in 1824 and 1825.* London (1828). [D]

7. KIRKWOOD, JAMES P. *Report on the Filtration of River Waters, for the Supply of Cities, as Practiced in Europe.* D. Van Nostrand, New York (1869). [B]

8. BOOTH, T. Discussion of "Antwerp Water Works," by WILLIAM ANDERSON. *Proc. Inst. C.E.*, Part II, 72:78 (1882–83). [E]

9. GOETZ, EUGEN. Filtration for Public Water Supplies With Especial Reference to the Double Filtration Plant at Bremen, Germany. *Trans. A.S.C.E.*, 53:210–217 (1904). [E]

10. CARRIERE, J. E. De Voorbehandeling Bij Langzame Zandfiltratie van Rivierwater. Korthuis, 's-Gravenhage, Netherlands (n.d.). [B]

11. LÜSCHER, O. Personal Letters. Zurich, Switzerland (November 1939 & March 1940).

12. DAVIDSON, J. R. Personal Letters. London (1935–39).

13. BARNES, A. A. Personal Letter. Birmingham, England (May 18, 1942).

14. BIRMINGHAM (England) WATER DEPARTMENT. *Description of Water Supply From Wales.* Birmingham (1908). [X]

15. VANLOAN, SETH M. Personal Letters. Philadelphia, Pa. (1937, 1939, 1940).

16. CLARK, HARRY W. Double Sand Filtration of Water at South Norwalk, Conn.—Removal of Organisms, Tastes and Odors. *Jour. N.E. W.W.A.*, 30:86–100 (1916).

17. BRACKEN, ELMER F. Personal Letter. Norwalk, Conn. (1940).

18. FIELD, FREDERIC E. Personal Letter. Montreal, Que. (1933 & 1939).

CHAPTER X

Drifting-Sand Rapid Filters

1. BOLLMANN FILTER GESELLSCHAFT. Personal Letters. (January 24 & March 9, 1934).

2. ANON. A Large Drifting Sand Water Filtration Plant for Toronto, Ont. *Eng. News*, 71:1446–1448 (1914). [E]

3. ANON. Drifting Sand Water Filters for Toronto, Ont. *Eng. News*, 76: 556–570 (1916). [E]

4. KIRKPATRICK, W. Personal Letter. Kingston, Jamaica, B.W.I. (October 20, 1939).

5. HANSON, TERENCE C. Personal Letter. Pernambuco, Brazil (February 28, 1940).

6. BAITY, H. C. Personal Letter. Rio de Janeiro, Brazil (1943).

7. MILLER, LEONARD. Personal Letter. London (January 20, 1942).

8. BRAMWELL, CHARLES B. Personal Letter. Belfast, Ireland (January 13, 1925).

9. ANON. Les Distributions d'Eau en France. L'Usine de Mézières-sur-Couesnon. (Alimentation en Eau de la Ville de Rennes.) *L'Eau* (Paris), pp. 31–34 (March 1939). [E]

Chapter XI

Natural Filters: Basins and Galleries

1. IMBEAUX, EDOUARD. *Annuaire Statistique et Descriptif des Distributions d'Eau de France, Algerie, Tunisie.* C. Dunod, Paris (3rd ed., 1931). [D]
2. D'AUBUISSON DE VOISINS, JEAN FRANCOIS. Histoire de l'Etablissement des Fontaines à Toulouse. *Histoire et Mémoires de l'Académie Royal des Sciences de Toulouse,* Series 2, Part 1, 2:159–400 (1823–27). [D]
3. KIRKWOOD, JAMES P. *Report on the Filtration of River Waters, for the Supply of Cities, as Practiced in Europe.* D. Van Nostrand, New York (1869). [B]
4. DAVIES, B. W. Personal Letters. Nottingham, England (1938–39).
5. HAWKSLEY, THOMAS. Description of Filter Basin and Gallery, Notingham. Vol. 2, pp. 52–53, *First Report of the Commission for Inquiring Into the State of Large Towns and Populous Districts.* Clowes & Sons, London (1844–45). [E]
6. CYRIL, WALMESLEY. Personal Letter. Perth, Scotland (n.d.).
7. KINGSBURY, FRANCIS H. Public Ground Water Supplies in Massachusetts. *Jour. N.E.W.W.A.,* 50: 149–196 (1936).
8. BURDICK, CHARLES B. Infiltration Galleries at the Des Moines, Iowa, Water Works. *Jour. N.E.W.W.A.,* 38:203–218 (1924).
9. DUMONT, ARISTIDE GEORGES. *Les Eaux de Nimes, de Paris et de Londres.* C. Dunod, Paris (1874). [D]
10. DEBAUVE, ALPHONSE ALEXIS & IMBEAUX, EDOUARD. *Distributions d' Eau.* Dunod, Paris (1905–06). [E]
11. GIESELER, E. A. A New Form of Filter Gallery at Nancy, France. *Eng. Rec.,* 51:148 (1905). [E]
12. FAWKES, A. W. ELLSON. River Water Supply for Moose Jaw, Saskatchewan. *American City,* 55:40 (1940). [D]
13. NICHOLS, WILLIAM RIPLEY. Filtration of Potable Water. p. 226, *9th Annual Report,* State Board of Health of Massachusetts, Boston (1878). [D]
14. FANNING, JOHN THOMAS. *A Practical Treatise on Water Supply Engineering.* New York (1877). [D]
15. ANON. Water Power in the United States. *Tenth Census of the United States (1880).* Govt. Printing Office, Washington, D.C. (1887). [B]
16. LAWRENCE, ROBERT L. Personal Letters. Nashville, Tenn. (1938).
17. BENN, GEORGE S. Personal Letters. Nashville, Tenn. (1938).
18. ANON. The Filter Galleries at Painesville, Ohio. *Eng. Rec.,* 43: 518 (1901). [D]
19. WARING, F. H. Personal Letter. Columbus, Ohio (April 3, 1940).
20. WATSON, JOHN. Abstraction of Water From a Stream by Infiltration. *Water & Water Eng.,* 34:151–152 (1932). [D]
21. ———. Personal Letters. Edinburgh, Scotland (1939).

Chapter XII

Plain Sedimentation

1. WEGMANN, EDWARD. The Water Works of Laodicea, Asia Minor. *Eng. Rec.,* 40:354–355 (1899). [E]
2. IDRISI, MUHAMMAD IBN MUHAMMAD AL. *Description de L'Afrique et de l'Espagne par Edrisī.* Tr. by R. DOZY & M. J. DE GOEJE. E. J. Brill, Leyde (1866). [D]

3. SHAW, THOMAS. *Travels or Observations Relating to Several Parts of Barbary and the Levant.* J. Ritchie, Edinburgh (1738). [D]

4. DAVIS, N. *Carthage and Her Remains.* Harper, New York (1861). [B]

5. CROES, JOHN JAMES ROBERTSON. The Water Works of Carthage. *Eng. Rec.*, 25:8 (1891). [E]

6. D'AUDOLLANT, AUGUSTE MARIE HENRI. *Carthage Remains: 146 Avant Jesus-Christ—698 Après Jesus-Christ.* Fontemoing, Paris (1901). [D]

7. IMBEAUX, EDOUARD. *Annuaire Statistique et Descriptif des Distributions d'Eau de France, Algerie, Tunisie.* C. Dunod, Paris (3rd ed., 1931). [D]

8. FRONTINUS, SEXTUS JULIUS. *The Two Books on the Water Supply of the City of Rome.* Tr. by CLEMENS HERSCHEL. Dana Estes & Co., Boston (1899). [D]

9. FABRETTI, RAPHA. *De Aquis et Aquaeductibus Veteris Romae. Dissertationes Tres.* I. B. Bufsotti, Rome (1680). [B]

10. VITRUVIUS, POLLIO. *The Ten Books on Architecture.* Tr. by M. H. MORGAN. Harvard Univ. Press, Cambridge (1914). [D]

11. CAESAR, C. JULIUS. *Caesar's Commentaries on the Gallic War, With the Supplementary Books Attributed to Hirtius, Including the Alexandrian War.* Tr. by W. A. McDEVITTE & W. S. BOHN. American Book Co., New York (190?). [G]

12. CORFIELD, WILLIAM HENRY. *Water and Water Supply.* D. Van Nostrand, New York (1875). [D]

13. ENGLEFIELD, HENRY C. On the Purification of Water by Filtration. *Jour. of Natural Philosophy, Chemistry & the Arts*, 9:95–97 (1804). [C]

14. STEIN, ALBERT. *Description of Works. Report of the Watering Committee to the Council of the Corporation of Lynchburg Relative to the Water Works, March 24, 1830.* Fletcher & Toler, Lynchburg, Va. (1830). [D]

15. *Manual of American Water Works.* Ed. by M. N. BAKER. Eng. News, New York (1891).

16. ARAGO, DOMINIQUE FRANCOIS. Report Made to the Academy of Sciences on the Filtering Apparatus of Henry de Fonvielle. Tr. by J. GRISCOM. *Jour. Franklin Inst.*, New Series, 22:206–212 (1838). [E]

17. BALDWIN, GEORGE R. The Quebec Water Works. (Extracts from the report of preliminary examination made by the engineer.) *Eng. News*, 5:148–150, 156–158, 163–166, 173–174 (1878). [E]

18. *Manual of American Water Works.* Ed. by M. N. BAKER. Eng. News, New York (1888).

19. CROES, JOHN JAMES ROBERTSON. *Statistical Tables From the History and Statistics of American Water Works.* Eng. News, New York (1885). [E]

20. PEARSON, GALEN W. Subsidence and Filtration of Western River Waters, and Availability of Natural Flow Into Filtering Wells and Galleries. *Proc. A.W.W.A.*, pp. 91–93 (1883).

21. PEARSONS, GALEN W. The Water Works of Kansas City. *Eng. News*, 18:328–329, 345–346, 380–400, 437, 454 (1887). [E]

22. KIERSTED, WYNKOOP. New Settling Basin of the Water Works of Kansas City. *Eng. News*, 67:371–373 (1912). [E]

23. GILKINSON, GEORGE F. Personal Letters. Kansas City, Mo. (January 1932 & June 1940).

24. SEDDONS, JAMES A. Clearing Water by Settlement: Observations and Theories. *Eng. News*, 22:607–609 (1889); 23:31–33, 98–99 (1890). [E]

25. HAZEN, ALLEN. On Sedimentation. *Trans. A.S.C.E.*, 53:45–71 (1904). [E]

26. U.S. WORKS PROGRESS ADMINISTRATION (New York City). *Bibliography on Coagulation and Sedimentation in Water & Sewage Treatment.* New York (1938). [B]

Chapter XIII

Coagulation: Ancient and Modern

1. ALPINO, PROSPERO. *Medicina Aegyptorum.* Venice (1591). [A]

2. D'ARCET, FELIX. Note Relative to the Clarification of the Water of the Nile and Water in General Which Holds Earthy Substances in Suspension. (Articles from the French Journals.) Tr. by J. GRISCOM. *Jour. Franklin Inst.,* New Series, 22:258–261 (1838). [E]

3. JOHNSTON, JAMES FINLAY WEIR. *The Chemistry of Common Life.* W. Blackwood & Sons, Edinburgh (1854–55). [D]

4. NAVARETTE, R. F. DOMINICK FERNANDEZ. *An Account of the Empire of China in a Collection of Voyages and Travels in Six Volumes.* Compiled by JOHN CHURCHILL. London (3rd ed., 1744–46). [D]

5. CLARK, EDWARD B. *William Sibert, The Army Engineer.* Philadelphia (1930). [X]

6. RUTTY, JOHN. *A Methodical Synopsis of Mineral Waters.* William Johnston, London (1757). [B]

7. BOSTOCK, JOHN. On the Spontaneous Purification of Thames Water. *Philosophical Trans., Royal Soc.,* Part 2, pp. 287–290 (1829). [D] *Philosophical Magazine, or Annals of Chemistry* (London), 7:268–271 (1830). [C]

8. BOOTH, ABRAHAM. *A Treatise on the Natural and Chymical Properties of Water and on Various Mineral Waters.* George Wightman, London (1830). [B]

9. ARAGO, DOMINIQUE FRANCOIS. Report Made to the Academy of Sciences on the Filtering Apparatus of Henry de Fonvielle. Tr. by J. GRISCOM. *Jour. Franklin Inst.,* New Series, 22:206–212 (1838). [E]

10. SIMPSON, JAMES. Discussion of "On the Supply of Water to the City of Glasgow," by D. MACKAIN. *Proc. Inst. C.E.,* 2:136–138 (1842–43). [E]

11. GRAHAM, THOMAS. Chemical Report on the Supply of Water to the Metropolis. *Jour. Chem. Soc.,* 4:375–413 (1852). [C]

12. SPENCE, PETER. Personal Letter. London (April 30, 1937).

13. GENIEYS, RAYMOND. Clarification et Dépuration des Eaux. *Annales des Ponts et Chaussées* (Paris), Series I:56–768 (1835). [D]

14. ANON. Pure Water. *Repertory of Patent Inventions,* 7:305 (1829). [D]

15. JEUNNET, M. C. Action Clarifiante de l'Alum sur les Eaux Bourbeuses. *Le Moniteur Scientifique Journal* (Paris), 7:1007–1009 (1865). [I]

16. AUSTEN, PETER T. & WILBER, FRANCIS A. Report of the Purification of Drinking Water by Alum. p. 141, *Annual Report of the State Geologist of New Jersey, 1884.* Trenton (1885). [X]

17. NETHERLANDS' COMMISSION. Report on Water Purification. Abstracted, *Chem. News* (London), 19:239 (1869). [X]

18. ANON. Das Wasserwert de Stadt. Groningen in Holland. *Glaser's Annalen für Gewerbe und Bauwesen,* 11:71–77 (1882). [D]

19. HAZEN, ALLEN. *The Filtration of Public Water Supplies.* J. Wiley & Sons, New York (1st ed., 1895). [D]

20. ANON. The Water Filtration Plant at Leeuwarden, Holland. *Eng. News,* 29:220–221 (1893). [E]

21. SCHOOLCRAFT, HENRY ROWE. *A View of the Lead Mines of Missouri.* C. Wiley & Co., New York (1819). [E]

22. AUSTEN, PETER T. The Purification of Water by Alum. *Scientific American Suppl.,* 21:8782 (1886). [D]

23. LEISEN, THEODORE. Personal Letters. Omaha, Neb. (1931–32).
24. WALKER, ISAAC S. Personal Letter. Philadelphia, Pa. (September 7, 1936).
25. EYER, FRANK A. Personal Letter. Harrisburg, Pa. (1936).
26. (a) MARVIN, F. O. Personal Letter. Lawrence, Kan. (June 28, 1900).
    (b) HOOD, W. C. Personal Letter. Lawrence, Kan. (1907).
    (c) HATCHER, M. P. Personal Letter. Kansas City, Mo. (1936).
    (d) BOYCE, EARNEST. Personal Letter. Lawrence, Kan. (1936).
27. REYER, GEORGE. Personal Letter. Nashville, Tenn. (October 19, 1931).
28. *McGraw Waterworks Directory.* McGraw Pub. Co., New York (1915).
29. DELAFIELD, CLARENCE. The Vicksburg Settling Basins. *Trans. A.S.C.E.*, 21:88–91 (1889). [E]
30. SMITH, J. B. Personal Letter. Vicksburg, Miss. (October 5, 1940).
31. KENDALL, THEODORE REED. Six-Hundred-Dollar Investment Continues $3600 Annual Savings. *American City*, 52:12:45 (1937). [D]
32. BRADY, JOSEPH. The Sandhurst Water Supply, Victoria, Australia. *Proc. Inst. C.E.* (London), 56:134 (1887–89). [E]
33. NICHOLS, WILLIAM RIPLEY. *Water Supply Considered Mainly From a Chemical and Sanitary Standpoint.* J. Wiley & Sons, New York (1883). [E]
34. HOUSTON, ALEXANDER CRUIKSHANK. *8th Report on Research Work*, Metropolitan Water Board, London (1912). [D]
35. ———. Search for Certain Pathogenic Microbes in Raw River Water and in Crude Sewerage. *9th Research Report*, Metropolitan Water Board, London (1913). [D]
36. LONDON METROPOLITAN WATER BOARD. *25th Annual Report.* London (1930). [X]

37. *The Water Engineer's Handbook and Directory, 1929–39.* Water & Water Eng., London (1939). [B]
38. WATTIE, ELSIE & CHAMBERS, CECIL W. Relative Resistance of Coliform Organisms and Certain Enteric Pathogens to Excess-Lime Treatment. *Jour. A.W.W.A.*, 35: 709–720 (1943).
39. SPENCER, THOMAS. On the Supply and Purification of Water. *Jour. Chem. Soc.*, pp. 83–85 (1859). [X]
40. KIRKWOOD, JAMES P. *Report on the Filtration of River Waters, for the Supply of Cities as Practiced in Europe.* D. Van Nostrand, New York (1869). [B]
41. GREAT BRITAIN RIVERS POLLUTION COMMISSION (1868). Domestic Water Supply of Great Britain. *6th Report.* George Edward Eyre & William Spottiswoode, London (1874). [E]
42. ANDERSON, WILLIAM. The Antwerp Waterworks. *Proc. Inst. C.E.* Part 2, 72:24–83 (1882–83). [E]
43. OGSTON, GEORGE HENRY. The Purification of Water by Metallic Iron in Mr. Anderson's Revolving Purifiers. *Eng. News*, 14:147–150 (1885). [E]
44. ANON. A So-called Rusting Tank for Producing a Coagulant for Water Purification. *Eng. News*, 52:266 (1904). [E]
45. EYROLLES, LEON. Extrait du Rapport, H. Chabal & Cie. Exposition Coloniale Internationale, Paris (1901). [X]
46. IMBEAUX, EDOUARD. *Annuaire Statistique et Descriptif des Distributions d'Eau de France, Algerie, Tunisie.* C. Dunod, Paris (3rd ed., 1931). [D]
47. ANON. The Relative Values of Ozone and Slow Filtration as a Means of Purifying Water. *Eng. News*, 43:92 (1900). [E]
48. PHILLIPS, FRANCIS C. Notes on the Purification of Allegheny River Water by the Anderson Process. *Proc. Eng. Soc. of Western Pa.*, pp. 27–40 (1893). [E]

49. PELIGOT, EUGENE. Études sur le Composition des Eaux. *Comptes Rendus Hebdomadaires des Séances,* 58:729–738 (1864). [D]
50. GARDNER, L. H. Discussion of "Methods of Filtration of Water in Use in the United States and Europe," by JOHN JAMES ROBERTSON CROES. *Proc. A.W.W.A.,* p. 74 (1883).
51. ———. The Clarification and Purification of Public Water Supply. *Scientific American Suppl.,* 20:8146 (1885). [D]
52. DRAGENDORFF, G. Purification of Water Containing Organic Impurities. *Pharmaceutical Jour.,* 3rd Series, 2:1049 (1872). [D]
53. GWINN, DOW R. The Use of Iron as a Coagulant in Connection With Mechanical Filtration. *Proc. A.W.W.A.,* pp. 94–99 (1900).
54. WOLMAN, ABEL; DONALDSON, WELLINGTON; & ENSLOW, LINN H. Recent Progress in the Art of Water Treatment. *Jour. A.W.W.A.,* 22:1161–1177 (1930).

CHAPTER XIV

Disinfection

1. HERODOTUS. *History.* English version, ed. by GEORGE RAWLINSON. D. Appleton & Co., New York (1859). [D]
2. HIPPOCRATES. *The Genuine Work of Hippocrates.* Tr. from the Greek by FRANCIS ADAMS. Williams & Wilkins Co., Baltimore (1939). [D]
3. PLACE, FRANCIS EVELYN. Water Cleansing by Copper. *Jour. Preventive Medicine* (London), 13:379 (1905). [A]
4. SUS'RUTA. The Hindu System of Medicine According to Sus'ruta. Tr. from the Sanskrit by UDOY CHAND DUTT. Vol. 95, Chap. 45, pp. 215–222, *Bibliotheca Indica, Royal Asiatic Society of Bengal.* J. W. Thomas, Calcutta (1883). [D]
5. PLINY THE ELDER. *The Natural History of Pliny.* Tr. by JOHN BOSTOCK & H. T. RIPLEY. (Bohn Classical Library) H. G. Bohn, London (1855–57). [D]
6. PLUTARCH. *Vitae Parallelae.* The tr. called Dryden's, corrected from the Greek and rev. by A. H. CLOUGH. Little, Brown & Co., Boston (1859). [D]
7. SEMPLE, ELLEN CHURCHILL. Domestic and Municipal Waterworks in Ancient Mediterranean Lands. *Geog. Review,* 21:466–474 (1931). [D]
8. AEGINETA, PAULUS. *The Seven Books of Paulus Aegineta.* Tr. from the Greek with a commentary by FRANCIS ADAMS. In *Greeks, Romans and Arabians on All Subjects Connected With Medicine and Surgery.* Sydenham Society, London (1844–47). [D]
9. IBN SINA, HUSSAIN IBN ABD ALLAH. *A Treatise on the Canon of Medicine of Avicenna.* Incorporating a tr. of the "First Book" by G. CAMERON GRUNER. Luzab & Co., London (1930). [D]
10. BOERHAAVE, HERMANN. *Elements of Chemistry.* Tr. from the original Latin by TIMOTHY DALLOW. J. & F. Pemberton, London (1735). [D]
11. DUNGLINSON, ROBLEY. *Human Health; or Influence of Atmosphere, Food . . . on the Healthy Man.* Philadelphia (1844). [D]
12. RIDEAL, SAMUEL. *Water and Its Purification.* C. Lockwood & Son. London (1902). [D]
13. ——— & ERIK. *Chemical Disinfection and Sterilization.* E. Arnold & Co., London (1921). [D]
14. BECHMANN, M. Purification of Water for Domestic Use. French Practice. *Proc. A.S.C.E.,* Part D, 44:183–190 (1904). [E]

15. IMBEAUX, EDOUARD. *Annuaire Statistique et Descriptif des Distributions d'Eau de France, Algerie, Tunisie.* C. Dunod, Paris (3rd ed., 1931). [D]
16. HINMAN, JACK J., JR. Water Supply in the Field. Vol. 1, No. 4, *Univ. of Iowa Studies in Medicine,* Iowa City (1918). [D]
17. LONDON METROPOLITAN WATER BOARD. *29th Annual Report.* London (1934). [X]
18. RIDEAL, SAMUEL & BAINES, E. The Suggested Use of Copper Drinking Vessels as a Prophylactic Against Water-Borne Typhoid. *Royal San. Inst. Jour.* (London), 25:591–595 (1904). [E]
19. KRAEMER, HENRY. Use of Copper in Destroying Typhoid Organisms and the Effects of Copper on Man. *Amer. Jour. Pharmacy*, 77:265–281 (1905). [K]
20. (a) AMY, JOSEPH. *Nouvelles Fontaines Domestiques, Approuvées par l'Academie Royale des Sciences.* J. B. Coignard, Paris (1750). [B]
    (b) ———. *Suite du Livre Intitulé "Nouvelles Fontaines Filtrantes Approuvées par l'Academie Royal des Sciences."* Antoine Boudet, Paris (1754). [B]
21. MOORE, GEORGE THOMAS & KELLERMAN, KARL F. A Method of Destroying or Preventing the Growth of Algae and Certain Pathogenic Bacteria in Water Supplies. *Bulletin No. 64.* Bureau of Plant Industry, U.S. Dept. of Agriculture, Govt. Printing Office, Washington, D.C. (1904). [D]
22. ———. Copper as an Algicide and Disinfectant in Water Supplies. *Bulletin No. 76.* Bureau of Plant Industry, U.S. Dept. of Agriculture, Govt. Printing Office, Washington, D.C. (1905). [D]
23. JACKSON, DANIEL D. The Purification of Water by Copper Sulfate. *Eng. News*, 54:307–308 (1905). [E]
24. MOORE, GEORGE THOMAS; ET AL. The Use of Copper Sulfate and Metallic Copper for the Removal of Organisms and Bacteria From Drinking Water. *Jour. N.E.W.W.A.*, 19:474–582 (1905).

25. BROWN, C. ARTHUR. The Practical Sterilization of Public Water Supplies by Means of Copper-Iron Sulfate and Filtration. *Proc. A.W.W.A.*, pp. 239–252 (1905).
26. NAGELI, CARL VON. Über Oligodynamischen Erscheinungen in Lebenden Zellen. *Schweizerische Naturforschenda Gesellschaft*, 33:1–59 (1893). [D]
27. THIELE, HERMANN & WOLF, K. Über die Bakterienschädigenden Einwirkungen der Metalle. *Archive für Hygiene*, 34:43–70 (1899). [D]
28. SACL, P. Über die Keim Kotende Firnwirkung von Metallen Oligodynamische Wirkung. *Klinische Wochenschaft* (Vienna), 30:714 (1917). [X]
29. SUCKLING, E. V. The Sterilization of Water by Catadyn Silver. *Water & Water Eng.* (London), 34:15–18 (1932). [E]
30. MOISEEV, S. V. The Sterilization of Water by Silver Coated Sand. *Jour. A.W.W.A.*, 26:217–238 (1934).
31. JUST, J. & SZNIOLIS, A. Germicidal Properties of Silver in Water. *Jour. A.W.W.A.*, 28:492–506 (1936).
32. ANON. Silver Process Effective in Water Sterilization. *Chem. & Metallurgical Eng.*, 41:372 (1934). [C]
33. SHAPIRO, ROBERT & HALE, FRANK E. An Investigation of the Katadyn Treatment of Water. *Jour. N.E. W.W.A.*, 51:113–124 (1937).
34. DORROH, J. H. The Design and Operation of Swimming Pools. *Jour. A.W.W.A.*, 27:100–107 (1935).
35. *Manual of Water Quality and Treatment.* American Water Works Assn., New York (1940).
36. RACE, JOSEPH. *Chlorination of Water.* J. Wiley & Sons, New York (1918). [E]
37. HOOKER, ALBERT HUNTINGTON. *Chloride of Lime in Sanitation.* J. Wiley & Sons, New York (1913). [E]
38. AMERICAN PUBLIC HEALTH ASSN. COMMITTEE ON SEWAGE DISPOSAL. Unpublished report for 1933. [X]

BIBLIOGRAPHY—CHAPTER XIV

39. Crimp, William Santo. *Sewage Disposal Works.* Griffin, London (1894). [E]
40. Fuller, George Warren. *Sewage Disposal.* McGraw Hill Book Co., New York (1911). [E]
41. Metcalf, Leonard & Eddy, Harrison P. Disposal of Sewage. Vol. III, *American Sewerage Practice.* McGraw-Hill Book Co., New York (1914–15). [D]
42. Anon. A Review of Attempts to Use Electricity in Sewage and Water Purification. *Eng. News,* 67: 534–536 (1912). [E]
43. *Patents for Inventions—Class 46: Filtering and Otherwise Purifying Liquids.* H.M. Stationery Office, London (1676–1866). [E]
44. Baker, M. N. *Sewage Purification in America.* Eng. News, New York (1893). [X]
45. Bache, R. Mead. Possible Sterilization of City Water. *Am. Philosophical Soc. Proc.,* 29:26–39 (1891). [D]
46. Anon. The Electrical Purification of Water. *Eng. Rec.,* 29:110 (1894). [E]
47. Gayol, Roberto. *Enforme el Sistema de Saneamiento por Medio de Soluciones de Chloruros Electrolizados.* A. N. Ayuntamiento, Mexico City (1893). [X]
48. Imbeaux, Edouard. Personal Letters. Nancy, France (August 20 & 30, 1934).
49. Drown, Thomas M. Electrical Purification of Water. *Jour. N.E.W.W.A.,* 8:183 (1893–94). [E]
50. Frankland, Edward. On the Metropolitan Water Supply During the Year 1866. *Chem. News* (London), 15:151 (1867). [C]
51. Wanklyn, J. Alfred. Report on the Organic Pollution of Drinking Waters. *British Med. Jour.,* p. 351 (March 30, 1872). [X]
52. Anon. Second Report Dealing With the Atkins Moulded-Carbon Filter. *British Med. Jour.,* p. 338 (September 21, 1872). [X]
53. Jewell, William M. Personal Letters. Chicago (1933–35).
54. Fuller, George Warren. *Report on the Investigation Into the Purification of the Ohio River Waters at Louisville, Ky.* D. Van Nostrand, New York (1898). [B]
55. Vaughan, Victor C. Report on Analyses of Water From the Adrian (Mich.) Water Co. Before and After Passing Through a Jewell Mechanical Filter, April, 1899. From papers of John A. Cole, Consulting Engineer to the Adrian Water Co. (1899). [X]
56. Smart, F. B. Personal Letters. Adrian, Mich. (1934–35).
57. Whipple, George C. Disinfection as a Means of Water Purification. *Proc. A.W.W.A.,* pp. 266–288 (1906).
58. Boby, William. The Early Use of Chlorine. *Jour. A.W.W.A.,* 23:283–284 (1931).
59. ———. Personal Letter. London (April 1931).
60. Houston, Alexander Cruikshank. *Studies in Water Supply.* Macmillan, London (1913). [E]
61. Johnson, George A. Hypochlorite Treatment of Public Water Supplies—Its Adaptability and Limitations. *Am. Jour. Pub. Health,* 1: 562–574 (1911). [E]
62. (a) Leal, J. L. The Sterilization Plant of the Jersey City Water Supply Co. at Boonton, N.J. *Proc. A.W.W.A.,* pp. 100–109 (1909).

    (b) Fuller, George Warren. Description of the Process and Plant of the Jersey City Water Supply Co. for the Sterilization of the Water of the Boonton Reservoir. *Proc. A.W.W.A.,* pp. 110–134 (1909).

    (c) Johnson, George A. Description of Methods of Operation of the Sterilization Plant of the Jersey City Water Supply Company at Boonton, N.J., and Discussion of Results of Analyses of Raw and Treated Water, With Notes on the Cost of the Treatment. *Proc. A.W.W.A.,* pp. 135–162 (1909).

63. LEAL, J. L.; FULLER, GEORGE WARREN; & JOHNSON, GEORGE A. Defendants' Testimony. Jersey City vs. Jersey Water Supply Co. In Chancery of New Jersey. (February–October 1909). [X]

64. ANON. The Hypochlorite Plant. *Eng. News*, 63:686 (1910); 64:483 (1910). [E]

65. GORMAN, ARTHUR E. Chronology of the Bubbly Creek Chlorination Plant. Based on data in his file. [X]

66. JENNINGS, CHARLES A. Personal Letters. (September–October 1936).

67. OTIS, JOHN C. The Poughkeepsie Water Works. *Jour. N.E.W.W.A.*, 23:283–301 (1909).

68. POUGHKEEPSIE BOARD OF PUBLIC WORKS. *Annual Report*. Poughkeepsie, N.Y. (1909). [E]

69. WEST, F. D. Disinfecting 200,000,000 Gallons of Water a Day—Experience With Chloride of Lime and Liquid Chlorine at Torresdale Filtration Plant (Philadelphia). *Jour. A.W.W.A.*, 1:403–455 (1914).

70. BALDWIN, ROBERT T. History of the Chlorine Industry. *Jour. Chem. Education*, 4:313–319 (1927). [C]

71. KIENLE, JOHN A. The Use of Liquid Chlorine for Sterilizing Water. *Proc. A.W.W.A.*, pp. 267–289 (1913).

72. U.S. PUBLIC HEALTH SERVICE. National Census of Water Treatment Plants of the United States. *Water Works Eng.*, 96:63–117 (1943).

73. ANON. *Water Sterilization by Gaseous Chlorine*. Paterson Engineering Co., Ltd., London (1928). [E]

74. LONDON METROPOLITAN WATER BOARD. *31st Annual Report*. London (1936). [E]

75. ANON. Chlorination of Water Supplies. (British Waterworks Assn. Questionnaire.) *The Surveyor*, 97:518 (1940). [E]

76. BUNAU-VARILLA, PHILIPPE. *L'Auto Javellization Imperceptible*. Paris (1926). [X]

77. WEST, FRANCIS D. Personal Letter. Arlington, N.J. (1931).

78. MCCONNEL, S. Verdunization. *Water & Water Eng.*, 37:310–311 (1935). [E]

79. IMHOFF, KARL. Personal Letter. Essen, Germany (April 20, 1941).

80. BAYLIS, JOHN ROBERT. *Elimination of Taste and Odor in Water*. McGraw-Hill Book Co., New York (1935). [E]

81. SOPER, GEORGE A. The Purification of Drinking Water by the Use of Ozone. *Eng. News*, 42:250–253 (1899). [E]

82. ANON. Sterilization of Water by Ozone. *Eng. News*, 42:340 (1899). [E]

83. ANON. Relative Value of Ozone and Slow Sand Filtration as a Means of Purifying Water. *Eng. News*, 43:92–95 (1900). [E]

84. SIMIN, NICHOLAS. Ozonized Air. *Proc. A.W.W.A.*, p. 41 (1901).

85. ANON. Production and Utilization of Ozone With Especial Reference to Water Purification. *Eng. News*, 63:488–496 (1910). [E]

86. ANON. Ozone Water Purification Plant at St. Petersburg, Russia. *Eng. Rec.*, 63:473 (1911). [E]

87. ANON. Ozone Plant at St. Petersburg, Russia. *Eng. News*, 66:783 (1911). [E]

88. ANON. Report on the Filtration of the Croton Water Supply, New York City. *Eng. News*, 58:561–562 (1907). [E]

89. HEAD, WILLIAM C. Personal Letter. Ann Arbor, Mich. (1933).

90. PRYER, R. W. Water Purification by Ozone, With Report on the Ann Arbor Plant. *Jour. Ind. & Eng. Chem.*, 6:797–800 (1914). [C]

91. CASWELL, H. H. Development of Ann Arbor Water Supply. *Jour. A.W.W.A.*, 32:1577–1586 (1940).

92. RAY, D. Personal Letter. Lindsay, Ont. (August 29, 1933).

BIBLIOGRAPHY—CHAPTER XIV

93. WALDEN, A. E. & POWELL, S. T. Sterilization of Water. *Proc. A.W.W.A.*, pp. 128–149 (1910).
94. POWELL, S. T. The Use of Ozone as a Sterilizing Agent for Water Purification. *Jour. N.E.W.W.A.*, 29: 87–93 (1915).
95. ———. Personal Letter. Baltimore, Md. (1933).
96. COX, CHARLES R. Significant Experiences in the Treatment of Water in New York State. *Jour. N.E. W.W.A.*, 53:444–457 (1939).
97. ANON. Milwaukee Testing Station Results on Filtration of Water. *Eng. News-Rec.*, 85:257–259 (1920).
98. JENKS, STANLEY. Personal Letter. Philadelphia, Pa. (October 7, 1941).
99. HANN, VICTOR. Ozone Treatment of Water. *Jour. A.W.W.A.*, 35:585–591 (1943).
100. BARTUSKA, JAMES F. Ozonation at Whiting, Ind. *Jour. A.W.W.A.*, 33: 2035–2050 (1941).
101. CONSOER, ARTHUR WARDEL & NELLIS, JAMES C. Ozone Reduces Water Odors and Tastes. *Eng. News-Rec.*, 127:367–369 (1941). [E]
102. ANON. Huntington Pumping Station of the South Staffordshire Waterworks Co. *Eng.*, 143:651–653, 711–713 (1937). [E]
103. WHITSON, M. T. B. The Treatment of Water by Ozone. Paper read before Manchester and District Assn. of Inst. C.E. (March 20, 1940). [X]
104. MORGANS, G. EWART. Personal Letter. London (May 23, 1940).
105. FRENCH FOREIGN OFFICE. Information obtained by French Consul at Philadelphia (1940). [X]
106. GURY, A. *L'Eau à Paris*. (Extraits d'un Manuscrit, écrit en 1913.) Paris (1939). [Mimeographed Abstract.] [E]
107. IMHOFF, KARL. Personal Letter. Essen, Germany (April 20, 1941).
108. GRANT, KENNETH C. Sterilization of Polluted Water by Ultra-violet Ray. *Eng. News*, 64:275 (1910). [E]
109. ANON. Sterilization of Polluted Water by Ultra-violet Rays at Marseilles, France. *Eng. News*, 64:633 (1910). [E]
110. CLEMENCE, WALTER. The Water Supply of Marseilles and Trials of Filtering and Sterilizing Apparatus. *Eng.*, 91:106–108, 139–142 (1911). [E]
111. ANON. Another French Quest for the Last Germ in City Water Supplies. *Eng. News*, 66:686 (1911). [E]
112. SMITH, A. T. Ultra-violet Rays Finish Treatment of Henderson (Ky.) Water Supply. *Eng. News-Rec.*, 79: 1021–1022 (1917). [E]
113. BLOCHER, JOHN MILTON. The Purification of Water by Ultra-violet Radiation. *Jour. A.W.W.A.*, 21: 1361–1372 (1929).
114. WARING, F. H. Personal Letter. Columbus, Ohio (October 18, 1939).
115. BOYCE, EARNEST. Personal Letter. Lawrence, Kan. (October 23, 1939).
116. SHAW, FRANK R. Personal Letters. Chicago (1933 & 1940).
117. WALDEN, A. E. & POWELL, S. T. Sterilization of Water by Ultra-violet Rays. *Proc. A.W.W.A.*, pp. 341–346 (1911).
118. VON RECKLINGHAUSEN, MAX. Purification of Water by Ultra-violet Rays. *Jour. A.W.W.A.*, 1:565–588 (1914).
119. ———. Ultra-violet Rays for Water Purification. *Jour. N.E.W.W.A.*, 29:202–213 (1915).
120. SPENCER, R. R. Ultra-violet Rays; Their Advantages and Disadvantages in the Purification of Drinking Water. *Jour. A.W.W.A.*, 4:172–182 (1917).
121. FAIR, GORDON M. Elements of the Theory Underlying the Disinfection of Water by Ultra-violet Light. *Jour. A.W.W.A.*, 7:325–342 (1920).
122. PERKINS, ROGER G. & WELCH, HENRY. Sterilization of Water by Ultra-violet Light as Emitted by the Carbon Arc. *Jour. A.W.W.A.*, 22:959–967 (1930).

CHAPTER XV

Distillation

1. ARISTOTLE. *De Generatione Animalium.* Tr. by ARTHUR M. PLATT. Clarendon Press, Oxford(1910). [G]
2. ST. BASIL. *Letters and Select Works.* (In Select Library of Nicene and Post-Nicene Fathers of the Christian Church, Second Series, 8:75.) Christian Literature Co., New York (1895). [D]
3. JABIR IBN HAYYAN. *The Works of Geber Englished by Richard Russell, 1678.* New ed. with introduction by E. J. HOLMYARD. J. M. Dent & Sons, London (1928). [B]
4. HALES, STEPHEN. *An Account of Some Attempts to Make Distilled Water Wholesome, Containing Useful and Necessary Instructions for Such as Undertake Long Voyages at Sea. Showing How Sea-Water May Be Made Fresh and Wholesome; and How Fresh-Water May Be Preserved Sweet.* R. Mauly, London (1739). [D]
5. GESNER, CONRAD. *A Newe Booke of Destyllation of Waters, Called the Treasure of Evonymous.* London (1565). [X]
6. HAWKINS, RICHARD. *The Observations of Sir Richard Hawkins in His Voyage Into the South Sea in the Year 1593.* Hakluyt Society, London (1847). [D]
7. BACON, FRANCIS. *Sylva Sylvarum, or a Natural History in Ten Centuries.* W. Lee, London (1670). [D]
8. GLAUBER, JOHANN RUDOLF. *The Works of the Highly Experienced and Famous Chymist.* Tr. by C. PARKE. London (1689). [D]
9. FRENCH, JOHN. *The Art of Distillation, or a Treatise of the Choicest Spagyricall Preparations Performed by Way of Distillation.* London (1651). [J]
10. BOYLE, ROBERT. *Letter to Dr. JOHN BEAL concerning fresh water made out of sea water; appended is an account of the Honourable ROBERT BOYLE's way of examining waters as to freshness and saltness.* London (1683). [X]
11. GREW, NEHEMIAH. *New Experiments and Observations Concerning Sea-Water Made Fresh.* London (1684). [J]
12. *Abridgements of Specifications Relating to Purifying and Filtering Water, Including the Distillation of Sea-Water to Produce Fresh Water, (A.D. 1675–1866).* George Edward Eyre & William Spottiswoode, London (1876).
13. HUNTER, FREDERIC MERCER. *An Account of the British Settlement of Aden in Arabia.* London (1877). [D]
14. News Note. *The Engineer* (London), 144:739 (1927). [X]
15. EHLERS, V. M. Personal Letter. Austin, Texas (September 6, 1939).

CHAPTER XVI

Aeration in Theory and Practice

1. ATHENAEUS OF NAUCRATIS. *The Deipnosophists.* Tr. by CHARLES BURTON GULICK. (Loeb Classical Library) W. Heinemann, London (1927). [D]
2. COLUMELLA, LUCIUS JUNIUS MODERATUS. *Of Husbandry.* A. Millar, London (1745). [J]
3. CHARAKA. *Charaka-Samhita.* Tr. by AVINASH CHANDRA KAVIRATNA. Calcutta (1888–1912). [D]
4. PLINY THE ELDER. *The Natural History of Pliny.* Tr. by JOHN BOSTOCK & H. T. RIPLEY. (Bohn

Classical Library) H. G. Bohn, London (1855-57). [D]

5. IBN SINA, HUSSAIN IBN ABD ALLAH. A Treatise on the Canon of Medicine of Avicenna. Incorporating a tr. of the "First Book" by G. CAMERON GRUNER. Luzab & Co., London (1930). [D]

6. BACON, FRANCIS. *Sylva Sylvarum, or a Natural History in Ten Centuries.* W. Lee, London (1670). [D]

7. WINTRINGHAM, CLIFTON. *A Treatise of Endemic Diseases. Different Nature of Airs, Waters.* Grace White, York, Eng. (1718). [B]

8. DAWNE, DARBY [EDWARD BAYNARD]. *Health, a Poem.* London (1716). [J]

9. ARMSTRONG, JOHN. *The Art of Preserving Health; a Poem.* London (1744). [J]

10. PLÜCHE, NOEL ANTOINE. *Spectacle de la Nature; or Nature Displayed.* Tr. from the original French by SAMUEL HUMPHREYS. J. & F. Pemberton, London (1736). [D]

11. *Encyclopedia Britannica.* A. Bell & C. MacFarquhar, Edinburgh (3rd ed., 1797). [D]

12. HALES, STEPHEN. 1. An Account of the Great Benefit of Blowing Showers of Fresh Air up Through Distilling Liquors. 2. Ventilators in Slave and Other Transport Ships. 3. Account of Some Trials to Cure the Ill Taste of Milk and to Sweeten Stinking Water. *Philosophical Trans. Royal Soc.,* 49:312-335, 339-347 (1755-56). [D]

13. PARKES, EDMUND ALEXANDER. *A Manual of Practical Hygiene, Prepared Especially for Use in the Medical Service of the Army.* J. Churchill & Sons, London (1864). [D]

14. SINCLAIR, JOHN. *The Code of Health and Longevity.* A. Constable & Co., Edinburgh (1807). [B]

15. AUSTEN, C. E. Communication Describing a Russian Aerator. *Proc. Inst. C.E.,* 27:46-47 (1867). [X]

16. HUMBER, WILLIAM. *A Comprehensive Treatise on the Water Supply of Cities and Towns With Numerous Specifications of Existing Waterworks.* London (1876). [D]

17. ANON. Memoir of Bryan Donkin. *Proc. Inst. C.E.,* Part 1, 115:391 (1893-94). [X]

18. DON, JOHN. *The Filtration and Purification of Water for Public Supply.* Inst. Mech. Engrs., London, pp. 51-55 (1909).

19. TORONTO COMMITTEE ON FIRE, WATER AND GAS. *Proceedings in Connection With the Supply of Water to the City.* Toronto, Ont. (1854). [X]

20. ANON. Water Power in the United States. *Tenth Census of the United States (1880).* Govt. Printing Office, Washington, D.C. (1887). [B]

21. *Manual of American Water Works.* Ed. by M. N. BAKER. Eng. News, New York (1897).

22. TUBBS, J. NELSON. *Annual Report.* Executive Board, Rochester, N.Y., Water Works (April 5, 1886). [X]

23. FISHER, EDWIN A. Personal Letters. Rochester, N.Y. (November 1940 & March 1941).

24. ACKERMAN, J. WALTER. Personal Letters. Utica, N.Y. (October 28 & November 7, 1941).

25. ANON. Water Purification in America—Aeration and Continuous Sand Filtration at Ilion, N.Y. *Eng. News,* 31:466-469 (1894). [E]

26. BABCOCK, STEPHEN E. Aeration of Water Supplies by Natural Canals and Low Dams. *Jour. N.E.W.W.A.,* 3:35-44 (1888-89).

27. *Manual of American Water Works.* Ed. by M. N. BAKER. Eng. News, New York (1889).

28. CODD, W. F. Water Purification in America—The Filter Bed of the Wannacomet Water Co., Nantucket, Mass. *Eng. News,* 31:336 (1894). [E]

29. WESTON, ARTHUR D. Personal Letter. Boston, Mass. (November 21, 1939).

30. ROBINSON, J. H. Personal Letter. Nantucket, Mass. (December 13, 1939).
31. DROWN, THOMAS M. Effect of Aeration of Natural Waters. pp. 384–394, *Annual Report*, Massachusetts State Board of Health (1891). [X]
32. ———. The Aeration of Natural Waters. *Jour. N.E.W.W.A.*, 7:96 (1892–93).
33. CARROLL, EUGENE. Personal Letter. Butte, Mont. (November 21, 1935).
34. CHARLES, S. A. My Experience in Filtration and Aeration. *Proc. A.W.W.A.*, pp. 71–77 (1898).
35. HAZEN, ALLEN. *The Filtration of Public Water Supplies*. J. Wiley & Sons, New York (1st ed., 1895). [B]
36. ANON. Removal of Iron From Water From a Filter Gallery at Reading (Mass.). *Eng. News*, 36:348 (1896). [E]
37. BANCROFT, LEWIS M. The Iron Removal Plant at Reading, Mass. *Jour. N.E.W.W.A.*, 11:295–300 (1896–97).
38. LEEDS, ALBERT RIPLEY. Aeration of Water. *Trans. A.S.C.E.* (1895). [X]
39. ———. Purification of the Water Supplies of Cities. *Jour. Franklin Inst.*, 123:93–107 (1887). [E]
40. PHILADELPHIA WATER WORKS. *Annual Report*. Philadelphia (1886). [X]
41. LEEDS, ALBERT RIPLEY. Mechanical Aeration of Water. *Stevens Indicator*, 9:309–317 (1892). [E]
42. BRUSH, CHARLES B. Discussion of "Algae and the Purity of Water Supplies," by GEORGE W. RAFTER. *Trans. A.S.C.E.*, 21:519–522 (1889).
43. ———. Aeration on a Gravity Water Supply. *Proc. A.W.W.A.*, pp. 73–76 (1891).
44. ———. Testimonial. *Catalog*, New York Filter Co., New York (1893).
45. COWLES, M. W. Personal Letters. Hackensack, N.J. (1935 & 1936).
46. NELSON, S. L. Testimonial, dated January 14, 1888. *Catalog*, New York Filter Co., New York (1893).
47. ANON. Water Purification at Norfolk (Va.). *Eng. News*, 43:346–348 (1900). [E]
48. FITZGERALD, R. W. Personal Letter. Norfolk, Va. (December 3, 1935).
49. CRANCH, E. T. Personal Letter. New Rochelle, N.Y. (January 15, 1941).
50. BOLLING, GEORGE E. Personal Letter. Brockton, Mass. (January 21, 1941).
51. GIBSON, J. E. Personal Letters. Charleston, S.C. (January & October 1941).
52. WICKERSHAM, L. E. Personal Letter. New York (1941).
53. ANON. Those New Filters. An Official Test at the Belleville Water Works. *Sanitary Era*, 1:89 (1886). [E]
54. ANON. Seventeen Cities and Towns That Now Have Plants in Operation. Purified Water by the Hyatt Pure Water System. *Sanitary Era*, 1:248 (1887). [E]
55. ANON. Still More Pure-Water Towns. *Sanitary Era*, 1:241 (1887). [E]
56. ANON. The Hyatt Water Filtering Plant at Long Branch (N.J.). *Eng. News*, 20:280 (1888). [E]
57. METCALF, LEONARD. Aerator for Mechanical Filter Plant at Winchester (Ky.). *Eng. News*, 45:410–411 (1901). [E]
58. CLARK, HARRY W. Double Sand Filtration of Water at South Norwalk, Conn.—Removal of Organisms, Tastes and Odors. *Jour. N.E.W.W.A.*, 30:86–100 (1916).
59. PIRNIE, MALCOLM. Effluent Aerators Control Mechanical Filters. *Eng. News-Rec.*, 99:376–380 (1927). [E]
60. ANDERSON, HERMAN. Aeration in Water Treatment. *Jour. N.E.W.W.A.*, 46:373–378 (1932).
61. ROE, FRANK C. Aeration of Water by Air Diffusion. *Jour. A.W.W.A.*, 27:897–904 (1935).

62. THOMPSON, LEONARD N. Problems of St. Paul Water Supply. *Jour. A.W.W.A.*, 23:202–210 (1931).
63. THUMA, ROSS A. Aeration With Compressed Air for Removing Odors. *Jour. A.W.W.A.*, 24:682–691 (1932).
64. BRUSH, WILLIAM W. Operation of New York's Water Supply. *Water Works & Sewerage*, 78:97–99 (1931).
65. Census of Municipal Purification Plants in the United States, 1930–31. American Water Works Assn., New York (1933).
66. U.S. PUBLIC HEALTH SERVICE. National Census of Water Treatment Plants of the United States. *Water Works Eng.*, 96:63–117 (1943).
68. BAYLIS, JOHN ROBERT. *Elimination of Taste and Odor in Water*. McGraw-Hill Book Co., New York (1935). [E]

CHAPTER XVII

Algae Troubles and Their Conquest

1. LEEUWENHOEK, ANTONY VAN. Observations, Communicated in a Dutch Letter Concerning Little Animals by Him Observed in Rain-, Well-, Sea-, and Snow Water, as Well as in Water Wherein Pepper Had Been Infused. *Philosophical Trans., Royal Soc.*, 10–13:821–831 (1676). [D]
2. KING, EDWARD. Several Observations and Experiments on the Animalcula in Pepper Water. *Philosophical Trans., Royal Soc.*, 17:861–865 (1693). [E]
3. HARRIS, JOHN. Some Microscopical Observations of Vast Numbers of Animalcula Seen in Water. *Philosophical Trans., Royal Soc.*, 19:254–259 (1696). [D]
4. RUTTY, JOHN. *A Methodical Synopsis of Mineral Waters*. William Johnston, London (1757). [B]
5. PRIESTLEY, JOSEPH. *Experiments and Observations on Different Kinds of Air*. London (1779–86). [G]
6. INGENHOUSZ, JAN. Nouvelles Experiences et Observations sur Divers Objets de Physique. *Monthly Review*, 3:201–208 (1790). [D]
7. DWIGHT, TIMOTHY. *Travels in New England and New York*. W. Baynes & Sons, London (1823). [D]
8. D'AUBUISSON DE VOISINS, JEAN FRANCOIS. Histoire de l'Etablissement des Fontaines à Toulouse. *Histoire et Mémoires de l'Académie Royal des Sciences de Toulouse*. Series 2, Part 1, 2:159–400 (1823–27). [D]
9. HAWKSLEY, THOMAS. Description of Filter Basin and Gallery, Nottingham. Vol. 2, pp. 52–53, *First Report of the Commission for Inquiring Into the State of Large Towns and Populous Districts*. Clowes & Sons, London (1844–45). [E]
10. BATEMAN, J. F. Discussion of "Experiments on Removal of Organic and Inorganic Substances in Water," by EDWARD BYRNE. *Proc. Inst. C.E.*, 26:544 (1866–67). [X]
11. BOLTON, FRANCIS. *London Water Supply*. International Health Exhibition, 1884. Clowes & Sons, London (1884). [B]
12. LINDLEY, WILLIAM. *London Water Supply*. International Health Exhibition, 1884. Clowes & Sons, London (1884). [B]
13. JERVIS, JOHN B. & JOHNSON, WALTER B. Report on Additional Water Supply. *Report of the Commissioners to Examine the Sources From Which a Supply of Pure Water May Be Obtained for the City of Boston*. Boston (1845). [D]
14. HORSFORD, E. N. & JOHNSON, CHARLES T. Report on Tastes and Odors. *Annual Report*. Cochituate Water Board, Boston (1854). [X]

15. DAVIS, JOSEPH P. Bad Tastes in Chestnut Hill Reservoir. *Annual Report.* Cochituate Aqueduct Board, Boston (1876). [X]

16. REMSEN, IRA. On the Impurity of the Boston Water Supply. *City of Boston Document No. 143.* Boston (November 12, 1881). [X]

17. FITZGERALD, DESMOND. Brief Summaries of Studies of Micro-organisms, Color and Filtration. *Annual Reports.* Boston Water Board (1888-94). [D]

18. ALBANY WATER BOARD. *Annual Reports.* Albany, N.Y. (1851–). [X]

19. WURTZ, HENRY. Analyses of Water Supply of Trenton, N.J. *Report of President and Directors,* Trenton Water Works. Philips, Trenton (1856). [D]

20. TORREY, JOHN. Discussion of Purity of Water Supplies. *Trans. A.S.C.E.,* 21:555-557 (1889). [E]

21. MCALPINE, WILLIAM JARVIS. Preliminary Report on Proposed Water Supply for Schenectady. (August 1867). [Unpublished Report.]

22. SPRINGFIELD WATER COMMISSIONERS. *Annual Reports.* Springfield, Mass. (1872-74). [X]

23. ———. *Special Report on the Improvement of the Present Water Supply and an Alternative New Independent System, April 14, 1902.* Including reports by P. M. BLAKE and by the STATE BOARD OF HEALTH. Springfield, Mass. (1902). [D]

24. SPRINGFIELD WATER SUPPLY COMMITTEE. *Report of Engineers to Special Water Supply Commission, March 23, 1904.* Springfield, Mass. (1904). [D]

25. POUGHKEEPSIE WATER BOARD. *Annual Report.* Poughkeepsie, N.Y. (1875). [X]

26. HUDSON WATER BOARD. *Annual Reports.* Hudson, N.Y. (1874-75). [X]

27. FORBES, FAYETTE F. A Study of Algae Growth in Reservoirs and Ponds. *Jour. N.E.W.W.A.,* 4:196-210 (1889-90).

28. ———. The Relative Taste and Odor Imparted to Water by Some Algae and Infusoria. *Jour. N.E.W.W.A.,* 6:90-97 (1891-92).

29. ———. Covered Reservoir at Brookline, Mass. *Jour. N.E.W.W.A.,* 8:113-117 (1893-94).

30. ———. The Appearance of *Chara Fragilis* in the Reservoir of a Public Water Supply. *Jour. N.E.W.W.A.,* 10:252-254 (1895-96).

31. BAKER, M. N. The Brookline Water Works and F. F. Forbes. *Jour. N.E.W.W.A.,* 46:77-92 (1932).

32. LEEDS, ALBERT RIPLEY; LECONTE, L. J.; & GARDNER, L. H. Report of Committee on Animal and Vegetable Growths Affecting Water Supplies. *Proc. A.W.W.A.,* pp. 27-45 (1891).

33. ALLEN, CHARLES P. Discussion of "Report of Committee on Animal and Vegetable Growths Affecting Water Supplies." *Proc. A.W.W.A.,* pp. 35-36 (1896).

34. MOORE, GEORGE THOMAS & KELLERMAN, KARL F. A Method of Destroying or Preventing the Growth of Algae and Certain Pathogenic Bacteria in Water Supplies. *Bulletin No. 64.* Bureau of Plant Industry, U.S. Dept. of Agriculture, Govt. Printing Office, Washington, D.C. (1904). [D]

35. ———. Copper as an Algicide and Disinfectant in Water Supplies. *Bulletin No. 76.* Bureau of Plant Industry, U.S. Dept. of Agriculture, Govt. Printing Office, Washington, D.C. (1905). [D]

36. (a) ANON. Preventing the Growth of Algae in Water Supplies. *Eng. News,* 51:496-499 (1904). [E]

(b) MOORE, GEORGE THOMAS; ET AL. The Use of Copper Sulfate and Metallic Copper for the Removal of Organisms and Bacteria From Drinking Water. *Jour. N.E.W.W.A.,* 19:474-582 (1905).

(c) JACKSON, DANIEL D. The Purification of Water by Copper Sulfate. *Eng. News,* 54:307-308 (1905).

(d) CAIRD, JAMES M. Some Experi

ments With Copper Sulfate. *Proc. A.W.W.A.*, pp. 253–256 (1905).
(e) ———. Copper Sulfate Results. *Proc. A.W.W.A.*, pp. 249–265 (1906).

37. HALE, FRANK E. Control of Microscopic Organisms in Public Water Supplies. *Jour. N.E.W.W.A.*, 44:361–385 (1930).

38. LOVEJOY, W. H. Algae Control by Creating Turbidity at Louisville (Ky.). *Eng. News-Rec.*, 101:505–507 (1928). [E]

39. TRIMBLE, EARLE J. Personal Letters. Ilion, N.Y. (August & September 1942).

40. CARROLL, EUGENE. Personal Letter. Butte, Mont. (December 1935).

41. GOUDEY, R. F. New Method of Copper Sulfating Reservoirs. *Jour. A.W.W.A.*, 28:163–179 (1936).

42. STEWART, E. P. Copper Sulfate Applied as a Spray for Algae Control. *Water Works Eng.*, 94:617 (1941).

Chapter XVIII

Softening

1. SHAW, PETER. *Chemical Lectures, Publickly Read at London, 1731 and 1732*. T. & T. Longman, London (1734). [D]

2. HOME, FRANCIS. *Experiments on Bleaching*. A. Kincaid & A. Donaldson, Edinburgh (1756). [B]

3. RUTTY, JOHN. *A Methodical Synopsis of Mineral Waters*. William Johnston, London (1757). [B]

4. CAVENDISH, HENRY. Experiments on Rathbone-Place Water. *Philosophical Trans., Royal Soc.*, 57:92–100 (1767). [D]

5. HENRY, THOMAS. On the Preservation of Sea Water From Putrefaction by Means of Quicklime. *Manchester Literary & Philosophical Soc. Memoirs*, 1:41–54 (1785). [D]

6. ———. An Account of a Method of Preserving Water at Sea From Putrefaction, and of Restoring to the Water Its Original Purity and Pleasantness, by a Cheap and Easy Process. *Manchester Literary & Philosophical Soc. Memoirs* (1785). [X]

7. DAVY, EDMUND. Experiments Made Upon the Hard Water at Black Rock, Near Cork. *Philosophical Magazine & Journal* (London), 52:3–8 (1818). [D]

8. BOOTH, ABRAHAM. *A Treatise on the Natural and Chymical Properties of Water and on Various Mineral Waters*. George Wightman, London (1830). [B]

9. CLARK, THOMAS. *A New Process for Purifying the Waters Supplied to the Metropolis by the Existing Water Companies*. R. J. E. Taylor, London (1841). [D]

10. ———. On Means Available to the Metropolis and Other Places for the Supply of Water Free From Hardness, and From Organic Impurity. *Jour. of the Soc. of Arts* (London), 4:429–433 (1856). [D]

11. GRAHAM, THOMAS; MILLER, W. A.; & HOFFMANN, A. W. *Report by the Government Commission on the Chemical Quality of the Supply of Water to the Metropolis*. General Board of Health, London (1851). [X]

12. CLARK, THOMAS. Evidence. *First Report of the Commission for Inquiring Into the State of Large Towns and Populous Districts*. Clowes & Sons, London (1844–45). [E]

13. HUMBER, WILLIAM. *A Comprehensive Treatise on the Water Supply of Cities and Towns With Numerous Specifications of Existing Waterworks*. London (1876). [D]

14. HOMERSHAM, SAMUEL COLLETT. Evidence given March 12, 1868. pp. 210–216, *Annual Report*. Great Britain Rivers Pollution Commission. London (1869). [E]

15. ANON. A Large Base-Exchange Softening Plant. *Water & Water Eng.* (London), 37:245 (1935). [E]

16. GREAT BRITAIN RIVERS POLLUTION COMMISSION (1868). *Report of the Commissioners to Inquire Into the Best Means of Preventing the Pollution of Rivers.* George Edward Eyre & William Spottiswoode, London (1870–74). [E]

17. ANON. Why Is Hard Water Unfit for Domestic Purposes? (Papers on Water No. 1) *Harper's New Monthly Magazine,* 1:50–52 (1850). [D]

18. LATHAM, BALDWIN. Softening of Water. *Eng. News,* 11:65–66; 81–82 (1881). [X]

19. PORTER, J. H. The Porter-Clark Process. *Jour. Soc. Chem. Industry* (London), 3:51–55 (1884). [C]

20. ANON. Water Softening Plant at Southampton. *Eng.,* 53:318–319 (1892). [E]

21. FULLER, GEORGE WARREN. Water Softening for Municipal Supplies. *Proc. A.W.W.A.,* pp. 113–138 (1906).

22. ARCHBUTT, LEONARD. The Softening of Water by the Archbutt-Deeley Process. *Eng. News,* 40:403–405 (1898). [E]

23. HOOVER, CHARLES P. Water Supply and Treatment. *Bulletin No. 211.* National Lime Assn., Washington, D.C. (1938). [B]

24. ———. Review of Lime-Soda Water Softening. *Jour. A.W.W.A.,* 29:1687–1696 (1937).

25. ANON. *Statistical Tables of American Water Works.* American Soft Water Co., New York (1887). [X]

26. ANON. Water Purification as Originated in the United States by Companies Now Included in the New York Filter Co. *Trade Catalog,* New York Filter Co., New York (1893). [X]

27. ANON. Water Filter and Lime Catcher for Locomotives. *Railway Age,* p. 279 (May 23, 1879). [X]

28. HANDY, JAMES O. Water Softening. *Eng. News,* 51:500–508 (1904). [E]

29. GERRISH, WILLIAM B. The Municipal Water Softening Plant at Oberlin, Ohio. *Jour. N.E.W.W.A.,* 19:422–436 (1905).

30. GRAF, AUGUST V. Personal Letter. St. Louis, Mo. (1940).

31. MAIGNEN, P. A. Discussion of "Water Softening for Municipal Supplies," by GEORGE WARREN FULLER. *Proc. A.W.W.A.,* pp. 127–138 (1906).

32. GOODELL, J. E. Personal Letters. Lancaster, Pa. (November 1939 & February 1940).

33. HOOVER, CHARLES P. Personal Letter. Columbus, Ohio (February 5, 1940).

34. GREGORY, JOHN H. The Improved Water and Sewage Works of Columbus, Ohio. *Trans. A.S.C.E.,* 67:206–224 (1910). [E]

35. HOOVER, CHARLES P. The Treatment of the Water Supply of the City of Columbus, Ohio. *Trans. A.S.C.E.,* 93:1505–1526 (1929). [E]

36. JENSEN, J. ARTHUR. Water Softening at Minneapolis. *Jour. A.W. W.A.,* 30:1547–1568 (1938).

37. THUMA, ROSS A. St. Paul Softens Water Cheaply. *Eng. News-Rec.,* 127:80–82 (1941). [E]

38. SPAFFORD, H. A. & KLASSEN, C. W. Illinois' Experiences in Lime Softening With Short-Time Upward-Flow Clarification. *Jour. A.W.W.A.,* 31:1734–1754 (1939).

39. BEHRMAN, A. S. & GREEN, W. H. Accelerated Lime-Soda Water Softening. *Ind. Eng. Chem.,* 31:128 (1939).

40. SPAULDING, CHARLES H. & TIMANUS, C. S. The New Water Purification Plant for Springfield, Illinois. *Jour. A.W.W.A.,* 27:327–336 (1935).

41. SPAULDING, CHARLES H. Conditioning of Water Softening Precipitates. *Jour. A.W.W.A.,* 29:1697–1707 (1937).

42. ANON. The "Excess Lime" Method of Disinfecting and Softening Water. *Eng. News,* 67:1087–1088 (1912). [E]

43. ANON. Sterilization of Water Supply by Free Lime. *Municipal Review* (London), 1:94 (1930). [D]
44. HOOVER, CHARLES P. & SCOTT, RUSSELL D. The Use of Lime in Water Purification. *Eng. News*, 72:586–590 (1914). [E]
45. OLSON, H. M. Benefits and Savings From Softened Water for Municipal Supply. *Jour. A.W.W.A.*, 31:607–639 (1939).
46. ———. Development and Practice of Municipal Water Softening. *Jour. A.W.W.A.*, 37:1002–1012 (1945).
47. U.S. DISTRICT COURT (Western District of New York). *Opinion of Judge R. Hazel*. (Finding Refinite Co.'s Zeolite Water Softener an Infringement of the Permutit Co.'s Patent.) Evening Post Job Printing Office, New York (n.d.). [B]
48. SIMIN, BORIS N. Water Softening by Means of Zeolith. *Proc. A.W.W.A.*, pp. 347–352 (1911).
49. JACKSON, DANIEL D. Water Softening by Filtration Through Artificial Zeolite. *Jour. A.W.W.A.*, 3:423–433 (1916).
50. *The Water Engineer's Handbook and Directory for 1939.* Water & Water Eng., London (1939). [B]
51. MOSES, H. E. Personal Letter. Harrisburg, Pa. (1938).
52. MONTGOMERY, JAMES M. & AULTMAN, WILLIAM W. Water-Softening and Filtration Plant of the Metropolitan Water District of Southern California. *Jour. A.W.W.A.*, 32:1–24 (1940).
53. STREETER, H. W. On Water Purification and Treatment. *Census of Municipal Water Purification Plants in the United States, 1930–31.* American Water Works Assn., New York (1933).
54. U.S. PUBLIC HEALTH SERVICE. National Census of Water Treatment Plants of the United States. *Water Works Eng.*, 96:63–117 (1943).
55. OLSON, H. M. Census of U.S. Municipal Water Softening Plants *Jour. A.W.W.A.*, 33:2153–2193 (1941)
56. ———. 1944 Census of U.S. Municipal Water Softening Plants. *Jour. A.W.W.A.*, 37:585–592 (1945).
57. ANON. Statistics of Water Works. *Canadian Engineer—Water & Sewage*, Annual Directory Number (April 1941). [X]
58. COLLINS, W. D.; LAMAR, W. L.; & LOHR, E. W. The Industrial Utility of Public Water Supplies in the United States, 1932. *U.S. Geological Survey Water Supply Paper No. 658.* Govt. Printing Office, Washington, D.C. (1934). [D]
59. COLLET, HAROLD. *Water Softening and Purification. The Softening and Clarification of Hard and Dirty Waters.* E. & F. N. Spon, London (1896). [B]
60. *Manual of Water Quality and Treatment,* American Water Works Assn., New York (1940).

CHAPTER XIX

Cause and Removal of Color

1. BOSTOCK, JOHN. On the Spontaneous Purification of Thames Water. *Philosophical Trans., Royal Soc.*, Part 2, pp. 287–290 (1829). [D]; *Philosophical Magazine, or Annals of Chemistry* (London), 7:268–271 (1830). [C]
2. WEST, WILLIAM. Practice and Philosophical Observations on Natural Waters. *Jour. Royal Inst. Great Britain* (London), 1:38–46 (1830–31). [D]
3. BEETZ, W. On the Colour of Water. *The London, Edinburgh & Dublin Philosophical Magazine,* 4th Series, 24:218–224 (1862). [D]
4. SPRING, WALTHERE. Oeuvres Complètes. *Société Chimique de Bel-*

gique, 1:659–900 (1914); 2:1815–1824 (1923). [D]

5. SWAN, CHARLES W.; WARD, EDWARD S.; & BOWDITCH, H. P. Report on the Sanitary Qualities of the Sudbury, Mystic, Shawshine and Charles River Waters. *City of Boston Document No. 102.* Boston (1874). [X]

6. FITZGERALD, DESMOND. Brief Summaries of Studies of Micro-organisms, Color and Filtration. *Annual Reports.* Boston Water Board (1888–94). [D]

7. ANON. Water Purification at Norfolk (Va.). *Eng. News,* 43:346–348 (1900). [E]

8. ANON. The Decolorization of Water by the Excess-Coagulation Method at Springfield, Mass. *Eng. News,* 70:974–975 (1913). [E]

CHAPTER XX

Iron and Manganese Removal

1. SALBACH, B. *Berichte über die Erfahrungen bei Wasserwerken und Wasserversorgungen.* Dresden (1893). [X]

2. WESTON, ROBERT SPURR. Some Recent Experiences in the Deferrization and Demanganization of Water. *Jour. N.E.W.W.A.,* 28:27–59 (1914).

3. HAZEN, ALLEN. *The Filtration of Public Water Supplies.* J. Wiley & Sons, New York (1st ed., 1895). [B]

4. ANON. Removal of Iron From Artesian Well Water by Mechanical Filtration. *Eng. News,* 35:364–367 (1896). [E]

5. ANON. Removal of Iron From a Filter Gallery at Reading, Mass. *Eng. News,* 36:348–351 (1896). [E]

6. ANON. Iron Removal From Ground Water at Far Rockaway, N.Y., by Slow Sand Filtration. *Eng. News,* 43:238 (1900). [E]

7. STERLING, CLARENCE J., JR.; & BELKNAP, JOHN B. Iron in the Ground Water Supplies of Massachusetts. *Jour. Boston Soc. C.E.,* 19:1–20 (1932). [X]

8. BEAUCHEMIN, A. O. French Indo-China Supply Secured Under American Methods. *Water Works Eng.,* 89:67–71 (1936).

9. CHARLES, R. S., JR. Pressure Filters for Iron Removal. *Jour. A.W.W.A.,* 30:1507–1513 (1938).

10. LEY, PETER. A Unique Iron Removal Plant. *Jour. A.W.W.A.,* 30:1493–1506 (1938).

11. KETEL, J. Personal Letter. Zutphen, Holland (February 6, 1940).

12. BARBOUR, FRANK A. Decarbonation and Removal of Iron and Manganese From Ground Water at Lowell, Mass. *Jour. A.W.W.A.,* 4:129–164 (1917).

13. ZAPFFE, CARL. The History of Manganese in Water Supplies and Methods for Its Removal. *Jour. A.W.W.A.,* 25:655–676 (1933).

14. ANON. Manganese and Iron Removal Plant for Lincoln, Nebraska. *Eng. News-Rec.,* 114:246–247 (1935). [E]

15. DAVIS, D. E. Personal Letter. Pittsburgh, Pa. (November 14, 1938).

16. U.S. PUBLIC HEALTH SERVICE. National Census of Water Treatment Plants of the United States. *Water Works Eng.,* 96:63–117 (1943).

Chapter XXI

Taste and Odor Control

1. RACE, JOSEPH. Chlorination and Chloramine. *Jour. A.W.W.A.*, 5: 63–82 (1918).
2. HAROLD, C. H. H. Further Investigations Into the Sterilization of Water by Chlorine and Some of Its Compounds. *Jour. Roy. Army Corps* (London), 45:190; 251; 350; 429 (1925).
3. ADAMS, B. A. The Chloramine Treatment of Pure Water. *Medical Officer* (London), 5:55 (1926).
4. MCAMIS, J. W. Prevention of Phenol Taste With Ammonia. *Jour. A.W.W.A.*, 17:341–350 (1927).
5. SPAULDING, CHARLES H. Preammoniation at Springfield, Illinois. *Jour. A.W.W.A.*, 21:1085–1096 (1929).
6. RUTH, EDWARD D. The Elimination of Taste and Odor in the Water Supply of Lancaster, Pennsylvania. *Jour. A.W.W.A.*, 23:396–399 (1931).
7. HARRISON, LOUIS B. Chlorphenol Tastes in Waters of High Organic Content. *Jour. A.W.W.A.*, 21:542–549 (1929).
8. LAWRENCE, W. C. Studies in Water Purification Processes at Cleveland, Ohio. *Jour. A.W.W.A.*, 23:896–902 (1931).
9. BRAIDECH, MATTHEW M. The Ammonia-Chlorine Process as a Means for Taste Prevention and Effective Sterilization. *Ohio. Conf. Water Purif.*, 9:67 (1930).
10. HOWARD, NORMAN J. & THOMPSON, RUDOLPH E. Chlorine Studies and Some Observations on Taste-Producing Substances in Water, and the Factors Involved in Treatment by the Super- and De-chlorination Method. *Jour. N.E.W.W.A.*, 40: 276–296 (1926).
11. ———. Discussion of "Chlorine Studies and Some Observations on Taste-Producing Substances in Water, and the Factors Involved in Treatment by the Super- and De-chlorination Method." *Jour. N.E. W.W.A.*, 41:59–62 (1927).
12. SCOTT, R. D. Research on Removal of Phenolic Tastes in Public Water Supplies. *Ohio Conf. Water Purif.*, 8:60 (1929).
13. GERSTEIN, H. H. Chlorophenol Tastes and Abnormal Absorption of Chlorine in the Chicago Water Supply. *Jour. A.W.W.A.*, 21:346–357 (1929).
14. FABER, HARRY A. Super-chlorination Practice in North America. *Jour. A.W.W.A.*, 31:1539–1560 (1939).
15. GRIFFIN, A. E. Reaction of Heavy Doses of Chlorine in Various Waters. *Jour. A.W.W.A.*, 31:2121–2129 (1939).
16. HASSLER, WILLIAM W. The History of Taste and Odor Control. *Jour. A.W.W.A.*, 33:2124–2152 (1941).
17. BAYLIS, JOHN ROBERT. *Elimination of Taste and Odor in Water.* McGraw-Hill Book Co., New York (1935).
18. ———. The Activated Carbons and Their Use in Removing Objectionable Tastes and Odors From Water. *Jour. A.W.W.A.*, 21:787–814 (1929).
19. IMHOFF, K. & SIERP, F. Active Charcoal Filter for Improving Taste of Potable Water. *Gas- und Wasserfach* (Berlin), 72:465 (1929).
20. SIERP, F. Improvement of the Smell and Taste of Drinking Water. *Technisches Gemeindeblatt* (Berlin), 32:153; 165; 186 (1929); abstracted *Jour. A.W.W.A.*, 23:462–463 (1931).
21. SPALDING, GEORGE R. Discussion of "Successful Super-chlorination and De-chlorination for Medicinal Taste of a Well Supply, Jamaica, N.Y." by FRANK E. HALE. *Jour. A.W. W.A.*, 23:384–386 (1931).

Chapter XXII

Medication by Means of the Water Supply

1. LITTLE, BEEKMAN C. Iodine Treatment of Water for Prevention of Goiter. *Jour. A.W.W.A.*, 10:556–558 (1923).
2. ———. Sodium Iodide Treatment of Rochester's Water Supply. *Jour. A.W.W.A.*, 12:68–86 (1924).
3. GOLER, G. W. Personal Letter. Rochester, N.Y. (December 29, 1931).
4. JOHNSON, A. M. Personal Letter. Rochester, N.Y. (September 27, 1933).
5. SHERMAN, HENRY A. Personal Letter. Sault Ste. Marie, Mich. (October 18, 1933).
6. SMITH, ALFRED E. Personal Letter. Derbyshire, England (January 29, 1934).
7. JOHNSON, H. M. Personal Letter. Anaconda, Mont. (September 5, 1939).
8. ———. Goiter and the Public Water Supply. *Jour. A.W.W.A.*, 16:205–206 (1926).
9. MELLEN, ARTHUR F. The Present Status of the Use of Iodides in the Minneapolis Water Supply. *Jour. A.W.W.A.*, 16:715–729 (1926).
10. ———. Personal Letter. Minneapolis, Minn. (September 15, 1933).
11. BOLT, R. A. & WOLMAN, ABEL. Treatment of Water With Iodide for the Prevention of Simple Goiter. Chap. XII, *Water Works Practice*. A Manual issued by the American Water Works Assn. Williams & Wilkins, Baltimore (1925).
12. OLSON, H. M. Personal Letters. Wadsworth, Ohio (February 17, 23 & 27, 1945).
13. DEAN, H. TRENDLEY. Domestic Water and Dental Caries. *Jour. A.W.W.A.*, 35:1161–1186 (1943).
14. KNAPP, HAROLD J. The Public Health Significance of Dental Deficiencies. *Jour. A.W.W.A.*, 35:1187–1190 (1943).
15. AST, DAVID B. A Program of Treatment of Public Water Supply to Correct Fluoride Deficiency. *Jour. A.W.W.A.*, 35:1191–1197 (1943).
16. WOLMAN, ABEL. What Are the Responsibilities of Public Water Supply Officials in the Correction of Dental Deficiencies? *Jour. A.W.W.A.*, 35:1198–1200 (1943).
17. AST, DAVID B. The Caries-Fluorine Hypothesis and a Suggested Study to Test Its Application. *Public Health Reports* (June 4, 1943).
18. COX, CHARLES R. Personal Letter. Albany, N.Y. (March 5, 1945).

INDEX

Editor's Note: The following is essentially an index of people and places. All men and localities mentioned in the text and footnotes are cited below together with the number of the page upon which any single mention appears or continuous reference begins.

No attempt has been made to provide a thorough subject index, but those subjects which are mentioned in the chapter titles and main subheadings of the text have been included as guides to the various sections of the book.

A

Abadie [of Toulouse, Fr.], 274
Abel, Frederick, 316
Aberdeen, Scot., 313
Abraham [ozonation], 353
Activated carbon treatment, 454
Adams, B. A., 452
Adams, Francis, 323
Adams, H. J., 212
Aden, Arabia, 360
Adrian, Mich., *ft.* 330, 335
Aegineta, Paulus, 7, 322, 450
Aeration; 361, 450
    forced, 377
    induced-air, 385
Albany, N.Y.; algae, 398, 401
    aniline plant, 190
    filtration, 144, 161, 264
Albany Steam Trap Co., 213
Albi, Fr., 281
Alborchetti, Alfonso, 13
Alexander the Great, 321
Alexandria, Egypt, 5, 9, 291
Algae control, 391, 451
Algiers, Algeria, 307
Allegheny, Pa., 317
Allen, Charles P., 412
Alpino, Prospero, 24, 300
Alston [Dr.], 24
Alton, Ill., 143
Altona, Ger., 133, 259
Alvord, Burdick & Howson, 213, 238, 280
American Filter Co., 205, 209

American Ozone Co., 349
American Soft Water Co., *ft.* 428
American Water Softener Co., 193, 223, 227
American Water Works Assn.; literature, 390, 439
    meetings, 458, 460
    organization, 141
American Water Works & Electric Co., 387
Amiens, Fr., 343
Amsterdam, Neth., 318, 445
Amy, Joseph; 21, 324
    filter patents, 30, 248, 253
Anaconda, Mont., 457
Anderson, Adam, 279
Anderson, H. B., *ft.* 223
Anderson, Herman, 389
Anderson, William, 313, 315, 319
Anderson, Ind., 325
Anderson Process; Belgium, 316
    France, 63
    U.S., 153, 246, 374, 376
Andrews, Horace, 167
Andrews, William D. & Bro., 169
Angers, Fr., 133, 279
Anglicus, Johannes, 357
Anklam [of Berlin, Ger.], *ft.* 445
Ann Arbor, Mich., 348
Anna, Ill., 433
Annapolis, Md., 143
Antiochus, 5, 286
Antwerp, Belg., 190, 258, 309, 315
Arabia, 7

511

Arago, Dominique Francois; coagulation, 300, 303, 305, 332
 filtration, 44, 46, 133
 sedimentation, 295
Archbutt, Leonard, 427
Archbutt-Deeley Process, 427
Aristotle, 4, 321, 357
Armour, Dole & Co., 217
Armstrong, John, 363
Asbury Park, N.J., 225, 263, 445
Ashby, A., 426
Ashby, Eng., 427
Ashton-under-Lyne, Eng., 352
Ast, D. B., 461
Athenaeus of Attilia, 5
Athenaeus of Naucratis, 4
Athens, Ga., 199
Athol, Mass., 216
Atkins, F. H., 255, 427
Atkins, William G., 427
Atkins Filter & Mfg. Co., 427
Atlanta, Ga., 194, 265
Atlantic Highlands, N.J., 225, 263, 445
Augusta, Ga., 296
Augusta, Me., 199
Austen, C. E., 368
Austen, Peter T., 307, 310
Auvergne, Fr., 18
Avalon, Pa., 437
Avicenna, 7, 323, 362
Avignon, Fr., 343
Ayr, Scot., 95, 249

B

Babcock, Stephen E., 373
Bache, R. Mead, 330
Bacon, Francis; aeration, 362, 450
 distillation, 357
 filtration, 9, 253
Badger, Ernest F., 348
Bahlman, Clarence, 244
Baiae, It., 6
Bailey, Alexander Mabyn, 360
Bailey, George H., 190
Bailey, George I., 172, 174
Bailey, J. W., 397
Baines, E., 324
Baity, H. C., 271
Bakenhus, Reuben E., 231
Baldwin, George R., *ft.* 132, 295
Baldwin, Loammi, *ft.* 95, 129, 131, *ft.* 295
Ball, Phineas, 384, 404

Baltimore, Md., 348, 398
Barker & Wheeler, 161
Barnes, A. A., 262
Barry [filtration], 39
Bartlesville, Okla., 251
Barwise [iodization], 457
Basil [Saint], 357
Bateman, J. F., 395
Baudet [Mayor, of Chateaudun, Fr.], 62
Bauscaren, George F., 243
Bay City, Mich., 453, 454
Baylis, John R., 345, 390, *ft.* 415, 442, 454
Baynard, Edward, 363
Beaver Falls, Pa., 349
Bechmann, M., 324
Beck, Romeyn, 163
Beetz, Wilhelm von, 440
Behrman, A. S., 433
Belfast, Ire., 265, 271
Belgium, 190, 258, 308
Belgrand, Eugène, 35, 37, 43, 51
Belknap, Mass., 446
Bell, Andrew James, 256
Belleville, Ill., 193, 194, 387
Benzenburg, George H., 243
Berea, Ohio, 355
Berlin, Ger., 133, 188, 269, 396
Bernard [of Paris, Fr.], 51
Bertrand, Jean, 25
Bible, 3, 418
Birmingham, Eng., 80, 262
Bischof, Gustav, 258, 313, 315, 319
Bishop, H. K., 157
Black Rock, Ire., 419
Blackburn, Eng., 252
Blake, Percy M., 216, 408
Blanken, H., 347
Blankenberg, Belg., 346
Blessing, James H., 213
Blunt [ultra-violet], 354
Boby, William, 336
Boerhaave, Hermann, 20, 323, 450
Bohman, H. P., 348
Bolling, Charles E., 129, 130
Bolling, George E., *ft.* 384
Bollmann [of Hamburg, Ger.], 265
Bollmann Filter-Gesellschaft, 266, 271
Bologna, It., 281
Bolt, R. A., 458
Bolton, Francis, 396
Bolton, Eng., 304
Boog, Robert, 77

INDEX 513

Boonton, N.J., 326, 336
Booth, Abraham, 116, 303, 309, 313, 419
Bordeaux, Fr., 295, 343
Bostock, John, 302, 440
Boston, Eng., ft. 106
Boston, Mass.; 129, 131, 144
  algae, 396, 405
  coagulation, 317
  color removal, 441
Boulton, Matthew, 79, 273
Boutelle, Thomas N., 143, 250
Bowman, C. H., 160
Boyle, Robert, 359
Bracken, Elmer F., 264
Brackett, F. W., & Co., 252
Braidech, Matthew M., 453
Brainerd, Minn., 447
Brampton, Ont., 271
Bramwell, Charles B., 271
Brande, William Thomas, 105, 107
Bremen, Ger., 259, 436
Breslau, Ger., 445, 447
Brest, Fr., 38
Brewster, N.Y., 330
Bridge, J. Howard, 347
Bridgeton, N.J., 446
Brighton, Eng., 352
Bristowe, Leonard H., 209
Brito, Saturnino de, 271
Brockton, Mass., 143, 220, 382, 384
Brookline, Mass., 280, 411, 447, 451
Brooklyn, N.Y., 334
Brown, C. Arthur, 325, 430
Brownell [electrolysis], 232, 335
Brownsville, Tex., 389, 451
Brunswick, Me., 199
Brush, Charles B., 201, 380, 384
Brussels, Belg., 343
Bryan, William B., 261
Bubbly Creek Filters, 336, 339
Buchan, William, 25
Buchanan, Robertson, 84
Budapest, Hung., 281
Bull, William B., 219, 320
Bunau-Varilla, Philippe, 343
Bunsen, Robert Wilhelm, 440
Burdick, Charles B., 213
Burgess, Edward, 398
Burlington, Iowa, 143, 250
Burnt Mills, Md., 78
Butler, Thomas, 23
Butte, Mont., 376, 413
Byrne, Edward, 119, 368

C

Caesar, Caius Julius, 5, 9, 291
Caird, James M., 160, 371
Cairo, Ill., 211
Calcutta, India, 315
Camden, Eng., 426
Canada; 23
  aeration, 370
  algae, 406
  coagulation, 309
  filtration, 125, 179, 264, 269
  iron removal, 446
  ozonation, 348
  softening, 428, 438
Canal de l'Ourcq, 45, 61, 318
Candy, F. P., 370, 437, 446
Canterbury, Eng., 423
Capellen, F. W., 207
Carbon; see also Charcoal
  activated, 454
Carcassonne, Fr., 281
Caries, dental, 460
Carmichael, Henry, 198, 199
Carpenter, George W., 144, 163, 172, 401
Carpenter, Lewis V., 240
Carriere, J. E., 259
Carroll, Eugene, 376, 414
Carthage, Mo., 143
Carthage, N. Africa, 13, 286
Cashmore, H. M., 80
Caterham, Eng., 423
Cavallo, Tiberius, 27, 450
Cavendish, Henry, 415, 418
Chabal, H., 88, 257, 261
Chadwick, Edwin, 120
Chalon, Fr. Indo-China, 446
Chambers, Cecil, 313
Chambers, Ephraim, 21
Chambers, John, 240
Champaign, Ill., 202, 381, 428
Chandler, Charles F., 164, 167
Chandler, H. C., 177
Chaplin, Alexander, 367
Charancourt, Francois-Gregoire de Bourbon, 36, 274
Charcoal; see also Carbon
  purification with, 26
Charles II of England, 359
Charles, S. A., 376
Charleston, S.C., 384
Charleston, W.Va., 193
Charlottenburg, Ger., 346, 445

Chartres, Fr., 353
Chase, John C., 332
Chateaudun, Fr., 62
Chatin [iodine], 459
Chattanooga, Tenn., 202, 221
Chelsea Water Works Co., 47, 84, 99, 119, 421
Cheltenham, Eng., 342
Chesborough, I. C., 166, 377
Chesbrough, E. S., 145
Chester, Pa., 311
Chicago, Ill., 145, 336, 339
Chilton, James R., 402
China, 24, 301, 322
Chisholm, C. A., 385
Chloride Process Co., 205
Chlorination, 326, 453
Christchurch, Eng., 336
Churchill, John, 24
Cincinnati, Ohio; 130
  coagulation, 320
  filtration, 140, 146, ft. 227, 237, 243
  sedimentation, 297
  softening, 432
Clarification, 40, 313
Clark, Charles, 367
Clark, Harry W., 232, 264, 389
Clark, Patrick, 183, 190
Clark, Thomas, 332, 415, 420, 434
Clark, William, 181
Cleaveland, W. F., 384
Clegg, Robert, 367
Clemence, Walter, 57, 355
Cleveland, Grover, 166
Cleveland, Ohio, 453
Clinton, Iowa, 143
Clow, James B., & Sons, 217
Clow, W. E., 217
Coagulation, 299
Cochrane, John, 87, 88
Codd, W. F., 375
Cohoes, N.Y., 175
Cole, Thomas A., 154
Collier, Joseph, 248, 255
Collin [engr., Carthage aqueduct], 290
Collingwood, Francis, 371
Collins [aeration], 379, ft. 380
Collins, W. D., 439
Colne Valley Water Co., 352, 423, 437
Color removal, 440
Columbus, Ohio, 143, 280, 431, 435
Columella [aeration], 362
Compagnie Dufaud, 37

Compagnie Francais, 49, 51
Compagnie Générale des Eaux, 18, 54, 63
Conger, W. H., 155
Consoer, Arthur W., 351
Conti, Nicolo de, 13
Continental Filter Co., 202, 223, 445
Cook, J. D., 181, 191, 249, 255
Copper; disinfection by, 324
Copper sulfate treatment, 412
Corbett, Joseph, ft. 61
Corfield, William Henry, 121, 292
Cork, Ire., 419
Coryat, Thomas, 13
Cotelle, Theodore, 367
Cowles, M. W., 381
Cox, C. R., 463
Cranch, E. T., 383
Cranston Hill Water Works Co.; 106, 110, 249, 273
  filtration, 80, 84, ft. 112
Creber, W. F. H., 67
Creston, Iowa, 221
Cresy, Edward, 117
Crimp, William Santo, ft. 327
Crocker Filtering Co., 193
Croes, John James Robertson; 201, ft. 250
  aeration, 383
  filtration, 137, 172, 249
  sedimentation, 289
Croker, Temple Henry, 21
Cronstedt [zeolite], 436
Crouy, Henry de, 255
Cuchet [filtration], 39
Cumberland Mfg. Co., 197, 231
Cyrus the Great, 4, 321

D

Daily, J. M., 348
Dankers, Jasper, 162
D'Arcet, Felix, 300
Darcy, Henry; filter patent, 42, 58, 179
  writings, 55, 88, 98
Darling, Edwin, 250
Darnall, C. R., 341
D'Aubuisson de Voisins, Jean Francois, 274, 277, 395, 451
d'Audollant, Auguste Marie Henri, 290
Davenport, Iowa, 211, 212
Davidson, J. R., 103
Davies, B. W., 277
Davis, Chester B., 209

Davis, J., 328
Davis, James L., 128, 130
Davis, Joseph P., 397
Davis, N., 287
Davis, W., 148, 410
Davy, Edmund, 419
Deacon, Martin, 269
Dean, H. Trendley, 460
Debauve, Alphonse Alexis, 281
de Brito, Saturnino, 271
Decatur, Ill., 280
de Conti, Nicolo, 13
DeCosta, J. D., 195
de Crouy, Henry, 255
Deeley, R. M., 427
Defiance, Ohio, 428
de Fonvielle, see Fonvielle
DeFrise [ozonation], 353
Delafield, Clarence A., 311
de La Hire, Philippe, 19
Delbrück [filtration], ft. 45, ft. 49, 51, 114, 257
Delhi, N.Y., 348
Deluc, Jean Andre, 293
De Normandy, Alphonse René le Mire, 367
Dental caries, 460
Denver, Colo., 412
Denver, Pa., 349, 351
Derby, Eng., 262, 279, 427
Derbyshire, Eng., 457
de Saussure, Horace Benedict, 293
Des Moines, Iowa, 280, 282
Desportes [of Paris, Fr.], 49
Detroit, Mich., 147
Deutsch, Claude, 199
Deutsch, William M., 199
DeVarona, I. M., 347
De Veau, John G., 383, 384
Devonshire, Easton, 316
D'Heureuse [aeration], ft. 378, 379, ft. 380
Diball, Tex., 360
Dickinson, Robert, 367
Diderot, Denis, 21
Digby, Kenelm, 9
Dijon, Fr., 58
Diophanes, 3
Disinfection, 321
Distillation, 357
Dobell [electrolysis], 328
Dodd, Ralph, 89, 294
Doherty, Frank, 431

Dominiceti, Bartholomew, 360
Don, John, 370
Donaldson, Wellington, 320
Dongola, Anglo-Egypt. Sudan, 301
Donkin, Bryan, 368
Dorr Co., 239
Dorroh, J. H., 326
Douglass, David Bates, 163
Downs [ultra-violet], 354
Dragendorff, G., 319
Dresden, Ger., 281, 436, 445
Drown, Thomas M., 152, 331, 361, 375, 377
Dryden, John, ft. 69
Dublin, Ire., 120, 133, 418
Ducommun, J., 43, 49
Ducommun, Théophile, 44
Dufaud Compagnie, 37
Dunglinson, Robley, 42, 132, 323, 327, 366
Dunkinfield, Eng., 352
Dunkirk, Fr., 57, 99
Dupin, Charles, 84
Dupuit, Jules, ft. 60
Durbin, W. H., 203
Duyk, Maurice, 336, 338
Dwight, Timothy, 393

E

Earl, George G., ft. 205, 207
East Jersey Water Co., 226, 227, 234, 337
East London Water Works Co., 89, 261
Easton, Mass., 143
Eddy, Harrison P., ft. 327
Eddy, Robert Henry, 144
Edgeworth, Pa., 448
Edinburgh, Scot., 90, 133, 294
Edmonton, Alta., 438
Edmunds, J., 368, 386
Egypt, 2, 5, 24
Ehlers, V. M., 360
Ehrenberg, C. F., 397
Eichorn [zeolite], 436
Electro Bleaching Gas Co., 341
Electrolysis, 330
Elgin, Ill., 209, 211
Elisha, 3
Elizabeth, N.J., 125, 141
Ellms, Joseph W., 231, 348
Elmira, N.Y., 141, 221, 370, 451
Elrod, Henry E., 389
Emigh, John, 156
Emmery, H. C., 44

Emmons, Stephen, 328
Engelfield, Henry C., 293
England, *see* Great Britain
Enslow, Linn H., 320
Essen, Ger., 281
Etobicoke Twp., Ont., 446
Evanston, Ill., 240, 413
Everett, Chester M., 154, 175
Excess lime treatment, 434
Exeter, N.H., 202

F

Faber, Harry A., 453
Fabretti, Rapha, 292
Fanning, John Thomas, 137, 167, 282
Far Rockaway, N.Y., 445
Farlow, W. G., 398, 406
Feldhaus, Franz Maria, 27
Fell, Richard, 367
Ferrand, Nicolas, 35, 366
Filter cisterns, 13
Filter fountains, 37, 51
Filter galleries, 83, 85, 273
Filter ships, 27, 36
Filter wash; reverse-flow, 57, 67
 sand-transfer, 265
Filtration; 20, 29, 64, 125, 179, 248, 253, 273
 domestic, 67
 drifting sand, 265
 gravity, 43
 hydraulics of, 58
 mechanical, 64, 183
 multiple, 11, 253
 natural, 45, 273
 non-submerged, 60
 pressure, 44, 46
 rapid, 179, 265
 slow sand, 63, 125
 upward-flow, 67, 248
Finistère, Fr., 354
FitzGerald, Desmond, 377, 396, 399, 400, 442
Fitzgerald, R. W., 383
Fitzgerald, Robert, 11, 358
Flad, Henry, 181, 249
Fleming, Sandford, *ft.* 370
Fluorination, 460
Fonvielle, Louis-Charles-Henri de, 44, 46, 60, 98, 133, 181
Forbes, Fayette F., 396, 412, 451
Fort Myer, Va., 341
Fort Wayne, Ind., 280

Fowler, Charles E., 150, 151, 410
Fox, C. J., 345
France; disinfection, 343, 353
 filtration, 29, 190, 272
 sedimentation, 295
Frankfort, N.Y., 373
Frankfurt am Main, Ger., 183, 188
Frankland, Edward, 119, 315, 332, 424
Frankland, Percy F., 122, 140
Fraser, James, 367
Fremay, Fr., 343
French, John, 358
Friedrichs, Carl C., Jr., *ft.* 209
Fries, Va., 356
Frontinus, Sextus Julius, 6, 290
Fteley, Alphonse, 371
Fuller, F. L., 377
Fuller, George Warren; *ft.* 327, 427
 Boonton, N.J., 338
 Cincinnati, Ohio, 243
 Little Falls, N.J., *ft.* 227, 228
 Louisville, Ky., 140, 228, 237, 242, 333
 Springfield, Mass., 408

G

Gadeeschen, Johannes, 357
Gale, James M., 87, 88, 99
Gale, William, 87
Galen, 6, 300
Gans, Robert, 435
Gardner, L. H., 183, 246, 318, 319
Garonne River, 274, 295
Gärtner, August, 62
Geber [Jabir ibn Hayyan], 7, 357
Geneva, Switz., 257, 343
Genieys, Raymond, 42, 44, 304
Genoa, It., 133, 279
Germany, 188, 445
Gerrish, W. B., 429
Gerson, Caesar, 255
Gerson, G. D., 256
Gerstein, H. H., 453
Gesner, Conrad, 357
Gianibelli, Federigo, 8
Gibb, Alexander, 81
Gibb, John, 40, 77, 294
Gibson, J. E., 385
Gilbertini, Dario, 122
Gill, Henry, 188
Gill, W., 130
Gill Hill, Kan., 356
Gillespie, Richard, 84
Girard, Pierre Simon, 36, 37

INDEX

Glasgow, Scot.; 45, 105, 110, 132
  aeration, 368, 369, 451
  filtration, 64, 80, 249, 257, 273
  sedimentation, 294
Glasgow Water Works Co., 80, 81, 273
Glauber, Johann Rudolf, 9, 358
Glens Falls, N.Y., 165
Goetz, Eugen, 62, 259, *ft.* 445
Goiter, treatment of, 456
Golden, Colo., 250
Goler, G. W., 456
Goodell, J. E., 431
Gorbals Gravitation Water Co.; aeration, 368, 369, 451
  filtration, 80, 87, 99
Gore, William, 269
Gorman, Arthur E., 339
Goudey, R. F., 414
Gowing, E. H., 159
Graf, August V., 246, *ft.* 310
Graff, Frederic, 130, 144, 377
Graffincourt, Cadet von, *ft.* 49
Graham, Thomas, 65
Grand Forks, N.D., 161
Grand Junction Water Works Co., 103
Grange-over-Sands, Eng., 252
Grant, Hugh J., *ft.* 331
Grant, Kenneth C., 354
Gray, Samuel M., 408, 432
Great Barrington, Mass., 280
Great Britain; 21
  coagulation, 302
  disinfection, 327, 342, 352
  filtration, 64, 248, 261, 269
  iron removal, 446
  sedimentation, 294
  softening, 415, 437
Greece, 3
Green, W. H., 433
Green Island, N.Y., 280
Greenalch, Wallace, 174, *ft.* 215
Greene, George S., 147
Greenock, Scot.; *ft.* 110
  color removal, 442
  filtration, 64, 91, 93, 249
Greenville, Tenn., 453
Greenville, Tex., 211
Greenwich, Conn., 194, *ft.* 216, 387
Gregory, John H., 432
Grew, Nehemiah, 360
Griffin, A. E., 453
Groningen, Neth., 309
Gros-Cailloux, Fr., 44, 49

Gunning, J. W., 318
Gury, A., 37, 353
Gwinn, Dow R., 320

H

Hackensack, N.J., 379
Hackensack Water Co., 379, 381, 454
Hadrian [Emperor], 287, 290
Hague (The), Neth., 445
Hale, Frank E., 326, 413
Hales, Stephen; aeration, 25, 363, 367, 450
  coagulation, 20
  distillation, 357, 359
Hall, Frank, 217
Halske [ozonation], 347
Hamburg, Ger., 445
Hamilton, Ont., 125, 279
Hamm, Ger., 454
Hancock, J. C., 406
Hand, Henry Y., *ft.* 370
Handy, James O., 429
Hannan, Frank, 62, 308, *ft.* 441, 442, *ft.* 445
Hannibal, Mo., 141
Happey [of Paris, Fr.], 40, 294, 366
Hardness scale, 421
Harold, C. H. H., 324, 342, 353, 452
Harris, C. J., 427
Harris, John, 392
Harris, John T., 334
Harris Magneto-Electric Filter Co., 231, 233, 334
Harrisburg, Pa., 264
Harrison, Charles Weightman, 327
Harrison, Louis B., 453, 454
Hartford, Conn., 398
Hartman, Gerda (Miss), *ft.* 1
Hassler, William W., 453
Haussmann, Georges Eugène, 43
Hawkins, Isaac, 116
Hawkins, Richard, 357
Hawks, A. McL., 251
Hawksley, Thomas, 120, 277, 395
Hawley, John B., *ft.* 35
Hazen, Allen; Albany, N.Y., 161, 173
  coagulation, 309, 317
  filtration, 140, 409
  Poughkeepsie, N.Y., 153
  sedimentation, 298
  writings, *ft.* 60, 133, 346, 445
Hazen & Whipple, 154, 175, 389
Hazen, Whipple & Fuller, 154

Heberden, William, 24
Helmreich, L. W., 374
Hempel, Johanna (Mrs.), 67, 71
Henderson, Charles B., 213
Henderson, Ky., 355
Henly, Eng., 24
Henry, Thomas, 415, 419
Hering, Rudolph, 237, 429
Hering & Fuller, 338, 339, 347, 432
Hermany, Charles A., 228, 235, 242, 332
Hermite, M. H., ft. 327, ft. 331
Hero of Alexandria, 3
Herodotus, 4, 321
Herschel, Clemens, 6, 172, 291, 332
Herscher, Charles, 323
Hill, Hibbert, 231
Hill, Nicholas S., Jr., 176, 428
Hinckley, H. V., ft. 334
Hinman, Jack J., Jr., 324
Hippocrates, 4, 321, 449
Hirtius, 5, 292
Hobart, Ind., 348, 349, ft. 351, 450
Hoboken, N.J., 380
Hodkinson, George F., 192, ft. 209, 219, ft. 223, 227
Hoffman [of Leipzig, Ger.], ft. 49
Holland, 20, 258, 308, 318
Hollis, F. S., 399, 442
Holly, Birdsill, 181
Holyoke, Mass., 398, 405
Home, Francis, 22, 23, 415
Homersham, Samuel Collett, 420, ft. 421, 422, 435
Hooker, Albert Huntington, ft. 327
Hoosick Falls, N.Y., 280
Hooten, Eng., 436
Hoover, Charles P., 428, 431, 435, 439
Hope, Thomas C., 90, 294
Horsford, E. N., 397
Horsley, John, 98, 425
Horton, Robert E., 176
Horton, Theodore, 163, 174, 176, 178
Horton, Kan., 355
Houston, Alexander Cruikshank, 304, 313, 336, 434, 452
Howard, Norman J., 452, 453
Huddleston, Thomas, 252
Hudson, N.Y., 136, 155, 411
Hudson Aqueduct Co., 155
Hudson River, 148, 155, 162, 403
Hughes, Samuel, 118
Huizinga [Prof., Univ. of Groningen], 309

Hull, B. H., 249
Hull, Eng., 359
Humber, William, 58, 368, 422
Humphrey, W. E., 263
Hungerford [mechanical filter], 347
Hunt, Charles W., 383
Hunter, Dard, 18
Huntingdon, Pa., 263
Huntington, W.Va., 240, 413
Huy, H. F., 341
Hyatt, Alpheus, 398
Hyatt, Isaiah Smith; 183, 205
  coagulation, 310, 318, 319
Hyatt, John Wesley; 183
  aeration, 368, 376, 377, 385, 389, 450
  filtration, 171, ft. 200, 249, 265, 272
Hyatt Pure Water Co., 202, 205, ft. 216, 230

I

Idrisi, Muhammad ibn Muhammad al, 287
Ilion, N.Y., 161, 373, 413
Illinois Central Railroad, 428
Imbeaux, Edouard; chlorination, 331, 343
  filtration, 57, 274, 281
  sedimentation, 290
Imhoff, Karl, 343, 354
India, 1, 222, 301, 321, 324
Indianapolis, Ind., 280
Infiltration basins, 25, 273
Ingenhousz, Jan, 393
Iodization, 456
Iola, Kan., 311
Iron; coagulation with, 313
  removal of, 445
Iselin, Adrian, 384
Ivry, Fr., 261

J

Jackson, Daniel D., 325, 436
Jacksonville, Fla., 374
Jacksonville, Ill., 398
Jamaica, N.Y., 446
Janet, M. L., 61
Japan, 21
Jefferson City, Mo., 374
Jegau [of Nantes, Fr.], 55
Jennings, Charles A., 339
Jersey City, N.J., 326, 336
Jervis, John B., 396
Jeunnet, M. C., 307

INDEX

Jewell, Ariel Clyde, 219
Jewell, Ira H., 217, 232
Jewell, O. H., & Sons, 219
Jewell, O. H., Filter Co., 219, 231
Jewell, Omar H., 205, 217, 329
Jewell, William M.; chlorination, 330, 332, 335
 coagulation, 320
 filtration, 217, 232
Jewell Export Filter Co., 222, ft. 223, 259, 263
Jewell Pure Water Co.; 202, ft. 330
 Davenport, Iowa, 213
 Ottumwa, Iowa, 217
 St. Johnsbury, Vt., 160
 Somerville, N.J., 193
 Terre Haute, Ind., 204
 Vicksburg, Miss., 312
Johnson, Charles T., 397
Johnson, George A.; chlorination, 336, 338, 339, ft. 340
 filtration, 231, 245
 softening, 432
Johnson, H. M., 457
Johnson, J. B., 374
Johnson, J. Y., 345
Johnson, John E., 202
Johnson, Richard, 367
Johnson, Walter R., 396
Johnson, William S., 264, 389
Johnston, James F. W., 301
Joinville le Pont, Fr., 346
Just, J., 325

K

Kalm, Pehr, 162
Kansas City, Mo., 296, 311
Katadyn, Inc., 326
Keene, N.H., 398
Kellerman, Karl F., 325, 412
Kent, Eng., 24, 302
Kent Waterworks Co., 422
Keokuk, Iowa, 250
Keyport, Mass., 445
Kienle, John A., 341
Kiersted, Wynkoop, 296
Kilmarnock, Scot., 99
King, Edward, 391
Kingman, Horace, ft. 384
Kingsbury, Francis H., 280
Kingsford, Thomas, ft. 146
Kingsford Starch Works, 146
Kingston, Jamaica, B.W.I., 271

Kingston, N.Y., 463
Kingston, Ont., 125, 141
Kirkwood, James P.; Albany, N.Y., 164
 Cincinnati, Ohio, 146, 244
 European cities, 55, 133, 258, 277, 279, 281, 282
 Hudson, N.Y., 136
 Lawrence, Mass., 136
 Lowell, Mass., 136
 Poughkeepsie, N.Y., 136, 148
 St. Louis, Mo., 146, 230, 245, 315, 320
Kirsanaff, Russ., 436
Kittaning, Pa., 263
Klassen, C. W., 433
Knapp, Harold J., 461
Koch, Robert, 123, 168, ft. 445
Koenigsburg, Ger., 445
Koyl, C. Herschel, ft. 433, 435
Kraemer, Henry, 324
Krause, G. A., 326
Kühne, W., 327

L

Lagane [of Toulouse, Fr.], 274
La Hire, Philippe de, 19
Lake Forest, Ill., 221
Lamar, W. L., 439
Lambeth Water Works Co., 60, 113
Lancashire, Eng., 64, 65, 90, 106
Lancaster, Pa., 263, 431, 453
Lanning, J. H., 144
Laodicea, Syria, 286
Lassone [of Versailles, Fr.], 37
Latham, Baldwin, 425, 434, 439
Laurens, Iowa, 437
Lawlor, Thomas F., 149
Lawrence, Robert L., Jr., 284
Lawrence, W. C., 453
Lawrence, William, ft. 433
Lawrence, Mass., 136, 161, 173, 230, 264, 280
Lawrence Experiment Station, ft. 61, ft. 112, 125, 139, 161, 232
Lawton, R. W., 222
Leal, John L., 337
Lee, James, 80, 98
Leeds, Albert Ripley; aeration, 185, 201, 345, 361, 377, 388, 390, 450
 Albany, N.Y., 170
 chlorination, 328
 National Filter Co., 199, 205, 207
Lees [of Paris, Fr.], 46
Leeuwarden, Neth., 309

Leeuwenhoek, Antony van, 391
Leghorn, It., 133, 258
Leicester, Eng., 133, 262
Leipzig, Ger., 445
Leisen, Theodore A., 236, 241, 251, 311, 317
Leningrad, Russ., see St. Petersburg
Lesage, Louis, 147
Leupold [of Bordeaux, Fr.], 295
Levy, E. C., 130
Lewiston, Me., 250
Lexington, Ky., 376
Libourne, Fr., 63
Lime; clarification with, 313
  excess, 434
Lincoln, Eng., ft. 106, 336, 338
Lincoln, Neb., 447
Lincolnshire, Eng., 106
Lind, James, 21, 366, 450
Lindley, William, 346, 396
Lindsay, Ont., 348
Linside Bleach Works, 80
Lisbon, Port., 343
Little Falls, N.J., 226, 227, 234
Little Falls, N.Y., 373, 451
Little Rock, Ark., 211
Littlewood [aeration], 363
Liverpool, Eng., 133, 426
Llambias, G. A., ft. 207
Lochridge, Elbert E., 408, 410
Locke [Dr., of Cincinnati, Ohio], 228
Lohr, E. W., 439
London, Eng.; 41, 64, 133
  algae, 395
  chlorination, 342
  coagulation, 303
  filtration, 89, 91, 99, 123, 257, 263
  purification, 8, 24
  sedimentation, 294
  softening, 418, 420, 434
London, Ont., 446
London Bridge Water Works Co., 84
Long, Charles R., 229, 236
Long Beach, Ind., 348, 349, ft. 351, 444, 450
Long Branch, N.J., 194, 387, 388
Los Angeles, Calif., 414, 438
Loudon, J. C., ft. 95
Loudonville, N.Y., 176
Louisiana, Mo., 225
Louisville, Ky.; algae, 413
  filtration, 140, ft. 227, 228
  chlorination, ft. 330, 332

Lovejoy, W. H., 238, 413
Lowell, Mass., 136, 147, 280, 447
Lowitz, George Moritz, 26
Lowitz, Johann Tobias, 26, 38, ft. 76, 132, 449
Ludlow, C. W., 191
Ludlow, William M., 377
Lüscher, O., 259
Lussac, Gay, 46
Lycurgus, 4
Lynchburg, Va., 130, 294
Lynn, Mass., 398
Lyons, Fr., 133, 279, 281, 292, 343

M

MacAlister, James, 95
MacDougal Polarite System, 233
Mackain, Donald, 81
Macon, Ga., 199
Magdeburg, Ger., 369
Magues [of Toulouse, Fr.], 275
Maidstone, Eng., 336
Maignen, P. A., 431
Malkah, Tunisia, 289
Mallet, Charles-Francois, 45, 81, 86, ft. 95, 106
Manchester, Eng.; filters, 67, 89, 105, 110
  ozonation, 352
  softening, 25, 421
Manchester, N.H., 146
Manganese removal, 446
Manning, Tex., 360
Marboutin, M., 61
Mareschal et Compagnie, 51
Mark [electrolysis], 335
Marlborough, Mass., 394
Marmier [ozonation], 353
Marseilles, Fr., 57, 60, 133, 354
Marshalltown, Iowa, 143
Marsigli, Luigi Ferdinando, 19, 253
Martin, Edward A., 330
Mascott, Tenn., 356
Mason, Amassa, 188
Mason, William P., 168
Massachusetts State Board of Health [see also Lawrence Experiment Station], 300
Massey-Mainwaring, W. F. B., 368, 386
Mather & Platt, 427
Matthews, William [Hydraulia], 17, 42, 46, 87, 116
Matthews, William [Supt., Southampton Waterworks], 427

INDEX

Mattice, Charles R., 225
Maudslay, Henry, 367
Maurras, André Eustache Gratien Auguste, 98, 114
Mayfield Print Works, 421
McAlpine, William Jarvis; Albany, N.Y., 144, 164
  Hudson, N.Y., 155
  Oswego, N.Y., 146
  Schenectady, N.Y., 146, 404
McAmis, J. W., 453
McCannon, William, 148
McConnel, S., 343
McCurdy [aeration], 379, *ft.* 380
McGowan [Dr., Lincoln, Eng.], 336
McKees Rocks, Pa., 437
Mederer von Wuthwehr, 27, 248
Medication, 456
Medlock, Henry, 313, 314, 317
Mellen, Arthur F., 458
Memphis, Tenn., 448
Mentone, Fr., 354
Merrick [of Philadelphia, Pa.], 377
Merthyr Tydfil, Wales, 265, 269
Metcalf, Leonard, *ft.* 327
Middelkerke, Belg., 336, 338
Middleborough, Mass., 447
Middletown, N.Y., 225
Mills, Hiram F., 139
Milwaukee, Wis., 348
Minneapolis, Minn., 222, *ft.* 223, 432, 458
Miquel, P., 61
Mirabeau, Honoré Gabriel Victor Riquetti de, 37
Mississippi River; 310
  New Orleans, 183, 185, 208, 234, 319
  St. Louis, 246
Mobile, Ala., 130
Moiseev, S. V., 325
Möller, K., 256
Monckton, E. H. C., 345
Monmouth, N.J., 388
Montbruel, Jean Baptiste Molin de, 35, 366
Monte Carlo, Fr., 343
Montfort, Denis, 39
Montpellier, Fr., 343
Montreal, Que., 147, 264
Moore, George Thomas, 325, 409, 412
Moose Jaw, Sask., 281
Morgans, G. Ewart, 352
Morison, Samuel L., 222

Morison-Allen Co., 219, 220
Morison-Jewell Filtration Co., 219, 221
Morse, Jedediah, 163
Morse, Robert, 78
Moses, 3
Moses, H. E., 437
Mouchet, H., 61
Mt. Clemens, Mich., 209, 211
Mt. Vernon, N.Y., 161
Mowatt, Magnus, *ft.* 113
Mulkey Salt Co., 460
Mylne, William, 86

N

Nageli, Carl von, 325
Nancy, Fr., 281, 353
Nantes, Fr., 29, 54, 133
Nantucket, Mass., 161, 374
Nashville, Tenn.; 130
  coagulation, 311
  filtration, 280, 283, 284
National City Water Works Co., 296
National Water Purifying Co.; 199, *ft.* 378, 385
  Champaign-Urbana, Ill., 381, 428
  New Orleans, La., 205
Navarette, R. F. Dominick Fernandez, 302
Nellis, James C., 351
Nelson, S. L., 202, 381, *ft.* 428
Nero, 6, 322
Netherlands, 20, 258, 308, 318
Neuburger, Albert, 6
New Britain, Conn., 405
New England Water Works Assn., 141
New Haven, Conn., 398
New Milford, Conn., 249
New Milford, N.J., 380, 381
New Orleans, La.; coagulation, 183, 318, 319, 320
  filtration, 205, 231
  reservoir, 130
  softening, 432
New River, 38, 342, 421, 440
New River Water Works Co., 114, 302
New Rochelle, N.Y., 383
New York, N.Y.; aeration, 390
  algae, 398, 404
  disinfection, 330, 338, 347
  sewage treatment, *ft.* 328
New York Continental Jewell Filtration Co., 222, 223, 226

New York Filter Mfg. Co.; 194, *ft.* 197, 199, 202, 205, 209, 213, 215, 221, *ft.* 378, 385
   Albany, N.Y., 174
   Elgin, Ill., 211
   New Rochelle, N.Y., 383
Newark, Eng., 279
Newark, N.J., 279
Newark Filtering Co.; 183, 189, 201, 209
   Albany, N.Y., 170, 171, 187
   Somerville, N.J., 191
Newburgh, N.Y., 414, 463
Newcastle, Eng., 417
Newmarket, Ont., 446
Newport, R.I., 183, 190
Newton, Isaac, 440
Newton, Mass., 280
Niagara Falls, N.Y., 220, 221, 341
Nice, Fr., 63, 353
Nichols, William Ripley; algae, 398, 405
   coagulation, 313
   filtration, 136, 152, 189, 281
Nicollet [Fr. scientist], 19
Nieuwersluis, Neth., 347
Nile River, 5, 292, 300
Nimes, Fr., 281
Norfolk, Va., 382, 444
Norman, Bradford, 190
Norman, George H., 190
Normandy, Alphonse René le Mire de, 367
Northern Illinois Water Corp., 212
Northern Indiana Public Service Co., 350
Norwich, Conn., 398
Norwood Engineering Co., 198, 339
Nottingham, Eng., 262, 277, 395
Nye, Walter B., 197

O

Oakland, Calif., 195
Oberlin, Ohio, 429, 434
Odor control, 361, 391, 449
Oesten [deferrization], *ft.* 445
Ogdensburg, N.J., 348
Ogston, George Henry, 316
Ohio River, 228, 243, 334
Ohio Salt Co., 459
Ohio Valley Water Co., 437
Oklahoma City, Okla., *ft.* 334
Olson, H. M., 436, 437, 438
Omaha, Neb., 297, 311
Ornstein, Georg, 341

Osbridge [aeration], 366
Oshawa, Ont., 271
Oshkosh, Wis., 199
Ostend, Belg., 338
Oswego, N.Y., 146
Otis, Daniel C., 181
Ottawa, Ont., 452
Otto, M. P., 347, 353
Ottumwa, Iowa, 217
Ozonation, 343
Ozone Co. of America, 348
Ozone Processes, Inc., 349, 350, 351

P

Painesville, Ohio, 284
Paisley, Scot., 64, 77, 95, 98, 110, 249, 257, 294
Palladius, Rutilius Taurus Aemilianus, 4
Palmer [electrolysis], 232, 335
Paris, Fr.; coagulation, 304
   disinfection, 343, 353
   filtration, 19, 21, 29, 37, 45, 63, 114, 261
   sedimentation, 294
Parker [Adm., Royal Navy], 76
Parker, Horatio N., 399
Parkes, Edmund Alexander, 366
Parmelee, Charles L., 223, 226, 228, 231
Parmentier [clarification], 40
Parr, S.C., 356
Parthenay, Fr., 324
Passaic River, 334
Paterson, William, *ft.* 69, 342
Paul [filtration], 257
Pauwells [of Dunkirk, Fr.], 58
Pawtucket, R.I., 250
Paxamus, 3
Peacock, James, 60, 64, 67, 93, 129, 181, 227, 248
Pearson, Galen W., 296
Peligot, Eugene, 318
Pennell, Reginald Humphrey Lee, 252
Pennsylvania Maignen Filtration Co., 263, 431
Percival, Thomas, 25
Perkins, C. E., 251
Permutit Softening, 436
Pernambuco, Braz., 265, 271
Perrysburg, Ohio, 355
Persia, 4, 7
Perth, Scot., 133, 277, 279
Peters [of Zurich, Switz.], 259

Petersburg, Va., 130
Peyre, Francois, Jr., 367
Philadelphia, Pa.; aeration, 377
  coagulation, 317
  disinfection, 330, 340, 341, 347, 349
  filtration, 144, 263
Piefke, C., 124, *ft.* 445
Pierson, George H., 191
Pillsbury, J. L., 431
Pirnie, Malcolm, 154, 389, 451
Pittsburgh, Pa., 144
Pittsburgh Testing Laboratory, 429
Place, Francis Evelyn, 1, 321, 324
Plato, 4
Pliny the Elder, 3, 6, 322, 362
Plüche, Noel Antoine, 20, 363
Plummer, M. W., 414
Plumstead, Woolwich & Charlton Consumers' Water Co., 422
Plutarch, 4, 322
Poccianti, Paschal, 258
Pocock, Lewis, 327
Pola, It., 331
Poland, 446
Poole, B. A., *ft.* 351
Pope, F. L., 328
Port Chester, N.Y., *ft.* 216, 387
Porter, Herbert, 426
Porter, John Henderson, 426
Portius, Lucas Antonius, *see* Porzio
Porzio, Luc Antonio, 11, 22, 27, 34, 248, 253, 258
Poughkeepsie, N.Y.; aeration, 389, 451
  algae, 410
  coagulation, 317
  disinfection, 340
  filtration, 136, 137, 148, 173, 264
Powers, J. J., *ft.* 328
Pozelius, A. F., *ft.* 203
Priestley, Joseph, 392
Proskauer, B., *ft.* 445
Providence, R.I.; aeration, 154, 389, 451
  filtration, 140, 143, 222
  iron removal, 447, 448
Puech, Armand, 57, 63, 88, 257, 261, 307
Puech-Chabal Aerators, 369
Puech-Chabal Filters, 63, 257, 261, 317, 355

Q

Quai des Celestin Filters, 38, 40, 132, 294, 366, 451
Quarnier, J.-F.-E., *ft.* 49

Quebec, Que., 295
Quincy, Ill., 320
Quindaro, Kan., 296

R

R.U.V. Co., Inc., 356
Race, Joseph, *ft.* 327, 340, 341, 452
Radcliffe, Eng., 67
Rafter, George W., 396
Rahway, N.J., 183, 185, 190, 197, 389
Rand, J. B. G., 149, 156, 411
Ransome ver Mehr Co., 269
Rapin, René, 11
Rapp [filtration], 190
Raritan, N.J., 191, 265, 310
Rasis, *see* Rhazes
Raymond [of Toulouse, Fr.], 275
Reading, Eng., 88, 257, 259
Reading, Mass., 280, 376, 445
Real, Graf von, *ft.* 49
Réaumur, René Antoine Ferchault de, 19, 31
Recife, Braz., 265, 271
Rees, Abraham, 116
Reeves, W., 257
Reeves Patent Filters Co., 257
Reisert Rapid Filter, 446
Remsen, Ira, 398
Rennes, Fr., 272
Reynolds, Schuyler, 162
Rhazes [Mohammedan physician], 323
Rheims, Fr., 343
Rich Hill, Mo., 193
Richmond, Va., 125, 127, 249, 294
Riddell, Ernest H., 209
Rideal, Erik, 323, 324
Rideal, Samuel, 124, 323, 324
Rider, J. B., 375
Rider, William B., 141
Rikkers, M., 190
Roberts Filter Mfg. Co., 208
Robins, George, 117, 327
Robinson, Andrew J., 181
Robinson, J. H., 375
Rochester, N.Y.; aeration, 371, 451
  iodization, 456, 458
  paper mill, 190
Rochon, M., 19, 38
Rock Island, Ill., 220
Rockland, Ont., 271
Rogers Park, Ill., 209, 211
Rome, It., 3, 6, 13, 290
Rossman, Clark G., 157

Rotterdam, Neth., 259
Rouart-Herscher & Co., 323
Rouen, Fr., 355
Ruhl [of Karlsruhe, Ger.], 189
Russell, George, 367
Russell, S. Bent, 236
Russia, 222, 368, 436
Ruth, Edward D., 453
Rutland, Vt., 280
Ruttan, H. N., 429
Rutty, John; 24
   algae, 392
   coagulation, 302
   softening, 415, 417
Rye, N.Y., ft. 216, 387

S

Saigon, Fr. Indo-China, 446
St. Basil [distillation], 357
St. Helens, Eng., 427
St. Johnsbury, Vt., 158, 250
St. Johnsbury Aqueduct Co., 160
St. Lawrence River, 23, 147, 309
St. Louis, Mo.; coagulation, 311, 317, 318, 319, 320
   filtration, 133, 146, 245
   sedimentation, 297
   softening, 430
St. Nazaire, Fr., 343
St. Paul, Minn., 390, 432
St. Petersburg, Russ., 347, 368
Salbach, B., 309, 445
Salford, Eng., ft. 61, 352
Sandhurst, Austral., 313
San Diego, Calif., 414
Santa Monica, Calif., ft. 334
Sault Ste. Marie, Mich., 457
Saunders, Charles, 23, 309
Saussure, Horace Benedict de, 293
Sawyer, Joseph B., 147
Sawyer, W. H., 328
Scheerer, Carl J. A., 318
Schenectady, N.Y., 146, 280, 404
Schiedam, Neth., 258, 309, 347
Schoolcraft, Henry Rowe, 310
Scotland, see Great Britain
Scott, Russell D., 435, 453, 454
Scotte, J. P., 25
Scowden, Theodore R., 228
Seddons, James A., 246, 297
Sedgwick, William T., 466
Sedimentation, 286

Seine River, 31, 35, 36, 39, 318
Selinsgrove, Pa., 311
Sellers, George H.; aeration, 376
   coagulation, 314, 317
   filtration, 249, 251
Semple, Ellen Churchill, 322
Senegal, Fr. W. Africa, 23, 313
Senegal River, 25
Sennar, Anglo-Egypt. Sudan, 301
Seville, Sp., 343
Shapiro, Robert, 326
Shaw, Frank H., 356
Shaw, Peter, 415, ft. 417
Shaw, Thomas, 287, 289
Shaws Water Joint Stock Co., 93
Shedd, J. Herbert, 147
Sheerness, Eng., 359
Sheffield, Eng., 262, 438
Sherman, Henry A., 457
Shield, H., 120
Shirley, Mass., 280
Sibert, William, 302
Sidney, Ohio, 211
Siemens, Werner von, 347, 353
Sierp, F., 454
Silver; disinfection by, 325
Simco, Ont., 446
Simin, Boris N., 436
Simin, Nicholas, 346, 436
Simpson, Arthur Telford, 113
Simpson, Charles Liddell, ft. 103, 113, ft. 209
Simpson, Clement P., ft. 103, 113
Simpson, Edward P., 113
Simpson, James; 60, 64, 81, 91, 132, 294
   Chelsea filters, 47, 99
   coagulation, 303
Simpson, James, Jr., ft. 103, 113, 421
Simpson, John, 113
Simpson, Thomas, 84, 113
Simpson, Thomas B., 113
Sinclair, John, 77, 367
Singh, Kundan Jugendia Pal, 1
Sloper, B. G., 98, 114
Sluyter, Peter, 162
Smart, F. B., 335
Smith, Angus, 118
Smith, J. Waldo, 228
Smith, James, 38
Smith, Marsden C., 131
Smith-Cuchet-Montfort Filter, 38, 47
Socrates, 4

INDEX

Softening, 23, 415
Somerville, N.J., 183, 190, 193, 197
Somerville & Raritan Water Co., 191, 197, 265, 310
Soper, George A., 232, 344, 346
Souchon [filtration], 51, 114
South Bethlehem, Pa., 263
South Norwalk, Conn., 141, 264, 388
South Staffordshire, Eng., 352
Southend, Eng., 313, 434
Southampton, Eng., 426
Southport, Eng., 314
Spafford, H. A., 433
Spalding, George R., 454
Spaulding, Charles H., 433, 453
Spence, Peter, ft. 209, 304
Spencer, Thomas, 313, 314, 319, 443
Spring, Walthère, 441
Springfield, Ill., 280, 284, 433, 453
Springfield, Mass., 161, 404, 444, 452
Stalybridge, Eng., 352
Stamford, Eng., ft. 106, 258, 315
Stamford, N.Y., 225
Stanley, F. T., 405
Starnes, J. S., 367
Stearns, Frederick P., 173
Steelton, Pa., 264
Stein, Albert, 127, 249, 294
Stephenson, Joseph, 255
Sterilization, 321
Sterling, Mass., 446
Stevens, Harold C., 232
Stillwater, Minn., 250
Stirrat, James, 88
Stockbridge, Mass., 141
Stockport, Eng., 443
Stonehaven, Scot., 284
Storer, J., 367
Storm Lake, Iowa, 251
Storrow, Charles S., ft. 95, 132
Stover, Frederick H., 239
Stow, John, 8
Streator, Ill., 211, 212
Streeter, H. W., 438
Strype, J., 8
Suez Canal, 360
Superchlorination, 453
Surbiton, Eng., 112
Swadlincote, Eng., 427
Switzerland, 293
Syracuse, N.Y., 414
Szniolis, A., 325, 446

T

Tacoma, Wash., 251, 317, 369, 374
Taste and odor control, 361, 391, 449
Taunton, Mass., 280
Taylor [of Paris, Fr.], 46
Taylor, W. E., 203
Telford, Thomas; 105, 110, 273, 420
  filtration, 81, 257
  sedimentation, 294
Ten Eyck, Philip, 163, 401
Terre Haute, Ind., 203, 221
Thames River, 22, 41, 84, 103, 120, 139, 257, 303, 342, 421, 423
Thames Water Co., 103
The Hague, Neth., 445
Theophrastus, 362
Theos, Antiochus, 286
Thom, Robert; 47, 58, 60, 64, 91, 110, 129, 132, 181, 227, 249
  color removal, 442
  filter details, 93
Thomas, George E., 341
Thompson [zeolite], 436
Thompson, R. W. S., 262
Thurston, Robert H., ft. 379
Tidd, M. M., 250
Tighe, James L., 216
Tillmanns, J., 62
Timanus, C. S., 433
Toledo, Ohio, 143, 255
Tomlinson, Charles, 30, ft. 98, 114, 117
Toronto, Ont.; aeration, ft. 370
  filtration, 265, 269, 280
  superchlorination, 452, 453
Torrey, John, 402, 404
Toulon, Fr., 353
Toulouse, Fr.; algae, 395, 451
  filtration, 36, 45, 133, 274, 279
Tours, Fr., 57
Trailigaz Filter, 63, 272
Traube [chlorination], 331
Trenton, N.J., 145, 403
Trimble, Earle J., 373, 413
Tritton, Henry, ft. 49, 179
Troy, N.Y., 165, 170, 175
Tubbs, J. Nelson, 371
Tunis, Tunisia, 389, 390
Tunkhannock, Pa., 193
Turn-Over Filter Co., 271

U

Ultra-violet; disinfection by, 354
Uneek Filter, 271
United States; aeration, 370, 390
  algae, 396, 406
  coagulation, 309
  disinfection, 328, 348
  filtration, 125, 179, 248, 263
  iron removal, 445, 447
  sedimentation, 294
  softening, 428, 435
U.S. Filtering & Purifying Co., 317
United Water Improvement Co., 347
Urbana, Ill., 381, *ft.* 428
Ure, Andrew, 83, 117, 327
Utica, N.Y., 372, 451

V

Vaillard-Desmoraux Apparatus, 323
Van Brunt, Cornelius, 410
Van Calker [Prof., Univ. of Groningen], 309
Van der Made [ozonation], 353
Van Loan, Seth M., 341, 349
Vater, Abraham, 21
Vaughan, Victor C., 335
Vedel [filtration], 53
Venice, It., 13
Verdun, Fr., 343
ver Mehr, John, Engineering Co., 269
ver Mehr, Ransome, Co., 269
Vermeule, C. C., 158
Versailles, Fr., 37
Vichy, Fr., 281
Vicksburg, Miss., 311
Vienna, Aust., 279
Villefranche-sur-Mer, Fr., 353
Vincennes, Ind., 225
Viribent [of Toulouse, Fr.], 275
Vitruvius, Pollio, 3, 292
Vogt Bros. Mfg. Co., 389
von Beetz, Wilhelm, 440
von Graffincourt, Cadet, *ft.* 49
von Nageli, Carl, 325
von Real, Graf, *ft.* 49
von Wuthwehr, Mederer, 27, 248
Vosmaer System, 347

W

Wakefield, Eng., 133, 314, 443
Walcott, William, 11, 358, 359
Wales, *see* Great Britain
Walker, Alexander T., 88, 257, 259
Walker, Leslie G., 259
Walker, Walter E., 154
Wallace & Tiernan Co., Inc., 341
Waltham, Mass., 280, 282
Wanklyn, J. Alfred, 121, 332
Ward, D. B., 152
Wardle, J. McClure, 158
Warmouth Sugar Refinery, 190
Warren, John E., 185, 195, 377, 445
Warren, Joseph A., 197
Warren, S. D., & Co., 195
Warrington, Eng., 395
Warsaw, Pol., 222
Washington, D.C., 326
Waterhouse-Forbes Apparatus, 323
Watervliet, N.Y., 175
Watier [of Nantes, Fr.], 55
Watson, Harold E., 190
Watt, James, 79, 83, 273
Wattie, Elsie, 313
Waukegan, Ill., 389, 451
Way, J. Thomas, 436
Wayne, Pa., 263
Webster, W., *ft.* 327, 328
Wedgwood, Josiah, 71
Weehawken, N.J., 380
Weir [of Glasgow, Scot.], 86
Weitzel, G., 147
Wellesley, Mass., 280
Wells, J. H., 213
Welsbach Street Illuminating Co., 349
West, Francis D., 343
West, William, 440
West Cheshire, Eng., 437
West Chester, Pa., 296
West Palm Beach, Fla., 389
Western Filter Co., 212, 231
Western New York Water Co., 341
Westervelt, Albert C., 183
Westinghouse-Cooper-Hewitt Co., 354
Weston, Arthur D., 375
Weston, Edmund B., 140, 222
Weston, Robert Spurr; color, 442
  demanganization, 447
  filtration, 208, 216, 231
  iron removal, 446
  ozonation, 346
Weyl, Theo, 317, 346
Wheeler, William, 388
Whipple, George C.; algae, 399, 409
  chlorination, 153, 336, 340
  ozonation, 347

Whitfield, George H., 131
Whiting, Ind., 349, 350, 450
Whitinsville, Mass., 280
Whitman, Requardt & Smith, 176
Whitson, M. T. B., 353
Whittier, Charles T., 232
Wickersham, L. E., *ft.* 203
Wiergate, Tex., 360
Wilber, Francis A., 307, 310
Wilkes-Barre, Pa., 143, 220
Willcomb, George E., 175, *ft.* 177
Williams [of Marlborough, Mass.], 394
Williams, John, 257
Williams Bay, Wis., 433
Williamson, David, 223
Williamson, David Charles, 223
Williamson, James E., 223
Williamson, L. L., 203
Willisden, Eng., 426
Wilmington, Del.; aeration, 376
 chlorination, 341
 coagulation, 317
 filtration, 251, 263
Wilson, James B., 238
Wilson, John, 38
Winchester, Ky., 388, 451
Winnipeg, Man., 202, 428
Wintringham, Clifton, 362
Wisbeach, Eng., 314, 443
Witt, Henry, 119

Wittstein [color], 441
Wolman, Abel, 320, 458, 462
Wood, A. B., *ft.* 207, *ft.* 209
Wood, H. J., 373
Woodhead, Sims, 336
Woolf, Albert E., *ft.* 327, 330, *ft.* 331, 338
Worcester, Eng., 317
Worcester, Mass., 405
Wormsley [Prof., Columbus, Ohio], 431
Worthen, William E., 170
Worthington, H. R., 191
Worthington-Simpson Co., 113
Wurtz, Henry, 403
Wuthwehr, Mederer von, 27, 248
Wylie [filtration], 252
Wyomissing, Pa., 437

X, Y, Z

York, Eng., 133, 263
York, Pa., 398
Yorkton, Sask., 446
Young, David, 89
Younger, Richard, 38
Youngstown, Ohio, 435

Zapffe, Carl, 447
Zeolite treatment, 435
Zurich, Switz., 259
Zutphen, Neth., 258, 446

CPSIA information can be obtained at www.ICGtesting.com
Printed in the USA
LVOW04s1025010914

401754LV00003B/294/P